Bayes Linear Statistics

WILEY SERIES IN PROBABILITY AND STATISTICS

Established by WALTER A. SHEWHART and SAMUEL S. WILKS

Editors: *David J. Balding, Peter Bloomfield, Noel A. C. Cressie, Nicholas I. Fisher, Iain M. Johnstone, J. B. Kadane, Geert Molenberghs, Louise M. Ryan, David W. Scott, Adrian F. M. Smith*
Editors Emeriti: *Vic Barnett, J. Stuart Hunter, David G. Kendall, Jozef L. Teugels*

A complete list of the titles in this series appears at the end of this volume.

Bayes Linear Statistics

Theory and Methods

Michael Goldstein and David Wooff
Durham University, UK

BICENTENNIAL
1807
WILEY
2007
BICENTENNIAL

Telephone (+44) 1243 779777

Email (for orders and customer service enquiries): cs-books@wiley.co.uk
Visit our Home Page on www.wileyeurope.com or www.wiley.com

Reprinted July 2007

Other Wiley Editorial Offices

John Wiley & Sons Inc., 111 River Street, Hoboken, NJ 07030, USA

Jossey-Bass, 989 Market Street, San Francisco, CA 94103-1741, USA

Wiley-VCH Verlag GmbH, Boschstr. 12, D-69469 Weinheim, Germany

John Wiley & Sons Australia Ltd, 42 McDougall Street, Milton, Queensland 4064, Australia

John Wiley & Sons (Asia) Pte Ltd, 2 Clementi Loop #02-01, Jin Xing Distripark, Singapore 129809

John Wiley & Sons Canada Ltd, 6045 Freemont Blvd, Mississauga, Ontario, L5R 4J3, Canada

Wiley also publishes its books in a variety of electronic formats. Some content that appears in print may not be available in electronic books.

Library of Congress Cataloging-in-Publication Data

Goldstein, Michael, 1949-
Bayes linear statistics : theory and methods / Michael Goldstein and David A. Wooff.
p. cm.
Includes bibliographical references and index.
ISBN-13: 978-0-470-01562-9 (cloth : alk. paper)
ISBN-10: 0-470-01562-4 (cloth : alk. paper)
1. Bayesian statistical decision theory. 2. Linear systems.
3. Computational complexity. I. David A. Wooff, II. Title.
QA279.5.G65 2007
519.5′42 – dc22

2006033618

British Library Cataloguing in Publication Data

A catalogue record for this book is available from the British Library

ISBN 978-0-470-01562-9 (HB)

Typeset in 10/12pt Times by Laserwords Private Limited, Chennai, India
Printed and bound in Great Britain by TJ International, Padstow, Cornwall
This book is printed on acid-free paper responsibly manufactured from sustainable forestry in which at least two trees are planted for each one used for paper production.

To our families:
Beverley, Ayanna, Damani, Timothy, and Chica.
Val, Ben, Dad, and my mother Mary

Contents

Preface

How should we use data to help us analyse our beliefs? This book is concerned with the subjectivist analysis of uncertainty, and develops methods that combine prior judgements with information derived from relevant data. Whenever we move from broadly data-focused questions, such as 'Does this data set suggest that a certain medical treatment is likely to be effective?', to broadly decision-motivated questions, such as 'Are we sufficiently confident, given all that we know about this treatment, to recommend its widespread use?', then we must make such a synthesis of data with more generalized forms of information. Because we may find this hard to achieve, we need some methodology to help us. This methodology should be clear, helpful, logically well founded and tractable.

The Bayesian approach to statistics is the natural methodology for this purpose. This approach treats all uncertainties within a common probabilistic framework, combining the different sources of information using the rules of probability. This approach has a sound logical foundation and a well-developed methodology and is popular and successful in many areas of application.

However, in large-scale applications, the Bayesian approach can easily become the victim of its own ambition. Representing all uncertainties in probabilistic form is a daunting task for complicated problems. This is partly because of the intrinsic difficulties in judging the value of each relevant source of knowledge. However, in large part, the task is difficult because the Bayesian approach requires us to specify our uncertainties to an extreme level of detail. In practice, it is usually beyond our ability to make meaningful specifications for our joint probability distributions for multiple outcomes.

If we do wish to follow a broadly Bayesian path, then we must either choose to make specifications that do not correspond to our actual uncertainties or be more modest about our ability to render our beliefs in probabilistic form. If the data are plentiful and unambiguous in their message or if the problem is not sufficiently important to merit careful analysis, then little harm is done by somewhat misrepresenting our beliefs. However, when the issue is important and data are less plentiful, then we must be more careful and honest. When we cannot make full belief specifications, we require alternative methods that respect the limitations on our abilities to specify meaningful beliefs and allow us to conduct partial analyses strictly in terms of the actual limited aspects of our beliefs that we are able to specify.

The Bayes linear approach offers an appropriate methodology for this purpose. By making expectation, rather than probability, the primitive quantity for the quantification of uncertainty, we move from a position that requires full probability specification to a less demanding position in which we specify directly whichever collection of expectation statements we feel are most relevant for expressing and modifying our beliefs in the light of observation.

To use such collections of expectation statements effectively, we must rebuild our approach to the analysis of uncertainty. That is the purpose of this book. We derive from first principles the Bayes linear approach to statistics, developing the methodology from practical, theoretical, and foundational viewpoints. Our approach is subjectivist and emphasizes the twin roles of interpretative measures to help us understand the implications of our collections of belief statements and diagnostic measures to help uncover serious conflicts between the various aspects of our specifications and observations. Modelling proceeds through direct specification of beliefs over observable quantities, exploiting second-order exchangeability. Bayes linear graphical models simplify belief specification and analysis and provide the natural setting for graphical displays to highlight the key features of the analysis.

Work on the Bayes linear approach has been going on for many years. Indeed, in preparing papers and talks, in writing software, and in writing this book we have been involved in this development throughout our academic lives. The approach is sufficiently mature to merit a detailed presentation. However, this book is in no way intended to be an exhaustive treatment of Bayes linear methodology. It is not even a complete account of our own work in this field, let alone a guide to all of the enormous volume of other work that has been done from a moment-based perspective. Rather, this work is a self-contained development of the basic features of the approach, based around the starting point that expectation is the natural primitive concept for the theory, and developing the practical and logical implications of this view in a unified way.

We trust that this work may be of interest and value to those who share our view, and we would be pleased to convert readers to this way of thinking. However, we also hope that readers with different foundational opinions will find value in exploring the implications of alternative views, both out of intellectual interest and because the approach yields a variety of simple and powerful methods that may give additional insights into their own procedures. From such perspectives, the Bayes linear approach may be viewed as achieving '90% of the answer for 10% of the effort'. This is not a recommendation that we should be lazy, but rather recognition that, when even 10% of the effort is a substantial amount of work, we are much more likely to be able to carry out a careful, thoughtful, and successful analysis if we concentrate our efforts where they will have the greatest effect. Whatever viewpoint we may have, it is important to understand that we often do have simple alternatives that we may make use of to avoid becoming overwhelmed by the complexities of more traditional analyses.

This book is suitable for a graduate readership. Most of the book is also suitable for a final-year undergraduate course. Some of the material has been used within the final-year undergraduate Mathematics programme at Durham University.

The index provided at the end of this book is almost entirely a pointer to coverage of theoretical material and definitions. Generally, examples directly follow the theory and so illustrate the material soon after its definition. We use a number of running examples for this purpose: an index of such examples is given in Appendix B. Appendix A lists the notation used in this book, together with a page reference to the main, or first, definition. The Bayes linear programming language [B/D] was used for most of the calculations needed; see Appendix C.

Writing this book has been both a pleasurable and a frustrating experience: pleasurable because the approach is both powerful and elegant and it has been very rewarding to revisit favourite ideas and to build them into a unified whole, adding illustrations and much extra material to consolidate our treatment; frustrating because limitations of space and time inevitably mean that in many places our discussion is curtailed while there still is much to be said. We are very grateful to our colleagues who have endured so pleasantly all our accounts of these ideas, to all our collaborators who have contributed so much to the development of the approach, to Wiley for their near infinite patience in waiting for our manuscript, and to our families for their love and support.

Michael Goldstein and David Wooff
Durham, August 2006

1

The Bayes linear approach

The subject of this book is the qualitative and quantitative analysis of our beliefs, with particular emphasis on the combination of beliefs and data in statistical analysis. In particular, we will cover:

(i) the importance of partial prior specifications for problems which are too complex to allow us to make meaningful full prior specifications;

(ii) simple ways to use our partial prior specifications to adjust our beliefs given observations;

(iii) interpretative and diagnostic tools that help us, first, to understand the implications of our collections of belief statements and, second, to make stringent comparisons between what we expect to observe and what we actually observe;

(iv) general approaches to statistical modelling based upon partial exchangeability judgements;

(v) partial graphical models to represent our beliefs, organize our computations and display the results of our analysis.

Our emphasis is methodological, so that we will mostly be concerned with types of specification and methods of analysis which are intended to be useful in a wide variety of familiar situations. In many of these situations, it will be clear that a careful, quantitative study of our beliefs may offer a valuable contribution to the problem at hand. In other cases, and in particular in certain types of problem that are conventionally treated by statisticians, the status of a belief analysis may be more controversial. Therefore, we shall begin our account by giving our views as to the role of the analysis of beliefs in such problems, and then briefly discuss what we perceive to be the strengths and weaknesses of the traditional Bayesian approach to belief analysis. We will briefly describe some of the distinctive features of Bayes

Bayes Linear Statistics: Theory and Methods M. Goldstein and D.A. Wooff

linear analysis, give an overview of the contents of this book and introduce the methodology by example.

1.1 Combining beliefs with data

To introduce our approach, compare the following examples. First, we test an individual for precognitive powers, and observe correct guesses in ten out of ten flips of a fair coin. Secondly, we test a promising new treatment against a current treatment for a disease, and observe that the new treatment outperforms the current treatment in each of ten trials on carefully matched pairs of patients.

The two experiments have, in a sense, yielded the same data, namely ten successes in ten binary trials. However, in the first case, most people would be intrigued but remain unconvinced that precognition had been demonstrated, whereas in the second case most people would be largely convinced of the efficacy of the new treatment. Such disagreements that might arise in the above analyses would be based, in the first case, on the extent of our predisposition to accept the existence of psychic powers, and, in the second, on possible medical grounds that we might have to be suspicious of the new treatment. Thus, similar data in different experiments may lead to different conclusions, when judged by the same person, and the same data may lead to different conclusions when judged by different people. In the above cases, the differences in the conclusions arise from differences in beliefs, either over the a priori plausibility of the hypotheses in the two experiments, or disagreements between individual beliefs as to the a priori plausibility of the hypothesis in a given experiment. More generally, people may disagree as to the relevance of the data to the conclusions or to any other feature of the probabilistic modelling required to reach a given conclusion.

Statistical theory has traditionally been concerned with analysing evidence derived from individual experiments, employing seemingly objective methods which lead to apparently clear-cut conclusions. In this view, the task of the statistician is to analyse individual data sets and, where necessary, pass the conclusions of the analysis to subject area specialists who then try to reach substantive conclusions. This viewpoint has the apparent virtue of turning statistics into a well-defined technical activity, which can be conducted in comparative isolation from the difficulties involved in making practical decisions. For example, in each of the two experiments above we may agree that, given a certain null hypothesis (no precognitive ability, no difference between treatments), the experiment has yielded a surprising result. This data analysis may be useful and revealing. However, as we have observed, such surprise may have different implications between experiments and between individuals. Ultimately, whether or not a particular data set suggests that a new treatment is better than the current treatment is only of interest if such consideration helps us to address the substantive question as to whether it is reasonable for us to believe and act as though the new treatment actually is better.

Such substantive analyses are much harder than the analysis of individual data sets, as they must confront and synthesize all of the evidence, including much that is fragmentary, contradictory, hard to find and difficult to assess, and for which there may be legitimate grounds for expert disagreement. However, these difficulties are unavoidable given that we want to reach substantive conclusions.

In practice, statisticians often do present themselves as addressing substantive issues, and are generally perceived as so doing by their clients. Indeed, the theory of statistical inference is generally formulated and perceived as an attempt to address substantive questions, but this may only be achieved within a traditional statistical analysis when the data set is sufficiently large and unambiguous as to overwhelm all other sources of prior information. When the statistical analysis is less clear-cut, it is necessary to synthesize the statistical results with all of the other considerations which might influence the substantive conclusions of the analysis. However, in current practice, this synthesis rarely takes place. As a result, the fate of far too many statistical analyses is to be accepted uncritically, or completely ignored, or treated in some other equally arbitrary fashion. The only way to avoid this fate is to frame the statistical analysis within the wider context with which the problem should be concerned, so that the purpose and construction of the analysis are directed at those things that we actually wish to know.

However, such a change in orientation requires a change in attitude and approach. Statisticians are used to being careful and precise in the collection and quantitative analysis of data. What we must further develop are the corresponding methods and skills for the specification and quantitative analysis of beliefs. As our beliefs are of fundamental interest, the study and refinement of these beliefs offer a central unifying principle for the bewildering variety of problems that we may confront when analysing uncertain situations.

The most fully developed methodology for such study is the Bayesian approach. We shall develop an alternative framework for the quantitative elicitation, analysis and interpretation of our beliefs, with particular emphasis on situations where our beliefs are at least partly influenced by statistical data. The framework is similar in spirit to the Bayes formalism. However, it differs in various important ways which are directed towards clearer and simpler analyses of beliefs, as, for reasons that we shall discuss in the next section, even the Bayesian approach can easily become, in practice, a methodology for using beliefs to analyse data, rather than a methodology for using data to analyse beliefs.

1.2 The Bayesian approach

Suppose that you visit a doctor, as you fear that you might have some particular disease, which you either have, event D, or you do not have, event D^c. The doctor gives you a test, which either is positive, event T, or not positive, event T^c. Before testing, you have a prior probability, $\mathrm{P}(D)$, that you have the disease. If you take the test, and the result is positive, then your conditional probability of the disease is given by Bayes' theorem as

$$P(D|T) = \frac{P(T|D)P(D)}{P(T|D)P(D) + P(T|D^c)P(D^c)}. \tag{1.1}$$

Using Bayes' theorem, we replace the question

- Does the data, i.e. the test, suggest that you have the disease?

with the substantive question

- Should you now believe that you have the disease?

The evidence provided by the data, in this case the **likelihood ratio**,

$$\frac{P(T|D)}{P(T|D^c)},$$

has been combined with the external evidence as to whether you have the disease, as summarized by the **prior odds ratio**,

$$\frac{P(D)}{P(D^c)},$$

to produce the composite conditional probability $P(D|T)$.

This form of argument dates back at least to the famous posthumously published essay of Thomas Bayes. At that time, probabilistic judgements were generally taken to be subjective quantifications of opinion. Subsequently, however, a different tradition arose, within which statisticians became reluctant to allow that a general statement, for example that a new treatment is better than a current treatment, could meaningfully be given a prior probability. As a result, use of the Bayes argument fell out of fashion, and probabilistic analysis was only deemed relevant within statistics to the extent that it applied to the outcomes of well-defined and repeatable sampling experiments.

While this may even now be a majority view, Bayes methods have recently grown again in popularity. This is partly due to the influence of decision analysis, in which the Bayes paradigm fits very naturally, and partly as a consequence of the critical re-examination of the logical, philosophical and practical basis of statistical procedures. The strengths of the Bayes approach are, first, that it appears to be more logical than most other approaches, replacing *ad hoc* methods with a unified methodology, and, second, that the approach may be used to address complex problems which cannot easily be considered within more traditional statistical paradigms. As a result, the approach has been judged to be successful in many applications, particularly where the analysis of data has been improved by combination with expert judgements.

However, perhaps because of the historical development, Bayes methods have themselves often been viewed as a sophisticated form of data analysis, so that much emphasis has been placed on 'objective Bayes methods' based on 'non-informative priors' and similar methods which are intended to extract information from a particular data set, without imposing any particular prior quantifications. Thus, there has developed a form of 'objective' Bayes methodology which is implicitly based around the idea that we may use beliefs to improve the analysis of data, in the sense that we may consider that data have a story to tell that is quite separate from the individual preconceptions that we may bring to the analysis. Such methods may be interesting, particularly for the analysis of large data sets, but they cannot address directly the substantive questions that concern us. To do so, we require the reverse process, namely to use data to help analyse beliefs. However, there is a fundamental difficulty in carrying out this program within the Bayes paradigm, namely that honest belief specification for large problems is usually very difficult.

Even in small problems, with few sources of uncertainty, it can be hard to distil all of our prior knowledge into a satisfactory full joint prior probability specification over all of the possible outcomes. In practical problems there may be hundreds of relevant sources of uncertainty about which we may make prior judgements. In such problems it is arguably impossible for us to carry out the Bayes programme, which requires us to specify meaningful probabilistic beliefs over collections of probability distributions over such high-dimensional structures. Even were we able to carry out such a full prior specification, we would usually find that the specification was too time-consuming and too difficult to check, document and validate to be worth the effort, unless we were working on questions that were of such importance that they justified the enormous expenditure of effort that is required simply to apply the paradigm in an honest fashion.

Even if we were able to make such high-dimensional specifications, the resulting Bayes analysis would often be extremely computer intensive, particularly in areas such as experimental design. Computational issues, while of great practical importance, are secondary to the fundamental difficulty of making meaningful high-dimensional prior probability specifications. However, such considerations do support the basic argument that we shall develop in this book, which is as follows.

The more complex the problem, the more we need help to consider the resulting uncertainties, but the more difficult it is to carry out a full Bayes analysis. Essentially, the Bayes approach falls victim to the ambition in its formulation. Often, the approach is considered to be a description of what a perfectly rational individual would do when confronted with the problem. The implication is that we should copy the behaviour of such an individual as closely as we can. However, as the complexity of problems increases, the disparity between the hypothetical abilities of the perfectly rational analyst and our actual abilities to specify and analyse our uncertainties becomes so wide that it is hard to justify the logical or practical relevance of such a formulation.

Therefore if, in complex problems, we are unable to make and analyse full prior specifications, it follows that we need to develop methods based around **partial** belief specification. We shall develop one such methodology, termed the **Bayes**

linear approach. The approach is similar in spirit to the full Bayes approach, and is particularly appropriate whenever the full Bayes approach requires an unnecessarily exhaustive description and analysis of prior uncertainty.

Depending on our viewpoint, we may view the Bayes linear approach either as offering a simple approximation to a full Bayes analysis, for problems where the full analysis would be too difficult or time-consuming, or as complementary to the full Bayes analysis, offering a variety of new interpretative and diagnostic tools which may be of value whatever our viewpoint, or as a generalization of the full Bayes approach, where we lift the artificial constraint that we require full probabilistic prior specification before we may learn anything from data.

1.3 Features of the Bayes linear approach

The following are important features of the Bayes linear approach.

1. The approach is subjectivist. We express our prior judgements of uncertainty in quantitative form, and adjust these uncertainties in the light of observation.
2. We use prior specifications which honestly correspond to prior beliefs. In order to do this, we must structure our analyses so that the prior specifications that we require are within the ability of the individual to make.
3. The approach is based on expectation rather than probability as a primitive. With expectation as a primitive, we may immediately obtain probabilities as expectations of indicator functions. With probability as a primitive, we need to determine all probabilities for a quantity before we may assess the expectation. Therefore, starting with expectation allows us to focus directly on the crucial uncertainties in the problem.
4. With expectation as a primitive, the fundamental object of interest is the collection of random quantities, which are naturally gathered into inner product spaces. Therefore, the resulting analysis follows from the geometric structure implied by the partial belief specification.
5. Beliefs are adjusted by linear fitting rather than conditioning. Therefore, the Bayes linear approach may be viewed as a simple and tractable approximation to a full Bayes analysis.
6. There are general temporal relationships between the adjusted beliefs created by linear fitting and our posterior beliefs. Full conditioning is a special case of linear fitting whose general temporal relation with posterior beliefs is no different than for any other linear fit. Therefore the full Bayes analysis may be also viewed as a particular special case of the Bayes linear approach.
7. As linear fitting is generally computationally simpler than full conditioning, we may often analyse complex problems, in particular those arising in experimental design, more straightforwardly than under the full Bayes counterpart.

8. We only specify beliefs over observable quantities, so that all of our belief statements can be given a direct, physical interpretation. We therefore construct underlying population models strictly by means of exchangeability judgements over observables, which is feasible precisely because we take expectation as the primitive for the theory.

9. Our aim is to develop improved assessments of belief. Partly, this is achieved by sensible processing of prior and data inputs. However, just as important is the qualitative interpretation of the belief adjustment. Therefore, we develop interpretative tools to identify which aspects of our prior judgements and the data are most influential for which aspects of our conclusions, so that we may judge whether or not our belief adjustments appear intuitively reasonable, and compare possible alternative adjustments, based for example on different sampling frames or experimental designs.

10. When we adjust our beliefs, we similarly need qualitative methods for interpreting the resulting collection of changes in belief. Therefore, we develop interpretative tools to summarize both the magnitude and the nature of the overall changes in belief, and to display conflict or consistency between the various sources of evidence which contribute to such changes.

11. Each belief statement made about an observable may be subsequently compared with the value of that observable. Stringent diagnostics are available to warn us of possible conflicts between our beliefs and reality.

12. There are important special cases, for example certain analyses for multivariate Gaussian models, where many aspects of the Bayes and the Bayes linear approaches correspond. Therefore, many of the interpretative and diagnostic tools that we describe will also be relevant for such analyses. Further, it is of general interest to separate those aspects of the Gaussian analysis which follow directly from the geometric implications of the second-order specification, from those aspects whose validity depends on the precise form of the Gaussian density function.

13. Much of the qualitative and quantitative structure of the Bayes linear analysis may be displayed visually using Bayes linear graphical models. These models aid the intuitive understanding of expected and observed information flow through complex systems, and also facilitate efficient local computation methods for the analysis of large systems.

1.4 Example

As a trailer for the ideas in the book we give the following example. The example is intended to convey the flavour of our approach, and so we refrain both from detailed exposition of the methodology and from deep analysis of the problem.

A factory produces two products. For planning purposes, the factory wishes to predict sales of the products in each period. In order to do this, various relevant information will be used, in particular the sales of the two products in the previous period. For this introduction, it will be sufficient to suppose that this is all that is explicitly used, though of course the judgements of the sales forecasters will be called on to formulate the prior beliefs.

For illustration, we shall imagine that sales at a time point soon to come are used to improve our understanding of sales at a more distant future time point. Thus, there are four quantities of interest: X_1 and X_2, the sales of products 1 and 2 at the first time point, and Y_1 and Y_2, the corresponding sales at the later time point.

For the simplest form of analysis that we shall describe, the sales forecaster first specifies prior expectations for the four quantities, together with a variance matrix over them. We will consider the problem of eliciting and specifying prior information in the form of expectations, variances, and covariances in Chapter 2. In the meantime, suppose that we have based our prior specifications on sample information from previous sales figures, and managerial judgements as to their relevance in the light of any special circumstances which may be felt appropriate to the current sales period.

1.4.1 Expectation, variance, and standardization

In this book, we assume basic knowledge of expectation, variance and covariance, and correlation. Suppose that X and Y are collections of m and n random quantities, respectively. The expectation for X is denoted by $\mathrm{E}(X)$, an $m \times 1$ vector with ith element $\mathrm{E}(X_i)$. The variance for X is denoted by $\mathrm{Var}(X)$, an $m \times m$ variance–covariance matrix with (i, i)th element $\mathrm{Var}(X_i)$ and with (i, j)th element giving the covariance between X_i and X_j, denoted by $\mathrm{Cov}(X_i, X_j)$. The covariance between X and Y is denoted by $\mathrm{Cov}(X, Y)$, an $m \times n$ covariance matrix with (i, j)th element $\mathrm{Cov}(X_i, Y_j)$. The correlation between X and Y is denoted by $\mathrm{Corr}(X, Y)$, an $m \times n$ correlation matrix with (i, j)th element $\mathrm{Corr}(X_i, Y_j)$, assuming finite non-zero variances $\mathrm{Var}(X_i)$ and $\mathrm{Var}(Y_j)$. We may find it helpful to refer to the standardized versions of quantities.

Definition 1.1 *For a random quantity X, we write the standardized quantity as*

$$\mathrm{S}(X) = \frac{X - \mathrm{E}(X)}{\sqrt{\mathrm{Var}(X)}}.$$

1.4.2 Prior inputs

Suppose that, in some appropriate units, the prior mean for each quantity is 100; the prior variance for X_1, X_2 is 25; the prior variance for the future sales Y_1, Y_2 is 100; and the prior correlation matrix over all four quantities is

	X_1	X_2	Y_1	Y_2
X_1	1.00	−0.60	0.60	−0.20
X_2	−0.60	1.00	−0.20	0.60
Y_1	0.60	−0.20	1.00	−0.60
Y_2	−0.20	0.60	−0.60	1.00

Thus, we might summarize our prior specifications as follows. We have the same expectation for sales for each product at each time point, but we are much less certain about the sales figures for the later time point. The correlation matrix specified expresses the belief that sales of each product are quite strongly positively correlated over the two time periods, but that the products are considered to compete and so sales of the two products are negatively correlated. Note that in this problem we do not complete the prior specification by choosing a prior joint probability distribution for these four quantities with the given mean and variance structure. Rather, our aim is to perform an analysis based solely on the partial prior specification that we have described.

We intend to use the sales at the first time point to improve our forecasts for sales at the later time point. Much of our approach deals with simultaneous analysis of **collections** of quantities, so, for convenience, we group together the two sales from the first time point into the collection $D = (X_1, X_2)$, and the two sales for the later time point into the collection $B = (Y_1, Y_2)$. There is no particular significance to the names B and D, except that we sometimes find it useful to retain D for a collection of 'data' quantities (i.e. quantities which we intend to observe, and so for which data will become available) and to retain B for a collection of 'belief' quantities (i.e. quantities that we wish to predict, and so for which we have prior beliefs followed by adjusted beliefs).

1.4.3 Adjusted expectations

There are many ways in which we might try to improve our forecasts for the collection B. A simple method, which exploits the prior mean and variance statements that we have made, is as follows. We can look among the collection of linear estimates, i.e. those of the form $c_0 + c_1X_1 + c_2X_2$, and choose constants c_0, c_1, c_2 to minimize the prior expected squared error loss in estimating each of Y_1 and Y_2. For example, we aim to minimize

$$\mathrm{E}([Y_1 - c_0 - c_1X_1 - c_2X_2]^2). \tag{1.2}$$

The choices of constants may be easily computed from the above specifications, and the estimators turn out to be

$$\mathrm{E}_D(Y_1) = 1.5X_1 + 0.5X_2 - 100, \tag{1.3}$$

$$\mathrm{E}_D(Y_2) = 0.5X_1 + 1.5X_2 - 100. \tag{1.4}$$

We call $\mathrm{E}_D(Y_1)$ the **adjusted expectation** for Y_1 given the information $D = [X_1, X_2]$. Similarly, $\mathrm{E}_D(Y_2)$ is the adjusted expectation for Y_2 given D. The adjusted expectations have a number of properties which we will come to below and in later chapters; in particular, they are themselves random quantities, so that they too have expectations, variances and so forth.

1.4.4 Adjusted versions

We will be concerned not only with the adjusted expectation for a quantity, but also with the residual component associated with it, which we call the **adjusted version** of the quantity. The adjusted version of Y given D is defined to be $\mathbb{A}_D(Y) = Y - \mathrm{E}_D(Y)$. In our example, the adjusted versions are

$$\mathbb{A}_D(Y_1) = Y_1 - (1.5X_1 + 0.5X_2 - 100), \tag{1.5}$$

$$\mathbb{A}_D(Y_2) = Y_2 - (0.5X_1 + 1.5X_2 - 100). \tag{1.6}$$

These adjusted versions have important roles to play in Bayes linear analysis in that they allow us to quantify the uncertainty expected to remain after an adjustment. A priori, we expect the residual component to be zero, $\mathrm{E}(\mathbb{A}_D(Y_i)) = 0$.

1.4.5 Adjusted variances

How useful are the adjusted expectations when judged as predictors? One way to assess how much information about the elements of B we gain by observing the elements of D is to evaluate the **adjusted variance** for each quantity. The adjusted variance for any quantity Y, given a collection of information D, is defined as

$$\mathrm{Var}_D(Y) = \mathrm{Var}(\mathbb{A}_D(Y)) = \mathrm{E}([Y - \mathrm{E}_D(Y)]^2),$$

being the minimum of the prior expected squared error loss in the sense of (1.2). This is a measure of the residual uncertainty, or, informally, the 'unexplained' variance, having taken into account the information in D. The portion of variation resolved is

$$\mathrm{Var}(Y) - \mathrm{Var}_D(Y) = \mathrm{Var}(\mathrm{E}_D(Y)).$$

For this example the adjusted variances are the same, so that we have

$$\mathrm{Var}_D(Y_1) = \mathrm{Var}_D(Y_2) = 60,$$

whereas we began with variances $\mathrm{Var}(Y_1) = \mathrm{Var}(Y_2) = 100$. Consequently, the value of observing sales at the first time point is to reduce our uncertainty about sales at the later time point by 40%. We typically summarize the informativeness of data D for any quantity Y by a scale-free measure which we call the **resolution** of Y induced by D, defined as

$$\mathrm{R}_D(Y) = 1 - \frac{\mathrm{Var}_D(Y)}{\mathrm{Var}(Y)} = \frac{\mathrm{Var}(\mathrm{E}_D(Y))}{\mathrm{Var}(Y)}. \tag{1.7}$$

In our example, the variance resolutions are $\mathrm{R}_D(Y_1) = \mathrm{R}_D(Y_2) = 0.4$. The resolution lies between 0 and 1, and in general, small (large) resolutions imply that the information has little (much) linear predictive value, given the prior specification.

In terms of the vector B, we began with a variance matrix $\mathrm{Var}(B)$ which we have decomposed into unresolved and resolved portions, each a matrix:

$$\mathrm{Var}(B) = \mathrm{Var}_D(B) + \mathrm{RVar}_D(B), \tag{1.8}$$

where $\mathrm{RVar}_D(B) = \mathrm{Var}(\mathrm{E}_D(B))$ is our notation for the resolved variance matrix for the adjustment of the collection B by the collection D, and equals the prior variance matrix for the adjusted expectation vector. The off-diagonal terms are **adjusted covariances** and **resolved covariances**. For example, the adjusted covariance between Y_1 and Y_2 given D is the covariance between the two residual components,

$$\mathrm{Cov}_D(Y_1, Y_2) = \mathrm{Cov}(\mathbb{A}_D(Y_1), \mathbb{A}_D(Y_2)),$$

and the resolved covariance is the change from prior to adjusted,

$$\mathrm{RCov}_D(Y_1, Y_2) = \mathrm{Cov}(Y_1, Y_2) - \mathrm{Cov}_D(Y_1, Y_2).$$

In our example, the decomposition (1.8) turns out to be

$$\mathrm{Var}(B) = \begin{bmatrix} 100 & -60 \\ -60 & 100 \end{bmatrix} = \begin{bmatrix} 60 & -60 \\ -60 & 60 \end{bmatrix} + \begin{bmatrix} 40 & 0 \\ 0 & 40 \end{bmatrix}.$$

The off-diagonal entries here show that $\mathrm{Cov}(Y_1, Y_2) = \mathrm{Cov}_D(Y_1, Y_2) = -60$, and that $\mathrm{RCov}_D(Y_1, Y_2) = 0$. It may seem a little puzzling that we do not seem to have resolved any of the covariance between Y_1 and Y_2. Indeed, the variance matrix for their adjusted versions is singular. We shall discover why this is so, and comment on it in more detail, later.

1.4.6 Checking data inputs

At some point, we may observe the values of D. In our case, suppose that the sales at the first time point turn out to be $x_1 = 109$ and $x_2 = 90.5$. (We follow convention in using lower case for observations and upper case for unknowns.) The first thing we do is to check that these observations are consistent with beliefs specified about them beforehand. A simple diagnostic is to examine the **standardized change** from the prior expectation to the observed value. In our example, the standardized changes are

$$\mathrm{S}(x_1) = \frac{x_1 - \mathrm{E}(X_1)}{\sqrt{\mathrm{Var}(X_1)}} = \frac{109 - 100}{\sqrt{25}} = 1.8, \tag{1.9}$$

$$\mathrm{S}(x_2) = \frac{90.5 - 100}{\sqrt{25}} = -1.9. \tag{1.10}$$

Each (squared) standardized change has prior expectation one. Informally, we might begin to suspect an inconsistency if we saw a standardized change of more than about two standard deviations; and be quite concerned to see standardized changes of more than about three standard deviations. We do not wish to give rigid rules or thresholds for interpreting these kinds of measure, as they are largely dependent on the context of the problem.

1.4.7 Observed adjusted expectations

When the data quantities are observed we may calculate the observed adjusted expectations. Replacing X_1, X_2 by $x_1 = 109$ and $x_2 = 90.5$ in (1.3) and (1.4), we obtain the following assessments:

$$\mathrm{E}_d(Y_1) = 1.5 \times 109 + 0.5 \times 90.5 - 100 = 108.75,$$

$$\mathrm{E}_d(Y_2) = 0.5 \times 109 + 1.5 \times 90.5 - 100 = 90.25.$$

We call these values **observed adjusted expectations**. Notice that our subscript notation uses lower case, $\mathrm{E}_d(\cdot)$, rather than upper case, $\mathrm{E}_D(\cdot)$ to indicate that the entire collection D has been observed to be d. The effect of the data here is to cause our expectations for future sales to follow a similar pattern, i.e. larger and smaller sales respectively in the two components.

1.4.8 Diagnostics for adjusted beliefs

It is valuable at this stage to check how different the observed adjusted expectation is from the prior expectation. A simple diagnostic is given by the change from prior to adjusted expectation, standardized with respect to the variance of the adjusted expectation. We have that $\mathrm{E}(\mathrm{E}_D(Y)) = \mathrm{E}(Y)$ for any Y and D. Thus, from (1.9), the standardized change is

$$\mathrm{S}(\mathrm{E}_d(Y)) = \frac{\mathrm{E}_d(Y) - \mathrm{E}(Y)}{\sqrt{\mathrm{Var}(\mathrm{E}_D(Y))}},$$

where the denominator in the standardization does not depend on the observed data. We call these standardized changes the **standardized adjustments**. In our example, they are:

$$\mathrm{S}(\mathrm{E}_d(Y_1)) = \frac{108.75 - 100}{\sqrt{40}} = 1.38, \qquad \mathrm{S}(\mathrm{E}_d(Y_2)) = \frac{90.25 - 100}{\sqrt{40}} = -1.54,$$

where in each case the squared standardized adjustment has prior expectation one. As such, the changes in expectation for sales at a future time point are 1.38 and 1.54 standard deviations, relative to variation explained, and so are roughly in line with what we expected beforehand.

1.4.9 Further diagnostics for the adjusted versions

As time progresses, we eventually discover actual sales, $y_1 = 112$ and $y_2 = 95.5$, of the two products. It is diagnostically important now to compare our predictions with what actually happened. There are two diagnostics to examine. First, we can compare a quantity's observation with its prior expectation, irrespective of the linear fitting on D. The standardized change in expectation for a quantity is given by (1.9). In our example, the standardized changes in expectation from prior to observed are $\mathrm{S}(Y_1) = (112 - 100)/10 = 1.2$ and $\mathrm{S}(Y_2) = -0.45$, so these future sales turned out to be consistent with our prior considerations.

A second diagnostic is given by examining the change from adjusted expectation to actual observation, relative to the associated adjusted variance, as this was the variation remaining in each Y_i after fitting on D, but before observing Y_1 and Y_2. By observing the actual sales values y_1, y_2, we observe the residual components, i.e. the adjusted versions $\mathbb{A}_D(Y_i) = Y_i - \mathrm{E}_D(Y_i)$. Given that they had prior expectation zero, we wish to see how far the adjusted versions have changed from zero, relative to their variances

$$\mathrm{Var}(\mathbb{A}_D(Y_i)) = \mathrm{Var}_D(Y_i).$$

The appropriate standardized change is thus

$$\mathrm{S}_d(y_i) = \mathrm{S}(\mathbb{A}_d(y_i)) = \frac{y_i - \mathrm{E}_d(Y_i)}{\sqrt{\mathrm{Var}_D(Y_i)}}.$$

In our example, the sales at the later time point, $y_1 = 112$, $y_2 = 95.5$, should be compared to the adjusted expectations $\mathrm{E}_d(Y_1) = 108.75$ and $\mathrm{E}_d(Y_2) = 90.25$, standardizing with respect to the adjusted variances:

$$\mathrm{Var}_D(Y_1) = \mathrm{Var}_D(Y_2) = 60.$$

We obtain

$$\mathrm{S}_d(y_1) = \frac{112 - 108.75}{\sqrt{60}} = 0.42 \quad \text{and} \quad \mathrm{S}_d(y_2) = \frac{95.5 - 90.25}{\sqrt{60}} = 0.68.$$

The squared standardized changes should again be about one, so our diagnostic checks suggest that both of our predictions were roughly within the tolerances suggested by our prior variance specifications. If anything, the adjusted expectations are, in terms of standard deviations, rather closer to the observed values than expected.

1.4.10 Summary of basic adjustment

Let us summarize our results so far in the form of tables, shown in Table 1.1. The analysis results in decomposing the sales quantities into two parts, the first of which comes from linear fitting on other quantities D, and the second of which is residual. Summary statistics are calculated for the original and component quantities; all summaries are additive over components, except for the standardized changes. We note that the diagnostics reveal nothing untoward: all the standardized changes are about in line with what was expected beforehand. In each case, the change from prior to adjusted expectation was slightly larger than expected, one up and one down; and in each case the standardized change from adjusted expectation to

Table 1.1 Adjusting future sales Y_1, Y_2 by previous sales: summary.

	Original	=	Adjusted expectation	+	Adjusted version
Quantity	Y_1	=	$\mathrm{E}_D(Y_1)$	+	$\mathbb{A}_D(Y_1)$
		=	$1.5X_1 + 0.5X_2 - 100$	+	$Y_1 - \mathrm{E}_D(Y_1)$
Prior expectation	$\mathrm{E}(Y_1)$	=	$\mathrm{E}(\mathrm{E}_D(Y_1)) = \mathrm{E}(Y_1)$	+	$\mathrm{E}(\mathbb{A}_D(Y_1)) = 0$
	100	=	100	+	0
Prior variance	$\mathrm{Var}(Y_1)$	=	$\mathrm{RVar}_D(Y_1)$	+	$\mathrm{Var}_D(Y_1)$
	100	=	40	+	60
Observed	y_1	=	$1.5x_1 + 0.5x_2 - 100$	+	$y_1 - \mathrm{E}_d(Y_1)$
	112	=	108.75	+	3.25
Standardized change	$\frac{y_1-\mathrm{E}(Y_1)}{\sqrt{\mathrm{Var}(Y_1)}}$		$\frac{\mathrm{E}_d(Y_1)-\mathrm{E}(Y_1)}{\sqrt{\mathrm{RVar}_D(Y_1)}}$		$\frac{y_1-\mathrm{E}_d(Y_1)}{\sqrt{\mathrm{Var}_D(Y_1)}}$
	1.2		1.38		0.42
Quantity	Y_2	=	$\mathrm{E}_D(Y_2)$	+	$\mathbb{A}_D(Y_2)$
		=	$0.5X_1 + 1.5X_2 - 100$	+	$Y_2 - \mathrm{E}_D(Y_2)$
Prior expectation	$\mathrm{E}(Y_2)$	=	$\mathrm{E}(\mathrm{E}_D(Y_2)) = \mathrm{E}(Y_2)$	+	$\mathrm{E}(\mathbb{A}_D(Y_2)) = 0$
	100	=	100	+	0
Prior variance	$\mathrm{Var}(Y_2)$	=	$\mathrm{RVar}_D(Y_2)$	+	$\mathrm{Var}_D(Y_2)$
	100	=	40	+	60
Observed	y_2	=	$0.5x_2 + 1.5x_2 - 100$	+	$y_2 - \mathrm{E}_d(Y_2)$
	95.5	=	90.25	+	5.25
Standardized change	$\frac{y_2-\mathrm{E}(Y_2)}{\sqrt{\mathrm{Var}(Y_2)}}$		$\frac{\mathrm{E}_d(Y_2)-\mathrm{E}(Y_2)}{\sqrt{\mathrm{RVar}_D(Y_2)}}$		$\frac{y_2-\mathrm{E}_d(Y_2)}{\sqrt{\mathrm{Var}_D(Y_2)}}$
	−0.45		−1.54		0.68

observed value was smaller than expected, and closer to the original prior expectation. Whether this should cause concern cannot be answered solely by examining single quantities using summaries such as these, useful though they are. In fact, we need also to analyse changes in our collection of beliefs, which we consider next.

1.4.11 Diagnostics for collections

We showed in §1.4.6 how we check individual data inputs by calculating standardized changes. To check a collection of data inputs, we need to make a basic consistency check, and if this is successful we proceed to calculate a global discrepancy. For the basic consistency check, recall that, for any random quantity X, if we specify $\text{Var}(X) = 0$ then we expect to observe $x = \text{E}(X)$: otherwise either the variance specification is wrong, or perhaps some error has occurred in collecting the data. For a collection (vector) of random quantities B, with observed value b, expectation $\text{E}(B)$, and variance matrix $\text{Var}(B)$, the basic consistency check is as follows. If $\text{Var}(B)$ is non-singular then the value of $b - \text{E}(B)$ is unconstrained, and the basic consistency check is passed. Otherwise, $\text{Var}(B)$ has one or more eigenvalues equal to zero. In this case, suppose that q is an eigenvector corresponding to a zero eigenvalue. Such eigenvectors identify linear combinations of the Bs having variance zero, as for each such eigenvector q, it is the case that $\text{Var}(q^T B) = 0$. Consequently, in the case of singularity the basic consistency check lies in verifying that $q^T b = q^T \text{E}(B)$ for every eigenvector q corresponding to a zero eigenvalue. Failure of the consistency check always corresponds to infinite values for the corresponding standardized changes. Following a successful basic consistency check, we calculate measures of discrepancy based on the Mahalanobis distance.

To return to checking data inputs, we are concerned with differences between a vector of data d and the vector of prior expectations $\text{E}(D)$. The variance matrix concerned here is

$$\text{Var}(D) = \begin{bmatrix} 25 & -15 \\ -15 & 25 \end{bmatrix},$$

which is full rank, so that the basic consistency check is passed. Next, for our measure of the difference between the data d and their prior expectations $\text{E}(D)$, we calculate the **discrepancy**, $\text{Dis}(d)$, as the Mahalanobis distance between d and $\text{E}(D)$:

$$\begin{aligned} \text{Dis}(d) &= (d - \text{E}(D))^T \text{Var}(D)^\dagger (d - \text{E}(D)) \\ &= \begin{bmatrix} 109 - 100 & 90.5 - 100 \end{bmatrix} \begin{bmatrix} 25 & -15 \\ -15 & 25 \end{bmatrix}^{-1} \begin{bmatrix} 109 - 100 \\ 90.5 - 100 \end{bmatrix} \\ &= 4.29. \end{aligned}$$

Here, $\text{Var}(D)^\dagger$ is the Moore–Penrose generalized inverse of $\text{Var}(D)$, equivalent to the usual inverse $\text{Var}(D)^{-1}$ when $\text{Var}(D)$ is full rank. The Moore–Penrose inverse is employed as we make no distinction between the handling of full rank and singular variance matrices: this is especially useful when analysing the structural implications of prior specifications. The discrepancy has prior expectation equal to the rank of the prior variance matrix $\text{Var}(D)$, which in our example has rank two. We thus obtain as a summary statistic of the discrepancy between the observed values and the prior specification, the **discrepancy ratio**,

$$\text{Dr}(d) = \frac{\text{Dis}(d)}{\mathbf{rk}\{\text{Var}(D)\}} = 2.15,$$

to be compared to its prior expectation of one. For single observations rather than collections, the discrepancies are just the squared standardized changes. None of these measures indicate any substantial problem with our prior formulation.

We showed in §1.4.8 how we calculate a standardized adjustment to check for a difference between an observed adjusted expectation and the corresponding prior expectation. As above, we obtain a global diagnostic by making a basic consistency check and then calculating a measure of discrepancy. The vectors to be compared are the observed adjustments, $\mathrm{E}_d(B)$, and their prior expectations, $\mathrm{E}(B)$. The variance matrix concerned is

$$\mathrm{Var}(\mathrm{E}_D(B)) = \mathrm{RVar}_D(B) = \begin{bmatrix} 40 & 0 \\ 0 & 40 \end{bmatrix},$$

which is full rank, so that the basic consistency check is passed. We obtain a global diagnostic for the observed adjustment by calculating the Mahalanobis distance between the observed adjusted expectations and the prior expectations, to give the **adjustment discrepancy**, $\mathrm{Dis}_d(B)$, where

$$\begin{aligned}\mathrm{Dis}_d(B) &= (\mathrm{E}_d(B) - \mathrm{E}(B))^T \mathrm{RVar}_D(B)^\dagger (\mathrm{E}_d(B) - \mathrm{E}(B)) \\ &= \begin{bmatrix} 108.75 - 100 & 90.25 - 100 \end{bmatrix} \begin{bmatrix} 40 & 0 \\ 0 & 40 \end{bmatrix}^{-1} \begin{bmatrix} 108.75 - 100 \\ 90.25 - 100 \end{bmatrix} \\ &= 4.29.\end{aligned}$$

As before, this discrepancy measure fails to suggest any substantial problem with our prior formulation.

For our final collection diagnostic of this section, we showed in §1.4.9 how to calculate the standardized change from observed adjusted expectation, $\mathrm{E}_d(Y_i)$, to actual observation y_i, where the standardization is with respect to the variance remaining in Y_i, $\mathrm{Var}_D(Y_i)$, before observing it. As above, we proceed to a global diagnostic where we wish to measure the discrepancy between the observed adjusted expectations $\mathrm{E}_d(B)$, and the actual observations $b = [y_1\ y_2]^T$, relative to the variance matrix $\mathrm{Var}_D(B)$. Another way of thinking about this is that we finally observe the adjusted versions $\mathbb{A}_D(B)$ and wish to see whether these observations are consistent with their prior variance–covariance specifications, $\mathrm{Var}(\mathbb{A}_D(B))$. For a basic consistency check, we have that

$$\mathrm{Var}(\mathbb{A}_D(B)) = \mathrm{Var}_D(B) = \begin{bmatrix} 60 & -60 \\ -60 & 60 \end{bmatrix}, \tag{1.11}$$

which is singular. There is one eigenvalue equal to zero, with corresponding eigenvector proportional to $[1\ 1]^T$. Consequently we have specified a variance of zero for

$$\begin{bmatrix} 1 & 1 \end{bmatrix}^T \begin{bmatrix} \mathbb{A}_D(Y_1) \\ \mathbb{A}_D(Y_2) \end{bmatrix} = \mathbb{A}_D(Y_1) + \mathbb{A}_D(Y_2),$$

and it is thus necessary to verify in this example that the observed adjusted versions sum to their expected value, which is zero. However, we see from Table 1.1 that the observed adjusted versions are 3.25 and 5.25, summing to $8.5 \neq 0$, so we have discovered a very serious flaw in our specification. In practice there is no point in proceeding further with the analysis. Had the basic consistency check not failed, we would have calculated the adjusted version discrepancy as

$$(b - \mathrm{E}_d(B))^T \mathrm{Var}_D(B)^\dagger (b - \mathrm{E}_d(B))$$

$$= \begin{bmatrix} 112 - 108.75 & 95.5 - 90.25 \end{bmatrix} \begin{bmatrix} 60 & -60 \\ -60 & 60 \end{bmatrix}^\dagger \begin{bmatrix} 112 - 108.75 \\ 95.5 - 90.25 \end{bmatrix} = 0.02.$$

1.4.12 Exploring collections of beliefs via canonical structure

To this point we have specified prior information, recorded some data, obtained predictions, calculated the value of the predictions, and compared expected to actual behaviour, largely focusing on the single quantities of interest, Y_1 and Y_2, the sales for two products at a future time point. Little of the analysis turned up anything surprising: changes in expectation were mostly about in line with what we expected. However, one of the diagnostics calculated for a collection revealed a very serious flaw, namely actual observations which should not have been possible given the prior specifications. This suggests, rightly, that our analysis should focus on analysing collections of beliefs, rather than on piecemeal analysis for single quantities. Further, to focus on collections of beliefs will allow us naturally to address many other relevant questions. For example, it reveals the implications of correlations between the collections of interest; it allows us to make global uncertainty and diagnostic assessments for entire collections or any sub-collections we choose; and it allows us easily to go beyond analysis of single quantities such as Y_1 and Y_2 to such quantities as total sales, $Y_1 + Y_2$, or the difference between sales, $Y_1 - Y_2$. Answering such questions is an important part of the Bayes linear approach.

It turns out, whether our interest is in making assessments for simple quantities such as Y_1, or for interesting linear combinations such as $Y_1 + Y_2$, or for global collections such as $B = [Y_1, Y_2]$, that for all such problems there is a natural reorganization which we may use to answer these questions directly. The reorganization arises by generating and exploiting an underlying **canonical structure**. This structure completely summarizes the global dynamics of belief adjustment for an analysis. For the two-dimensional problem, this amounts to finding the linear combinations of Y_1 and Y_2 about which D is respectively most and least informative, in the sense of maximizing and minimizing the variance resolution. In our example, these linear combinations have a particularly simple form; they are Z_1 and Z_2, where

$$Z_1 = 0.112(Y_1 + Y_2) - 22.361, \tag{1.12}$$

$$Z_2 = 0.056(Y_1 - Y_2). \tag{1.13}$$

For convenience, we have centred each Z_i so that it has prior mean zero, and scaled it so that it has prior variance one. We call Z_1 and Z_2 respectively the first and second **canonical directions**. Canonical directions are always uncorrelated. For our example, Z_1 is essentially a linear combination giving total sales, and Z_2 is the difference between sales. As far as the original sales quantities are concerned, they can be expressed in terms of the canonical quantities as

$$Y_1 = 4.472(Z_1 + 2Z_2) + 100,$$

$$Y_2 = 4.472(Z_1 - 2Z_2) + 100.$$

In addition to calculating the canonical directions, we also calculate their resolutions $\mathrm{R}_D(Z_1)$ and $\mathrm{R}_D(Z_2)$ from (1.7). We call these the **canonical resolutions**. The canonical directions and canonical resolutions together comprise the canonical structure. In our example, the resolutions in the canonical directions are $\mathrm{R}_D(Z_1) = 1$ and $\mathrm{R}_D(Z_2) = 0.25$. In the latter case, the implication is that the minimum variance resolution for **any** linear combination of the two unknown sales quantities is 0.25, i.e. by observing D we expect to 'explain' at least 25% of the variance for **all** linear combinations of our future sales quantities, Y_1 and Y_2.

The resolution of Z_1 turns out to be exactly 1. This means that, according to our prior specifications, there will be no uncertainty remaining in Z_1 once we have observed the previous sales X_1, X_2. This might appear to be good news: we are, after all, hoping to reduce our uncertainty about future sales by linear fitting on these two explanatory quantities. However, let us look a little more closely at the implications. Z_1 is proportional (except for a constant) to total sales: $Y_1 + Y_2 = 8.944Z_1 + 200$, so that one implication of our prior specification is that we shall have no uncertainty about $Y_1 + Y_2$ after we have observed X_1 and X_2. Indeed, as the adjusted expectations of Y_1, Y_2 are given above as $\mathrm{E}_d(Y_1) = 108.75$ and $\mathrm{E}_d(Y_2) = 90.25$ respectively, we shall apparently know certainly that $Y_1 + Y_2$ will be $108.75 + 90.25 = 199$. Did we really intend our prior specifications to contain the algebraic implication that we will 'know' total future sales in advance? Most likely we did not; and indeed later we actually observe total sales of $y_1 + y_2 = 112 + 95.5 = 207.5$, which flatly contradicts the prior specification, and which resulted in the failure of the consistency check in the previous section.

Now, what has led to this position? To find out, we obtain the adjusted expectations for the canonical quantities Z_1 and Z_2. For simplicity we introduce an obvious notation for the main sums and differences:

$$X^+ = X_1 + X_2, \quad X^- = X_1 - X_2, \quad Y^+ = Y_1 + Y_2, \quad Y^- = Y_1 - Y_2.$$

The adjusted expectations for the canonical quantities are:

$$\mathrm{E}_D(Z_1) = 0.224X^+ - 44.722, \tag{1.14}$$

$$\mathrm{E}_D(Z_2) = 0.056X^-.$$

The resolution $\mathrm{R}_D(Z_1) = 1$ corresponds to having an adjusted variance of zero for $\mathrm{E}_D(Z_1)$, shown as (1.14), so that the correlation between Z_1 (where $Z_1 \propto Y^+$)

and $\mathrm{E}_D(Z_1)$ (where $\mathrm{E}_D(Z_1) \propto X^+$) must be equal to one. Thus, X^+ and Y^+ have a prior correlation of one, and this explains why Y^+ becomes 'known' as soon as we observe x^+.

Now, while this was a logical consequence of our prior specification, it is quite possible that we had not realized, when we made our pairwise prior correlation specifications, that we were building such a strong degree of dependency between X^+ and Y^+. Indeed, it will usually be the case, particularly when we come to specify beliefs over large, complex and highly interdependent collections of quantities, that our initial prior specifications will have surprising and counter-intuitive consequences, which may cause us to reconsider the basis for our specifications. It is for this reason that it is vital to carry out a global analysis, by generating and examining the canonical structure, to ensure coherence and consistency over and between belief specifications and data. In particular, many defects are not discovered if we carry out analyses piecemeal – for example, nearly all of the analyses carried out in §1.4.3 to §1.4.10 are unremarkable when Y_1 and Y_2 are considered separately, but are revealed to be dubious when we analyse them as a collection. We did receive a hint of the underlying problem earlier, in §1.4.5, where we noticed the singularity in the adjusted variance matrix. Singularities showing up here are directly related to finding canonical resolutions equal to one.

In this particular example, the canonical quantities Z_1, Z_2 are the suitably centred and scaled versions of Y^+ and Y^-. Because of the symmetries involved in the prior specification, the **canonical data quantities** $\mathrm{E}_D(Z_1), \mathrm{E}_D(Z_2)$ are likewise the suitably centred and scaled versions of X^+ and X^-. Note that these are also uncorrelated. In later chapters we shall discuss in detail the use of such canonical structures and explain the relationship with classical canonical correlation analysis.

1.4.13 Modifying the original specifications

In this case, let us suppose that we reconsider our prior specifications. There are many changes that we might make. Suppose, for simplicity, that we decide not to change our prior means and variances for the four sales quantities, but just to weaken one or two of the correlations. In terms of the four sums and differences, the original prior correlation matrix was:

	X^-	X^+	Y^-	Y^+
X^-	1			
X^+	0	1		
Y^-	0.5	0	1	
Y^+	0	1	0	1

Inspecting the matrix, suppose we decide that it is appropriate to weaken the correlation between X^+ and Y^+ to 0.8. With this change, the prior correlation matrix over sales becomes

	X_1	X_2	Y_1	Y_2
X_1	1			
X_2	−0.60	1		
Y_1	0.56	−0.24	1	
Y_2	−0.24	0.56	−0.60	1

so that the actual effect is to decrease generally all the correlations between the sales quantities.

1.4.14 Repeating the analysis for the revised model

We now repeat our analysis with the modified belief specifications. The results are rather similar, and have similar interpretations. The adjusted expectations are now

$$\begin{aligned}\mathrm{E}_D(Y_1) &= 100 + 1.3(X_1 - 100) + 0.3(X_2 - 100), \\ \mathrm{E}_D(Y_2) &= 100 + 0.3(X_1 - 100) + 1.3(X_2 - 100),\end{aligned} \tag{1.15}$$

so that $x_1 = 109$, $x_2 = 90.5$ yields observed adjusted expectations of

$$\mathrm{E}_d(Y_1) = 108.85 \quad \text{and} \quad \mathrm{E}_d(Y_2) = 90.35.$$

These are about the same as for the original prior specifications. As before, the adjusted variances are the same for the two products,

$$\mathrm{Var}_D(Y_1) = \mathrm{Var}_D(Y_2) = 67.2,$$

so that the variance resolutions are 32.8%. Compared to the original specifications, the weakening of the underlying correlations leads to the explanatory quantities being less informative for future sales. The standardized changes in expectation (prior to adjusted) are $\mathrm{S}(\mathrm{E}_d(Y_1)) = 1.55$ and $\mathrm{S}(\mathrm{E}_d(Y_2)) = -1.69$, a little larger than before. Finally, when we observe $y_1 = 112$ and $y_2 = 95.5$, the standardized changes from adjusted expectation to observed are 0.38 and −0.63 respectively. Summaries of the basic adjustments are shown in Table 1.2.

In terms of the vector B, the decomposition of the prior variance matrix into unresolved and resolved portions is now (with the correlation matrices shown underneath),

$$\text{Variances:} \quad \begin{bmatrix} 100 & -60 \\ -60 & 100 \end{bmatrix} = \begin{bmatrix} 67.2 & -52.8 \\ -52.8 & 67.2 \end{bmatrix} + \begin{bmatrix} 32.8 & -7.2 \\ -7.2 & 32.8 \end{bmatrix},$$

$$\text{Correlations:} \quad \begin{bmatrix} 1 & -0.6 \\ -0.6 & 1 \end{bmatrix} \quad \begin{bmatrix} 1 & -0.79 \\ -0.79 & 1 \end{bmatrix} \quad \begin{bmatrix} 1 & -0.22 \\ -0.22 & 1 \end{bmatrix},$$

so that unlike for the first prior specification, there has been some alteration to the covariance structure for the residual portions of Y_1 and Y_2. The understanding of such changes to the covariance structure is a matter we defer until later.

Table 1.2 Adjusting future sales Y_1, Y_2 by previous sales: summary for the modified structure, giving expectations E(·), variances Var(·), and standardized changes S(·).

	Initial	=	Adjusted expectation	+	Adjusted version
	Y_1	=	$0.3X_1 + 1.3X_2 - 100$	+	$Y_1 - (0.3X_1 + 1.3X_2 - 100)$
Prior E(·)	100	=	100	+	0
Prior Var(·)	100	=	32.8	+	67.2
Data	112	=	108.85	+	3.15
Change S(·)	1.2		1.55		0.38
	Y_2	=	$0.3X_1 + 1.3X_2 - 100$	+	$Y_2 - (0.3X_1 + 1.3X_2 - 100)$
Prior E(·)	100	=	100	+	0
Prior Var(·)	100	=	32.8	+	67.2
Data	95.5	=	90.35	+	5.15
Change S(·)	−0.45		−1.69		−0.63

For the modified model, we recalculate the canonical structure. The two canonical directions are as in (1.12) and (1.13), with corresponding canonical resolutions $\mathrm{R}_D(Z_1) = 0.64$ and $\mathrm{R}_D(Z_2) = 0.25$. It follows that we expect to 'explain' 64% of the variation in the direction/linear combination $Z_1 \propto Y^+$, and this is the most we can learn about any linear combination of the two future sales quantities. Otherwise, the canonical structure is as before.

The canonical structure helps us to understand the implications of our belief specifications. There are two ideas. The first is that we examine the implications of our belief specifications as they affect variance reduction, and the second is that we do this globally, i.e. simultaneously over all linear combinations of interest, thereby taking account of the relationships expressed between the quantities being predicted. Our unknowns have been reorganized as a canonical structure which has two directions, scaled so that the prior variance in each is one, and so that the removed variance in each is the corresponding canonical resolution. Consequently we will talk of the global structure as having initial uncertainty $1 + 1 = 2$ and resolved uncertainty $0.64 + 0.25 = 0.89$, with resolution averaged over the structure evaluated as $0.89/2 = 0.445$. This single number, which we call the **system resolution** for our collection B of future sales quantities, is a simple quantification of the value of the information for the entire collection B. We treat the system resolution just as we treat resolutions for individual quantities such as Y_1. That is, a system resolution of zero implies that the information contains no potential to reduce uncertainties in the collection by linear fitting, whereas a system resolution of one implies that the information precisely identifies all the elements of the collection B. In this way we begin to distance ourselves from the idea that the individual quantities are the fundamentals of interest, and approach instead the idea that the **collections** constitute the fundamentals of interest. This blurring of the distinction between single quantities and collections of them has many advantages, particularly as the dimensionality of a problem increases.

1.4.15 Global analysis of collections of observations

In previous sections we saw that piecemeal analyses for individual quantities such as Y_1 provided little or no evidence of the serious flaws present in the prior belief specification; these flaws were revealed only by calculating and interpreting the underlying canonical structure. In a Bayes linear analysis we assess both the expected value of information sources and diagnostics (such as standardized changes) comparing expected to actual behaviour. Therefore, the question arises: is it sufficient to examine standardized changes for the single elements of a collection, or, analogous to the underlying canonical structure, is there a more informative underlying diagnostic structure? Recall that one motivation for calculating the canonical structure was to find the linear combination with maximum variance reduction. Suppose, analogously, that we calculate the linear combination Y^* with the largest squared change in expectation, relative to prior variance. For the observations $x_1 = 109$, $x_2 = 90.5$, this turns out to be

$$Y^* = 0.0478Y_1 - 0.0678Y_2 + 2.0000.$$

This linear combination, which has been centred so that it has prior expectation zero, has adjusted expectation $\mathrm{E}_D(Y^*) = 1.078$ and, for a reason we shall come to, a prior variance also of 1.078. Thus, the largest change in expectation from prior to adjusted for any linear combination of the future sales quantities is about $\sqrt{1.078} = 1.038$ prior standard deviations. It appears that the interplay between prior specifications and the data used to compute adjusted expectations is about as expected. As Y^* has been deliberately chosen to maximize the squared standardized change in expectation, we now describe how to assess the magnitude of the maximal change associated with it.

It turns out that Y^* has a unique and important role to play in Bayes linear analysis, and so we introduce a notation and a name for it. For a collection B being adjusted by a further collection D observed to be d, we call the linear combination in B with the largest standardized squared change in expectation the **bearing**, and we use the notation $\mathbb{Z}_d(B)$ for it. It is a simple linear combination of the quantities being predicted (here, Y_1 and Y_2), with the coefficients being functions of the data used to generate the observed adjusted expectation (here, x_1 and x_2). The bearing has two useful properties.

1.4.15.1 Summary of direction and magnitude of changes

The bearing summarizes the direction and magnitude of changes between prior and adjusted beliefs in the following sense: for any quantity Y constructed from the elements of the collection B, the change in expectation from prior to adjusted is equal to the prior covariance between Y and the bearing $\mathbb{Z}_d(B)$ so that $\mathrm{E}_d(Y) - \mathrm{E}(Y) = \mathrm{Cov}(Y, \mathbb{Z}_d(B))$. In our example it is simple to illustrate this result: we have

$$\mathbb{Z}_d(B) = 0.0478Y_1 - 0.0678Y_2 + 2,$$

so that

$$\begin{aligned}\mathrm{Cov}(Y_1, \mathbb{Z}_d(B)) &= \mathrm{Cov}(Y_1, 0.0478Y_1 - 0.0678Y_2 + 2)\\ &= 8.85 = 108.85 - 100\end{aligned}$$

and

$$\mathrm{Cov}(Y_2, \mathbb{Z}_d(B)) = -9.65 = 90.35 - 100.$$

Changes in expectation for other linear combinations, such as Y^+ and Y^-, are obtained as easily. For example,

$$\begin{aligned}\mathrm{E}_d(Y^+) &= \mathrm{Cov}(Y^+, \mathbb{Z}_d(B)) = -0.8,\\ \mathrm{E}_d(Y^-) &= \mathrm{Cov}(Y^-, \mathbb{Z}_d(B)) = 18.5.\end{aligned}$$

In particular, recalling that we noticed above that Y^* has a prior variance equal to its change in expectation, 1.078, we now observe that this is explained because

$$\mathrm{E}_d(\mathbb{Z}_d(B)) - \mathrm{E}(\mathbb{Z}_d(B)) = \mathrm{Cov}(\mathbb{Z}_d(B), \mathbb{Z}_d(B)) = \mathrm{Var}(\mathbb{Z}_d(B)).$$

1.4.15.2 Global diagnostic

The bearing provides a global diagnostic which gives a guide as to how well the data agree with the prior information. We have already seen that $\mathbb{Z}_d(B)$ is the linear combination having the largest squared change in expectation, relative to prior variance. We will call this change, which we have seen is just $\mathrm{Var}(\mathbb{Z}_d(B))$, the **size of the adjustment**, and introduce the notation $\mathrm{Size}_d(B)$ for it. It is natural to compare this maximum data effect with our expectation $\mathrm{E}(\mathrm{Size}_D(B))$ for it, where expectation is with respect to the data quantities and prior to them being observed. This expectation turns out to be

$$\mathrm{E}(\mathrm{Size}_D(B)) = \mathrm{E}(\mathrm{Var}(\mathbb{Z}_D(B))) = \sum_i \mathrm{R}_D(Z_i),$$

i.e. the sum of the canonical resolutions. In our example, the size of the adjustment and its prior expectation are

$$\mathrm{Var}(\mathbb{Z}_d(B)) = 1.078,$$

$$\mathrm{E}(\mathrm{Var}(\mathbb{Z}_D(B))) = 0.64 + 0.25 = 0.89.$$

For a simple global diagnostic we calculate $\mathrm{Sr}_d(B)$, the ratio of these quantities, which we call the **size ratio for the adjustment** of B given the observations $D = d$. In our example we obtain $\mathrm{Sr}_d(B) = 1.078/0.89 = 1.21$. This ratio has expectation one. Large size ratios indicate larger than expected changes in expectation, suggesting that the data are in sharp disagreement with our prior specifications. Small size ratios indicate smaller changes in expectation than expected and may imply that our prior variance specifications were too large. In our example, the size ratio is fairly close to one, suggesting little conflict between our prior information and the observations.

1.4.16 Partial adjustments

We have so far addressed the adjustments of both single quantities and collections by a single collection of information sources. We now move on to explore the partial effects and implications of individual pieces of information. In the following example, each 'information source' will be a single random quantity, but the approach works in just the same way when the individual information sources are themselves collections of quantities. Some of the reasons for studying partial adjustments are as follows. First, at the **design** stage, some of the information sources may be expensive to observe and so there may be advantages in excluding them as predictors if they are not individually valuable in helping to reduce variation in the unknowns. Secondly, at the **analysis** stage, it is valuable to know which aspects of the data have led us to our conclusions. Thirdly, at the **diagnostic** stage, adjustments are usually based on data from different sources which may or may not be in general agreement – for example, the data from one information source may suggest that an adjusted expectation should rise, whilst data from a different information source may suggest the reverse. In such cases it can easily happen that an overall adjustment appears quite plausible, but conceals surprising conflicts between different pieces of evidence. Bayes linear analysis permits us to explore the interactions between the various sources of beliefs and data in a way which highlights any such discordant features.

Key to understanding (linear) partial effects is the notion that one information source is often at least partly a surrogate for another information source. For example, if two vectors U and V are perfectly correlated in the sense that every linear combination constructed from the elements of U is perfectly correlated with some linear combination constructed from the elements of V, then we could essentially throw away V as U carries all the relevant information. Thus, when U and V are correlated, there will be a portion of V which is irrelevant when we also have U, and vice versa. We introduced in §1.4.4 the notion of, and notation for, the decomposition of a single random quantity into an adjusted expectation plus an adjusted version. We now extend this notation to vectors of random quantities. That is, we write

$$U = \mathrm{E}_V(U) + [U - \mathrm{E}_V(U)] = \mathrm{E}_V(U) + \mathbb{A}_V(U).$$

Informally, in a linear framework, $\mathrm{E}_V(U)$ and $\mathbb{A}_V(U)$ are respectively (1) the portion of the information source U that is also carried by V, and (2) the residual portion of U not duplicated by any part of the information source V.

Before we illustrate the Bayes linear approach to design via partial adjustment, it may be helpful to consider the usefulness of summaries of partial effects in the traditional context of stepwise linear regression. In stepwise regression the usual setting is that of one or more response variables with a large number of explanatory variables, where it is desired to determine a small subset of explanatory variables according to some criterion – such as the explanation of a given percentage of variation in the response variables. Two simple approaches to finding such a subset are forward selection and backward elimination. The former proceeds by beginning with an empty set of explanatory variables and then sequentially adding to this set the explanatory variables which are most helpful in explaining remaining variation in the response variables. The latter proceeds by taking the full set of explanatory variables and then sequentially removing those explanatory variables which are least helpful in explaining variation in the response variables. Both these notions have their analogues in Bayes linear methodology. With regard to forward addition of variables, the partial effect of interest is the extra percentage of variance explained in the response variables. With regard to backward deletion of variables, the partial effect of interest is the reduction in the explained variance of the response variables attributable to removing an explanatory variable.

In our example so far we have used our information sources X_1 and X_2 jointly as D to learn about future sales. Suppose now that we consider how important each is individually in predicting future sales. We adjust first by X_1 and then perform the **partial adjustment** by $\mathbb{A}_{X_1}(X_2)$, the adjusted version of X_2 given X_1, which is the portion of X_2 which has not already been contributed to the adjustment by X_1.

Details of the resulting variance resolutions are shown in Table 1.3. For example, when X_1 alone is used, the expected variance resolution in Y_1 is $\mathrm{R}_{X_1}(Y_1) = 0.3136$, rising to $\mathrm{R}_D(Y_1) = 0.3280$ when X_2 is also used. The **partial resolution** contributed by $\mathbb{A}_{X_1}(X_2)$ is thus, by subtraction,

$$\mathrm{R}_{\mathbb{A}_{X_1}(X_2)}(Y_1) = 0.0144.$$

Table 1.3 shows clearly that X_1 is mainly informative for Y_1, and that the residual portion of X_2 having taken into account X_1 has little extra explanatory power. For

Table 1.3 Variance implications for individual quantities and their collection.

	Resolution given X_1 $\mathrm{R}_{X_1}(\cdot)$	Resolution given X_1 and X_2 $\mathrm{R}_{X_1 \cup X_2}(\cdot)$	Partial resolution $\mathrm{R}_{\mathbb{A}_{X_1}(X_2)}(\cdot)$
Y_1	0.3136	0.3280	0.0144
Y_2	0.0576	0.3280	0.2704
B	0.1640	0.4450	0.2810

explaining variation in Y_2, the role is reversed. In the context of this example, if we were particularly interested in predicting sales of Y_1 rather than Y_2, and if X_2 was expensive to measure, we might decide at this stage not to bother observing X_2 but to depend on only observing X_1. Actual design decisions will depend on context and will take into account issues such as the expense of observing quantities such as X_1 and the utility of reducing variation in quantities such as Y_1. If we are concerned with explaining variation globally across the collection B, we notice that the variance resolutions are $\mathrm{R}_{X_1}(B) = 0.1640$ and $\mathrm{R}_{\mathbb{A}_{X_1}(X_2)}(B) = 0.2810$ respectively, indicating that both information sources are valuable.

Given data X_1 alone, the adjusted expectations are

$$\mathrm{E}_{X_1}(Y_1) = 1.12(X_1 - 100) + 100,$$

$$\mathrm{E}_{X_1}(Y_2) = -0.48(X_1 - 100) + 100.$$

Consequently, if we observe X_1 to be larger than expected, the expectations for Y_1 and Y_2 are revised upwards and downwards, respectively. These movements are due to the prior correlations shown in §1.4.13 in that X_1 is positively correlated with Y_1 and negatively correlated with Y_2. The actual observation $x_1 = 109$ gives adjusted expectations of $\mathrm{E}_{x_1}(Y_1) = 110.08$ and $\mathrm{E}_{x_1}(Y_2) = 95.68$. These are standardized changes of ± 1.8 standard deviations relative to the variances resolved.

If we now make the partial adjustment by X_2, or rather by the adjusted version $\mathbb{A}_{X_1}(X_2)$, we obtain **partial adjusted expectations** which provide the formulae to update the expectations from the current adjusted expectation (given only X_1) to that based on both X_1 and X_2. In doing so, it is helpful to introduce some extra notation. Let

$$\mathrm{E}_{[X_2/X_1]}(B) = \mathrm{E}_{X_1 \cup X_2}(B) - \mathrm{E}_{X_1}(B)$$

be the partial adjustment of B by X_2 given that we have already adjusted by X_1. Such partial adjustments necessarily have expectation zero. We find that

$$\mathrm{E}_{[X_2/X_1]}(Y_1) = 0.18(X_1 - 100) + 0.30(X_2 - 100),$$

$$\mathrm{E}_{[X_2/X_1]}(Y_2) = 0.78(X_1 - 100) + 1.30(X_2 - 100).$$

In this case, if we observe X_2 to be larger than expected, the partial change in expectation for both Y_1 and Y_2 is upward. As we did observe $x_2 = 90.5$, the partial change in expectation is $0.18(109 - 100) + 0.30(90.5 - 100) = -1.23$ for Y_1

Table 1.4 Exploring the implications of partial adjustment for Y_1 and Y_2.

	Results for Y_1		
	Prior	Given X_1	Given X_1 and X_2
Expectation	100.0	110.08	108.80
Variance	100.0	68.64	67.20
Total variance resolved		31.36	32.80
Change in expectation		10.08	−1.28
Change in variance resolved		31.36	1.44
Squared standardized change in expectation		3.24	1.05
	Results for Y_2		
	Prior	Given X_1	Given X_1 and X_2
Expectation	100.0	95.68	90.35
Variance	100.0	94.24	67.20
Total variance resolved		5.76	32.80
Change in expectation		−4.32	−5.33
Change in variance resolved		5.76	27.04
Squared standardized change in expectation		3.24	1.05

and $0.78(109 - 100) + 1.30(90.5 - 100) = -5.33$ for Y_2. These are standardized changes of 1.03 standard deviations relative to the respective resolutions in variance. A summary for the adjustments is given in Table 1.4. Overall, we notice that the expectation for Y_1 rose and then fell back slightly whilst the expectation for Y_2 fell and then fell again. None of the standardized changes are particularly large and we conclude that the magnitudes of the changes in expectation are in apparent agreement with the prior specification.

Because the initial data source X_1 is uncorrelated with the partial data source $\mathbb{A}_{X_1}(X_2)$, notice how the overall adjusted expectations for Y_1 and Y_2 given in (1.15) have been decomposed into additive initial and partial adjustments. That is, we have

$$\mathrm{E}_D(\cdot) = \mathrm{E}_{X_1}(\cdot) + \mathrm{E}_{[X_2/X_1]}(\cdot).$$

1.4.17 Partial diagnostics

We saw in Table 1.4 that the expectation for Y_1 rose and then fell slightly, so that the two information sources, X_1 and $\mathbb{A}_{X_1}(X_2)$, might be said to have contradictory implications for Y_1, whereas the two information sources are apparently complementary as far as Y_2 is concerned. Obviously we can make similar judgements for whichever quantities are of interest, such as total future sales Y^+, but it is simpler to calculate a global summary of the implication of two sources of information. Recall that in §1.4.15 we introduced the **bearing for the adjustment** to summarize the magnitude and direction of changes in expectation implied by a data source. For a partial adjustment we calculate the **bearing for the partial adjustment**, which summarizes the magnitude and direction of changes in expectation implied by the additional partial information source.

In our example, the initial bearing given data x_1, the partial bearing given extra data $\mathbb{A}_{x_1}(x_2)$, and the overall bearing given all the data $d = x_1 \cup x_2$, are

$$\begin{aligned}
\text{Initial:} \quad \mathbb{Z}_{x_1}(B) &= 0.1170(Y_1 - 100) + 0.0270(Y_2 - 100) \\
\text{Partial:} \quad \mathbb{Z}_{[X_2/X_1]}(B) &= -0.0692(Y_1 - 100) - 0.0948(Y_2 - 100) \\
\text{Overall:} \quad \mathbb{Z}_d(B) &= \mathbb{Z}_{x_1}(B) + \mathbb{Z}_{[X_2/X_1]}(B) \\
&= 0.0478(Y_1 - 100) - 0.0678(Y_2 - 100).
\end{aligned}$$

As in §1.4.15, each bearing is associated with a **size ratio** measuring the discrepancy between data and belief specifications taken as a whole across the collection being adjusted. In this example, the size ratios for the initial, partial, and overall adjustments are respectively 3.24, 1.05, and 1.21. None of these, each of which has prior expectation unity, appears particularly large or disturbing, and we might conclude that the changes in expectation implied by the data are in general agreement with the prior specifications.

As a change in expectation for any quantity such as Y_1 can be represented as a covariance between that quantity and a bearing, we also note that the implications of the two data sources for changes in expectation are opposite: typically positive for the first, and typically negative for the second. To formalize this idea, the most useful single summary is the correlation between the bearings for the two data sources, which we call a **path correlation**. In this example, it is

$$\text{PC}(x_1, \mathbb{A}_{x_1}(x_2)) = \text{Corr}(\mathbb{Z}_{x_1}(B), \mathbb{Z}_{[X_2/X_1]}(B)) = -0.3633.$$

The interpretation is that there is a very mild form of conflict between the two information sources.

We have already seen that the standardized changes in expectation at each stage for the two quantities are not too surprising in relation to the variance resolved at each stage. However, we should be aware that an overall adjustment by all the data can mask (either by cancelling out or by averaging) two surprising and/or contradictory changes in belief. As an illustration, we repeat the diagnostic analysis using the canonical structure for the data quantities, which we saw at the foot of §1.4.12 to be the current sales total and sales difference, X^+ and X^-. Thus, we reorganize the data sources to be these canonical data quantities, and use them to make predictions about future sales.

The analysis proceeds as described in previous sections, but we shall not detail it as our interest here is only in the diagnostic evidence. Suppose that we carry out an initial adjustment of B by X^+, and then a further partial adjustment by X^-, which is uncorrelated with X^+, so that we have $\mathbb{A}_{X^+}(X^-) = X^-$. We find that the bearings are

$$\begin{aligned} \text{Initial:} \quad & \mathbb{Z}_{x^+}(B) = -0.01(Y^+ - 200) \\ \text{Partial:} \quad & \mathbb{Z}_{x^-}(B) = 0.0578Y^- \\ \text{Overall:} \quad & \mathbb{Z}_d(B) = 0.0478(Y_1 - 100) - 0.0678(Y_2 - 100), \end{aligned}$$

so that there is a natural and straightforward correspondence between data sources and what the data source is informative for: previous total sales are informative for future total sales, and previous sales differences for future sales differences. Because of the uncorrelatedness of these quantities, observe for example that previous sales totals X^+ are valueless for making linear predictions about a future sales difference, Y^-. The overall bearing $\mathbb{Z}_d(B)$, which is of course the same however we reorganize the information sources, has a corresponding size ratio of 1.21. However, the size ratios for the initial and partial adjustments are respectively 0.0125 and 4.2781. The interpretation here is that the changes in expectation induced by the first data source, X^+, were surprisingly small compared to the expected level of variance explained, whereas the changes in expectation induced by the second data source, X^-, were perhaps disturbingly large. A plausible explanation would be that we overstated our prior variability for the sales totals, and that we understated variability for the sales differences, or perhaps that there are errors in the data. In such cases, we might choose to re-examine our prior specifications and the data. Note that, as will often be the case, diagnostic inspection based on the canonical structure gives a clearer picture of potential problems with the overall prior formulation than is obtained by inspection of the adjustments of the original quantities.

1.4.18 Summary

A good analysis of even simple problems such as these requires the knowledgeable use of effective tools. Our analysis here is incomplete as we have only introduced some of the basic machinery of the Bayes linear approach, and yet we have shown how fairly simple ideas and procedures lead directly into the heart of a problem, offering tools that work as well for collections as they do for single quantities, and that reveal quickly the important aspects of a combined belief and data structure. We could possibly have made a more detailed prior specification. However, by concentrating on the reduced belief specifications required for the second-order structure we have been able to apply a simple and efficient methodology under which we can control input requirements, and within which the implications of the belief specifications and any observations can be readily discerned. Various aspects of the Bayes linear analysis are thus revealed: straightforward specification of genuine beliefs, exploration of their implications, their adjustment using data, and diagnostics comparing expected to actual behaviour. This methodology works in essentially the same way as we increase the number of quantities in the problem, in which case we will find that the role of the canonical structure becomes increasingly important in clarifying the effects of complex belief adjustments.

1.5 Overview

The Bayes linear approach has been developed to the level where it is usable as a general framework within which to develop statistical methodology. As with any such methodology, much work may be required to bring the approach to bear on particularly challenging practical problems. However, the basic elements of the approach are sufficiently well developed to merit a unified exposition. Our intention, in this book, is to present in a systematic way those central methodological features that we consider to be both essential for and distinctive to the Bayes linear approach. Thus, we do not address the many aspects of belief specification, statistical modelling and data analysis which are common to our approach and other views of statistical analysis. Nor do we attempt to summarize all of the ways in which moment specification and analysis are currently exploited within statistical methodology. Instead, by concentrating on the essentials of the approach, we aim to give at least the outline of a unified methodology for belief analysis from a particular subjectivist viewpoint based on partial belief specification taking expectation as primitive. Whether we consider this approach as (the skeleton of) a complete methodology of itself or as part of a much larger toolkit of approaches to belief modelling and analysis will depend both on our philosophical viewpoint and on the types of problem which we wish to address.

The organization of this book is as follows. In Chapter 2, we introduce the ingredients which we will blend in later chapters, namely prior means, variances and covariances, assessed as primitive quantities. We give a brief introduction to the idea of expectation as primitive, and discuss, by example, some simple approaches to prior specification for means, variances and covariances.

The basics of our approach are threefold: (i) we specify collections of beliefs and analyse how we expect beliefs to change given our planned data collection; (ii) we collect information and analyse how our beliefs have actually changed; (iii) we compare, diagnostically, expected to actual changes in our beliefs. Step (i) is addressed in Chapter 3, where we explain the basic operations within our approach, namely the adjustment of collections of expectations and variances, by linear fitting on data. We develop the basic properties of belief adjustment and describe the natural geometric setting for the analysis. A general construction is introduced, namely the belief transform, for interpreting collections of belief adjustments through the eigenstructure of the transform.

We address steps (ii) and (iii) of our general approach in Chapter 4, which is concerned with interpretation and diagnostic evaluation of the observed belief adjustment given data. In particular, we describe the construction and interpretation of the bearing for a belief adjustment, which is a form of linear likelihood for the analysis, which summarizes the overall direction and magnitude of a collection of adjustments.

Usually, our information comes from different sources: for example, there may be different time points, different populations, different types of quantity. It is useful to identify how much information we expect from each source, and then to consider whether the various data sources are giving consistent or a contradictory information. In Chapter 5, we apply the three-step programme – (i) interpret expected adjustments, (ii) interpret actual adjustments, (iii) compare actual to interpreted effects – when the data have been divided into portions. We therefore consider partial belief adjustments and develop the corresponding partial belief transforms and partial bearings for an adjustment carried out in stages.

Exchangeability (the property that beliefs over a collection of objects would not be affected by permutation of the order of the objects) is a fundamental subjective judgement underlying many statistical applications. In principle, exchangeability judgements allow us to carry out statistical modelling purely in terms of our judgements over observables. Unfortunately, in the usual Bayes formalism, this is very difficult, and exchangeability tends to be hidden from view. Because of our simplified approach to belief specification, however, it is both feasible and natural to build statistical models directly from second-order exchangeability judgements over observables. This process is covered in Chapter 6, where we develop and interpret the representation theorem for second-order exchangeable random quantities. Chapter 6 is also concerned with how to adjust beliefs over the resulting exchangeability models. We derive useful general results which greatly simplify the analysis of such models, through the special properties of the corresponding belief transforms. In Chapter 7, we extend such analyses to cover collections of data which are individually second-order exchangeable, and which satisfy natural second-order exchangeability relationships between each pair of collections. In Chapter 8, we address the issues that arise in learning about population variances from exchangeable samples.

To this point, we have treated a particular type of belief transform as our basic interpretative tool for analysing collections of belief changes. However, this type of transform is itself a special case of a much wider class of transforms, which are examined in Chapter 9, all of which are based on comparisons between collections of variance and covariance specifications. We give the general construction for such transforms, and illustrate the approach with various problems of comparison over models and designs.

Graphical models are a powerful tool for graphically representing and evaluating our beliefs. Bayes linear graphical models, covered in Chapter 10, perform this task for describing and manipulating our second-order specifications. We may also display quantitative information, expressing our three-step sequence – expected effects, observed effects and their comparison – in a natural way on the diagram. Thus, the diagrams express both the modelling and the analysis of beliefs. Further, the local computation properties of these models allow us to tackle large problems in a straightforward and systematic way.

In Chapter 12, we cover the technical material that we need for efficient implementation of the Bayes linear approach, assuming a somewhat higher level of knowledge of matrix algebra than in the rest of the book. The matrix algebra required is covered in Chapter 11.

2

Expectation

In a quantitative analysis of beliefs, we turn some aspects of prior judgements into numbers. Each such number is a statement of knowledge, and as such requires careful consideration. Therefore, we will often need to be modest and specify a relatively small and carefully chosen collection of quantitative judgements about aspects of the problem which are meaningful and clearly relevant to the solution of our problem.

In general, the level of detail at which we choose to describe our beliefs will depend on:

- how interested we are in the various aspects of the problem;
- our ability to specify each aspect of our uncertainty;
- the level of resources, in terms of time, money and effort that we are willing to expend on choosing, observing and analysing data relevant to the problem;
- how much detail is required from our prior specification in order to extract the useful information from the data.

This means that the analysis depends not only upon the observed data but also upon the level of detail to which we express our beliefs. In the Bayes linear approach, we concentrate upon the specification and analysis of those limited aspects of uncertainty which are essential to the question at hand. To allow such restricted specification, we choose expectation, rather than probability, as the primitive for quantifying aspects of prior beliefs. In this chapter, we discuss the role of expectation as a primitive concept, and illustrate the direct specification of expectations by example.

2.1 Expectation as a primitive

De Finetti (1974, 1975) gives a careful development of expectation as the basis for subjectivist theory. He offers the following intuitive operational interpretation

Bayes Linear Statistics: Theory and Methods M. Goldstein and D.A. Wooff

for expectation. Suppose that X is a random quantity, i.e. a numerical quantity whose value you do not presently know. Consider a ticket which will entitle you to a cash payoff of precisely X money units, when the value of X is revealed. Call $\mathrm{E}(X)$ the **expectation** of X, where $\mathrm{E}(X)$ is the fair price (as judged by you) for the ticket on X, meaning that you are indifferent between the sure gain $\mathrm{E}(X)$ and the random gain X.

There are two intuitive properties of price. First, the fair price should not be greater than the largest payoff from the ticket or less than the smallest payoff from the ticket, so that

$$\inf X \leq \mathrm{E}(X) \leq \sup X. \tag{2.1}$$

Secondly, suppose that you buy two tickets, one on X and one on Y. Together, you have paid $\mathrm{E}(X) + \mathrm{E}(Y)$ and bought a ticket on $X + Y$, suggesting that

$$\mathrm{E}(X + Y) = \mathrm{E}(X) + \mathrm{E}(Y). \tag{2.2}$$

The above arguments are informal, but sufficient for our purpose. We shall discuss how to make the argument more rigorous in §2.2. The crucial aspect for our purposes is that we consider expectations directly, as primitive quantities, rather than as derived quantities calculated from intermediary probability specifications. Different individuals may validly make different assessments for the expectation of a random quantity, as they will bring to bear different knowledge, experience, judgements and abilities. All that we require is that the collection of expectations should obey the fundamental properties of expectation, namely conditions (2.1) and (2.2).

For any event, H, we identify H with the corresponding indicator function $H = 1$ if H occurs, $H = 0$ otherwise. The probability of the event H is equivalently the expectation, $\mathrm{E}(H)$, of the indicator function for H, and we make no distinction between probability and expectation in what follows. De Finetti, in his development, uses the term **prevision** for probabilities and expectations, so that $\mathrm{P}(X)$ is an 'expectation' or a 'probability', depending on whether or not X is an indicator function. Our preference for using $\mathrm{E}(X)$ for both cases is solely due to the more widespread familiarity of the conventional expectation notation.

In certain applications, it may be both feasible and sensible to make a full probability specification over a partition of possibilities. In such cases, from (2.2), directly assessed expectations will agree with those expectations calculated by the usual formulae (at least for finite partitions). However, there are many other ways in which we may make expectation statements, some of which we shall describe in this chapter, and we shall usually develop alternative methods of direct prior specification, particularly for complex problems. The practical superiority of expectation over probability as the fundamental quantification of belief is that with expectation as fundamental we can make as many or as few belief specifications as we deem appropriate, whereas with probability as fundamental we are forced to make all probability statements over a partition before we may make any expectation statements.

We may similarly consider conditional expectation as a 'called off' fair price. Your **conditional expectation** for X given event H, namely $\mathrm{E}(X|H)$, is the fair price for a ticket which pays X money units if H occurs, while, if H does not occur, the ticket is discarded, and your price is refunded.

The following coherence argument gives the familiar formula for conditional expectation. Observe first that the ticket pays X if H occurs, $\mathrm{E}(X|H)$ otherwise. Compare this ticket with a ticket which pays XH, with fair price $\mathrm{E}(XH)$, or equivalently pays X if H occurs and zero otherwise. Both tickets pay the same if H occurs, so that the difference in the fair price for the two tickets is the fair price for the ticket which pays $\mathrm{E}(X|H)(1-H)$, which has fair price $\mathrm{E}(X|H)(1-E(H))$. Equating these fair prices gives the required formula, namely

$$\mathrm{E}(XH) = \mathrm{E}(X|H)\mathrm{E}(H). \tag{2.3}$$

When X is also an indicator function, then (2.3) gives the usual form for conditional probability, namely

$$\mathrm{P}(A|B) = \frac{\mathrm{P}(A \cap B)}{\mathrm{P}(B)}.$$

In subsequent chapters, we shall contrast belief updating by conditioning with more general linear forms of belief adjustment.

2.2 Discussion: expectation as a primitive

In this section, we briefly consider some of the general issues which arise when treating expectation as a primitive.

Note, first, that our argument for linearity of expectation, namely property (2.2), was heuristic rather than formal, as there is no compelling reason why the sum of a collection of fair prices need be a fair price for the sum. We may address this in various ways. The simplest is to restrict attention entirely to small money payoffs, for which the equivalence may seem reasonable. Alternatively, we may recognize that the equivalence is only precise if all payoffs are in units of utility and develop the ideas of utility and expectation together. This approach is logically superior, but might seem somewhat laborious as we do not intend to exploit utility considerations in the subsequent development. A further approach is to consider payoffs in a currency for which the linearity of fair prices seems intuitively more persuasive. Probability currency is particularly appropriate for this purpose. This currency consists of tickets in a raffle with a fixed prize, so that the various combinations of payoffs do not affect the prize to be won, but only the probability of winning the prize; for a discussion of probability currency, see, for example, Walley (1991).

We may feel that the concept of a fair price is itself vague, and that it is unclear what, if any, are the unfortunate consequences of deviating from the linearity of such prices. De Finetti gives a tighter operational definition for expectation, using a quadratic scoring rule as follows. Your **expectation** $\mathrm{E}(X)$ for X is the value that you would specify if you were subsequently to incur a penalty score L given by

$$L = c[X - \mathrm{E}(X)]^2, \tag{2.4}$$

where c is a constant defining the units of loss. This definition appears to be very different from the fair price definition for expectation in the preceding section. However, we may see, informally, that the definitions are equivalent as follows. You prefer penalty $c[X - \mathrm{E}(X)]^2$ to any other penalty $c(X - \mathrm{E}(X) - d)^2$. Therefore you prefer the difference to zero, and expanding both penalties, this corresponds to preferring $c(\mathrm{E}(X) + d/2)$ to cX if $d > 0$, and vice versa if $d < 0$. Therefore, defining $\mathrm{E}(X)$ through (2.4) is equivalent to making $\mathrm{E}(X)$ your fair price, subject to the same qualifications on the additivity of penalties that we made above in our discussion of (2.2).

Conditional expectation can be similarly defined in terms of a 'called off' quadratic penalty score, as follows. $\mathrm{E}(X|H)$ is the value that you would specify if the penalty L given by (2.4) were replaced by the penalty

$$L_H = cH[X - \mathrm{E}(X|H)]^2. \tag{2.5}$$

While operational rules do suggest ways to measure expectations, their primary function is simply to clarify that expectation is a measurable aspect of your beliefs in contrast to the relative frequency theory of probability, in which probability is claimed to be an intrinsic, but unobservable, property of a repeatable experimental set-up. The operational measurements also identify the properties that such expectations should satisfy.

We have already introduced two measuring devices and we will suggest various other approaches to prior specification below. It is an interesting philosophical and psychological question as to whether each approach is really measuring the same underlying judgements, but for our purposes we will take the pragmatic approach of using whichever methods seem best suited to the problem at hand.

Similarly, we do not address the important question of imprecision in our judgements, namely that, as we have observed in the example, each operational procedure might, in a particular problem, lead to a range of judgements. For example, you might be sure that your fair price for a ticket on X lay in some interval, but be unwilling to narrow that interval to a single point. Discussion of such judgements, from somewhat different viewpoints, may be found in Lad et al. (1992) and Walley (1991). Such interval judgements may be valid and useful, but we do not consider them here because our interest is primarily in describing simple analyses which may give insights into complex, high-dimensional problems, while interval-based analyses are usually considerably more difficult to carry out. Thus, we prefer to make precise expectation specifications, carry out our analysis and then, if appropriate, carry out a sensitivity analysis showing how our conclusions might change under alternative specifications, and we will describe methodology for this purpose in our general development.

By what criteria can you judge the reasonableness of a collection of expectation assessments? Suppose that you specify expectations $\mathrm{E}(X_1), \ldots, \mathrm{E}(X_r)$. De Finetti offers the following coherence criterion.

Your assessments $\mathrm{E}(X_1), \ldots, \mathrm{E}(X_r)$ are **coherent** if there are no other choices $x_1, \ldots, x_r$ with the property that, for each possible collection of outcomes for $X_1, \ldots, X_r$, we have

$$\sum_i c_i (X_i - x_i)^2 < \sum_i c_i [X_i - \mathrm{E}(X_i)]^2.$$

In other words, you do not have a preference for a given overall random penalty if you have the option of choosing an alternative overall penalty which will certainly be smaller, whatever the outcome. The necessary and sufficient condition for a collection of expectations, $\mathrm{E}(X_1), \ldots, \mathrm{E}(X_r)$, to be coherent is that the point $(\mathrm{E}(X_1), \ldots, \mathrm{E}(X_r))$ lies in the closed convex hull of the set of possible values of the random vector $(X_1, \ldots, X_r)$ in r-dimensional Euclidean space. This condition has, as an immediate corollary, the condition (2.1). Applying the condition to the vector $(X_1, \ldots, X_r, \sum_{i=1}^r a_i X_i)$ gives the general form of (2.2), namely

$$\mathrm{E}\left(\sum_{i=1}^r a_i X_i\right) = \sum_{i=1}^r a_i \mathrm{E}(X_i). \tag{2.6}$$

The relation (2.3) for conditional expectations follows similarly from the coherence condition of avoiding sure loss.

Examining (2.6), observe that we have finite additivity for expectations. Whether we should impose further constraints of countable additivity on these expectations is an interesting and somewhat controversial question. However, in this book, we are largely interested in methodology for analysing finite collections of expectations of bounded random quantities, so this distinction will not be important.

2.3 Quantifying collections of uncertainties

The most basic features of our uncertainty that we might express are the following:

- judgements as to the magnitudes of various quantities;
- some degree of confidence in the judgements of magnitude;
- some expression as to how strongly judgements about the quantities are interrelated, so that observation on some of the quantities may be used to modify judgements on other quantities.

Our framework is as follows. We begin by supplying a list $C = [X_1, \ldots, X_k]$ of random quantities, for which we shall quantify aspects of our uncertainty. We term C the **base** for our analysis.

1. For each $X_i \in C$, we specify the expectation, $\mathrm{E}(X_i)$, giving a simple quantification of our belief as to the magnitude of X_i.

2. For each $X_i \in C$, we specify the variance, $\mathrm{Var}(X_i)$, quantifying our uncertainty or degree of confidence in our judgements of the magnitude of X_i.

3. For each pair $X_i, X_j \in C, i \neq j$, we specify the covariance, $\mathrm{Cov}(X_i, X_j)$, which expresses the relationship between judgements on the quantities, namely the extent to which observation on X_j may influence our belief as to the size of X_i.

The only restrictions that we require on the elements of C are that all means and variances be finite. These expectations and variances are specified directly. In the remainder of this chapter, we shall discuss, by example, how to make such prior specifications. We control the level of detail of our investigations by our choice of the collection C. For example, in a problem of medical diagnosis, we might just list certain critical symptoms, or include any number of further symptoms and explanatory factors. Clearly, such a collection will be fluid and elements will be added and removed for many practical reasons. However, it simplifies our exposition to suppose that at any particular time we have chosen a particular collection C and made the corresponding direct first- and second-order belief specifications.

We may choose to include in C various powers or other transforms of various observable quantities, so that our choice of C both summarizes the level of detail for our beliefs that we have chosen to describe, and places limits on the conclusions that we may draw. The most detailed collection that we could possibly select would consist of the indicator functions for all of the combinations of possible values of all of the possible quantities of interest. With this choice of C, we obtain a full probability specification over an implied outcome space. On occasions, this special case may be appropriate. Often, in complex problems, a joint probability distribution may serve as a convenient qualitative approximation for various aspects of the quantitative beliefs that we might express. However, in general, full probabilistic specification is unwieldy as our fundamental expression of the actual belief statements that we choose to make, in that it requires such an extremely large number of statements of knowledge, expressing judgements to such a fine level of detail, that typically we would have neither the interest nor the ability to make most of these judgements in a meaningful way. Therefore, for large problems we will often restrict attention to small sub-collections of this maximal collection. One of the themes of this book is that it is useful to have the choice of working explicitly with small collections of careful belief specifications rather than being forced to specify in an artificial manner very large collections of pseudo-belief statements.

As the number of elements increases, even the full second-order specification may become difficult. However, we only require full second-order specification over C in order to be able to consider how observation of the values of any sub-collection, E say, affects judgements over any other sub-collection, F say. It may be that we do not really need to assess how each subset E affects each F, but rather that we can split C into (not necessarily disjoint or exhaustive) sub-collections $C_i = (E_i, F_i), i = 1, \ldots, r$, say, and consider only how judgements over each F_k are affected by information represented in E_k, in which case we would only need to specify covariances over the pairs (F_k, E_k). What we lose by such simplification is the ability to consider the impact of the larger data set E upon our beliefs over the full collection F. Such trade-offs are unavoidable if we wish to control the complexity of the specification process, rather than pretending that we can actually quantify every judgement that we could conceivably make. Thus, in our approach, we select one or several bases C_i, chosen because we feel, a priori, that the benefits from carrying out the full second-order specification over each sub-collection outweigh the efforts of making these prior specifications.

2.4 Specifying prior beliefs

This work is concerned with the quantitative analysis of beliefs. Our methods take as a starting point that a preliminary quantification of beliefs has been made. We will be concerned with the interpretation, diagnostic evaluation and visual representation of the implications of our belief statements. All of these analyses may cause us to reassess our prior evaluations, and thus we should properly view prior specification as an iterative process, supported by all of the tools that we shall develop. However, it is useful to give at least some introductory discussion of the first stages in prior quantification. Therefore, we shall touch briefly on various guiding principles, which we believe to be helpful in carrying out this process. These principles are most easily described in a practical context. Therefore, we will describe in some detail how the specification might be carried out for a particular example, which we now introduce. Various themes that we shall develop in later chapters will be introduced informally in this account.

2.4.1 Example: oral glucose tolerance test

Suppose that a doctor works at a clinic where patients are frequently diagnosed for diabetes. One of the diagnostic tests that has been used is the oral glucose tolerance (OGT) test. In this test, a patient fasts for 12 hours, usually overnight. The blood glucose level is then measured (in mmol/litre). The patient takes a glucose solution (and nothing else) and the blood glucose level is measured again after 2 hours. The level may also be measured at various intervening times, but, during the period in which the data that we shall use were collected, in the late 1980s, the World Health Organization 'diagnosis' depended only upon these two values, with responses above certain thresholds suggesting diabetes. A summary of the possible diagnoses for given blood glucose levels is shown in Table 2.1.

Now, let us suppose that the clinic is occasionally involved in the diagnosis of diabetes in elderly people (say, over 60). The doctor might find it somewhat suspicious that the diagnostic levels that are set for the OGT test do not incorporate

Table 2.1 Oral glucose tolerance tests: diagnosis thresholds *(circa 1980s)*.

Blood glucose level mmol/litre	Diagnosis for fasting measurement	Blood glucose level mmol/litre	Diagnosis for 2-hour measurement
under 7	Healthy	under 7	Healthy
more than 7	Diabetes	7 to 10	Impaired glucose tolerance
		more than 10	Diabetes

any form of correction for age, as it would be a reasonable a priori assumption that, even for healthy elderly people, the sugar might take longer to be absorbed into the blood than for healthy young people. She might suspect that the empirical calibration of the test was based on experiments on comparatively young people, so that the test would tend to misclassify healthy elderly people as diabetic.

Suppose that the doctor is sufficiently curious about this that she decides to try to form a reasoned judgement about OGT test levels for elderly people. Suppose further that when she asks around, she finds that nobody else seems to know anything about the calibration of the OGT test on healthy elderly people. She discovers that there is a large literature on the test, which she finds to be somewhat confusing (as she is a non-specialist), contradictory (as experimental values quoted in different studies do not quite match up), and largely irrelevant (as everything that she can find relates to younger people anyway). Furthermore, as she suspected, she can find nothing on the calibration of the test for elderly people. This example is intended to be purely illustrative. However, the numbers that we shall use are derived from an experiment that was carried out by a clinician who was motivated by the scarcity of available information at the time of the experiment concerning the calibration of the OGT test for elderly people (see Wickramasinghe et al. 1992), and so this analysis is intended to be plausible for the individual at that time.

Suppose that the doctor is herself elderly. Indeed, suppose she has actually retired but still helps out at the clinic on a voluntary basis. Perhaps this is why she is sensitive to the interpretation of standard medical procedures for old people, and also why she has time to reflect about what she is doing. The simplest way that she can see to get any information on the effect of the OGT test on healthy elderly people is to administer the test on herself. This will provide minimal information but at least it offers a starting point for any further investigations. The simplicity of this scenario allows us to introduce some basic ideas in a straightforward manner.

In this example, we consider how the doctor can express in quantitative form her partial knowledge about the responses of elderly people to the OGT test. The values that we shall specify are in no way intended to represent best expert elicitations, but simply to illustrate the types of values that might be expressed by someone who has access to certain limited information about the quantities. In subsequent chapters, we will suggest ways in which she might modify her beliefs in the light of her own, and other, responses to the test.

2.5 Qualitative and quantitative prior specification

Suppose that we are considering various questions, for which it will be relevant to quantify aspects of our uncertainties. This quantification is an iterative procedure. One way to represent this procedure is through the following stages.

1. Identify those quantities for which beliefs are to be specified: wherever possible, such quantities should be directly observable, though we may need to introduce essential modelling or explanatory quantities which are unobservable. Loosely, there will be two types of quantities, namely primary quantities for which the quantification is of direct interest, and secondary quantities which are of interest mainly in providing information which is relevant to the quantification of beliefs about the quantities of primary interest.
2. Consider what relevant information is available for these quantities.
3. Develop a qualitative representation of beliefs about the quantities, expressing, as far as possible, the sources of uncertainty and the linkages between our prior information and the quantities of interest. This representation may be supported through graphical modelling.
4. Develop the statistical relationships between data and beliefs implied by the graphical model using basic considerations of exchangeability, temporal development and so forth.
5. Specify all numerical uncertainties (in our case, means, variances, and covariances), exploiting the graphical, statistical representation.
6. Check quantifications for consistency, plausibility and coherence.

We have two goals in this process. First, we aim to describe honestly our uncertainties as reached after careful consideration of the available prior evidence. In particular, we wish to lay out our prior reasoning in a sufficiently transparent manner that the plausibility of our reasoning can be judged, potential weaknesses in our argument can be identified, and the range of plausible disagreement over the various steps in the prior construction may be quantified. Secondly, we wish our overall specification to be sufficiently extensive to cover all the important aspects of the problem under investigation.

In practice, these goals may be contradictory, given constraints of time, resources and ability. Often, we must compromise between a very careful specification over limited aspects of a problem which are comparatively straightforward to consider and a somewhat rougher elicitation over a much wider collection of measures which we would like to include in the analysis. Methodology based on partial aspects of prior specification, by reducing the complexity of the prior elicitation, may help to limit this conflict, but a substantial element of individual judgement will always remain.

We now illustrate how we might carry out the above program of qualitative and quantitative prior specification in the context of our example.

2.6 Example: qualitative representation of uncertainty

2.6.1 Identifying the quantities of interest

While there are many aspects of the OGT test that we might examine, there are two obvious quantities of concern, namely the fasting and 2-hour glucose levels that the doctor will observe upon herself. Call these D_0 and D_2. The intention is to observe D_0 and D_2, and so to modify beliefs about similar readings on other healthy elderly patients.

In order that we may consider clearly defined observable quantities, we suppose that the doctor imagines a thought experiment in which an elderly person is chosen at random from the local population of healthy elderly people with no history of, nor any family history of, diabetes. In the thought experiment, she envisages giving the OGT test to the chosen individual and measuring G_0, the fasting glucose level, and G_2, the 2-hour glucose level. These values are, of course, hypothetical. However, they are meaningful to consider – indeed, the doctor may already be considering whether she should carry out an experiment to measure these quantities on a group of such individuals. Thus, the base for the analysis is the collection

$$C = [G_0, G_2, D_0, D_2].$$

The doctor will specify prior means, variances, and covariances for each member of C.

2.6.2 Identifying relevant prior information

The doctor considers what she knows about the various quantities. As we have supposed, she is able to find no information that is directly related to the responses of elderly people to the OGT test. Thus she decides to find out what she can about a group which has been widely studied, namely healthy younger people. Again, suppose that she finds the literature somewhat unclear and is further concerned that there might even be regional differences in test responses.

To simplify the account, suppose that the doctor finds details of a particular experiment which was actually carried out in her locality, in which the OGT test was administered to 15 healthy young people. All that is quoted in the paper are the sample summary statistics, which are as follows (in mmol/litre):

- The sample mean for the 15 observations of fasting glucose level is 4.16.
- The sample mean for the 15 observations of 2-hour glucose level is 5.5.
- The sample standard deviation for the 15 observations of fasting glucose level is 0.726.
- The sample standard deviation for the 15 observations of 2-hour glucose level is 0.949.
- The sample correlation is 0.422.

(Again, while the example is illustrative, these are the summary statistics, with a certain amount of rounding, from the study which we have already mentioned, as performed by Wickramasinghe et al. (1992); analysis of the data is given in Farrow and Leyland (1991).)

These figures give the doctor some ideas about the effect of the OGT test upon an 'average', healthy young person. She now considers what information she has about responses of elderly people. All she has to go on here are qualitative judgements about the similarities between younger and older individuals, coupled with certain implicit negative information.

For example, if average responses for elderly people were so large that there was hardly any overlap with scores for young people, then this would, arguably, have already been noticed and joined the medical folklore. A similar constraint is given by the diagnostic limits for the OGT test as given by the World Health Organization, which states that values over 10 mmol/l after 2 hours suggest diabetes, while values between 7 and 10 mmol/l after 2 hours suggest impaired glucose tolerance. (This latter diagnosis presupposes that the fasting level was under 7 mmol/l. Fasting levels over 7 mmol/l are automatically taken to suggest diabetes.) Again, the doctor doubts that most healthy elderly people would be classified diabetic by the OGT test, as this would be likely to be noticed, but she finds it quite plausible that many healthy elderly people would be classified in the intermediary category of impaired glucose tolerance.

2.6.3 Sources of variation

We construct a qualitative representation of uncertainty by considering the various sources of variation for our problem. Suppose that the doctor reasons that her uncertainty for each of G_0 and G_2 can usefully be thought of as deriving from three main sources.

1. Her judgements are based in part upon her prior judgement as to typical responses for healthy young people. She is uncertain as to the value of such a typical response.

2. She is uncertain as to the magnitude of the difference between typical responses for a healthy young person and a healthy elderly person.

3. She is uncertain as to how much the actual glucose values that she will observe will differ from a typical response for this quantity, due simply to differences between the individuals in the healthy elderly population.

Suppose that as a simplification the doctor makes the judgement that she has nothing useful to say about how these three sources of uncertainty might be interrelated. She therefore decides to consider these three aspects of her uncertainty separately and then combine them to give her overall uncertainty for G_0 and G_2.

2.6.4 Representing population variation

For our statements of belief to be meaningful and honest, we prefer well-defined observable quantities for which to make expectation statements, even if we might have to resort to thought experiments to define these quantities. The need for such quantities is equally important when we come to consider such abstractions as a 'typical' healthy elderly person.

What thought experiment addresses the idea of a 'typical' response? We suggest the following. Imagine that the OGT test was performed on a very large sample of healthy young people and a similarly large sample of healthy elderly people, say several thousand for each group. Suppose that in this thought experiment we evaluated the sample averages for each group. Call Y_0, Y_2 the sample averages for fasting and 2-hour glucose levels for the young people and E_0, E_2 the sample averages of these quantities for the elderly people. Given a large sample size, uncertainty about the values of these quantities due to sampling fluctuation will be negligible in comparison to uncertainty about average behaviour in the population. These quantities therefore may be used to express beliefs about typical responses. (This is an informal expression of **second-order exchangeability**, which will be considered in detail in a subsequent chapter, where we shall make precise the relationship between such averages from large samples and the limiting concept of the corresponding population means.)

2.6.5 The qualitative representation

2.6.5.1 Uncertainty for G_0

We have suggested that the doctor might envisage the magnitude of fasting glucose G_0 as being comprised of three contributing effects. Relating our thought experiment to these three effects, she is uncertain about:

- typical young responses, as expressed by uncertainty as to the value of Y_0;
- the difference between typical responses for the young and the elderly, as expressed by uncertainty as to the value of $C_0 = E_0 - Y_0$;
- the difference between the typical elderly response and the response of a particular elderly person as expressed by uncertainty as to the value of $R_0 = G_0 - E_0$.

As such, we write

$$G_0 = R_0 + C_0 + Y_0.$$

The doctor judges that there are no relationships that she wishes to specify between these three terms and so she assigns zero covariance for each pair. Therefore she can partition the expectation and variance of G_0 as

$$\mathrm{E}(G_0) = \mathrm{E}(R_0) + \mathrm{E}(C_0) + \mathrm{E}(Y_0),$$

$$\mathrm{Var}(G_0) = \mathrm{Var}(R_0) + \mathrm{Var}(C_0) + \mathrm{Var}(Y_0).$$

Thus she assigns the mean and variance for G_0 by first assigning a mean and variance for each of the three terms on the right of the above equations.

2.6.5.2 *Uncertainty for G_2*

As above, the doctor writes

$$G_2 = R_2 + C_2 + Y_2,$$

where $C_2 = E_2 - Y_2$ and $R_2 = G_2 - E_2$. Again, she assigns zero covariance between each pair. Therefore she partitions the expectation and variance of G_2 as

$$\mathrm{E}(G_2) = \mathrm{E}(R_2) + \mathrm{E}(C_2) + \mathrm{E}(Y_2),$$

$$\mathrm{Var}(G_2) = \mathrm{Var}(R_2) + \mathrm{Var}(C_2) + \mathrm{Var}(Y_2).$$

2.6.5.3 *Covariance between G_0 and G_2*

The doctor has partitioned G_0 and G_2 as $R_0 + C_0 + Y_0$ and $R_2 + C_2 + Y_2$, respectively. She decides that her judgements on individual variation (as expressed by the R components), typical differences between young and elderly (as expressed by the C components) and typical values for the young (as expressed by the Y components) are unrelated. She therefore sets all the corresponding covariances between the three collections of quantities to zero. Equivalently, she decides that all of her covariance between G_0 and G_2 can be attributed to the covariances between each of the pairs of quantities in the decomposition, i.e. that

$$\mathrm{Cov}(G_0, G_2) = \mathrm{Cov}(R_0, R_2) + \mathrm{Cov}(C_0, C_2) + \mathrm{Cov}(Y_0, Y_2).$$

2.6.5.4 *Uncertainties for D_0 and D_2*

The doctor partitions D_0 and D_2 as

$$D_0 = Z_0 + C_0 + Y_0,$$

$$D_2 = Z_2 + C_2 + Y_2,$$

where C_0, C_2, Y_0, and Y_2 are as before and Z_0, Z_2 are the corresponding quantities to R_0 and R_2, namely $Z_0 = D_0 - E_0$ and $Z_2 = D_2 - E_2$, the individual discrepancies between the doctor's readings and the average readings for healthy elderly people at the two time points. As above,

$$\mathrm{E}(D_0) = \mathrm{E}(Z_0) + \mathrm{E}(C_0) + \mathrm{E}(Y_0),$$

$$\mathrm{E}(D_2) = \mathrm{E}(Z_2) + \mathrm{E}(C_2) + \mathrm{E}(Y_2),$$

$$\mathrm{Var}(D_0) = \mathrm{Var}(Z_0) + \mathrm{Var}(C_0) + \mathrm{Var}(Y_0),$$

$$\mathrm{Var}(D_2) = \mathrm{Var}(Z_2) + \mathrm{Var}(C_2) + \mathrm{Var}(Y_2),$$

$$\mathrm{Cov}(D_0, D_2) = \mathrm{Cov}(Z_0, Z_2) + \mathrm{Cov}(C_0, C_2) + \mathrm{Cov}(Y_0, Y_2).$$

2.6.5.5 *Covariance between $[G_0, G_2]$ and $[D_0, D_2]$*

The doctor sees no reason why there should be any correlation between either of the pair (Z_0, Z_2) and either of the pair (R_0, R_2). Thus she sets all correlations between the two pairs equal to zero. This immediately determines the covariances between (D_0, D_2) and (G_0, G_2). We have

$$\text{Cov}(D_0, G_0) = \text{Cov}(Z_0 + C_0 + Y_0, R_0 + C_0 + Y_0) = \text{Var}(C_0) + \text{Var}(Y_0)$$

and similarly

$$\text{Cov}(D_0, G_2) = \text{Cov}(D_2, G_0) = \text{Cov}(C_0, C_2) + \text{Cov}(Y_0, Y_2),$$

$$\text{Cov}(D_2, G_2) = \text{Var}(C_2) + \text{Var}(Y_2).$$

2.6.5.6 *Comparing $[R_0, R_2]$ and $[Z_0, Z_2]$*

The only differences between the variances and covariances for $[G_0, G_2]$ and for $[D_0, D_2]$ arise from differences in the variance and covariance assessments for $[R_0, R_2]$ and for $[Z_0, Z_2]$. In this account we will suppose that the doctor sees no reason why her responses would be more or less variable than those of a randomly selected healthy elderly patient. Thus she sets

$$\text{Var}(R_0) = \text{Var}(Z_0), \quad \text{Var}(R_2) = \text{Var}(Z_2), \quad \text{Cov}(R_0, R_2) = \text{Cov}(Z_0, Z_2),$$

implying that

$$\text{Var}(D_0) = \text{Var}(G_0), \quad \text{Var}(D_2) = \text{Var}(G_2), \quad \text{Cov}(D_0, D_2) = \text{Cov}(G_0, G_2).$$

Of course, in practice there are reasons why a retired doctor might consider that she was different from a typical elderly person. We have identified the doctor's view of herself as typical of her group for three reasons. First, we have probably said enough already about how beliefs might be specified in this example, so to avoid reader burn-out it seems prudent to cut this discussion short. Secondly, we do not have any data on the effect of the OGT test on retired doctors, but we do have values that we shall use for the doctor's response based on tests on typical samples of healthy elderly people, so we would rather make our protagonist typical anyway. Finally, it seems reasonable that after reflection the doctor might consider that the factors which control the reactions to the test were sufficiently complex that she would be wary of ascribing much difference between her uncertainties for D_0, D_2 and for G_0, G_2, based on what might be purely superficial distinguishing characteristics. This assessment of her mean and variance structure as numerically equal to that of all other members of the group is a further example of a second-order exchangeability specification.

2.6.6 Graphical models

Already, for our simple problem, we have introduced a variety of quantities for which we have made highly structured qualitative judgements. In order to keep

Figure 2.1 The doctor's graphical model.

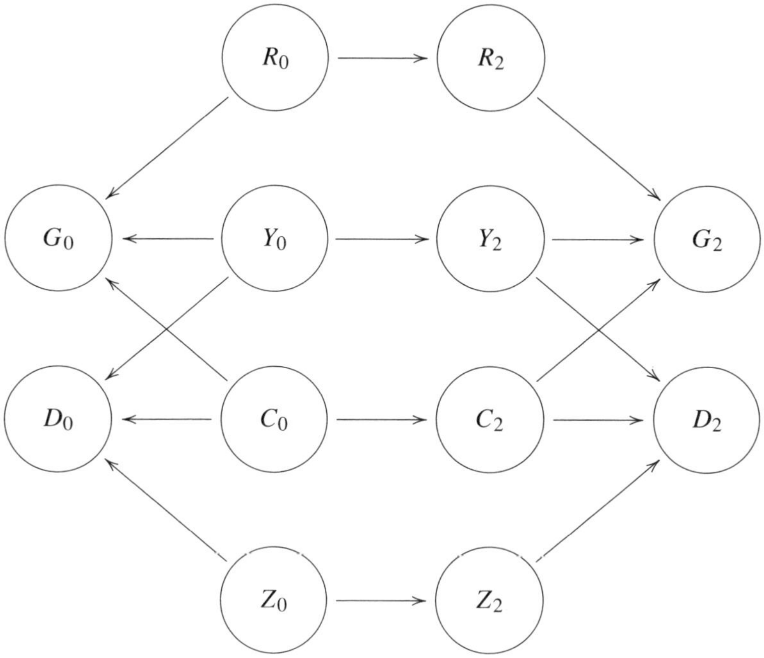

track of these relationships, and to communicate to others the qualitative structure of our argument, it is helpful to have simple graphical representations of these judgements. **Bayes linear graphical models** produce very useful pictures for this purpose. The model for the specification that we have made is given in Figure 2.1. In this picture, each node represents the corresponding random quantity. Arcs express predictive relationships in the sense that if there is no arc between a pair of nodes then, informally, the corresponding random quantities are uncorrelated given their parents. In many problems, we construct our qualitative representation of uncertainty directly by drawing the corresponding diagram. We shall describe the construction, interpretation, and use of such diagrams in Chapter 10.

Informally, from the diagram, R_0, Y_0, C_0, Z_0 are mutually uncorrelated (they are not linked by arcs and have no common ancestor group). Similarly R_2, Y_2, C_2, Z_2 are uncorrelated, and the only links between the two groups are the correlations between the pairs (R_0, R_2), (Y_0, Y_2), (C_0, C_2), and (Z_0, Z_2), as these are the only arcs between the two collections. Finally, G_i is influenced by R_i, Y_i, C_i and D_i is influenced by Y_i, C_i, Z_i (for $i = 0, 2$), and there are no other relationships between the quantities.

2.7 Example: quantifying uncertainty

2.7.1 Prior expectations

2.7.1.1 *Expectation for D_0 and G_0*

We have the qualitative representation

$$\mathrm{E}(G_0) = \mathrm{E}(R_0) + \mathrm{E}(C_0) + \mathrm{E}(Y_0).$$

The doctor now assigns expectations for each term on the right-hand side of the above equation.

Y_0 The sample mean for the responses of the 15 young people is 4.16. The doctor has no knowledge which would cause her to raise or lower this value, and so sets her prior expectation for Y_0 equal to 4.16.

C_0 While the doctor considers that there might be a difference between fasting glucose levels for young and elderly people, she also regards the 12 hours of fasting as a sufficiently long period to suggest that age-related slowness of response to blood sugar levels should not be a relevant factor. Therefore, she sees no reason a priori for C_0 to be positive or negative and she assigns a zero expectation for C_0.

R_0 The expectation of R_0 is zero, as the typical elderly response E_0 is simply an average of individual responses, each with the same prior expectation as G_0.

G_0 and D_0 The prior expectation for G_0, and for D_0, is therefore 4.16.

2.7.1.2 *Expectation for D_2 and G_2*

This is assessed as for D_0 and G_0.

Y_2 The doctor sets her prior expectation equal to 5.5, the sample mean for the 2-hour responses of the 15 young people.

C_2 The belief that blood sugar takes longer to absorb for elderly than for young people suggests that the prior expectation for C_2 should be positive. The doctor doubts that average responses for healthy elderly people would be over the 'impaired glucose tolerance level' of 7 mmol/l. Therefore she judges her prior mean for C_2 to lie between zero and 1.5 mmol/l. She judges her uncertainty to be roughly symmetric across this interval and so she selects the mid-point, which she rounds up to 0.75, as her prior mean for C_2.

R_2 The expectation of R_2 is zero, as for R_0.

G_2 and D_2 The prior expectation for G_2, and for D_2, is therefore 6.25.

2.7.2 Prior variances

2.7.2.1 *Variance of G_0 and D_0*

In the qualitative representation,

$$\mathrm{Var}(G_0) = \mathrm{Var}(R_0) + \mathrm{Var}(C_0) + \mathrm{Var}(Y_0).$$

Variances are now assigned for each term of the above equation.

Y_0 The doctor reasons that her judgement is based on a sample mean with variance 0.035, being the sample variance divided by the sample size. This does not contradict anything that she can think of. Therefore, as her numerical judgement is based strictly on this survey (to keep the argument in this chapter simple), she identifies her uncertainty for Y_0 with the sample variance of the estimate for this quantity, which corresponds very roughly to Bayes updating with a vague prior. She takes $\mathrm{Var}(Y_0)$ to be 0.05, which is the sample variance of the sample mean with about a 50% mark-up for natural scepticism as to possible flaws in the experiment.

C_0 The doctor reasons that she has no particular a priori reason to expect C_0 to be large. She would be mildly surprised to find that E_0 differed from Y_0 by more than 0.75 and very surprised to find that E_0 differed from Y_0 by more than 1.5 mmol/l. It seems appropriate to her to view these values as roughly the one and two standard deviation points of her implicit 'probability distribution' for C_0, so she sets $\mathrm{Var}(C_0)$ to be 0.57.

R_0 The sample variance for the young healthy individuals is about 0.5, which suggests a prior variance for individual variation for young people as 0.5. The doctor sees no a priori reason why elderly healthy people should be more or less variable than young healthy people, after a long fasting period, and so she decides to set $\mathrm{Var}(R_0)$ at 0.5 as well.

G_0 and D_0 These variances are the sum of the above terms, i.e. 1.12.

2.7.2.2 *Variance of G_2 and D_2*

We make a similar evaluation for $\mathrm{Var}(G_2)$ and of $\mathrm{Var}(D_2)$, as follows.

Y_2 This is assigned in a similar way to $\mathrm{Var}(Y_0)$. As the reported sample variance for the 2-hour glucose values is roughly twice that for fasting glucose, $\mathrm{Var}(Y_2)$ is assessed as twice $\mathrm{Var}(Y_0)$, i.e. 0.1.

C_2 The doctor sees no reason why C_2 should be negative. She expects C_2 to be positive, but doubts whether the difference would be so large as to classify the majority of healthy elderly people as having impaired glucose tolerance.

Thus, she judges a value of 1.5 mmol/l as high for this quantity, as this value would push many elderly patients into the impaired glucose tolerance group. A value near zero she judges unlikely, but not impossible. She judges that C_2 is most likely to be in the interval 0.5 to 1.0, and equally likely to be smaller or larger than this interval. Thus she sets her prior median to be 0.75, her central 50% prior interval to be [0.5, 1.0], and her lower and upper 25% intervals to be $[-0.5, 0.5)$ and $(1.0, 2.0]$. She then decides that her beliefs about C_2 may be roughly described by the following probability density function:

$$p(x) = \begin{cases} 0.25, & -0.5 \leq x < 0.5, \\ 1, & 0.5 \leq x \leq 1.0, \\ 0.25, & 1.0 < x \leq 2.0. \end{cases}$$

A probability distribution such as this will not exactly represent the doctor's beliefs. For example, there is no sharp discontinuity in beliefs at 0.5 or 1.0, nor such a clear cut-off at −0.5 and 2. Perhaps she might smooth the distribution or peak it in the middle. Certainly, this would be important were she to conduct a full probabilistic analysis. However, the calculations that we shall make will not be unduly sensitive to mild smoothing. Therefore, we will leave the density specification in this form to emphasize both that probabilistic specification is a very useful intermediary device for converting generalized feelings of uncertainty into expectation-type statements and that it is a matter of subjective judgement as to when the specification has been made in sufficient detail to give sensible values for the various expectation statements that we require. In this case, $\text{Var}(C_2)$ is set equal to the variance of the above distribution, which is $1/3$. Observe that this value is somewhat smaller than $\text{Var}(C_0)$, as the doctor considers that C_2 is very likely to be positive, and she has a plausible upper bound to limit the magnitude of this quantity.

Note that the approach of successively dividing the region of possible values into equal probability sub-intervals can be very helpful in prior quantification, as in many situations we may feel more comfortable in judging two events as having equal probability than we would in judging more general relative magnitudes of probabilities. The prior quantiles that we specify may be exploited in various ways. First, we may simply evaluate the maximum and minimum variances consistent with the given assessments, to give general guidance as to the range of allowable specifications that we might make. Secondly, we might fit standard forms of distribution such as the normal or log-normal, and assess the corresponding variances. Thirdly, as in our illustration, we might specify a prior density which captures roughly the qualitative shape of our prior beliefs, in order to assess the variance.

In any given assessment, we may even use all three approaches, preferably within some convenient computer-based elicitation environment. A typical elicitation tool would allow us to sketch our prior density directly on the screen or to fit and display simple standard parametric families to our chosen quantiles. For example, in the above analysis, the doctor has specified a distribution on the interval $(-0.5, 2)$. A simple elicitation tool might transform the interval to $(0, 1)$ and display, for example, the beta distribution with the same mean and variance as derived above (in this case a beta distribution with both parameters equal to $59/32$).

R_2 This is set by the judgement that there should be more variation in the responses of elderly than of young people after 2 hours, as there should be a general slowing down with age of the speed with which blood sugar is absorbed, but this decline is unlikely to be uniform over individuals. The sample standard deviation in the young group was about 0.95. Doubling this value for elderly patients would suggest a noticeable proportion of healthy, elderly people exceeding the critical value of 10. This seems a little high so instead she raises the standard deviation by about 50%, setting $\text{Var}(R_2) = 2.0$.

G_2 and D_2 These variances are the sum of the above terms, namely 2.43.

2.7.3 Prior covariances

Y_0 and Y_2 The doctor decides that, because of the relatively high accuracy with which Y_0 and Y_2 are determined, her correlation between Y_0 and Y_2 is negligible.

C_0 and C_2 Her prior covariance between C_0 and C_2 is relatively high, as if she discovered that elderly people had considerably higher (or lower) fasting glucose levels than their young counterparts, this would strongly suggest large differences for the 2-hour levels. She sets the correlation between C_0 and C_2 to be 0.7, giving $\text{Cov}(C_0, C_2) = 0.30$.

Note that she may support this assessment by considering the quantity $C_2 - C_0$. Either by direct assessment, or by forming prior quantiles for this quantity, she may form a judgement for $\text{Var}(C_2 - C_0)$. As she has already assessed $\text{Var}(C_0)$, $\text{Var}(C_2)$, this gives an indirect evaluation for $\text{Cov}(C_0, C_2)$. Let us suppose that this assessment is consistent with the value given above.

R_0 and R_2 The study on young patients quoted a sample correlation of 0.422 between fasting and 2-hour glucose levels. In our notation, this gives a sample estimate for the correlation between R_0 and R_2 for the young group. The doctor sees no persuasive reason to raise or lower the correlation for the elderly group. Taking the same value, rounded, for the elderly controls, she assigns $\text{Cov}(R_0, R_2) = 0.42$.

G and D From the qualitative representation, we therefore determine that

$$\text{Cov}(G_0, G_2) = \text{Cov}(D_0, D_2) = 0.72, \quad \text{Cov}(D_0, G_0) = 0.62,$$

$$\text{Cov}(D_0, G_2) = \text{Cov}(D_2, G_0) = 0.30, \quad \text{Cov}(D_2, G_2) = 0.43.$$

2.7.4 Summary of belief specifications

In our example the doctor has specified the following expectations:

$$E(G_0) = E(D_0) = 4.16,$$
$$E(G_2) = E(D_2) = 6.25.$$

She has specified the following variances and covariances:

$$\text{Var}(G_0) = \text{Var}(D_0) = 1.12,$$
$$\text{Var}(G_2) = \text{Var}(D_2) = 2.43,$$
$$\text{Cov}(G_0, D_0) = 0.62,$$
$$\text{Cov}(G_2, D_2) = 0.43,$$
$$\text{Cov}(G_0, D_2) = \text{Cov}(G_2, D_0) = 0.3,$$
$$\text{Cov}(G_0, G_2) = \text{Cov}(D_0, D_2) = 0.72.$$

2.8 Discussion: on the various methods for assigning expectations

It is mistaken to suppose that, because prior expectation statements are subjective, they are also largely arbitrary. On the contrary, they are no more arbitrary than any other form of reasoning. We value quantifications of belief as we value any argument, namely to the extent to which the assessments are developed in a clearly and carefully reasoned manner. In our example, to keep the discussion fairly simple, we have cut short the specification process at various stages, in that far more information is available on all aspects of the prior quantification than we have made use of, and even the information that we have described could doubtless have been analysed more carefully. However, this only serves to emphasize that prior specification is not primarily a psychological issue, but depends instead upon the careful consideration of the available information.

An element of arbitrariness does enter when we turn our generalized reasoning about our uncertainties into precise values for our probabilities and expectations. For example, the doctor in our account above was fairly casual in rounding up or down her expectation judgements. In a more careful and detailed elicitation, she would have paid attention to the maximum and minimum values for each of her quantitative assessments which were consistent with her heuristic arguments, and thus found bounds for each of her composite expectation and variance specifications. Such bounds may be exploited in various ways. For example, we may carry out a sensitivity analysis on the conclusions of any subsequent analysis based on variation of the prior specifications over the region that we have identified, or, at the least, over some subset of this region based on identification of the most crucial aspects of our beliefs. Alternatively, we might consider our prior beliefs to be imprecise so that we could use the variation over allowable choices of prior inputs to determine the imprecision in our posterior conclusions; for a detailed treatment of the role of imprecision in probabilistic analysis, see Walley (1991).

There are no strict rules that we can give for how to quantify prior beliefs because in every case it will be a personal judgement as to what are the relevant features of the information about the situation, and how such information should be turned into quantitative specifications. However, there are various techniques that are useful to help us to turn our generalized qualitative knowledge into numerical evaluations. In the example, we have employed a variety of methods. In particular:

1. studying summary statistics from samples in related populations;
2. setting rough bounds, often from negative inferences from specialized knowledge;
3. specifying probability quantiles;
4. specifying probability distributions consistent with those quantiles;
5. identifying one and two standard deviation intervals, from which variances can be judged;
6. introducing underlying 'population means';
7. partitioning variances and covariances into terms corresponding to uncorrelated components;
8. assessing a covariance by considering the variance of the difference of the corresponding quantities;
9. constructing graphical models to display the qualitative relationships between the quantities of interest.

While all of the above are methods that we have often found helpful in specifying our beliefs, they are in no way intended to be exhaustive of the general approaches which have been put forward for the elicitation of probabilities and expectations. It is an interesting and important question as to which are the most appropriate methods of prior specification for any problem. We have emphasized in our account the scientific as opposed to the psychological basis for the specification, and suggested ways in which the various ingredients going into a prior judgement might be separated out and analysed with as much care as possible. However, even for simple problems, we do not have unlimited time or resources, and we must draw a line somewhere; this itself is a matter of judgement.

Is it meaningful to ask whether the doctor has got the numbers correct. Well, she could probably give better numbers if she thought more deeply and consulted more 'experts' – at least the numbers would be better in the twin senses that she might feel more confident in her assertions and she might feel that her assertions were more firmly grounded in reality. However, we should also recognize that the collection of arguments that the doctor has brought to bear even in our simplified account is probably more stringent than many people ever bring to bear on any argument about anything in their entire lifetime. It is reasonable to expect that the process of laying out our beliefs in numerical form, and carefully arguing the value for each number in our prior specification, should help us to make better judgements because we cannot escape into unexamined generalities, and we are forced to consider how the various aspects of our beliefs relate to each other.

However, we only gain this benefit if the numbers that we specify reflect meaningful judgements. In particular, it is often difficult to strike an appropriate balance between the desire to specify uncertainties about a large number of aspects of a problem (in order to bring to bear as many relevant sources of information as we can) and the need to specify only those uncertainties that we have the ability and patience to evaluate meaningfully.

We must recognize that we may not have the capability to investigate our problems as deeply as we might ideally wish. Even in our example problem, which was chosen to be as small as possible while still retaining some genuine content, and even with a bare minimum level of detail in the doctor's specifications, there is still a substantial effort required to produce the numbers that we require, provided that a genuine attempt is made to produce these values in a thoughtful manner. As problems become more complex, the gap between a 'perfect argument' and the belief quantifications that we can genuinely produce grows ever wider. Thus, our priority is to construct methods which are as simple as possible and allow us to utilize whatever limited aspects of our prior judgements we are able to specify, without imposing the pretence that a much wider class of hypothetical quantifications, for example all of the quantifications that are required for a many-dimensional prior probability distribution, have also been assessed.

In this chapter, we have just introduced some basic issues that arise in quantifying beliefs. A good starting point for investigating the literature on elicitation is the discussion meeting on this topic organized by the Royal Statistical Society. Kadane and Wolfson (1998), O'Hagan (1998), Craig et al. (1998) and the accompanying discussion give a variety of viewpoints and an extensive list of references for pursuing work in this area; for a more recent overview, see Garthwaite et al. (2005) and references therein. The paper by Farrow (2003) is interesting in giving constructive methods for building large subjective covariance structures.

3

Adjusting beliefs

Suppose that we have specified means, variances, and covariances for a collection of random quantities. If the values of some of the quantities become known, then this will cause us to reassess our judgements over the remaining quantities. In this chapter, we describe the adjustment of expectations and variances by linear fitting on observed quantities. We derive the basic properties of such belief adjustments and discuss the foundational interpretation of this form of analysis. We describe the canonical analysis for a collection of belief adjustments. This analysis summarizes all of the implications of the implied changes in beliefs, by the construction of the belief transform associated with the adjustment, and has a natural geometric interpretation.

3.1 Adjusted expectation

We have a collection, C, of random quantities, for which we have specified prior means, variances, and covariances. Suppose now that we observe the values of a subset, $D = \{D_1, \ldots, D_k\}$, of C. We intend to modify our beliefs about various quantities, $B = \{B_1, \ldots, B_r\}$, in C, given the values of the collection D.

A simple method by which we can modify our prior expectation statements is to evaluate the adjusted expectation for each quantity. The **adjusted expectation** of a random quantity X, given observation of a collection of quantities D, written $\mathrm{E}_D(X)$, is the linear combination

$$\mathrm{E}_D(X) = \sum_{i=0}^{k} h_i D_i$$

which minimizes

$$\mathrm{E}\left(\left[X - \sum_{i=0}^{k} h_i D_i\right]^2\right) \tag{3.1}$$

Bayes Linear Statistics: Theory and Methods M. Goldstein and D.A. Wooff

over all collections $h = (h_0, h_1, \ldots, h_k)$, where D_0 is the unit constant, i.e. $D_0 = 1$. $\mathrm{E}_D(X)$ is also called the **Bayes linear rule** for X given D.

The minimization in (3.1) is determined by the prior mean, variance, and covariance specifications for X and the vector $D = (D_1, \ldots, D_k)$. We make the following definition.

Definition 3.1 *The* ***adjusted expectation*** *of a random quantity X, given observation of a collection of quantities D, written* $\mathrm{E}_D(X)$, *is*

$$\mathrm{E}_D(X) = \mathrm{E}(X) + \mathrm{Cov}(X, D)\mathrm{Var}(D)^{\dagger}(D - \mathrm{E}(D)). \tag{3.2}$$

(Observe that we follow the simplification, here and subsequently, of using the same notation to refer to the set of quantities $D = \{D_1, \ldots, D_k\}$ when signifying the collection of quantities used for an adjustment, as in $\mathrm{E}_D(X)$, and to signify the vector whose elements are $(D_1, \ldots, D_k)$, as in the right-hand side of (3.2).)

The matrix $\mathrm{Var}(D)^{\dagger}$ in (3.2) is the Moore–Penrose generalized inverse, namely the generalized inverse constructed strictly from the space of positive eigenvectors. When $\mathrm{Var}(D)$ is non-singular, $\mathrm{Var}(D)^{\dagger} = \mathrm{Var}(D)^{-1}$ is simply the usual matrix inverse.

When $\mathrm{Var}(D)$ is invertible, we may show that (3.2) minimizes (3.1), by observing that, for any choice $h_* = (h_1, \ldots, h_k)$, the value of h_0 minimizing $\mathrm{E}([X - h_*^T D - h_0]^2)$ is $h_0 = \mathrm{E}(X - h_*^T D) = \mathrm{E}(X) - h_*^T \mathrm{E}(D)$, and with this choice for h_0,

$$\mathrm{E}([X - h_*^T D - h_0]^2) = \mathrm{Var}(X - h_*^T D) = \mathrm{Var}(X) + h_*^T \mathrm{Var}(D) h_* - 2h_*^T \mathrm{Cov}(D, X).$$

Setting the derivative of this relation to zero gives (3.2).

An alternative indirect derivation of (3.2) follows by checking that, with $\mathrm{E}_D(X)$ as defined by (3.2), we have

$$\mathrm{E}([X - \mathrm{E}_D(X)]D_i) = 0, \quad i = 0, 1, \ldots k, \tag{3.3}$$

so that, for any scalars $b_0, b_1, \ldots, b_k$, and corresponding linear combination $D_b = \sum_{i=0}^{k} b_i D_i$,

$$\mathrm{E}([X - \mathrm{E}_D(X) - D_b]^2) = \mathrm{E}([X - \mathrm{E}_D(X)]^2) + \mathrm{E}(D_b^2),$$

from which it is immediate that $\mathrm{E}_D(X)$ minimizes (3.1). A full direct derivation for (3.2) in the general case is given in §12.4.

3.2 Properties of adjusted expectation

Property 3.2 *Adjusted expectation obeys the following properties, which follow directly from* (3.2).

3.2.1: *Adjusted expectation is* ***linear***. *For any quantities* X_1, X_2 *and constants* a_1, a_2 *we have*

$$\mathrm{E}_D(a_1 X_1 + a_2 X_2) = a_1 \mathrm{E}_D(X_1) + a_2 \mathrm{E}_D(X_2). \tag{3.4}$$

3.2.2: *Adjusted expectation is* ***conglomerable****, which means that expectations over adjusted expectations yield prior expectations, namely, for any* X,

$$\mathrm{E}(\mathrm{E}_D(X)) = \mathrm{E}(X). \tag{3.5}$$

Definition 3.3 *We define the* ***adjusted version*** *of* X *given* D, $\mathbb{A}_D(X)$, *to be the 'residual' form*

$$\mathbb{A}_D(X) = X - \mathrm{E}_D(X). \tag{3.6}$$

Property 3.4 *Adjusted versions obey the following properties:*

3.4.1:

$$\mathrm{E}(\mathbb{A}_D(X)) = 0; \tag{3.7}$$

3.4.2:

$$\mathrm{Cov}(\mathbb{A}_D(X), D) = 0, \tag{3.8}$$

3.4.3:

$$\mathrm{Cov}(\mathbb{A}_D(X), \mathrm{E}_D(X)) = 0. \tag{3.9}$$

3.3 Adjusted variance

Definition 3.5 *The* ***adjusted variance****, of* X *given* D, *denoted* $\mathrm{Var}_D(X)$, *is defined to be*

$$\mathrm{Var}_D(X) = \mathrm{E}([X - \mathrm{E}_D(X)]^2) = \mathrm{Var}(\mathbb{A}_D(X)). \tag{3.10}$$

Substituting for $\mathrm{E}_D(X)$ from (3.2), the value of $\mathrm{Var}_D(X)$ is determined by our prior variances and covariances as

$$\mathrm{Var}_D(X) = \mathrm{Var}(X) - \mathrm{Cov}(X, D)\mathrm{Var}(D)^\dagger\mathrm{Cov}(D, X). \tag{3.11}$$

We write X as the sum of the two uncorrelated components

$$X = \mathbb{A}_D(X) + \mathrm{E}_D(X),$$

so that we can split $\mathrm{Var}(X)$ as

$$\mathrm{Var}(X) = \mathrm{Var}(\mathbb{A}_D(X)) + \mathrm{Var}(\mathrm{E}_D(X)). \tag{3.12}$$

Definition 3.6 *The* ***variance of*** X ***resolved by*** D, $\mathrm{RVar}_D(X)$, *is defined as*

$$\mathrm{RVar}_D(X) = \mathrm{Var}(\mathrm{E}_D(X)) = \mathrm{Cov}(X, D)\mathrm{Var}(D)^\dagger\mathrm{Cov}(D, X). \tag{3.13}$$

We therefore write the variance partition for X as

$$\mathrm{Var}(X) = \mathrm{Var}_D(X) + \mathrm{RVar}_D(X). \tag{3.14}$$

Definition 3.7 *A simple scale-free quantification of the effect of an adjustment is the* ***resolution****,* $\mathrm{R}_D(X)$*, defined as*

$$\mathrm{R}_D(X) = \frac{\mathrm{RVar}_D(X)}{\mathrm{Var}(X)} = 1 - \frac{\mathrm{Var}_D(X)}{\mathrm{Var}(X)}. \tag{3.15}$$

$\mathrm{R}_D(X)$ lies between zero and one. If $\mathrm{R}_D(X)$ is near zero then either the collection D is not expected to be informative for X, relative to our prior knowledge about X, or our beliefs have not been specified in sufficient detail to exploit the information contained in D.

Definition 3.8 *We define the* ***adjusted covariance****,* $\mathrm{Cov}_D(X, Y)$*, to be*

$$\begin{aligned}\mathrm{Cov}_D(X, Y) &= \mathrm{Cov}(\mathbb{A}_D(X), \mathbb{A}_D(Y))\\ &= \mathrm{E}([X - \mathrm{E}_D(X)][Y - \mathrm{E}_D(Y)])\\ &= \mathrm{Cov}(X, Y) - \mathrm{Cov}(X, D)\mathrm{Var}(D)^{\dagger}\mathrm{Cov}(D, Y),\end{aligned}$$

and similarly the ***resolved covariance****,* $\mathrm{RCov}_D(X, Y)$*, to be*

$$\begin{aligned}\mathrm{RCov}_D(X, Y) &= \mathrm{Cov}(\mathrm{E}_D(X), \mathrm{E}_D(Y))\\ &= \mathrm{Cov}(X, D)\mathrm{Var}(D)^{\dagger}\mathrm{Cov}(D, Y).\end{aligned}$$

3.4 Interpretations of belief adjustment

We now discuss the various interrelated interpretations of adjusted expectations and variances. First, if we take a Bayesian view based on complete probabilistic specification of all uncertainties, then we may view adjusted expectations as offering simple tractable approximations to their full Bayes counterparts, which are useful in problems which are sufficiently complex that the full specification and analysis would be too time-consuming. For example, in many problems arising in Bayesian experimental design, we must compare each of a large collection of possible designs, where each design must be evaluated by a preposterior analysis based on the expected value of information provided by samples from that design. Such assessments are notoriously computer-intensive, and the Bayes linear counterpart to the full Bayes analysis may be the only version of the design choice problem which is tractable. In addition, the Bayes linear analysis leads to various interpretative measures and diagnostic tests which offer insights which will be relevant to any full Bayes analysis. The Bayes linear 'approximation' is exact in certain cases. The most important is when we adjust on the indicator functions for an event partition, as we shall discuss below. More generally, the approximation is exact whenever the posterior mean is linear in the elements of the conditioning set, for example when all of the quantities are jointly normal. Because of the importance of both the normal form and the mean square approximation, the form of the Bayes linear estimator has appeared widely in the literature. Of particular importance are the papers by Stone (1963) and Hartigan (1969), which are among the first to discuss the role of such assessments in Bayes analysis in the context of partial prior specification.

An alternative interpretation is to view the quantity $\mathrm{E}_D(X)$ as an 'estimator' of the value of X, which combines the data with simple aspects of our prior beliefs in an intuitively plausible manner and which leads to a useful methodology. Further, on occasions, our prior judgements may be constructed by simple computations on available data (for example, using sample variance matrices as proxies for prior variance matrices, which may be an acceptable approximation when analysing a large data set). In such cases, adjusted expectation can be viewed as complementary to certain standard estimators in multivariate analysis.

However, usually we do not view expectation as an estimate. For example, probabilities are expectations of the corresponding indicator functions, but we rarely view the probability as an estimate for the indicator function. Just as we view expectation as a primitive, we may similarly view adjusted expectation as a basic quantification of certain further aspects of our beliefs. Indeed, we have already observed that, in de Finetti's formal development of expectation, the principle operational definition that he offers is that our expectation for X is the value x which we would choose under penalty (2.4). In this view, adjusted expectation simply expresses the extension of our choice of preferences from the certain choice x to the random choice

$$L_D = c\left[X - \sum_{i=0}^{k} x_i D_i\right]^2. \tag{3.16}$$

A particular case of interest is when the collection D **represents a partition**, i.e. $\{D_1, \ldots, D_k\}$ are the indicator functions for a partition, so that each D_i is 1 or 0, and $\sum_i D_i = 1$. In this case, expanding the penalty (3.16) gives the equivalent form,

$$L_D = \sum_{i=1}^{k} cD_i(X - x_i)^2. \tag{3.17}$$

By comparison with the operational definition of conditional expectation, as the value chosen to minimize the score (2.5), we see that each x_i should be chosen as the corresponding conditional expectation, so that

$$\mathrm{E}_D(X) = \sum_{i=1}^{k} \mathrm{E}(X|D_i)D_i. \tag{3.18}$$

Therefore, in the special case where D represents a partition, the adjusted expectation for X is numerically equal to the conditional expectation for X. Under this view, adjusted expectation is a natural generalization of conditional expectation, where we drop the restriction that we must only 'condition' on the indicator functions for a partition.

In line with our various interpretations of belief adjustment, we may give corresponding interpretations to adjusted variance. We may view $\mathrm{Var}_D(X)$ as:

- a simple, easily computable upper bound on full Bayes preposterior risk, under quadratic loss, for any full prior specification consistent with the given mean and variance specifications;
- the 'mean squared error' of the estimator $\mathrm{E}_D(X)$;
- a primitive expression, interpreted as we would a prior variance, but applied to the 'residual variation' when we have extracted the variation in X 'accounted for' by D.

We have described three alternative views of adjusted expectation, each of which has merit in certain contexts and reflects various contrasting views that may be held concerning the revision of beliefs. Our concern in this book is to describe the practical machinery of our approach. Therefore, for the most part, we will move between these three interpretations, viewing adjusted expectation as an intuitively plausible numerical summary statement about our beliefs given the data, based on certain clearly defined aspects of our prior beliefs. As with any other formal analysis that we might carry out, adjusted expectations offer logical information in quantitative form which we may use as we deem appropriate to improve our actual posterior judgements.

However, there remain various important questions concerning the relationship between belief adjustment based upon partial prior specification and the coherent revision of beliefs. There are various foundational arguments to suggest why we should view adjusted expectation as a primitive, the precise sense in which adjusted expectation may be viewed as an 'estimator', and the general properties which may be claimed for the estimate. Further, such arguments reverse our first interpretation above by identifying a full Bayes analysis as a simple special case of the general analysis which we advocate. A full foundational analysis would take us beyond the intended scope of this book. We shall content ourselves here with a brief description of the general relationship between belief adjustment and belief revision which underlies our approach.

3.5 Foundational issues concerning belief adjustment

Suppose that we are now making expectation statements and that we intend to revise these statements at some future time point, t say. Part of the information that we shall use to revise our beliefs is the observation of the collection $D = \{D_1, \ldots, D_k\}$.

To link belief statements at different times requires some form of temporal coherence condition. The condition that we shall employ is the **temporal sure preference condition**. The temporal sure preference condition is as follows: if it is logically necessary that we will prefer a certain small random penalty U to W at some given future time, then we should not now have a strict preference for penalty W over U. To separate changes in belief from changes in utility, we may suppose that the penalties are paid in probability currency, for example in numbers of tickets for a lottery with a fixed prize. With this penalty scale, our preferences obey the expectation preference property, namely that preferring penalty A to penalty B is equivalent to assigning $\mathrm{E}(A) < \mathrm{E}(B)$, as expectation for the penalty corresponds to probability of the reward.

The reasons why we base our development on the requirement that future sure preference should be respected by today's preference are as follows. First, the principle does appear to be reasonable in a very wide range of problems of uncertainty. Secondly, temporal sure preference is a very weak requirement on our current preferences, as we only require that it should apply to sure preferences which are a very small sub-collection of our future preferences. Further, temporal sure preference may be viewed as a natural extension to the familiar principle of avoiding sure loss, based, in this case, in accumulating penalties over time, in the sense that if we break temporal sure preference then we appear to be happy to pay to switch between penalties A and B now, in the certain knowledge that we will subsequently be prepared to pay again to switch back to the penalty we currently hold. Temporal sure preference is not a basic rationality criterion, but rather an operationally testable property of our current beliefs about our future beliefs, which we will often find it reasonable to accept. It is the minimal principle which is sufficient to derive an operational account of the inferential content of the subjectivist theory, for individuals with limited abilities to enumerate future possibilities and to specify beliefs over such possibilities. When we are unwilling, in particular circumstances, to accept the principle, then we must either modify aspects of our beliefs, or develop from first principles any links that we are prepared to assert between current and future beliefs.

We use the temporal sure preference principle as follows. Suppose that $D = (D_1, \ldots, D_k)$ is a vector of random quantities, whose values will definitely be a part of the information set which is known to us by time t, at which time we shall declare a revised expectation, $\mathrm{E}_t(X)$, for X. From our operational definition for expectation, at time t, we must prefer $L_t = c[X - \mathrm{E}_t(X)]^2$ to any other penalty of this form, and in particular L_t will be our preferred choice of penalty of the form

$$L(h, h_0, h_1, \ldots, h_k) = c\left[X - h\mathrm{E}_t(X) - \sum_i h_i D_i\right]^2.$$

From the temporal sure preference principle, L_t must be our preferred choice of penalty of form $L(h, h_0, h_1, \ldots, h_k)$ now, and by the expectation preference property, we must therefore currently assess that

$$\mathrm{E}([X - \mathrm{E}_t(X)]^2) = \inf_{h,h_0,h_1,\ldots,h_k} \mathrm{E}\left(\left[X - h\mathrm{E}_t(X) - \sum_i h_i D_i\right]^2\right).$$

Therefore, $\mathrm{E}_t(X)$ is the Bayes linear rule for X in the collection

$$(\mathrm{E}_t(X), D_1, \ldots, D_k),$$

which implies the following relationships between our adjusted expectation, our actual posterior expectation, and the value of the outcome X, namely

$$X = \mathrm{E}_t(X) + R_t(X), \quad \mathrm{E}_t(X) = \mathrm{E}_D(X) + S_t(X), \tag{3.19}$$

where $R_t(X)$, $S_t(X)$ have zero expectation and are uncorrelated with each other and with all the elements of D. Therefore, $\mathrm{E}_D(X)$ is informative for $\mathrm{E}_t(X)$ in precisely the same way that $\mathrm{E}_t(X)$ is informative for X. In this sense, $\mathrm{E}_D(X)$ may be viewed as a prior inference for the actual posterior judgement that we shall make, having observed various data, of which the collection D is known to be a part. There is no implication that $\mathrm{E}_D(X)$ will fully express our genuine revised belief concerning the expectation of X. Rather, a certain portion of the variance in $\mathrm{E}_t(X)$ is resolved by assessment of $\mathrm{E}_D(X)$, and the ratio $\mathrm{Var}(\mathrm{E}_D(X))/\mathrm{Var}(\mathrm{E}_t(X))$ is a simple measure of the value of the belief adjustment in revising our beliefs.

We may interpret conditional probabilities in just this way, from (3.18), namely as prior inferences for posterior beliefs via the equation

$$\mathrm{E}_t(X) = \sum_i \mathrm{E}(X|D_i)D_i + S_t(X), \tag{3.20}$$

where now $\mathrm{E}(S_t(X)|D_i) = 0$, for all i. Thus, it is not usually incoherent to appear to behave a priori as though we shall use Bayes conditioning to update our beliefs. However, we need an additional and much stronger temporal principle in order to move from zero conditional expectations for each S_t, as given in (3.20), to an a priori belief that each such quantity is identically zero, as required in standard interpretations of Bayesian conditioning. Such a principle must rely on the notions that we may anticipate, a priori, all possible outcomes that we might observe, that we may, at this time, quantify our posterior beliefs given each such combination of outcomes, and that we will in no way change these assessments, for example by further reflection, before we observe the conditioning events. Unlike the temporal sure preference principle which is weak enough to apply with great generality, this Bayesian temporal principle is so demanding that it will rarely, if ever, apply to the real problems that we face. Thus, it is preferable to view conditioning as a special case of prior inference, while recognizing that in certain simple standard situations we may be able to treat conditional and posterior expectations as though they were the same with only small loss of precision.

In summary, our view is that posterior judgements will always be subjective, for the same reason that prior judgements are subjective, namely that the full reasoning that we shall bring to bear is too complex to allow of a complete logical description in advance. The relations between actual posterior revisions and belief analysis based on partial prior specifications are stochastic, rather than deterministic, and statements of the form (3.19) encapsulate all that can be said about the relationship between the formal (full or linear) Bayes inference and our actual posterior beliefs.

3.6 Example: one-dimensional problem

We begin with a simple hypothetical problem in which we intend using one quantity X, as yet unobserved, to help reduce uncertainty about another quantity Y. To carry out the Bayes linear analysis we need expectations and variances for X and Y, together with a covariance between them. Suppose that these are $\text{Var}(X) = \text{Var}(Y) = 1$, $\text{Cov}(X, Y) = 0.6 = \text{Corr}(X, Y)$, $\text{E}(X) = 2$, and $\text{E}(Y) = 1$. By (3.2), Y has adjusted expectation

$$\begin{aligned}\text{E}_X(Y) &= \text{E}(Y) + \frac{\text{Cov}(X, Y)}{\text{Var}(X)}(X - \text{E}(X)) \\ &= 1 + 0.6(X - 2) \\ &= 0.6X - 0.2.\end{aligned}$$

The conglomerability property (3.5) is easily verified:

$$\text{E}(0.6X - 0.2) = 1 = \text{E}(Y).$$

If instead we had specified correlations of 0.8 or 0.4 between X and Y, we would have arrived at $\text{E}_X(Y) = 0.8X - 0.6$ or $\text{E}_X(Y) = 0.4X + 0.2$. Thus, viewing $\text{E}_X(Y)$ as an estimator of Y, observe that this estimator is weakly or strongly dependent on X according to the degree of relationship expressed beforehand, as the prior correlation between Y and X.

The adjusted variance for Y is given by (3.11) as

$$\text{Var}_X(Y) = \text{Var}(Y) - \frac{\text{Cov}(X, Y)^2}{\text{Var}(X)} = 1 - 0.6^2 = 0.64.$$

Thus in this case the adjusted variance depends only on the magnitude of the correlation between X and Y: all the uncertainty in Y will be removed when $\text{Corr}(X, Y) = 1$ (as then Y is linearly equivalent to X), whereas X is useless for adjusting Y when X and Y are uncorrelated. The resolved variance here is $\text{RVar}_X(Y) = 0.36$, and as the prior variance was 1, the resolution – loosely the proportion of prior variance explained – is $\text{R}_X(Y) = 0.36$ also.

In this first example the adjustment of Y by X has resulted in the following partition into adjusted expectation (estimator) and residual quantity, with properties:

$$\begin{aligned}Y &= \text{E}_X(Y) && + Y - \text{E}_X(Y) \\ &= 0.6X - 0.2 && + (Y - 0.6X + 0.2) \\ &= \text{Adjusted expectation} && + \text{Residual quantity}\end{aligned}$$

$$\begin{aligned}\text{E}(Y) &= \text{E}(\text{E}_X(Y)) && + \text{E}(Y - \text{E}_X(Y)) \\ &= \text{E}(Y) && + 0\end{aligned}$$

$$\begin{aligned}\mathrm{Var}(Y) &= \mathrm{Var}(\mathrm{E}_X(Y)) && + \mathrm{Var}(Y - \mathrm{E}_X(Y))\\ &= \mathrm{RVar}_X(Y) && + \mathrm{Var}_X(Y)\\ &= 0.36 && + 0.64\\ &= \text{Resolved variance} && + \text{Adjusted variance}\end{aligned}$$

3.7 Collections of adjusted beliefs

We have described how to adjust prior expectations for a single quantity, X, using observations on a collection $D = \{D_1, \dots, D_k\}$. We evaluate adjusted expectations, adjusted versions, and adjusted variances for a collection $B = \{B_1, \dots, B_r\}$ of elements in the same way. We consider B, D as vectors, of dimension r and k, respectively.

Definition 3.9 *The* ***adjusted expectation*** *for collection B given collection D is calculated componentwise as in* (3.2)*, giving*

$$\mathrm{E}_D(B) = \mathrm{E}(B) + \mathrm{Cov}(B, D)\mathrm{Var}(D)^{\dagger}(D - \mathrm{E}(D)). \tag{3.21}$$

Definition 3.10 *We define the* ***adjusted version*** *of the collection B given D to be the 'residual' vector*

$$\mathbb{A}_D(B) = B - \mathrm{E}_D(B). \tag{3.22}$$

Property 3.11 *The properties of adjusted expectations for a random vector are as for a single quantity:*

3.11.1: *For any conformable matrices A_1, A_2 and random vectors B_1, B_2,*

$$\mathrm{E}_D(A_1B_1 + A_2B_2) = A_1\mathrm{E}_D(B_1) + A_2\mathrm{E}_D(B_2). \tag{3.23}$$

3.11.2:

$$\mathrm{E}(\mathrm{E}_D(B)) = \mathrm{E}(B) \tag{3.24}$$

so that

$$\mathrm{E}(\mathbb{A}_D(B)) = 0, \tag{3.25}$$

the r-dimensional null vector.

3.11.3:

$$\mathrm{Cov}(D, \mathbb{A}_D(B)) = 0 \tag{3.26}$$

so that

$$\mathrm{Cov}(\mathrm{E}_D(B), \mathbb{A}_D(B)) = 0, \tag{3.27}$$

the $r \times k$ null matrix.

Therefore, just as for a single quantity X, we partition the vector B as the sum of two uncorrelated vectors, namely

$$B = \mathrm{E}_D(B) + \mathbb{A}_D(B), \tag{3.28}$$

so that we may partition the variance matrix of B into two variance components

$$\mathrm{Var}(B) = \mathrm{Var}(\mathrm{E}_D(B)) + \mathrm{Var}(\mathbb{A}_D(B)). \tag{3.29}$$

We call

$$\mathrm{RVar}_D(B) = \mathrm{Var}(\mathrm{E}_D(B))$$

the **resolved variance matrix** for B by D. We call

$$\mathrm{Var}_D(B) = \mathrm{Var}(\mathbb{A}_D(B))$$

the **adjusted variance matrix, for** B **by** D. $\mathrm{Var}_D(B)$ is calculated as in (3.11), namely

$$\mathrm{Var}_D(B) = \mathrm{Var}(B) - \mathrm{Cov}(B, D)\mathrm{Var}(D)^{\dagger}\mathrm{Cov}(D, B), \tag{3.30}$$

so that

$$\mathrm{RVar}_D(B) = \mathrm{Cov}(B, D)\mathrm{Var}(D)^{\dagger}\mathrm{Cov}(D, B). \tag{3.31}$$

We define the **adjusted covariance matrix**, $\mathrm{Cov}_D(B_1, B_2)$, to be

$$\begin{aligned}\mathrm{Cov}_D(B_1, B_2) &= \mathrm{Cov}(\mathbb{A}_D(B_1), \mathbb{A}_D(B_2))\\ &= \mathrm{E}([B_1 - \mathrm{E}_D(B_1)][B_2 - \mathrm{E}_D(B_2)]^T)\\ &= \mathrm{Cov}(B_1, B_2) - \mathrm{Cov}(B_1, D)\mathrm{Var}(D)^{\dagger}\mathrm{Cov}(D, B_2),\end{aligned}$$

and similarly the **resolved covariance matrix**, $\mathrm{RCov}_D(B_1, B_2)$, to be

$$\begin{aligned}\mathrm{RCov}_D(B_1, B_2) &= \mathrm{Cov}(\mathrm{E}_D(B_1), \mathrm{E}_D(B_2))\\ &= \mathrm{Cov}(B_1, D)\mathrm{Var}(D)^{\dagger}\mathrm{Cov}(D, B_2).\end{aligned}$$

3.8 Examples

3.8.1 Algebraic example

To introduce the issues involved in carrying out analyses for higher-dimensional problems, consider the following hypothetical problem in which there are two unknowns Y_1 and Y_2 of interest, and two as yet unobserved quantities X_1 and X_2 which we could use to learn about Y_1 and Y_2. We gather these quantities into the collections (vectors) $B = [Y_1, Y_2]$ and $D = [X_1, X_2]$. For convenience we will specify for each of the four quantities a variance of one and an expectation of zero. We will also suppose that $\mathrm{Cov}(Y_1, Y_2) = v$, where $|v| < 1$; $\mathrm{Cov}(X_1, X_2) = u$, where $|u| < 1$; and that each pair (Y_i, X_j) has the same covariance, $\mathrm{Cov}(X_i, Y_j) = \rho$.

Notice, before we carry out any adjustments, that we cannot distinguish (as far as the specified beliefs are concerned) between Y_1 and Y_2 or between X_1 and X_2, so that we expect a large degree of symmetry in the analysis. The vector of expectations and the joint covariance (and in this case correlation) matrix over these quantities are

$$\mathrm{E}\left(\begin{bmatrix} D \\ B \end{bmatrix}\right) = \mathrm{E}\left(\begin{bmatrix} X_1 \\ X_2 \\ Y_1 \\ Y_2 \end{bmatrix}\right) = \begin{bmatrix} 0 \\ 0 \\ 0 \\ 0 \end{bmatrix}, \tag{3.32}$$

$$\mathrm{Var}\left(\begin{bmatrix} D \\ B \end{bmatrix}\right) = \begin{bmatrix} \mathrm{Var}(D) & \mathrm{Cov}(D, B) \\ \mathrm{Cov}(B, D) & \mathrm{Var}(B) \end{bmatrix} = \mathrm{Var}\left(\begin{bmatrix} X_1 \\ X_2 \\ Y_1 \\ Y_2 \end{bmatrix}\right) = \begin{bmatrix} 1 & u & \rho & \rho \\ u & 1 & \rho & \rho \\ \rho & \rho & 1 & v \\ \rho & \rho & v & 1 \end{bmatrix}. \tag{3.33}$$

Prior to any adjustment, we must first ensure that the belief specifications are coherent: generally this means that we must ensure that the joint variance–covariance matrix over all quantities is non-negative definite. In order to do this we employ the following theorem.

Theorem 3.12 *The matrix* (3.33) *is non-negative definite if and only if the following three properties hold:*

3.12.1: $\mathrm{Var}(D)$ *is non-negative definite;*

3.12.2: $\mathrm{Cov}(D, B) \in \mathbf{range}\{\mathrm{Var}(D)\}$*;*

3.12.3: $\mathrm{Var}(B) - \mathrm{Cov}(B, D)\mathrm{Var}(D)^{-}\mathrm{Cov}(D, B)$ *is non-negative definite for any choice of generalized inverse for* $\mathrm{Var}(D)$.

In general, to avoid cluttering the flow of the statistical treatment we defer the matrix algebra required for implementation of Bayes linear methods to Chapter 11, and the implementation itself to Chapter 12, and make forward references to them as necessary. Thus, the general form for checking non-negative definiteness conditions for partitioned matrices can be found as Theorem 11.35.

To return to our example, the first condition is trivially satisfied by design. Indeed, as we have specified $|u| < 1$, we can take the inverse of $\mathrm{Var}(D)$ in subsequent equations, rather than a generalized inverse. The second condition, which requires that $\mathrm{Cov}(D, B)$ be in the linear span of the columns of $\mathrm{Var}(D)$ (see Definition 11.23), is also satisfied in this example as $\mathrm{Var}(D)$ has full column rank. The third condition is satisfied when

$$\begin{bmatrix} 1 & v \\ v & 1 \end{bmatrix} - \begin{bmatrix} \rho & \rho \\ \rho & \rho \end{bmatrix} \begin{bmatrix} 1 & u \\ u & 1 \end{bmatrix}^{-1} \begin{bmatrix} \rho & \rho \\ \rho & \rho \end{bmatrix} = \begin{bmatrix} 1 - \frac{2\rho^2}{1+u} & v - \frac{2\rho^2}{1+u} \\ v - \frac{2\rho^2}{1+u} & 1 - \frac{2\rho^2}{1+u} \end{bmatrix} \tag{3.34}$$

is non-negative definite. This reduces to the requirement that

$$|\rho| \leq \frac{1}{2}\sqrt{(1+u)(1+v)}. \tag{3.35}$$

By (3.21), which is the vector analogue of (3.2), the collection B has adjusted expectation

$$\mathrm{E}_D(B) = \mathrm{E}(B) + \mathrm{Cov}(B, D)\mathrm{Var}(D)^{\dagger}(D - \mathrm{E}(D))$$

$$\text{i.e.} \quad \begin{bmatrix} \mathrm{E}_D(Y_1) \\ \mathrm{E}_D(Y_2) \end{bmatrix} = \begin{bmatrix} 0 \\ 0 \end{bmatrix} + \begin{bmatrix} \rho & \rho \\ \rho & \rho \end{bmatrix} \begin{bmatrix} 1 & u \\ u & 1 \end{bmatrix}^{-1} \left(\begin{bmatrix} X_1 \\ X_2 \end{bmatrix} - \begin{bmatrix} 0 \\ 0 \end{bmatrix} \right)$$

$$= \begin{bmatrix} \frac{\rho}{1+u}(X_1 + X_2) \\ \frac{\rho}{1+u}(X_1 + X_2) \end{bmatrix}. \tag{3.36}$$

Thus, Y_1 and Y_2 have the same adjusted expectation – hardly surprising, because of the symmetry amongst the prior specifications that we have noted. Observe that increasing or decreasing the correlation between the Y_i, X_j pairs results in a straightforward strengthening or weakening of how changes in X_1 and X_2 affect the estimators.

By (3.30), which is the vector analogue of (3.11), the collection B has adjusted variance matrix

$$\mathrm{Var}_D(B) = \mathrm{Var}(B) - \mathrm{Cov}(B, D)\mathrm{Var}(D)^{\dagger}\mathrm{Cov}(D, B).$$

In fact, we have already calculated this matrix in (3.34) as

$$\mathrm{Var}_D(B) = \begin{bmatrix} \mathrm{Var}_D(Y_1) & \mathrm{Cov}_D(Y_1, Y_2) \\ \mathrm{Cov}_D(Y_2, Y_1) & \mathrm{Var}_D(Y_2) \end{bmatrix} = \begin{bmatrix} 1 - \frac{2\rho^2}{1+u} & v - \frac{2\rho^2}{1+u} \\ v - \frac{2\rho^2}{1+u} & 1 - \frac{2\rho^2}{1+u} \end{bmatrix}. \tag{3.37}$$

Observe that one of the conditions for non-negative definiteness of the joint variance–covariance matrix is thus equivalent to ensuring that the adjusted variance matrix is non-negative definite.

The implication of the adjustment for reducing uncertainty in Y_1, Y_2 is as follows. We began with variances of one for Y_1 and Y_2, and the expected value of fitting on the data quantities X_1 and X_2 is to reduce each of these variances to $1 - 2\rho^2/(1+u)$. As the prior variances are one in each case, the resolutions are the same as the resolved variances. For example,

$$\mathrm{R}_D(Y_1) = 1 - \frac{\mathrm{Var}_D(Y_1)}{\mathrm{Var}(Y_1)} = \frac{2\rho^2}{(1+u)}. \tag{3.38}$$

Consequently, large proportions of variation in the Y_js would be explained for large ρ and small u. This is the case when there is little overlap in the two data sources, and when the data sources are strongly correlated with the Y_js. Correspondingly, small proportions of variation in the Y_js would be explained for small ρ and large u. The adjusted covariances are changed by the same amounts: from v to $v - 2\rho^2/(1+u)$. Notice, therefore, that the adjusted correlation matrix (i.e. the adjusted variance matrix in correlation form) is

$$\begin{bmatrix} 1 & \frac{v(1+u)-2\rho^2}{(1+u)-2\rho^2} \\ \frac{v(1+u)-2\rho^2}{(1+u)-2\rho^2} & 1 \end{bmatrix}.$$

The correlation between Y_1 and Y_2, given the data quantities, can be stronger or weaker than at the outset, depending on the sign of v and the magnitude of ρ.

3.8.1.1 *Degeneracy in the variance matrices*

We are careful throughout this book to allow variance matrices exhibiting degeneracies – in other words, we allow that there may be one or more (but not all!) linear combinations of the random quantities having variance zero. In this example we have insisted that $|u| < 1$, $|v| < 1$ for algebraic convenience. However, let us examine briefly what happens in the case of degeneracy, which occurs – for example – when we allow $u = 1$. In this case we would have a non-negative definite variance matrix

$$\mathrm{Var}(D) = \begin{bmatrix} 1 & 1 \\ 1 & 1 \end{bmatrix},$$

corresponding to which the linear combination $X_1 - X_2$ has variance zero. The Moore–Penrose generalized inverse (see Lemma 11.9) is

$$\mathrm{Var}(D)^{\dagger} = \frac{1}{4}\begin{bmatrix} 1 & 1 \\ 1 & 1 \end{bmatrix}.$$

It is then straightforward to check for non-negative definiteness as we did in (3.34): the condition we obtain is that we need

$$|\rho| \le \frac{1}{\sqrt{2}}\sqrt{(1+v)},$$

suggesting that (3.35) can be extended to handle $|u| \le 1$. This is indeed so, but to make certain we need also to have checked Property 3.12.2 for this degenerate case. To do so, we need to make sure that

$$[I - \mathrm{Var}(D)\mathrm{Var}(D)^{\dagger}]\mathrm{Cov}(D, B) = 0$$

(see Lemma 11.28), and this is true in this case, establishing the result.

What happens if we employ an alternative generalized inverse? One such alternative is

$$\mathrm{Var}(D)^{-} = \frac{1}{4}\begin{bmatrix} -7 & 1 \\ -7 & 17 \end{bmatrix}.$$

Suppose that we form adjusted expectations as in (3.36), but using this generalized inverse. We then obtain

$$\mathrm{E}_D(Y_1) = \mathrm{E}_D(Y_2) = \frac{1}{2}\rho(-7X_1 + 9X_2). \tag{3.39}$$

Now compare (3.36) and (3.39): we have

$$\text{Moore–Penrose:} \qquad \mathrm{E}_D(Y_i) \propto (X_1 + X_2), \tag{3.40}$$

$$\text{an alternative:} \qquad \mathrm{E}_D(Y_i) \propto (X_1 + X_2) - 8(X_1 - X_2). \tag{3.41}$$

Thus, for the Moore–Penrose inverse, the adjusted expectation only depends on $(X_1 + X_2)$, the single non-zero eigenvector of $\mathrm{Var}(D)$. For the alternative generalized inverse (and, indeed, any other choice of generalized inverse) the adjusted expectation depends on both eigenvectors.

The two forms (3.40), (3.41) are numerically the same when all zero eigenvectors have observed values equal to their expectations, i.e. in cases where the data are consistent with their prior specifications, but otherwise the answers will be different. Consequently, the Moore–Penrose generalized inverse is the inverse which cannot lead to adjusting by 'impossible information' as, uniquely, it guarantees that adjusted expectations are formed from linear combinations with positive variance.

3.8.2 Oral glucose tolerance test

For this example we return to the oral glucose tolerance test problem introduced in Chapter 2. The problem is described in §2.4.1, the quantities of interest are described in §2.6.1, and the results of a process of belief elicitation for the problem are summarized in §2.7.4, and restated below, with the variance matrix also shown in correlation form for convenience. We gather the quantities of interest into the collections (vectors) $B = \{G_0, G_2\}$ and $D = \{D_0, D_2\}$.

$$\mathrm{E}(B) = \begin{bmatrix} \mathrm{E}(G_0) \\ \mathrm{E}(G_2) \end{bmatrix} = \begin{bmatrix} 4.16 \\ 6.25 \end{bmatrix}, \qquad \mathrm{E}(D) = \begin{bmatrix} \mathrm{E}(D_0) \\ \mathrm{E}(D_2) \end{bmatrix} = \begin{bmatrix} 4.16 \\ 6.25 \end{bmatrix}, \tag{3.42}$$

$$\mathrm{Var}(B) = \begin{bmatrix} \mathrm{Var}(G_0) & \mathrm{Cov}(G_0, G_2) \\ \mathrm{Cov}(G_0, G_2) & \mathrm{Var}(G_2) \end{bmatrix} = \begin{bmatrix} 1.12 & 0.72 \\ 0.72 & 2.43 \end{bmatrix}, \tag{3.43}$$

$$\mathrm{Var}(D) = \begin{bmatrix} \mathrm{Var}(D_0) & \mathrm{Cov}(D_0, D_2) \\ \mathrm{Cov}(D_0, D_2) & \mathrm{Var}(D_2) \end{bmatrix} = \begin{bmatrix} 1.12 & 0.72 \\ 0.72 & 2.43 \end{bmatrix}, \tag{3.44}$$

$$\mathrm{Cov}(B, D) = \begin{bmatrix} \mathrm{Cov}(G_0, D_0) & \mathrm{Cov}(G_0, D_2) \\ \mathrm{Cov}(G_2, D_0) & \mathrm{Cov}(G_2, D_2) \end{bmatrix} = \begin{bmatrix} 0.62 & 0.30 \\ 0.30 & 0.43 \end{bmatrix}, \tag{3.45}$$

$$\text{Corr}(B, B) = \text{Corr}(D, D) = \begin{bmatrix} 1 & 0.436 \\ 0.436 & 1 \end{bmatrix}, \tag{3.46}$$

$$\text{Corr}(B, D) = \begin{bmatrix} \text{Corr}(G_0, D_0) & \text{Corr}(G_0, D_2) \\ \text{Corr}(G_2, D_0) & \text{Corr}(G_2, D_2) \end{bmatrix} = \begin{bmatrix} 0.554 & 0.182 \\ 0.182 & 0.177 \end{bmatrix}. \tag{3.47}$$

Some of the principal features of the doctor's variance and covariance specifications are as follows.

- A person is diagnosed as having at least impaired glucose tolerance if the 2-hour blood glucose level is beyond the threshold of 7.0. For a typical elderly person, the prior expectation $\text{E}(G_2) = 6.25$ and variance $\text{Var}(G_2) = 2.43$ lead to crude two and three standard deviation intervals for G_2 of about $(3.13, 9.37)$ and $(1.57, 10.93)$ respectively. In each case, the upper boundary is well beyond the diagnostic threshold, reflecting the doctor's belief that the OGT test might misdiagnose many elderly patients.

- She is less sure about the 2-hour measurements than she is about the fasting measurements. Further, she has assigned a correlation of about 0.436 between G_0 and G_2 and between D_0 and D_2.

- Her covariances specified between B and D imply that G_0 and D_0 are moderately related (a correlation of about 0.554), whereas G_2 and D_2 are only weakly related, with the correlation between them being about 0.177. She has also specified about the same degree of relationship (a correlation of about 0.182) between D_0 and G_2.

- Informally, the strength of correlation between two quantities determines the degree to which one quantity can be used to help learn about another. We have here a moderate correlation, 0.554, between D_0 and G_0, and small correlations of 0.182 between D_0 and G_2 and 0.177 between D_2 and G_2, suggesting that the fasting measurement D_0 will be more informative than the 2-hour measurement D_2 in learning about both unknowns G_0 and G_2.

By (3.21), the collection B has adjusted expectation

$$\text{E}_D(B) = \text{E}(B) + \text{Cov}(B, D)\text{Var}(D)^{\dagger}(D - \text{E}(D))$$

$$\text{i.e.} \quad \begin{bmatrix} \text{E}_D(G_0) \\ \text{E}_D(G_2) \end{bmatrix} = \begin{bmatrix} 4.16 \\ 6.25 \end{bmatrix} + \begin{bmatrix} 0.62 & 0.30 \\ 0.30 & 0.43 \end{bmatrix} \begin{bmatrix} 1.12 & 0.72 \\ 0.72 & 2.43 \end{bmatrix}^{-1} \left(\begin{bmatrix} D_0 \\ D_2 \end{bmatrix} - \begin{bmatrix} 4.16 \\ 6.25 \end{bmatrix} \right)$$

$$= \begin{bmatrix} 0.5858D0 - 0.0501D2 + 2.0363 \\ 0.1904D0 + 0.1206D2 + 4.7047 \end{bmatrix}.$$

It can be difficult to tell whether a small coefficient truly indicates unimportance because of the different expectations and scalings of D_0 and D_2. For this reason, it can be useful to examine the **standardized adjusted expectations**,

$$\mathrm{E}_D(G_0) = (0.62)\mathrm{S}(D_0) - (0.08)\mathrm{S}(D_2) + 4.16,$$

$$\mathrm{E}_D(G_2) = (0.20)\mathrm{S}(D_0) + (0.19)\mathrm{S}(D_2) + 6.25,$$

where $\mathrm{S}(D_i)$ is shorthand for the standardized representation of D_i (see Definition 1.1). Now each coefficient multiplies a quantity that has expectation zero and variance unity, so that the coefficients are more readily comparable. (The constants added in each case are the initial expectations for G_0 and G_2.) We see that the adjusted expectation for G_0 depends essentially on D_0, plus a base value of 4.16; whereas the adjusted expectation for G_2 depends upon a rather larger base value, plus essentially an average of the before-and-after blood glucose readings.

By (3.30), the collection B has adjusted variance matrix

$$\mathrm{Var}_D(B) = \mathrm{Var}(B) - \mathrm{Cov}(B, D)\mathrm{Var}(D)^{\dagger}\mathrm{Cov}(D, B).$$

In terms of the basic variances and covariances, this is

$$\begin{aligned}\begin{bmatrix} \mathrm{Var}_D(G_0) & \mathrm{Cov}_D(G_0, Y_2) \\ \mathrm{Cov}_D(Y_2, G_0) & \mathrm{Var}_D(Y_2) \end{bmatrix} &= \begin{bmatrix} 1.12 & 0.72 \\ 0.72 & 2.43 \end{bmatrix} \\ &\quad - \begin{bmatrix} 0.62 & 0.30 \\ 0.30 & 0.43 \end{bmatrix} \begin{bmatrix} 1.12 & 0.72 \\ 0.72 & 2.43 \end{bmatrix}^{-1} \begin{bmatrix} 0.62 & 0.30 \\ 0.30 & 0.43 \end{bmatrix} \\ &= \begin{bmatrix} 0.7718 & 0.5658 \\ 0.5658 & 2.3211 \end{bmatrix}. \end{aligned} \tag{3.48}$$

Thus, adjusting uncertainty in G_0 reduces initial variance $\mathrm{Var}(G_0) = 1.12$ to adjusted variance $\mathrm{Var}_D(G_0) = 0.7718$; and reduces initial $\mathrm{Var}(D_2) = 2.43$ to adjusted variance $\mathrm{Var}_D(D_2) = 2.3211$. The variance resolutions are 31.09% and 4.48% respectively, so we expect that the value of the measurements that the doctor makes upon herself (D_0 and D_2) will be to remove about a third of the uncertainty in G_0, the fasting glucose level in a typical elderly person, but only a small fraction of the uncertainty in the typical elderly person's 2-hour glucose level.

In summary, the implication of the adjustment is to decompose each quantity into a residual component which we call the adjusted version and a fitted component which we call the adjusted expectation. As such, for each quantity we decompose the initial variation into portions remaining and resolved. The decompositions are shown in Table 3.1.

One of the questions of interest for this example is whether blood glucose levels act in the same way for healthy young and healthy elderly patients fed glucose as part of the OGT test. The suspicion is that glucose levels are generally higher for the elderly, and may also take longer to drop to the fasting level after ingestion of glucose, so that the 2-hour measurement may falsely indicate diabetes. This suggests that we examine the difference between the fasting measurement and the 2-hour measurement (i.e the blood glucose levels before and 2 hours after ingesting the

Table 3.1 Summary of decompositions.

G_0	=	Adjusted version	+	Adjusted expectation
	=	$G_0 - \mathrm{E}_D(G_0)$	+	$\mathrm{E}_D(G_0)$
	=	$G_0 - (0.56D_0 - 0.05D_2 + 2.04)$	+	$0.56D_0 - 0.05D_2 + 2.04$
$\mathrm{Var}(G_0)$	=	$\mathrm{Var}_D(G_0)$	+	$\mathrm{RVar}_D(G_0)$
	=	$\mathrm{Var}(G_0 - \mathrm{E}_D(G_0))$	+	$\mathrm{Var}(\mathrm{E}_D(G_0))$
1.12	=	0.77	+	0.35
100%	=	68.91%	+	31.09%
initial	=	remaining uncertainty	+	resolved uncertainty
$\mathrm{Var}(G_2)$	=	$\mathrm{Var}_D(G_2)$	+	$\mathrm{RVar}_D(G_2)$
2.43	=	2.32	+	0.11
100%	=	95.52%	+	4.48%

glucose). Thus we construct $G_h = G_2 - G_0$ and explore how we may use the doctor's own measurements to reduce uncertainties about it. We obtain $\mathrm{E}(G_h) = 2.09$, $\mathrm{Var}(G_h) = 2.11$, $\mathrm{Cov}(G_h, D_0) = -0.32$, and $\mathrm{Cov}(G_h, D_2) = 0.13$. A priori, a two standard deviation interval for G_h is given by 2.09 ± 2.91, corresponding to a tentative belief that the blood glucose level after 2 hours is still greater than the fasting level, but allowing for the possibility that we may be wrong. The adjusted expectation and standardized adjusted expectation for G_h are

$$\mathrm{E}_D(G_h) = \mathrm{E}(G_h) + \mathrm{Cov}(G_h, D)\mathrm{Var}(D)^{\dagger}(D - \mathrm{E}(D)) \tag{3.49}$$

$$= 2.09 + \begin{bmatrix} -0.32 & 0.13 \end{bmatrix} \begin{bmatrix} 1.12 & 0.72 \\ 0.72 & 2.43 \end{bmatrix}^{-1} \left(\begin{bmatrix} D_0 \\ D_2 \end{bmatrix} - \begin{bmatrix} 4.16 \\ 6.25 \end{bmatrix} \right) \tag{3.50}$$

$$= -0.3954D_0 + 0.1707D_2 + 2.6683 \tag{3.51}$$

$$= -0.4185\,\mathrm{S}(D_0) + 0.2660\,\mathrm{S}(D_2) + 2.09. \tag{3.52}$$

The adjusted variance for G_h is

$$\mathrm{Var}_D(G_h) = \mathrm{Var}(G_h) - \mathrm{Cov}(G_h, D)\mathrm{Var}(D)^{\dagger}\mathrm{Cov}(D, G_h)$$

$$= 2.11 - \begin{bmatrix} -0.32 & 0.13 \end{bmatrix} \begin{bmatrix} 1.12 & 0.72 \\ 0.72 & 2.43 \end{bmatrix}^{-1} \begin{bmatrix} -0.32 \\ 0.13 \end{bmatrix}$$

$$= 1.9613. \tag{3.53}$$

The adjusted variance of 1.9613 is hardly reduced from the prior variance of 2.11 – a reduction of only about 7% in fact – and so the data are not expected to be very informative for the difference between the two glucose measurements. Indeed, the two adjusted standard deviation interval around our prior expectation continues to include zero. The adjusted expectation is intuitively reasonable: larger than expected observed values of D_0 (D_2) will cause us to revise downwards (upwards) our expectations for the difference. We return, in §3.11.3, to why the data are expected to be so uninformative for G_h.

3.8.3 Many oral glucose tolerance tests

Using only a single pair of observations on herself, we have seen only a small fall in adjusted variance for the quantities of interest. Suppose the doctor now decides that she could take more observations into account, perhaps by taking measurements from a sample of further healthy elderly people, perhaps from patients attending her clinics for check-ups. Indeed, let us suppose that she takes a sample of n such people, that she labels her measurements as D_{i0} and D_{i2} for the ith person measured, and that she collects these into the vector D_i (any confusion between D_0, D_2, D_i should be resolved from context). Assuming that these n people form a random sample of the typical elderly people that she has in mind, expectation and variance–covariance specifications are as in (3.42) to (3.45). That is,

$$\mathrm{E}(D_i) = \begin{bmatrix} \mathrm{E}(D_{i0}) \\ \mathrm{E}(D_{i2}) \end{bmatrix} = \begin{bmatrix} 4.16 \\ 6.25 \end{bmatrix}, \tag{3.54}$$

$$\mathrm{Var}(D_i) = \begin{bmatrix} \mathrm{Var}(D_{i0}) & \mathrm{Cov}(D_{i0}, D_{i2}) \\ \mathrm{Cov}(D_{i0}, D_{i2}) & \mathrm{Var}(D_{i2}) \end{bmatrix} = \begin{bmatrix} 1.12 & 0.72 \\ 0.72 & 2.43 \end{bmatrix}, \tag{3.55}$$

$$\mathrm{Cov}(B, D_i) = \begin{bmatrix} \mathrm{Cov}(G_0, D_{i0}) & \mathrm{Cov}(G_0, D_{i2}) \\ \mathrm{Cov}(G_2, D_{i0}) & \mathrm{Cov}(G_2, D_{i2}) \end{bmatrix} = \begin{bmatrix} 0.62 & 0.30 \\ 0.30 & 0.43 \end{bmatrix}. \tag{3.56}$$

We also need to take into account covariances between different patients. For $i \neq j$,

$$\mathrm{Cov}(D_i, D_j) = \begin{bmatrix} \mathrm{Cov}(D_{i0}, D_{j0}) & \mathrm{Cov}(D_{i0}, D_{j2}) \\ \mathrm{Cov}(D_{i2}, D_{j0}) & \mathrm{Cov}(D_{i2}, D_{j2}) \end{bmatrix} = \begin{bmatrix} 0.62 & 0.30 \\ 0.30 & 0.43 \end{bmatrix}. \tag{3.57}$$

Suppose we now carry out the adjustment of B by this information. One approach is to arrange the n pairs of observations $D_{10}, D_{12}, \ldots, D_{n0}, D_{n2}$ into the $2n \times 1$ vector D^n. The $2 \times 2n$ covariance matrix $\mathrm{Cov}(B, D^n)$ and the $2n \times 2n$ variance matrix $\mathrm{Var}(D^n)$ are readily constructed from (3.54) to (3.57), and then we can apply the adjusted expectation and adjusted variance rules (3.21) to (3.31). Amongst the results we obtain are adjusted variances for G_0 and G_2 as the sample size increases from $n = 1$ to $n = 100$ patients. These adjusted variances, together with the adjusted variance for the 2-hour fasting difference, $G_h = G_0 - G_2$, are summarized in Figure 3.1. This shows that the extra reduction in variance in all three quantities is both quite small and very slow as more information arrives. The prior variance for G_h is $\mathrm{Var}(G_h) = 2.11$, which is expected to reduce to $\mathrm{Var}_{D^1}(G_h) = 1.9613$ for $n = 1$, $\mathrm{Var}_{D^2}(G_h) = 1.9031$ for $n = 2$, $\mathrm{Var}_{D^3}(G_h) = 1.8676$ for $n = 3$, and so forth, reducing to $\mathrm{Var}_{D^{100}}(G_h) = 1.6757$ for $n = 100$. Meanwhile, the adjusted variances for G_0 and G_2 converge swiftly to 0.50 and 2.00, respectively.

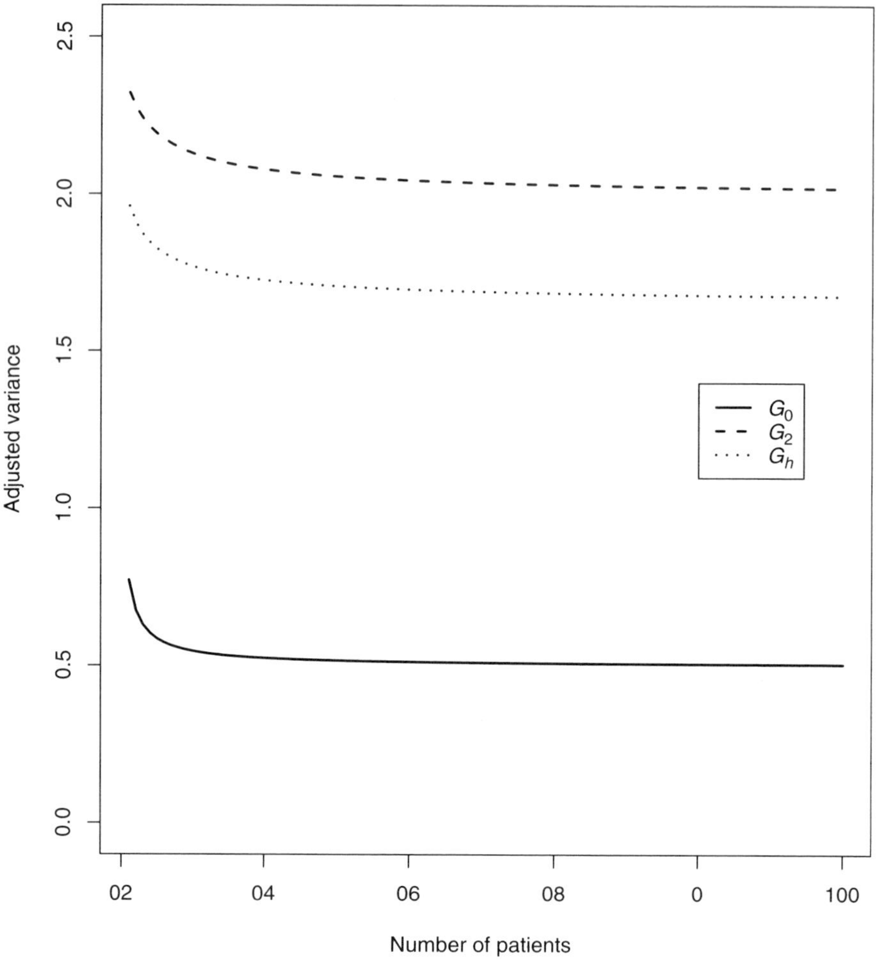

Figure 3.1 The fall in adjusted variance for G_0, G_2, and G_h as more information arrives.

As far as the last part of this example is concerned, a number of interpretational and methodological issues have begun to surface. We begin to see, for example, the notion of exploring the implications of differing sample sizes for adjustments, together with measuring the value of acquiring extra information, and explaining the rate at which evidence accumulates. The assessment of extra evidence is treated in detail in Chapter 5. Understanding the rate at which evidence accumulates will follow through calculating the underlying canonical structure for the problem: we begin to address this in the remainder of this chapter. In relation to increasing samples sizes, it would appear that the basic calculation of adjusted expectations and adjusted variances requires inverting here a $2n \times 2n$ matrix. Whilst this can be done, for larger sample sizes we quickly exceed the capacity of whatever computer we use to perform the calculations. It is clear from (3.54) to (3.57) that there is substantial symmetry amongst the belief specifications. It should come as no surprise that there is a way of exploiting such symmetries through the notion of exchangeability. We shall show in Chapter 6 how we fully exploit exchangeability, both to ease the computational burden and to help understand and utilize the underlying structure of a problem such as this. Amongst other things, we shall see, for this glucose-level example, that we need only calculate a 2×2 inverse to obtain full results for whatever sample size we desire.

In relation to the convergence of adjusted variances seen in Figure 3.1, notice that the limits are the values

$$\mathrm{Var}(G_0) - \mathrm{Cov}(D_0, G_0) = 1.12 - 0.62 = 0.50,$$

$$\mathrm{Var}(G_2) - \mathrm{Cov}(D_2, G_2) = 2.43 - 0.43 = 2.00.$$

We will see in general in Chapter 6 why we see such behaviour, which aspects of uncertainty are being resolved, and how this can be interpreted. Indeed, in §6.6 we shall return to this example and show how such convergence links to the idea of unresolvable residual variation for an individual, in the context of learning about a population mean.

3.9 Canonical analysis for a belief adjustment

We have described how to adjust a belief specification by linear fitting on data. In complex problems, it is often far from obvious how our collection of beliefs is affected by belief adjustment. Usually, therefore, it is important for us not only to obtain collections of adjusted values but also to understand and interpret the overall changes in belief over the whole collection of quantities of interest. Such an analysis, which we call **canonical analysis**, is helpful both in identifying the strengths and weaknesses of competing data sets that we may choose to collect and also in uncovering surprising and possibly unintended consequences of a particular belief adjustment that may cause us to re-examine our overall belief specification. In this section, we introduce some of the tools that we shall use for this purpose.

3.9.1 Canonical directions for the adjustment

When we evaluate a collection $\{\mathrm{E}_D(B_1), \ldots, \mathrm{E}_D(B_k)\}$ of adjusted expectations, we also implicitly evaluate the adjusted expectation and variance for each linear combination, $\sum_i h_i B_i$, of the elements of B, by the linearity of adjusted expectation (3.4), as

$$\mathrm{E}_D(h^T B) = \mathrm{E}_D\left(\sum_{i=1}^{r} h_i B_i\right) = \sum_{i=1}^{r} h_i \mathrm{E}_D(B_i) = h^T \mathrm{E}_D(B) \qquad (3.58)$$

and

$$\text{RVar}_D(h^T B) = h^T \text{RVar}_D(B)h. \tag{3.59}$$

For a general vector C, we denote by $\langle C \rangle$ the collection of linear combinations, $h^T C = \sum_i h_i C_i$, of elements of C. Often, we are principally interested in how our beliefs change over $\langle B \rangle$, given an adjustment by D. For example, if $B = \{B_1, \ldots, B_k\}$ represents a partition, then a typical element of $\langle B \rangle$ is a linear combination $a_1 B_1 + \ldots + a_k B_k$, i.e. the random quantity which takes value a_i if outcome B_i occurs. In this case, therefore, $\langle B \rangle$ is the collection of random variables defined over the partition, which will often be of more interest than are the individual partition probabilities.

One way to assess the information about $\langle B \rangle$ that we expect to receive by observing D is as follows. We first identify the particular linear combination $Y_1 \in \langle B \rangle$ for which we expect the adjustment by D to be most informative in the sense that Y_1 maximizes the resolution $\text{R}_D(Y)$ over all elements $Y \in \langle B \rangle$ with non-zero prior variance. From (3.15), this is equivalent to minimizing the ratio of adjusted to prior variance. We then proceed to identify directions for which we expect progressively less information. By analogy with similar types of canonical variable calculations in traditional multivariate analysis, we make the following definition.

Definition 3.13 *The* jth **canonical direction** *for the adjustment of* B *by* D *is the linear combination* Y_j *which maximizes* $\text{R}_D(Y)$ *over all elements* $Y \in \langle B \rangle$ *with non-zero prior variance which are uncorrelated a priori with* $Y_1, \ldots, Y_{j-1}$. *We scale each* Y_j *to have prior expectation zero and prior variance one. The values*

$$\lambda_i = \text{R}_D(Y_i) = \text{RVar}_D(Y_i) \tag{3.60}$$

are termed the **canonical resolutions**.

We will also refer to the canonical directions as **canonical quantities**. The number of canonical directions that we may define is equal to the rank,

$$r_B = \mathbf{rk}\{\text{Var}(B)\}, \tag{3.61}$$

of the variance matrix of the elements of B. The number of positive canonical resolutions depends on the covariance matrix between B and D.

Definition 3.14 *The number of positive canonical directions is*

$$r_{\mathbb{T}} = \mathbf{rk}\{\text{Cov}(D, B)\} \leq \min(\mathbf{rk}\{\text{Var}(B)\}, \mathbf{rk}\{\text{Var}(D)\}). \tag{3.62}$$

When this covariance matrix is zero, none of the canonical resolutions are non-zero and D is uninformative for linear adjustment of B. We will often be interested in finding the largest and the smallest resolutions given D for any element of $\langle B \rangle$, namely λ_1 for the element Y_1, and λ_{r_B} for Y_{r_B}.

We find the canonical quantities as follows. Suppose that $\text{Var}(B)$ is positive definite, and so has full rank. The canonical directions and resolutions are found by sequentially maximizing the ratio

$$\text{R}_D(h^T B) = \frac{h^T \text{RVar}_D(B) h}{h^T \text{Var}(B) h}. \tag{3.63}$$

As $\text{Var}(B)$ is positive definite, we may write $\text{Var}(B) = A^T A$, where A is invertible, so that we may rewrite (3.63) as

$$\text{R}_D(h^T B) = \frac{u^T M u}{u^T u}, \tag{3.64}$$

where $Ah = u$ and

$$M = (A^T)^{-1} \text{RVar}_D(B) A^{-1}.$$

It is straightforward to show, from (3.64), that the canonical resolutions are the eigenvalues, λ_i, of M, with corresponding canonical directions $h_i^T B$, where $h_i = A^{-1} u_i$ and u_i is the eigenvector of M corresponding to λ_i. Noting that u is an eigenvector of M with eigenvalue λ if and only if $h = A^{-1} u$ is an eigenvector of $\text{Var}(B)^{-1} \text{RVar}_D(B)$ with the same eigenvalue λ, we may therefore identify the canonical directions and resolutions with the eigenstructure of $\text{Var}(B)^{-1} \text{RVar}_D(B)$. We have shown the following.

Theorem 3.15 *The jth canonical resolution for the adjustment of B by D is the jth largest eigenvalue of the matrix*

$$\text{Var}(B)^{-1} \text{RVar}_D(B).$$

The jth canonical direction is the linear combination $h^T B$, where h is the eigenvector of this matrix corresponding to this eigenvalue, scaled to prior expectation zero and variance one.

When $\text{Var}(B)$ is not of full rank, then we may replace $\text{Var}(B)^{-1}$ by the corresponding generalized inverse. We discuss the more general case in §12.7.

3.9.2 The resolution transform

In the preceding section, we showed that the canonical directions and resolutions for the adjustment of B by D are given by the eigenstructure of a matrix $\text{Var}(B)^{\dagger} \text{RVar}_D(B)$. This matrix has a central role in Bayes linear statistics. In this section we explore some of its properties.

Definition 3.16 *The **resolution transform matrix** is defined as*

$$\begin{aligned} \mathbb{T}_{B:D} &= \text{Var}(B)^{\dagger} \text{RVar}_D(B) \\ &= \text{Var}(B)^{\dagger} \text{Cov}(B, D) \text{Var}(D)^{\dagger} \text{Cov}(D, B). \end{aligned} \tag{3.65}$$

When it is obvious from the context that the adjustment is evaluated over the collection B, then we will sometimes simplify the notation by removing the explicit dependence on B and denote the resolution transform by

$$\mathbb{T}_D = \mathbb{T}_{B:D}.$$

We can calculate the canonical directions $Y_1, \ldots, Y_{r_B}$ by finding the normed right eigenvectors of $\mathbb{T}_{B:D}$, which we write as $v_1, \ldots, v_{r_B}$, ordered by eigenvalues

$$1 \geq \lambda_1 \geq \lambda_2 \geq \ldots \geq \lambda_{r_B} \geq 0$$

and scaled, for each i, as $v_i^T \text{Var}(B) v_i = 1$, so that

$$Y_i = v_i^T (B - \text{E}(B)) \quad \text{and} \quad \text{Var}_D(Y_i) = 1 - \lambda_i. \tag{3.66}$$

In practice, this eigenstructure is often extracted from the equivalent form

$$\text{Var}(B)\mathbb{T}_{B:D} = \text{RVar}_D(B),$$

in order to avoid numerical problems which may arise where some or all of the variance specifications contain linear combinations with variance zero; see §12.7 for a full discussion. The resolution transform matrix has rank

$$\mathbf{rk}\{\mathbb{T}_{B:D}\} = r_{\mathbb{T}},$$

and this is the number of positive eigenvalues, and thus canonical resolutions, as noted in (3.62).

For any $U \in \langle B \rangle$, we can write $U = h^T B$ for some h. We may consider $\mathbb{T}_{B:D}$ as a linear operator over $\langle B \rangle$ by defining

$$\mathbb{T}_{B:D}(U) = (\mathbb{T}_{B:D} h)^T B. \tag{3.67}$$

The resolution transform is of intrinsic interest as the object which summarizes, through the eigenstructure, all of the effects of the belief adjustment over $\langle B \rangle$. Much of the interest of this transform derives from the following property.

Property 3.17 *For any $X = g^T B, U = h^T B \in \langle B \rangle$, we have*

$$\text{RCov}_D(X, U) = \text{Cov}(X, \mathbb{T}_{B:D}(U)). \tag{3.68}$$

Result (3.68) follows as

$$\begin{aligned}\text{Cov}(g^T B, \mathbb{T}_{B:D} h^T B) &= g^T \text{Var}(B)\mathbb{T}_{B:D} h \\ &= g^T \text{Var}(B)\text{Var}(B)^{\dagger}\text{Cov}(B, D)\text{Var}(D)^{\dagger}\text{Cov}(D, B)h \\ &= \text{RCov}_D(g^T B, h^T B).\end{aligned}$$

3.9.3 Partitioning the resolution

The collection $\{Y_1, Y_2, \ldots\}$ forms a mutually uncorrelated 'grid' of directions over $\langle B \rangle$, summarizing the effects of the adjustment. Each quantity, $X \in \langle B \rangle$, may be resolved along the canonical directions as

$$X - \mathrm{E}(X) = \sum_i \mathrm{Cov}(X, Y_i) Y_i. \tag{3.69}$$

From (3.68), we have

$$\begin{aligned} \mathrm{RVar}_D(X) &= \mathrm{Cov}(X, \mathbb{T}_{B:D}(X)) \\ &= \mathrm{Cov}\left(\sum_i \mathrm{Cov}(X, Y_i) Y_i, \sum_i \lambda_i \mathrm{Cov}(X, Y_i) Y_i\right) \\ &= \sum_i \lambda_i [\mathrm{Cov}(X, Y_i)]^2. \end{aligned} \tag{3.70}$$

We can exploit this as follows.

Definition 3.18 *For any $X \in \langle B \rangle$, the* ***resolution partition*** *for X is the decomposition of the overall resolution for X by D into the orthogonal resolutions accounted for by each of the canonical directions:*

$$\mathrm{R}_D(X) = \sum_i \mathrm{c}_i(X) \lambda_i, \tag{3.71}$$

$$\textit{where } \mathrm{c}_i(X) = [\mathrm{Corr}(X, Y_i)]^2 \tag{3.72}$$

$$\textit{and } \sum_i \mathrm{c}_i(X) = 1. \tag{3.73}$$

The resolution partition shows that we expect to learn most about those elements of $\langle B \rangle$ which have large correlations with those canonical directions with large resolutions.

Definition 3.19 *By analogy with the resolution for a single random quantity, we define the* ***resolved uncertainty for*** *$\langle B \rangle$ given adjustment by D to be*

$$\mathrm{RU}_D(B) = \sum_{i=1}^{r_B} \lambda_i = \mathbf{tr}\{\mathbb{T}_{B:D}\}. \tag{3.74}$$

The resolved uncertainty is the sum of the resolutions for any collection of r_B elements of $\langle B \rangle$ with prior variance one, which are a priori uncorrelated.

Definition 3.20 *We define the* ***system resolution for*** *B to be the average resolved uncertainty for the collection, namely*

$$\mathrm{R}_D(B) = \frac{\mathrm{RU}_D(B)}{r_B} = \frac{1}{r_B} \sum_{i=1}^{r_B} \lambda_i. \tag{3.75}$$

The system resolution provides qualitatively similar information for the collection B to that expressed by the resolution for a single quantity, X. $\mathrm{R}_D(B)$ is the average of the resolutions for each canonical direction, so that a value near one implies that we expect substantial reduction in variance for most elements of $\langle B\rangle$, while a value near zero indicates that there are a variety of elements for which the adjustment is not expected to be informative.

3.9.4 The reverse adjustment

We may similarly evaluate the canonical directions for the adjustment of D by B, based on the eigenstructure of

$$\mathbb{T}_{D:B} = \mathrm{Var}(D)^{\dagger}\mathrm{RVar}_B(D).$$

The relationship between the sets of canonical directions is analogous to that for the canonical variables constructed in a traditional canonical correlation analysis and is as follows.

Property 3.21 *If v_i is an eigenvector of $\mathbb{T}_{B:D}$ with eigenvalue λ_i, so that*

$$\mathbb{T}_{B:D}v_i = \mathrm{Var}(B)^{\dagger}\mathrm{Cov}(B, D)\mathrm{Var}(D)^{\dagger}\mathrm{Cov}(D, B)v_i = \lambda_i v_i, \tag{3.76}$$

then we must also have

$$\mathbb{T}_{D:B}\check{v}_i = \lambda_i \check{v}_i, \tag{3.77}$$

where

$$\check{v}_i = \mathrm{Var}(D)^{\dagger}\mathrm{Cov}(D, B)v_i. \tag{3.78}$$

Typically we scale each $\check{v}_i$ so that the resulting canonical direction has prior variance one. In summary, the canonical directions for the original adjustment, Y_i, and for the reverse adjustment, $\check{Y}_i$, are defined respectively as

$$Y_i = v_i^T(B - \mathrm{E}(B)) \tag{3.79}$$

$$\check{Y}_i = \frac{1}{\sqrt{\lambda_i}}v_i^T\mathrm{Cov}(B, D)\mathrm{Var}(D)^{\dagger}(D - \mathrm{E}(D)) \tag{3.80}$$

$$= \frac{1}{\sqrt{\lambda_i}}\mathrm{E}_D(Y_i). \tag{3.81}$$

Thus, the non-zero canonical resolutions of B by D and of D by B, namely the eigenvalues of $\mathbb{T}_{D:B}$ and $\mathbb{T}_{B:D}$, are the same, and the canonical directions for the adjustment of D by B are the scaled adjusted expectations for the $r_{\mathbb{T}}$ directions with $\lambda_i > 0$. For any directions with $\lambda_i = 0$, we have $\mathrm{E}_D(Y_i) = 0$. In particular, just as the quantities $\{Y_1, \dots, Y_{r_B}\}$ are mutually uncorrelated, so also are the quantities $\{\mathrm{E}_D(Y_1), \dots, \mathrm{E}_D(Y_{r_B})\}$ mutually uncorrelated, and each Y_i is uncorrelated with each $\mathrm{E}_D(Y_j)$, $j \neq i$, with

$$\mathrm{Cov}(Y_i, \check{Y}_j) = 0,\ i \neq j; \quad \mathrm{Cov}(Y_i, \check{Y}_i) = \sqrt{\lambda_i}. \tag{3.82}$$

3.9.5 Minimal linear sufficiency

Definition 3.22 *The collections Y_+ and $\check{Y}_+$, consisting respectively of eigenvectors of $\mathbb{T}_{B:D}$ and $\mathbb{T}_{D:B}$ corresponding to positive eigenvalues λ_i, are the* ***minimal linear sufficient*** *collections of elements of $\langle B\rangle$ and $\langle D\rangle$ for the adjustment of B by D. We call the collection $\check{Y}_+$ the* ***heart of the transform****, denoted $\mathbb{H}(D/B)$.*

The number of such eigenvectors corresponding to positive eigenvalues is $r_{\mathbb{T}}$. The property of minimal sufficiency follows because (i) the adjustment of B by $\check{Y}_+$ is identical to the adjustment by D and there is no proper subset of $\langle\check{Y}_+\rangle$ for which this is true, and (ii) the adjustment of Y_+ by D is numerically equivalent to the adjustment of B by D and there is no proper subset of Y_+ for which this is true, so that

$$\mathbb{T}_{B:D} = \mathbb{T}_{B:\mathbb{H}(D/B)}. \tag{3.83}$$

Therefore, we only need to measure the values in $\check{Y}_+$ and assess changes in expectation over Y_+, although the observed values of the eigenvectors Y^0 corresponding to zero eigenvalues may play a useful diagnostic role for our prior specifications. We let $\mathbb{H}^{\perp}(D/B)$ represent the orthogonal complement of $\mathbb{H}(D/B)$ in D, a space which is spanned by the eigenvectors corresponding to the eigenvectors Y^0.

3.9.6 The adjusted belief transform matrix

Sometimes, it is more convenient to make calculations based on an alternative transform, given as follows.

Definition 3.23 *The* ***adjusted belief transform matrix*** *for the adjustment of B by D is*

$$\mathbb{S}_{B:D} = \mathbb{I} - \mathbb{T}_{B:D},$$

where $\mathbb{I}$ is the identity matrix.

As with the resolution transform, when the collection B is obvious from context, we may simplify the notation and write

$$\mathbb{S}_D = \mathbb{S}_{B:D}.$$

For any $Y, U \in \langle B\rangle$, we have

$$\mathrm{Cov}_D(Y, U) = \mathrm{Cov}(Y, \mathbb{S}_{B:D}(U)), \tag{3.84}$$

so that $\mathbb{S}_{B:D}$ 'transforms' adjusted covariance to prior covariance. Observe that $\mathbb{S}_{B:D}$ and $\mathbb{T}_{B:D}$ have essentially the same eigenstructure, as W is an eigenvector of $\mathbb{T}_{B:D}$ with eigenvalue λ if and only if W is an eigenvector of $\mathbb{S}_{B:D}$ with eigenvalue $1-\lambda$.

3.10 The geometric interpretation of belief adjustment

While, for simplicity, we have presented the basic results on belief adjustment in matrix form, there is a natural underlying geometry for representing and analysing belief adjustment. The framework is as follows.

For any collection of individual random quantities C, we may construct the vector space where each $C_i \in C$ is a vector and finite linear combinations of vectors, $\sum_i h_i C_i$, are the corresponding combinations of random quantities. We denote this space as $\langle C \rangle$. Covariance imposes an inner product on $\langle C \rangle$, namely $(X, Y) = \text{Cov}(X, Y)$. Variance acts as a norm over this space, namely $\|X\|^2 = \text{Var}(X)$. As the unit constant has zero length under this norm, we identify all quantities which differ by a constant, or equivalently suppose that we have subtracted the prior expectation from each quantity and equate with zero all quantities with zero variance. The squared length of a vector therefore corresponds to the prior variance. As we adjust beliefs, the expected length of each vector is reduced by an amount equal to the resolved variance. If C is a finite collection, then we denote $\langle C \rangle$ with covariance inner product as $[C]$, the **(partial) belief structure over the base** C. If C is an infinite collection, then we define $[C]$ to be the closure of the corresponding inner product space over $\langle C \rangle$.

Orthogonality in $[C]$ corresponds to lack of correlation. For example, if E, F are the indicator functions corresponding to two events, then $(E, F) = 0$ if and only if E, F are independent. Two subspaces E and F are orthogonal, written $E \perp F$, if all elements of E are uncorrelated with all elements of F.

In this formalism, the adjusted expectation $\text{E}_D(X)$ for any $X \in [C]$ is the orthogonal projection of X into $[D]$, namely the element $Y \in \langle D \rangle$ minimizing $\|X - Y\|$. The adjusted variance is the squared distance between the element X and the subspace $[D]$, and the resolved variance is the squared length of the adjusted expectation. As $\text{E}_D(X)$ is the projection of X into D, we have

$$(X - \text{E}_D(X)) \perp [D], \tag{3.85}$$

and in particular $(X - \text{E}_D(X)) \perp \text{E}_D(X)$, so that

$$\|X\|^2 = \|X - \text{E}_D(X))\|^2 + \|\text{E}_D(X)\|^2,$$

which is the geometric form for the variance partition (3.14).

The projection $\text{E}_D(.)$ from B to D and the projection $\text{E}_B(.)$ from D to B are adjoint transformations, that is: for $X \in [B]$ and $Y \in [D]$ we have, from (3.85), that

$$(X, \text{E}_B(Y)) = (X, Y) = (\text{E}_D(X), Y).$$

Therefore, resolved covariance may be represented as

$$\text{RCov}_D(X, Z) = (\text{E}_D(X), \text{E}_D(Z)) = (X, \text{E}_B(\text{E}_D(Z))) = (X, \mathbb{T}_{B:D}(Z)), \tag{3.86}$$

where $\mathbb{T}_{B:D}$ is the resolution transform defined as the composition of the projection from $[B]$ to $[D]$ and the adjoint projection from $[D]$ to $[B]$, namely

$$\mathbb{T}_{B:D}(X) = \text{E}_B(\text{E}_D(X)). \tag{3.87}$$

$\mathbb{T}_{B:D}$ is a self-adjoint operator over $[B]$, as for each $X, Z \in [C]$,

$$(X, \mathbb{T}_{B:D}(Z)) = \mathrm{RCov}_D(X, Z) = (\mathbb{T}_{B:D}(X), Z).$$

The resolution matrix defined by (3.65) is a coordinate representation of $\mathbb{T}_{B:D}(\cdot)$. If

$$X = h_1 B_1 + \ldots + h_r B_r \in \langle B \rangle,$$
$$h = [h_1, \ldots, h_r]^T,$$
$$u = [u_1, \ldots, u_r]^T,$$

then

$$\mathbb{T}_{B:D}(X) = u_1 B_1 + \ldots + u_r B_r$$

if and only if $\mathbb{T}_{B:D} h = u$. In particular, $\mathbb{T}_{B:D} h = \lambda h$ if and only if we have $\mathbb{T}_{B:D}(X) = \lambda X$.

If r_B is finite, then, as $\mathbb{T}_{B:D}$ is self-adjoint, the operator has a full set of orthogonal unit eigenvectors $Y_1, \ldots, Y_{r_B}$, with eigenvalues

$$1 \geq \lambda_1 \geq \ldots \geq \lambda_{r_B} \geq 0.$$

The eigenvectors form an orthonormal basis for $[B]$. We may therefore express each $X \in [B]$ as $X = \sum_i (X, Y_i) Y_i$. Therefore, from (3.86),

$$\mathrm{RVar}_D(X) = \sum_i \lambda_i (X, Y_i)^2,$$

so that the eigenstructure of $\mathbb{T}_{B:D}$ has the properties that we have previously identified, and, in particular, the ith canonical direction is Y_i, the eigenvector of $\mathbb{T}_{B:D}$ corresponding to the ith largest eigenvalue, with canonical resolution λ_i.

We represent belief adjustment within this structure as follows.

Definition 3.24 *If we adjust each member of the collection $\{B\}$ by D, then we obtain a new collection $\{\mathbb{A}_D(B_1), \ldots, \mathbb{A}_D(B_k)\}$. The belief structure with this base is termed the **adjusted belief structure of** B **by** D and is written $[B/D]$.*

We may view $[B/D]$ as representing a belief structure over the linear space $\langle \mathbb{A}_D(B) \rangle$. However, it is also useful to view $[B/D]$ as an inner product space constructed over $\langle B \rangle$ but with the covariance inner product replaced by the adjusted covariance inner product

$$(X, Y)_D = \mathrm{Cov}_D(X, Y) = \mathrm{Cov}(\mathbb{A}_D(X), \mathbb{A}_D(Y)). \tag{3.88}$$

Let $\mathbb{I}$ represent the identity operator $\mathbb{I}(X) = X$. The adjusted belief structure may be represented by the adjusted belief transform

$$\mathbb{S}_{B:D} = \mathbb{I} - \mathbb{T}_{B:D},$$

using the relation, for all $X, Y \in [B]$, that

$$(X, Y)_D = (X, \mathbb{S}_{B:D}(Y)). \tag{3.89}$$

In this formalism, finite and infinite collections of quantities may be analysed in identical fashion. For example, conditioning for continuous probability measures may be expressed through projections between the corresponding belief structures. For this case, these are the Hilbert spaces of (equivalence classes of) functions (differing by a constant) defined over the underlying probability spaces which are square integrable with respect to the prior probability measure, where the inner product between two functions is covariance. As we are mainly interested, in this account, in properties of the adjustment of finite collections, we will not here consider those features which are specific to infinite collections of adjustments. General statements of all results from functional analysis that we use may be found, for example, in Bachman and Narici (1966).

3.11 Examples

3.11.1 Simple one-dimensional problem

We continue with our simple one-dimensional example first presented in §3.6. The results of the previous sections concern the implications of using one **collection** of quantities for learning about another **collection** of quantities, and are thus obvious and less interesting for one-dimensional problems. However, the basic definitions still apply and we will calculate them and discuss them briefly before passing on to the more interesting multivariate problems given in succeeding examples.

For a one-dimensional problem, the **resolution transform matrix** is, by (3.65), a single number:

$$\mathbb{T}_{Y:X} = \frac{\text{RVar}_X(Y)}{\text{Var}(Y)} = 0.36.$$

This 'matrix' has one eigenvalue, $\lambda_1 = 0.36$, and one eigenvector $v_1 = \alpha 1$, where α is any appropriate scalar. The single **canonical direction** in this example is thus $W_1 \propto \alpha Y + const$, where α is chosen so that W has prior variance one, i.e. $\alpha = 1/\sqrt{\text{Var}(Y)} = 1$, and where we arrange the canonical direction to have prior expectation zero. Thus, there is here a single canonical direction $W_1 = Y - 1$ with canonical resolution $\lambda_1 = 0.36$. A canonical resolution of 0.36 indicates that 36% of the uncertainty about Y (or any linear transformation of Y such as $3Y - 2$) is expected to be resolved by observing X.

3.11.2 Algebraic example

We return now to the simple hypothetical problem introduced in §3.8.1. By (3.65), the resolution transform matrix is

$$\mathbb{T}_{B:D} = \text{Var}(B)^{\dagger}\text{Cov}(B, D)\text{Var}(D)^{\dagger}\text{Cov}(D, B) \tag{3.90}$$

$$= \begin{bmatrix} 1 & v \\ v & 1 \end{bmatrix}^{-1} \begin{bmatrix} \rho & \rho \\ \rho & \rho \end{bmatrix} \begin{bmatrix} 1 & u \\ u & 1 \end{bmatrix}^{-1} \begin{bmatrix} \rho & \rho \\ \rho & \rho \end{bmatrix}$$

$$= \frac{2\rho^2}{(1+u)(1+v)} \begin{bmatrix} 1 & 1 \\ 1 & 1 \end{bmatrix}. \tag{3.91}$$

Here $\mathbb{T}_{B:D}$ is symmetric, although this will not usually be so, and has all its entries equal to $2\rho^2/[(1+u)(1+v)]$. The eigenvalues of (3.91) are

$$\lambda_1 = \frac{4\rho^2}{(1+u)(1+v)} \quad \text{and} \quad \lambda_2 = 0,$$

with corresponding eigenvectors

$$\psi_1 = \alpha_1 \begin{bmatrix} 1 \\ 1 \end{bmatrix} \quad \text{and} \quad \psi_2 = \alpha_2 \begin{bmatrix} 1 \\ -1 \end{bmatrix}, \tag{3.92}$$

for some appropriate α_1 and α_2. The resulting canonical directions are thus

$$W_1 = \psi_1^T B = \alpha_1(Y_1 + Y_2) \quad \text{and} \quad W_2 = \psi_2^T B = \alpha_2(Y_1 - Y_2).$$

We define the canonical directions to have prior variance one, so we need to choose α_1, α_2 so that $\text{Var}(W_1) = \text{Var}(W_2) = 1$, and so take $\alpha_1 = 1/\sqrt{2(1+v)}$ and $\alpha_2 = 1/\sqrt{2(1-v)}$. The quantities Y_1, Y_2 have prior expectation zero, so we require no extra centring. Therefore, the canonical resolutions and directions are

$$\lambda_1 = \frac{4\rho^2}{(1+u)(1+v)}, \qquad W_1 = \frac{1}{\sqrt{2(1+v)}}(Y_1 + Y_2) \tag{3.93}$$

$$\lambda_2 = 0, \qquad W_2 = \frac{1}{\sqrt{2(1-v)}}(Y_1 - Y_2). \tag{3.94}$$

Now let us interpret the canonical structure. The quantity in $[B]$ (i.e. the linear combination of Y_1, Y_2) with the greatest variance explanation, relative to prior variance, is the normed sum W_1. Furthermore, the proportion of variance explained is λ_1, which will be large when ρ is large and u and v are small – that is, when the X_is and Y_js are strongly correlated and when the X_i's and Y_j's have small variances. Similarly, the proportion of variance explained will be small when ρ is small and u and v are large – that is, when the X_is and Y_js are weakly correlated and when the X_is and Y_js have large variances. Notice that the largest possible reduction in variance is one, yielding the restriction

$$\lambda_1 \leq 1 \Rightarrow \frac{4\rho^2}{(1+u)(1+v)} \leq 1,$$

which we saw earlier in (3.35) as a condition ensuring coherence of the belief specifications.

The implication from considering the second canonical quantity, W_2, is that we learn least about the normed difference between Y_1 and Y_2. Furthermore, as the corresponding canonical resolution is $\lambda_2 = 0$, we can in fact learn nothing about this quantity: the data quantities $D = [X_1, X_2]$ are expected to be entirely uninformative for it. Even for such a simple two-dimensional problem as this, this structural implication of our belief specifications may have been far from obvious. It actually arises because the same correlation of ρ has been specified between each Y_i and X_j. To digress slightly, if we amend our covariance specifications by setting $\text{Cov}(Y_1, X_2) = \text{Cov}(Y_2, X_1) = 0$, then we find that the canonical quantities remain proportional to $Y_1 + Y_2$ and $Y_1 - Y_2$, but that the corresponding canonical resolutions become $\rho^2/[(1+u)(1+v)]$ and $\rho^2/[(1-u)(1-v)]$, respectively. The actual values are not as important here (the two sets of belief specifications are not directly comparable for this purpose) as the qualitative change in structure: the alternative belief specifications lead to a situation in which it **is** possible to reduce variation in all linear combinations constructed from Y_1 and Y_2.

To return to our original specifications, we have had the perhaps unwelcome news that we cannot learn about $Y_1 - Y_2$. The consequences for learning about other linear combinations such as Y_1 alone, or $2Y_1 - Y_2$, depend on how strongly correlated each such linear combination is with the canonical directions, because of (3.69) and (3.71). Take, for example, Y_1. Its covariances (and correlations, as all the quantities are standardized) with the canonical quantities are

$$\text{Cov}(Y_1, W_1) = \text{Cov}(Y_1, \alpha_1(Y_1 + Y_2)) = \sqrt{\frac{1+v}{2}}$$

$$\text{Cov}(Y_1, W_2) = \text{Cov}(Y_1, \alpha_2(Y_1 - Y_2)) = \sqrt{\frac{1-v}{2}}.$$

Therefore, by (3.69) we may write

$$Y_1 = \sqrt{\frac{1+v}{2}}W_1 + \sqrt{\frac{1-v}{2}}W_2, \tag{3.95}$$

and it is immediately clear that Y_1 is more strongly related to W_1 than to W_2 when v is positive.

By (3.71), the proportion of variance in Y_1 explained by the data quantities can be partitioned into additive contributions from each of the canonical quantities. We have here, by (3.72),

$$\text{c}_1(Y_1) = [\text{Corr}(Y_1, W_1)]^2 = \frac{1+v}{2}, \tag{3.96}$$

$$\text{c}_2(Y_1) = \frac{1-v}{2}. \tag{3.97}$$

Thus, by (3.71), the resolution in Y_1 by adjusting by D must be equal to

$$\begin{aligned} \mathrm{R}_D(Y_1) &= \mathrm{c}_1(Y_1)\lambda_1 + \mathrm{c}_2(Y_1)\lambda_2 \\ &= \frac{1+v}{2}\frac{4\rho^2}{(1+u)(1+v)} + \frac{1-v}{2} \times 0 \\ &= \frac{2\rho^2}{1+u}, \end{aligned} \tag{3.98}$$

as we saw in (3.38). The canonical analysis has thus revealed that the amount of variation in Y_1 accounted for is due to how much we learn in the canonical direction W_1, and to how strongly Y_1 and W_1 are correlated. There is no contribution to variance resolution in the second canonical direction W_2 as nothing can be learnt about this direction. Notice also that

$$\mathrm{R}_D(W_1) = \lambda_1 = \frac{4\rho^2}{(1+u)(1+v)} \geq \mathrm{R}_D(Y_1) = \frac{2\rho^2}{1+u} \geq \mathrm{R}_D(W_2) = \lambda_2 = 0, \tag{3.99}$$

illustrating that resolutions for all linear combinations of Y_1 and Y_2 are bounded by the maximal and minimal canonical resolutions for the adjustment.

The arguments work similarly for other quantities of interest, for example $Y^* = 2Y_1 - Y_2$. As we have

$$\begin{aligned} \mathrm{Corr}(Y^*, W_1) &= \sqrt{\frac{1+v}{2(5-4v)}} \\ \mathrm{Corr}(Y^*, W_2) &= 3\sqrt{\frac{1-v}{2(5-4v)}}, \end{aligned}$$

it follows from (3.71) that

$$\begin{aligned} \mathrm{R}_D(Y^*) &= \frac{1+v}{2(5-4v)}\frac{4\rho^2}{(1+u)(1+v)} + 3\frac{1-v}{2(5-4v)} \times 0 \\ &= \frac{2\rho^2}{(5-4v)(1+u)}. \end{aligned} \tag{3.100}$$

Compare (3.100) with (3.98). For positive v we expect to learn less about Y^* than we do about Y_1 because Y^* is more weakly related to the canonical direction W_1, the only direction in which we expect to learn anything. Notice how simple it is to deduce from the canonical structure the expected effects of belief adjustment for any linear combination of the adjusted quantities, simply through correlations with the canonical quantities and their resolutions.

A simple guide as to the value of the data quantities for explaining variation in the quantities in the collection of interest **taken as a whole** is given by the average canonical resolution (3.75). Such measures are especially useful when applied to very complicated models with many quantities where it is impractical to assess the effects of belief adjustment at the lowest level of individual quantity, and where it is efficacious to make assessments over possibly quite large collections of unknowns. In this simple hypothetical example, $\text{Var}(B)$ has rank two (there are two axes of variation), with canonical resolutions respectively $4\rho^2/[(1+u)(1+v)]$ and zero, so that the average resolution is $\text{R}_D(B) = 2\rho^2/[(1+u)(1+v)]$.

The canonical quantities may, of course, themselves be adjusted directly. Gather them into the collection $W = [W_1, W_2]$ and adjust by the collection D. We know already that the prior and resolved variance matrices are

$$\text{Var}(W) = \begin{bmatrix} 1 & 0 \\ 0 & 1 \end{bmatrix}, \qquad \text{RVar}_D(W) = \begin{bmatrix} \lambda_1 & 0 \\ 0 & \lambda_2 \end{bmatrix} = \begin{bmatrix} \frac{4\rho^2}{(1+u)(1+v)} & 0 \\ 0 & 0 \end{bmatrix}.$$

We can calculate the adjusted expectations directly, by (3.21), or by exploiting the linearity of adjusted expectation (3.23). We have from (3.93), (3.94),

$$W = \begin{bmatrix} \frac{1}{\sqrt{2(1+v)}} & \frac{1}{\sqrt{2(1+v)}} \\ \frac{1}{\sqrt{2(1-v)}} & -\frac{1}{\sqrt{2(1-v)}} \end{bmatrix} B. \tag{3.101}$$

Thus, recalling that we obtained $\text{E}_D(B)$ as (3.36),

$$\text{E}_D(W) = \begin{bmatrix} \frac{1}{\sqrt{2(1+v)}} & \frac{1}{\sqrt{2(1+v)}} \\ \frac{1}{\sqrt{2(1-v)}} & -\frac{1}{\sqrt{2(1-v)}} \end{bmatrix} \text{E}_D(B) \tag{3.102}$$

$$= \begin{bmatrix} \frac{1}{\sqrt{2(1+v)}} & \frac{1}{\sqrt{2(1+v)}} \\ \frac{1}{\sqrt{2(1-v)}} & -\frac{1}{\sqrt{2(1-v)}} \end{bmatrix} \begin{bmatrix} \frac{\rho}{1+u}(X_1 + X_2) \\ \frac{\rho}{1+u}(X_1 + X_2) \end{bmatrix} \tag{3.103}$$

$$\text{i.e.} \quad \begin{bmatrix} \text{E}_D(W_1) \\ \text{E}_D(W_2) \end{bmatrix} = \begin{bmatrix} \frac{2\rho}{(1+u)\sqrt{2(1+v)}}(X_1 + X_2) \\ 0 \end{bmatrix}. \tag{3.104}$$

Notice that $\text{E}_D(W_2) = 0$ is forced as we have already discovered that the data quantities X_1 and X_2 are useless for predicting W_2, and so the adjusted expectation cannot differ from the prior expectation of zero.

Just as the effects of belief adjustment on uncertainties can be deduced through the canonical quantities, so too can adjusted expectations for any linear combination of the quantities of interest be immediately deduced from the canonical structure. For example, we have from (3.95) that

$$Y_1 = \sqrt{\frac{1+v}{2}}W_1 + \sqrt{\frac{1-v}{2}}W_2$$

so that

$$\text{E}_D(Y_1) = \sqrt{\frac{1+v}{2}}\text{E}_D(W_1) + \sqrt{\frac{1-v}{2}}\text{E}_D(W_2)$$

$$= \sqrt{\frac{1+v}{2}} \mathrm{E}_D(W_1) \tag{3.105}$$

$$= \frac{\rho}{1+u}(X_1 + X_2),$$

as we found before in (3.36). Notice that only W_1 appears in (3.105), illustrating that only the canonical quantities which correspond to positive canonical resolutions are necessary to assess the belief adjustment over the collection B of interest. In this context, W_1 is **minimal linear sufficient** for the collection B.

In fact we can develop this idea of minimal linear sufficient collections (discussed in §3.9.5) further. Suppose that we reverse our notion of adjustment in this example and think instead of adjusting D by B (rather than B by D). The calculations are in general as straightforward as the other calculations in this example. However, in this case, thanks to the symmetry in the specifications, we can obtain all the results for the adjustment of D by B by taking the results of the adjustment of B by D, and simply swapping v with u and X with Y wherever they appear. For example, the canonical resolutions and canonical quantities for the reverse adjustment are (swapping u, v and X, Y in (3.93), (3.94)):

$$\check{\lambda}_1 = \frac{4\rho^2}{(1+u)(1+v)}, \qquad \check{W}_1 = \frac{1}{\sqrt{2(1+u)}}(X_1 + X_2), \tag{3.106}$$

$$\check{\lambda}_2 = 0, \qquad \check{W}_2 = \frac{1}{\sqrt{2(1-u)}}(X_1 - X_2). \tag{3.107}$$

Notice that the canonical resolutions for the two adjustments $\mathbb{T}_{B:D}$ and $\mathbb{T}_{D:B}$ must be identical, $\lambda_1 = \check{\lambda}_1, \lambda_2 = \check{\lambda}_2$, by (3.76).

It is clear so far from (3.36) and (3.104) that the adjusted expectations for any quantity constructed from the basic quantities of interest in B seem to depend only upon multiples of the sum $X_1 + X_2$, now identified as the sole canonical quantity $\check{W}_1$ for the reverse adjustment having a positive resolution. Indeed, this single quantity is minimal linear sufficient for the collection D for adjusting the collection B. As we have already discovered that W_1 is minimal linear sufficient for B, we conclude that all aspects of adjustment in either direction are carried solely by W_1, $\check{W}_1$, and covariances with the other quantities of interest. The canonical structure for the reverse adjustment can, in fact, be deduced directly from the canonical structure for the main adjustment. For example, by (3.81) we have

$$\begin{aligned}
\check{W}_1 &= \frac{1}{\sqrt{\lambda_1}} \mathrm{E}_D(W_1) \\
&= \sqrt{\frac{(1+u)(1+v)}{4\rho^2}} \frac{2\rho}{(1+u)\sqrt{2(1+v)}}(X_1 + X_2), \text{ by (3.93), (3.104)} \\
&= \frac{1}{\sqrt{2(1+u)}}(X_1 + X_2),
\end{aligned}$$

which we found in (3.106). That is, the canonical quantities for the reversed adjustment are the scaled adjusted expectations for the canonical quantities of the main adjustment.

The findings in this section trail some important ideas which we will develop later. Two important special cases are as follows. First, full exchangeability arises when $u = v = \rho$, so that this is the case where we sample two individuals from a population and where we want to predict values for two other individuals. There are obvious simplifications we can make to the formulae of this section in this case, but the feature that learning takes place via means remains key. We will explore some of the issues by extending this example to full exchangeability in §6.13. Secondly, this example provides a foretaste of analysing co-exchangeable samples from two populations. Again, we develop the example in this context more fully in §7.3, and again we will see that the relationships between the two samples comes through the corresponding sample means. Such mean linkage is a fundamental consequence of exchangeability which we will strongly exploit in subsequent chapters.

3.11.3 Oral glucose tolerance test

For this example we return to the oral glucose tolerance test problem for which we have already calculated adjusted expectations and adjusted variances in §3.8.2. The canonical directions are found by calculating the eigenstructure of the resolution matrix given by (3.65):

$$\begin{aligned}\mathbb{T}_{B:D} &= \begin{bmatrix}1.12 & 0.72\\0.72 & 2.43\end{bmatrix}^{-1}\begin{bmatrix}0.62 & 0.30\\0.30 & 0.43\end{bmatrix}\begin{bmatrix}1.12 & 0.72\\0.72 & 2.43\end{bmatrix}^{-1}\begin{bmatrix}0.62 & 0.30\\0.30 & 0.43\end{bmatrix}\\ &= \begin{bmatrix}0.3336 & 0.1345\\-0.0354 & 0.0050\end{bmatrix}.\end{aligned}$$

This resolution matrix is asymmetric, as will often be the case. The canonical quantities are constructed from the eigenvalues and **right** eigenvectors of this matrix. For practical implementations of the theory, it is more convenient to calculate the canonical structure from a symmetrized version of $\mathbb{T}_{B:D}$; details may be found in Chapter 12. The canonical quantities are found to be

$$W_1 = 1.0059G_0 - 0.1136G_2 - 3.4745, \tag{3.108}$$

$$W_2 = 0.3020G_0 - 0.7039G_2 + 3.1431, \tag{3.109}$$

with corresponding canonical resolutions

$$\lambda_1 = 0.3184, \quad \lambda_2 = 0.0202. \tag{3.110}$$

The overall resolution is $\mathrm{RU}_D(B) = 0.3184 + 0.0202 = 0.3386$, whereas there are two different axes of variation as the rank of the prior variance matrix over B is two. Thus, the resolution for the overall collection is

$$\mathrm{R}_D(B) = 0.3386/2 = 0.1694, \tag{3.111}$$

so that the collection of measurements D that the doctor makes upon herself is expected on average to reduce by about 17% uncertainties in the collection B of measurements for a typical elderly person. Notice that both canonical resolutions are positive, unlike the situation in the previous example, so that W_1 and W_2 together are necessary to constitute a minimal linear sufficient collection for B.

It is useful to portray the canonical quantities in terms of the standardized versions (Definition 1.1) of G_0 and G_2:

$$W_1 = 1.0645\,\mathrm{S}(G_0) - 0.1771\,\mathrm{S}(G_2), \tag{3.112}$$

$$W_2 = 0.3196\,\mathrm{S}(G_0) - 1.0972\,\mathrm{S}(G_2). \tag{3.113}$$

Thus, the linear combination of quantities in $\langle B\rangle$ about which we expect to learn most is W_1, and we expect to remove about 31.84% of our uncertainty in this direction. Any other linear combination of elements in $\langle B\rangle$ which is highly correlated with W_1 will likewise have a similar variance reduction. The second canonical quantity W_2 represents the direction in which we expect to learn least. Its resolution of only about 2.02% suggests that we learn almost nothing about both W_2 and linear combinations highly correlated with W_2. The magnitude of the coefficients in (3.112) indicates that the first canonical quantity is more strongly related to G_0 than to G_2, whereas the reverse is true for the second canonical quantity. Therefore, taking into account the canonical resolutions of about 31.84% and 2.02% respectively, the data quantities (D_0 and D_2, the doctor's own fasting and 2-hour measurements) are expected to be much more informative for G_0 (the typical elderly person's fasting glucose measurement) than for G_2 (the corresponding 2-hour measurement). This confirms what we saw in §3.8.2, where we calculated $\mathrm{Var}_D(G_0) = 31.09\%$ and $\mathrm{Var}_D(G_2) = 4.48\%$. Notice that these resolutions are necessarily bounded by the canonical resolutions:

$$\lambda_2 = 0.0202 < 0.0448, \quad 0.3109 < 0.3184 = \lambda_1.$$

As the second canonical resolution is so small, we conclude that, for the purpose of learning about G_0 and G_2, the information contained in $[D]$ is essentially one-dimensional: we reduce uncertainty only in the direction of W_1. Examination of the standardized form of the first canonical direction W_1 above shows that G_0 is the major component, whereas G_2 is the major component of W_2. Hence, we are learning mostly in the direction of G_0, and learning very little in the direction of G_2. This can be confirmed by examining the resolution partition, exploiting (3.71), which is as follows.

	Resolution partition		
Quantity	W_1	W_2	Total
G_0	0.3103	0.0005	0.3109
G_2	0.0263	0.0185	0.0448

This shows that the overall explained proportion of variance for G_0 is 31.09%, i.e. 31.03% + 0.06%, with most of the learning in the direction W_1, and a trivial contribution of 0.06% in the direction of W_2. The corresponding breakdown for G_2 is $\mathrm{R}_D(G_2) = 0.0448 = 0.0263 + 0.0185$, so that the two canonical directions offer modest but more balanced variance reductions for G_2.

Now we return briefly to the difference quantity $G_h = G_2 - G_0$ which we constructed in §3.8.2. It should now be clear that we could have deduced its adjusted expectation and variance without physically constructing it, using (3.69) and (3.71). We found earlier that G_h had variance 2.11 and adjusted variance 1.9613, so that the proportion of resolved variance is $\mathrm{R}_D(G_h) = 0.0705$. Now we can discover through the canonical structure just why we are doing so badly here. We have $\mathrm{Corr}(G_h, W_1) = -0.4107$ and $\mathrm{Corr}(G_h, W_2) = -0.9118$, so that there is a very strong correlation between G_h and the second canonical direction, about which we expect to explain very little. The resolution partition turns out to be, from (3.71),

$$\begin{aligned}\mathrm{R}_D(G_h) &= (-0.4107^2)0.3148 + (-0.9118^2)0.0202 \\ &= 0.0537 + 0.0168 = 0.0705, \end{aligned} \tag{3.114}$$

making it clear that most of what we learn about G_h comes through a quite tenuous link with the first canonical quantity, rather than through the much more strongly related second canonical quantity, which is relatively useless for prediction.

3.11.3.1 *Many tests*

Suppose, as in §3.8.3, that we intend to make multiple measurements from a large sample of similar individuals. As above, organize our two quantities of interest, G_0 and G_2, into the collection B, and arrange the sample of n measurements as the collection D^n, as in §3.8.3. This collection contains $2n$ elements. We can calculate the canonical structure for the adjustment of B by D^n for any desired value of n. The resolution transform remains a 2×2 matrix even though we are now adjusting by a $2n$-dimensional information space. We have already calculated the canonical structure for $n = 1$ above, as (3.108) and (3.110), so let us see what happens when we increase the sample size. For $n = 2$ we obtain

$$\begin{aligned} W_1 &= 1.0059G_0 - 0.1136G_2 - 3.4745, \\ W_2 &= 0.3020G_0 - 0.7039G_2 + 3.1431, \\ \lambda_1 &= 0.4071, \quad \lambda_2 = 0.0353; \end{aligned}$$

and for $n = 3$ we obtain

$$\begin{aligned} W_1 &= 1.0059G_0 - 0.1136G_2 - 3.4745, \\ W_2 &= 0.3020G_0 - 0.7039G_2 + 3.1431, \\ \lambda_1 &= 0.4488, \quad \lambda_2 = 0.0471. \end{aligned}$$

We can continue with larger n, but the main features are already obvious: the canonical directions seem to be the same, regardless of sample size, whilst the canonical resolutions gradually increase in magnitude. These features turn out to be the foundations of Bayes linear analysis for sampling from exchangeable populations. We will see in Chapter 6 why the canonical directions are indeed the same for all sample sizes; we will derive a formula which shows exactly how the corresponding canonical resolutions increase as the sample size increases; and we will explain how and why this core structure is so helpful in understanding the analysis of beliefs for such problems.

3.12 Further reading

Lindley (1965) and Savage (1971) are early and enormously influential developments of the basic principles of Bayesian statistics. Bernardo and Smith (1994), O'Hagan and Forster (2004) and Robert (2001) each give an excellent overview of the standard Bayesian approach to statistics. The seminal works by de Finetti (1974, 1975) develop the Bayesian approach from a viewpoint in which expectation rather than probability is the natural primitive. Lad (1996) is a valuable complement to these volumes, giving much useful background and deriving many implications of this approach. The role of subjectivity in Bayesian statistics remains controversial; for a current account of relevant issues, see Goldstein (2006) and the accompanying discussion.

Stone (1963) and Hartigan (1969) are important in the early treatment of the role of partial prior specification using moments in Bayes analysis. Mouchart and Simar (1980) contains a useful summary of basic least squares results for Bayesian analysis. The particular form of the Bayes linear development that we have described begins with Goldstein (1975a,b), where some of the basic properties of Bayes linear adjustment are described in geometric form. Goldstein (1974) considers Bayes linear adjustment when we only specify bounds for our prior variance specifications. This analysis is extended in Goldstein (1984) which considers how we may make Bayes linear assessments when our prior specification consists of probability quantiles for the variables. The canonical form for the belief adjustment is described in Goldstein (1981), and the structure of adjusted belief spaces is described in Goldstein (1988a). The relationship between the Bayes linear approach and traditional regression modelling is explored in Goldstein (1976), and this analysis is extended in Goldstein (1980).

The practical and philosophical issues arising in the use of the Bayes linear approach are discussed in general terms in Goldstein (1987a,b, 1994a). Goldstein (1999) gives a short overview of the Bayes linear approach. Wooff (1992), Goldstein and Wooff (1995) and Wooff and Goldstein (2000b) describe the Bayes linear approach as related to the computing language [B/D] which we have created for carrying out Bayes linear analyses, and which was used to carry out the calculations in this book.

The foundational treatment that we have described, based on temporal coherence relations between beliefs at different time points, is developed in Goldstein (1983b, 1985, 1986b, 1997). While this book is concerned with the analysis of finite collections of beliefs, this approach also casts light on problems arising in infinite collections; see, for example, Goldstein (2001) which demonstrates how the linear geometric formulation may help to clarify puzzling paradoxes arising in finitely additive specifications. In Goldstein and Shaw (2004), belief adjustment is generalized to Bayes linear kinematics which describe how adjustments may be determined when, rather than observing data, we simply change our prior judgements about the values that the data might take, allowing us to mix probabilistic and expectation-based calculations within the same analysis.

Because of the strong relationship between Gaussian models and Bayes linear analyses, much applied Gaussian work has direct relevance to the Bayes linear approach; for example, much of West and Harrison (1997) may be recast from a Bayes linear viewpoint. A particular area in which Bayes linear methods have been applied is the analysis of computer simulators for complex physical systems, in particular for problems with large input and output spaces; see Craig et al. (1996, 1997, 2001) and Goldstein and Rougier (2005). For an application of the Bayes linear approach in the water industry, for which much of the data comprises judgements by experts at various levels of detail, see O'Hagan et al. (1992). For further applications, see O'Hagan (1987), Kuo (1988), Mukhopadhyay and Vidakovic (1995), Wooff et al. (1998), Coolen et al. (2001) and Little et al. (2004).

4

The observed adjustment

We have explained how to construct adjusted expectations given a collection of observations. After we make the observations and evaluate these adjustments, we need informative summaries to help us to understand qualitatively the changes in our beliefs which are suggested by these adjustments. We must also consider whether the observations suggest that we should re-examine any aspects of our prior formulation. Therefore, we now develop interpretative and diagnostic methodology for analysing the observed adjustment. We begin by considering discrepancy measures between observations and expectations. We then apply these discrepancy measures for diagnostic evaluation of adjusted expectation, and develop the bearing, or linear likelihood, for the adjustment, to allow us to interpret the changes in beliefs implied by the adjustment.

4.1 Discrepancy

Each prior statement that we make describes our beliefs about some random quantity. If we observe this quantity, then we may compare what we expect to happen with what actually happens. For a single random quantity X, suppose that we have only specified, a priori, the quantities $\mathrm{E}(X)$ and $\mathrm{Var}(X)$. Then a simple comparison is as follows.

Definition 4.1 *Given the observed value $X = x$, we define the* ***standardized observation****, $\mathrm{S}(x)$, and the* ***discrepancy*** *between the observation and the prior assessment, $\mathrm{Dis}(x)$, as*

$$\mathrm{S}(x) = \frac{x - \mathrm{E}(X)}{\sqrt{\mathrm{Var}(X)}}, \tag{4.1}$$

$$\mathrm{Dis}(x) = [\mathrm{S}(x)]^2 = \frac{[x - \mathrm{E}(X)]^2}{\mathrm{Var}(X)}. \tag{4.2}$$

Bayes Linear Statistics: Theory and Methods M. Goldstein and D.A. Wooff

A very large value for $\mathrm{Dis}(x)$ might suggest, under some circumstances, that $\mathrm{E}(X)$ has been misspecified or the variability of X underestimated, or the value of x misrecorded. Similarly, a very small discrepancy might suggest an overestimate of the variability of X. How large or small such discrepancies must be to warrant attention, and how such misspecification should be interpreted, will depend on the context. For example, it is quite different to carry out diagnostic checking on our own judgements or on the judgements of others, particularly when issues such as the competence or even the probity of the analysis may be called into question. It is similarly quite different to carry out diagnostic checking on a small study, where all prior judgements have been carefully made by experts in the subject area, or to carry out such checking on the judgements for a large and complex study, for which many of the prior judgements have been reached by fairly crude heuristic arguments. In general, the role of simple, systematic diagnostics, which draw attention to surprising and anomalous features of a prior specification, increases in importance as the size and complexity of the study grow, and our confidence decreases that each prior judgement is the product of careful reflection and that each data value has been correctly recorded.

Therefore, we cannot give general thresholds for discrepancy measures, although there are certain simple heuristics which may sometimes be useful. For example, the so-called **three-sigma rule** (Pukelsheim 1994) states that for any unimodal continuous random quantity, $\mathrm{P}(|\mathrm{S}(X)| \leq 3) \geq 0.95$, which might, on occasion, suggest three standard deviations as a possible diagnostic threshold, particularly when the qualitative judgement of unimodality is considered to be applicable. For multimodal distributions, particularly with modes in the tails of the distributions, we would expect to observe correspondingly larger discrepancy values.

In general, the principal function of discrepancy measures is comparative. When we make many observations, then it is useful to have simple measures which direct our attention to whatever subsets of the observations are sufficiently aberrant from prior specification to merit careful examination.

4.1.1 Discrepancy for a collection

For a collection of quantities, the natural counterpart to the above discrepancy measure is as follows.

Definition 4.2 *For comparing an observed data vector* $D = d = (d_1, \ldots, d_k)$ *with the prior assessments* $\mathrm{E}(D)$, $\mathrm{Var}(D)$ *for the vector, calculate the* ***discrepancy*** *as*

$$\mathrm{Dis}(d) = (d - \mathrm{E}(D))^T \mathrm{Var}(D)^\dagger (d - \mathrm{E}(D)). \tag{4.3}$$

Note that discrepancy is additive over uncorrelated sub-collections of elements of D, i.e. if $D = (D_1, D_2)$ where $\mathrm{Cov}(D_1, D_2) = 0$, then having observed the two sub-vectors $d = (d_1, d_2)$, we have

$$\mathrm{Dis}(d) = \mathrm{Dis}(d_1) + \mathrm{Dis}(d_2). \tag{4.4}$$

The prior expected value of $\mathrm{Dis}(D)$ is given by

$$\begin{aligned}\mathrm{E}(\mathrm{Dis}(D)) &= \mathrm{E}((D-\mathrm{E}(D))^T\mathrm{Var}(D)^\dagger(D-\mathrm{E}(D)))\\ &= \mathrm{E}(\mathbf{tr}\{\mathrm{Var}(D)^\dagger(D-\mathrm{E}(D))(D-\mathrm{E}(D))^T\})\\ &= \mathbf{rk}\{\mathrm{Var}(D)\}. \end{aligned} \tag{4.5}$$

Definition 4.3 *We denote the normalized discrepancy by*

$$\mathrm{Dr}(d) = \frac{\mathrm{Dis}(d)}{\mathbf{rk}\{\mathrm{Var}(D)\}}, \tag{4.6}$$

termed the ***discrepancy ratio*** *for d, with prior expectation 1.*

A natural heuristic for examining discrepancy ratios is to assess the magnitude of $\mathrm{Dr}(d)$. This specification depends on the value of $\mathrm{Var}(\mathrm{Dis}(D))$. We may make this specification directly, for example from observation of similar discrepancies in related problems, or we may deduce this discrepancy from a specification of fourth moments over the elements of D. We might make this specification directly, or through some distributional approximation. For example, we might judge the quantities $D_1, D_2, \ldots, D_k$ to be approximately multivariate normal. Then $\mathrm{Dis}(D)$ has approximately a chi-squared distribution with $r = \mathbf{rk}\{\mathrm{Var}(D)\}$ degrees of freedom, with

$$\mathrm{E}(\mathrm{Dis}(D)) = r \quad \text{and} \quad \mathrm{Var}(\mathrm{Dis}(D)) = 2r.$$

Having obtained a value for $\mathrm{Var}(\mathrm{Dis}(D))$, we might refer to appropriate distributional tables, or we might choose a simple conservative bound using, for example, Chebyshev's inequality, which gives, for any k,

$$P\left(-k \le \frac{\mathrm{Dis}(D)-r}{\sqrt{2r}} \le k\right) \le 1-k^{-2},$$

which can be reorganized as

$$P\left(1-k\sqrt{\frac{2}{r}} \le \mathrm{Dr}(D) \le 1+k\sqrt{\frac{2}{r}}\right) \le 1-k^{-2}.$$

For example, setting $k = 3\sqrt{2}$ gives

$$P\left(1-\frac{6}{\sqrt{r}} \le \mathrm{Dr}(D) \le 1+\frac{6}{\sqrt{r}}\right) \le 0.9444 \tag{4.7}$$

approximately. In comparison, if we use the normal approximation not only to suggest a reasonable value for $\mathrm{Var}(\mathrm{Dis}(D))$ but also to generate a probabilistic specification for $\mathrm{Dis}(D)$, then we arrive approximately at

$$P\left(1-\frac{2.7}{\sqrt{r}} \le \mathrm{Dr}(D) \le 1+\frac{2.7}{\sqrt{r}}\right) = 0.9444.$$

Bounds such as these can be helpful in offering simple, somewhat *ad hoc*, heuristics for fast routine scanning of large belief specifications. In later sections, we will look in more detail at the construction and interpretation of different types of discrepancy measures.

4.1.2 Evaluating discrepancy over a basis

An equivalent method for evaluating the discrepancy starts by selecting any maximal collection of uncorrelated quantities $W_1, \ldots, W_r$, where r is the rank of $\mathrm{Var}(D)$, each with prior mean zero and variance one. In what follows, we will often use the particular choice $W^+ = (W_1, \ldots, W_r)$ which are the standardized principal components corresponding to the non-zero eigenvalues of $\mathrm{Var}(D)$. The principal components are such that $W_i = a_i^T(D - \mathrm{E}(D))$, where a_i is the eigenvector of $\mathrm{Var}(D)$ corresponding to the ith smallest non-zero eigenvalue, and a_i is scaled so that $\mathrm{Var}(W_i) = 1$; see, for example, Krzanowski and Marriott (1994). If we observe $W_i = w_i$, for $i = 1, \ldots, r$, then we may evaluate the discrepancy as

$$\mathrm{Dis}(d) = \sum_{i=1}^{r} w_i^2. \tag{4.8}$$

4.1.3 Discrepancy for quantities with variance zero

If $\mathrm{Var}(D)$ is not of full rank, then we may derive a further collection of $(k - r)$ linearly independent combinations $W^0 = (W_{r+1}, \ldots, W_k)$, each with prior mean and prior variance equal to zero. The discrepancy measure $\mathrm{Dis}(d)$ does not depend on the observed values w_i for these remaining components. We must carry out a separate consistency check that, for each $F \in \langle D \rangle$, the collection of linear combinations of the elements of D, with $\mathrm{Var}(F) = 0$, the observed value, $F = f$, is actually equal to the prior expectation $\mathrm{E}(F)$, and it is sufficient to check that $w_i = \mathrm{E}(W_i)$ for $i = r + 1, \ldots, k$. If this check fails, then we must reconsider the status of the zero variances that we have assigned and check the validity of the data measurements. In all of the analyses that we shall describe, we shall assume that quantities with zero variance pass this check and take observed values equal to their expectations.

4.2 Properties of discrepancy measures

Discrepancy may be interpreted as a summary measure over $\langle D \rangle$. We separate $\langle D \rangle$ into two subspaces, $\langle D \rangle^+$, $\langle D \rangle^0$, where $\langle D \rangle^+$ is the set of linear combinations of the elements of W^+, and $\langle D \rangle^0$ is the set of linear combinations of the elements of W^0 so that all elements of $\langle D \rangle^0$ have variance zero while all non-zero members of $\langle D \rangle^+$ have positive variance. For any $F = h^T D = \sum_i h_i D_i \in \langle D \rangle$, the observed value of F, when $D = d$, is the value $f = \sum_i h_i d_i$. We denote by $\langle d \rangle^+$ the collection of such linear combinations of elements of d corresponding to elements $h^T D \in \langle D \rangle^+$. For $F = h^T D \in \langle D \rangle^+$ with observed value f, we have

$$\mathrm{Dis}(f) = \frac{[h^T(d - \mathrm{E}(D)]^2}{h^T \mathrm{Var}(D) h}. \tag{4.9}$$

Suppose we want to find the element $f \in \langle d \rangle^+$ with maximum discrepancy. As $\mathrm{Var}(D)$ is non-negative definite, we may write $\mathrm{Var}(D) = CC^T$. Therefore, we can write Dis(f) as

$$\text{Dis}(f) = \frac{[u^T v]^2}{u^T u},$$

where $u = C^T h$, $v = C^\dagger(d - \text{E}(D))$. Therefore, Dis($f$) is maximized at $u = v$, or equivalently, when $h = \text{Var}(D)^\dagger(d - \text{E}(D))$. Substituting this value for h into (4.9) gives

$$\max_{f \in \langle d \rangle^+} \text{Dis}(f) = [d - \text{E}(D)]^T \text{Var}(D)^\dagger [d - \text{E}(D)] = \text{Dis}(d) \tag{4.10}$$

The element $\dot{W}_d \in \langle D \rangle^+$ whose observed value achieves this maximum discrepancy is given as follows.

Definition 4.4 *The* ***discrepancy vector*** *for d is given by* $\dot{W}_d$*, where*

$$\dot{W}_d = \dot{a_d}^T (D - \text{E}(D))$$

and

$$\dot{a_d} = \text{Var}(D)^\dagger (d - \text{E}(D)). \tag{4.11}$$

There is a natural sense in which the discrepancy vector summarizes all of the diagnostic information in the observation, which follows from the following geometric property. For any $F = h^T D \in \langle D \rangle$ we have

$$\begin{aligned} \text{Cov}(F, \dot{W}_d) &= h^T \text{Var}(D) \dot{a}_d \\ &= h^T (d - \text{E}(D)) = f - \text{E}(F). \end{aligned} \tag{4.12}$$

Thus, $\dot{W}_d$ is the element of $\langle D \rangle$ which expresses the direction and the magnitude of the observed vector in relation to the prior belief specifications, in the sense that

$$\text{Cov}(F, \dot{W}_d) = 0 \Leftrightarrow f = \text{E}(F). \tag{4.13}$$

We have therefore shown that there is a single 'direction', $\dot{W}_d$, in $\langle D \rangle$ with the property that all differences between observations and prior expectations are in this direction. Therefore, it will often be informative to identify this direction in order to understand how observations differ from expectations. In particular,

$$\text{Var}(\dot{W}_d) = \text{Dis}(d), \tag{4.14}$$

and scaling $\dot{W}_d$ by some multiple, α say, to give discrepancy vector $\alpha \dot{W}_d$ corresponds to multiplying every distance $f - \text{E}(F)$ by α. The general idea motivating this construction, namely to find a single random quantity which summarizes all of the diagnostic information for a collection of random quantities, is useful in a variety of contexts that we shall develop. In particular, a similar form of data reduction may be applied to adjusted beliefs, as we shall describe below. The geometric interpretation underlying all such representations is discussed in §4.10.

4.2.1 Evaluating the discrepancy vector over a basis

An equivalent method for constructing the discrepancy vector is to select any maximal collection of uncorrelated quantities, each with prior mean zero and variance one, for example the standardized principal components W^+. If we observe $W_i = w_i$, $i = 1, \ldots, r$, then we may construct the discrepancy vector as

$$\dot{W}_d = \sum_{i=1}^{r} w_i W_i, \tag{4.15}$$

so that $\text{Var}(\dot{W}_d) = \sum_{i=1}^{r} w_i^2$. Note, therefore, the equivalence of (4.8) and (4.14). Note also that if $\dot{w}_d$ is the observed value of $\dot{W}_d$ then, from (4.12) and (4.15), we again have

$$\text{Cov}(\dot{W}_d, \dot{W}_d) = \dot{w}_d - \text{E}(\dot{W}_d) = \sum_{i=1}^{r} w_i^2.$$

4.3 Examples

4.3.1 Simple one-dimensional problem

We continue with our simple one-dimensional example first presented in §3.6. There we found that our estimator for Y was $\text{E}_X(Y) = 0.6X - 0.2$, and we calculated a value for our expected uncertainty for Y after we see X.

Suppose we now observe X to be x. As soon as we make an observation, it is advisable to see whether it is consistent with the beliefs expressed about it. As $\text{E}(X) = 2$ and $\text{Var}(X) = 1$, the **standardized observation** and **discrepancy** are, by (4.1),

$$\text{S}(x) = x - 2$$

$$\text{Dis}(x) = (x - 2)^2.$$

For example, $\text{S}(4) = 2$ and $\text{Dis}(4) = 4$. Whether or not these measures lead us to doubt any of our specifications depends on the context. If we ourselves had carried out a very careful prior analysis, a standardized value of 2 might seem to us quite surprising, whereas the same standardized value obtained as part of a much larger and more complex prior specification exercise might be rather less surprising. The implication in the latter case may be that we need to check whether more careful attention should have been paid to the corresponding prior specification.

4.3.2 Detecting degeneracy

One of the first tasks to perform when data become available is to check the consistency of the observations with the prior specifications made about them, to reveal potential anomalies. As an example, suppose that we intend using four quantities $X = [X_1, X_2, X_3, X_4]$ to help us learn about some other quantities. Indeed, suppose that after a prior specification exercise we arrive at the following expectation vector and variance matrix:

$$\mathrm{E}(X) = \begin{bmatrix} 2 \\ 2 \\ 3 \\ -1 \end{bmatrix}, \quad \mathrm{Var}(X) = \begin{bmatrix} 3 & 3 & 3 & 0 \\ 3 & 7 & 3 & -4 \\ 3 & 3 & 23 & 0 \\ 0 & -4 & 0 & 4 \end{bmatrix}.$$

Suppose we now observe these quantities and calculate their individual standardized observations and discrepancies as

$$x = \begin{bmatrix} 2.5 \\ 2.6 \\ 3.9 \\ -1.2 \end{bmatrix}, \quad x - \mathrm{E}(X) = \begin{bmatrix} 0.5 \\ 0.6 \\ 0.9 \\ -0.2 \end{bmatrix}, \quad \mathrm{S}(x) = \begin{bmatrix} 0.29 \\ 0.23 \\ 0.19 \\ -0.10 \end{bmatrix}, \quad \mathrm{Dis}(x) = \begin{bmatrix} 0.08 \\ 0.05 \\ 0.04 \\ 0.01 \end{bmatrix}.$$

As such, the observations seem very much (perhaps a bit too much) in line with the prior specification. Let us, however, carry out a discrepancy analysis. To do so, we construct the principal components $W_1, \ldots, W_4$ of $\mathrm{Var}(X)$, scaled to have variance one and centred to have expectation zero. They are:

$$W_1 = 0.0338X_1 + 0.0421X_2 + 0.1960X_3 - 0.0084X_4 - 0.7480,$$
$$W_2 = 0.0768X_1 + 0.2479X_2 - 0.0738X_3 - 0.1711X_4 - 0.5992,$$
$$W_3 = 0.4372X_1 + 0.0845X_2 - 0.0784X_3 + 0.3527X_4 - 0.4554,$$
$$W_4 = 0.5774X_1 - 0.5774X_2 - 0.0000X_3 - 0.5774X_4 - 0.5774.$$

The corresponding observed values are:

$$w_1 = -0.2202,$$
$$w_2 = -0.1550,$$
$$w_3 = -0.1282,$$
$$w_4 = -0.0577.$$

Observe that $\mathrm{Var}(X)$ is not full rank: it has a null space $W^0 = (W_4)$. The principal component W_4, which is proportional to $X_1 - X_2 - X_4$, has zero variance: $\mathrm{Var}(W_4) = 0$. However, this is contradicted by the fact of the observed value of W_4 differing from its expected value. In terms of the original quantities, $\mathrm{Var}(W_4) = 0$ implies that we know $X_1 - X_2 - X_4$ to be equal to its expectation, calculated as $\mathrm{E}(X_1 - X_2 - X_4) = 1$. However, the observed data obey $x_1 - x_2 - x_4 = 1.1 \neq 1$, so that the observations are incompatible with their corresponding prior specifications. There are many possible reasons why such contradictions might occur:

- We might deliberately have constructed a prior variance matrix containing implicitly or explicitly zero variances, but find when we see the data that we were wrong.

- In constructing beliefs over many quite complicated quantities, we might not have realized that we were building in such a strong structural relationship.
- The data are not quite what they purport to be.

In some cases we can re-examine $\text{Var}(X)$ and decide whether $\text{Var}(W_i) = 0$ was intended as a logical constraint. In any case, when such contradiction occurs we would have no choice but to revise our prior specification.

For the range of $\text{Var}(D)$ we have $W^+ = (W_1, W_2, W_3)$ and the discrepancy over this space can be determined via (4.8) as

$$\text{Dis}(d) = \sum_{i=1}^{3} w_i^2 = (-0.2202)^2 + (-0.1550)^2 + (-0.0577)^2 = 0.0889.$$

This can be compared with its expected value of $\mathbf{rk}\{\text{Var}(D)\} = 3$, from which we conclude that the data do not appear strongly to contradict the prior specification over W^+: perhaps the data in these three dimensions are rather closer to the prior specification than would be expected.

4.3.3 Oral glucose tolerance test

For this example we return to the oral glucose tolerance test problem. Recall that our doctor is using herself as a guinea pig in order to learn about the validity of the OGT test as it is interpreted for healthy elderly patients. In earlier sections we obtained the adjusted expectation and adjusted variance corresponding to her prior specifications. Now let us suppose that she proceeds to take the test herself and records her blood glucose level before and 2 hours after ingesting the requisite dose of glucose. They turn out to be as follows. D_0 is observed to be $d_0 = 5.4$ and D_2 is observed to be $d_2 = 9.8$. (These values are genuine in the sense that they are taken from a randomly chosen healthy elderly person who did indeed take the OGT test in the study (Wickramasinghe et al. 1992).)

4.3.3.1 Internal data consistency

An obviously useful check is to examine whether the data are consistent with the beliefs expressed about them. The simplest way of doing this is to **standardize** the data by (4.1). For the data we saw, $d_0 = 5.4$ and $d_2 = 9.8$,

$$\text{S}(d_0) = \frac{d_0 - \text{E}(D_0)}{\sqrt{\text{Var}(D_0)}} = 1.17, \qquad (4.16)$$

$$\text{S}(d_2) = \frac{d_2 - \text{E}(D_2)}{\sqrt{\text{Var}(D_2)}} = 2.28. \qquad (4.17)$$

The corresponding **discrepancies** (4.2) are

$$\text{Dis}(d_0) = 1.17^2 = 1.37$$

$$\text{Dis}(d_2) = 2.28^2 = 5.19,$$

each having prior expectation 1. The observation d_2 might be a little suspect: the doctor's 2-hour reading is more than two standard deviations distant from her expectation. The fasting measurement d_0 is just over one standard deviation from her expectation. Notice that, according to the standard World Health Organization OGT test diagnoses shown in Table 2.1, the doctor would be classified as suffering from impaired glucose tolerance, and would only marginally escape a presumably incorrect diagnosis of diabetes. Informally, then, these measurements support the doctor's ideas about the OGT test and the elderly.

We can also check the data discrepancy over the collection D taken globally, by evaluating (4.3) and (4.6). The discrepancy for the collection D is

$$\text{Dis}(d) = \begin{bmatrix} 5.4 - 4.16 & 9.8 - 6.25 \end{bmatrix} \begin{bmatrix} 1.12 & 0.72 \\ 0.72 & 2.43 \end{bmatrix}^{-1} \begin{bmatrix} 5.4 - 4.16 \\ 9.8 - 6.25 \end{bmatrix} = 5.23, \tag{4.18}$$

with discrepancy ratio $\text{Dr}(d) = 5.23/2 = 2.61$ as $\text{Var}(D)$ has rank 2. The rule of thumb (4.7) suggests that ratios in the region of

$$1 + 6/\sqrt{\mathbf{rk}\{\text{Var}(D)\}} = 1 + 6/\sqrt{2} = 5.24$$

or above might cause alarm. Thus, the maximal data discrepancy across the collection of the measurements that the doctor makes upon herself is within what we would expect from her prior specifications. The corresponding single-quantity discrepancies were found to be 1.37 for d_0 and 5.19 for d_2, so it is clear that the observation d_2 contributes most to the discrepancy assessed for the entire collection. The linear combination having the maximal discrepancy is, by (4.11), the **discrepancy vector**

$$\begin{aligned} \dot{W}_d &= 0.2D_0 + 1.4D_2 - 9.6 \\ &= 0.22\,\text{S}(D_0) + 2.18\,\text{S}(D_2), \end{aligned} \tag{4.19}$$

which, as the larger coefficient is for D_2, reiterates that the largest discrepancy is mostly in the direction of D_2. Discrepancies for any linear combination in $\langle D \rangle$ can be deduced through this one vector. For example, if we want to consider the difference $D_h = D_2 - D_0$ between the doctor's fasting and 2-hour blood glucose level, we use the property (4.12) to calculate

$$\begin{aligned} d_h - \text{E}(D_h) &= \text{Cov}(D_2 - D_0, \dot{W}_d) \\ &= \begin{bmatrix} -1 & 1 \end{bmatrix} \begin{bmatrix} 1.12 & 0.72 \\ 0.72 & 2.43 \end{bmatrix} \begin{bmatrix} 0.2 \\ 1.4 \end{bmatrix} \\ &= 2.31, \end{aligned}$$

from which we can calculate that the discrepancy for D_h is

$$2.31^2/\text{Var}(D_h) = 2.53$$

as $\mathrm{Var}(D_h) = 2.11$, and consequently that the standardized value of d_h is $\sqrt{2.53} = 1.59$.

We could equivalently obtain the discrepancy vector via an orthonormal basis for $\mathrm{Var}(D)$, as discussed in §4.1.2. Such a basis is provided by the principal components of $\mathrm{Var}(D)$, scaled to have variance one and centred to have expectation zero:

$$W_1 = 0.2439D_0 + 0.5517D_2 - 4.4628, \qquad \text{observed value} \quad w_1 = 2.261,$$

$$W_2 = 1.0215D_0 - 0.4517D_2 - 1.4625, \qquad \text{observed value} \quad w_2 = -0.337.$$

For example, we may simply compute the discrepancy as

$$\sum_{i=1}^{2} w_i^2 = 2.261^2 + (-0.337)^2 = 5.23,$$

as in (4.18), and it is simple to obtain the discrepancy vector as

$$\dot{W}_d = w_1 W_1 + w_2 W_2,$$

giving (4.19). For a two-dimensional problem, there is a unique direction F uncorrelated with $\dot{W}_d$. For this direction we must have $f = \mathrm{E}(F)$. In this example, F is given by

$$F \propto w_2 W_1 - w_1 W_2 = -2.39D_0 + 0.84D_2 + 4.73.$$

In k dimensions, there will be $k - 1$ such uncorrelated directions, for each of which the observed value is equal to the expected value.

4.4 The observed adjustment

Suppose that we specify beliefs about a quantity, X, and then adjust these beliefs using a collection D. When we observe the data values,

$$D = d = (d_1, \ \ldots, \ d_k),$$

then we may evaluate the random quantity $\mathrm{E}_D(X)$ given in (3.2).

Definition 4.5 *The* ***observed adjusted expectation*** *is the value* $\mathrm{E}_d(X)$, *where*

$$\mathrm{E}_d(X) = \mathrm{E}(X) + \mathrm{Cov}(X, D)\mathrm{Var}(D)^{\dagger}(d - \mathrm{E}(D)), \tag{4.20}$$

with interpretation as described in §3.4 and §3.5.

4.4.1 Adjustment discrepancy

We may apply the standardized diagnostics of §4.1 to the observed random quantity $\mathrm{E}_d(X)$, giving the following diagnostics.

Definition 4.6 *The* ***standardized adjustment****,* $\mathrm{S}_d(X)$*, is*

$$\mathrm{S}_d(X) = \mathrm{S}(\mathrm{E}_d(X)) = \frac{\mathrm{E}_d(X) - \mathrm{E}(\mathrm{E}_D(X))}{\sqrt{\mathrm{Var}(\mathrm{E}_D(X))}} = \frac{\mathrm{E}_d(X) - \mathrm{E}(X)}{\sqrt{\mathrm{RVar}_D(X)}}, \tag{4.21}$$

and the corresponding ***adjustment discrepancy****,* $\mathrm{Dis}_d(X)$*, is*

$$\mathrm{Dis}_d(X) = \mathrm{Dis}(\mathrm{E}_d(X)) = \frac{[\mathrm{E}_d(X) - \mathrm{E}(X)]^2}{\mathrm{RVar}_D(X)} = [\mathrm{S}_d(X)]^2. \tag{4.22}$$

The value of $\mathrm{Dis}_d(X)$ may suggest that our beliefs about X appear to be more or less affected by the data than we had expected. Very large discrepancies may raise the possibility that we have been overly confident in describing our uncertainty or that there may be errors in our observations, while very small discrepancies may suggest that we have been overly modest in valuing our prior knowledge about the value of X. As with any discrepancy analysis, the investigation is exploratory, and is intended to identify those areas of the prior specification which might require further consideration, such as prior assumptions which were intended to simplify a complex assessment process, and which may have seemed innocuous at the time, but turn out to be grossly conflicting with the data outcomes.

More generally, such diagnostics provide us with qualitative and quantitative information. If our observations appear to change substantially certain aspects of our beliefs, then, usually, we would want to be aware of this. Even when no simple explanation of a possible discrepancy occurs to us, aspects of our beliefs which have changed by much less or more than we had expected will often be of intrinsic interest. Such diagnostics are of particular importance when we make large collections of belief adjustments, or we are checking belief adjustments made by someone else, so that we need simple, automatic methods to call our attention to particular aspects of the assessments which we might usefully re-examine. In later chapters, we will suggest ways to augment these exploratory measures by more inferential diagnostics.

4.4.2 Adjustment discrepancy for a collection

When we adjust $B = (B_1, \ldots, B_r)$, by a further collection D, we evaluate the **observed adjusted expectation vector** $\mathrm{E}_d(B) = (\mathrm{E}_d(B_1), \ldots, \mathrm{E}_d(B_r))$. We may therefore evaluate the discrepancy for the adjustment vector as follows.

Definition 4.7 *The* ***adjustment discrepancy*** *for B is evaluated as*

$$\mathrm{Dis}_d(B) = \mathrm{Dis}(\mathrm{E}_d(B)) = [\mathrm{E}_d(B) - \mathrm{E}(B)]^T \mathrm{RVar}_D(B)^{\dagger} [\mathrm{E}_d(B) - \mathrm{E}(B)], \tag{4.23}$$

with corresponding ***adjustment discrepancy vector***

$$\ddot{W}_d = \ddot{a}_d{}^T [\mathrm{E}_D(B) - \mathrm{E}(\mathrm{E}_D(B))] = \ddot{a}_d{}^T [\mathrm{E}_D(B) - \mathrm{E}(B)], \tag{4.24}$$

where

$$\ddot{a}_d = \mathrm{Var}(\mathrm{E}_D(B))^{\dagger} [\mathrm{E}_d(B) - \mathrm{E}(\mathrm{E}_D(B))] \tag{4.25}$$

$$= \mathrm{RVar}_D(B)^{\dagger} [\mathrm{E}_d(B) - \mathrm{E}(B)]. \tag{4.26}$$

The adjustment discrepancy has prior expectation

$$\mathrm{E}(\mathrm{Dis}_D(B)) = \mathbf{rk}\{\mathrm{RVar}_D(B)\},$$

which we can show to be equal to the column rank of $\mathrm{Cov}(D, B)$ (see §12.6), and is equal to $r_{\mathbb{T}}$, the number of positive canonical resolutions for the corresponding resolution transform $\mathbb{T}_{B:D}$. Thus, a normalized version is as follows.

Definition 4.8 *The* ***adjustment discrepancy ratio*** *for B having observed $D = d$ is evaluated as*

$$\mathrm{Dr}_d(B) = \frac{\mathrm{Dis}_d(B)}{r_{\mathbb{T}}}, \tag{4.27}$$

with prior expectation one.

We may use similar heuristic arguments to those supporting relations such as (4.7) to obtain corresponding guideline intervals such as

$$P\left(1 - \frac{6}{\sqrt{r_{\mathbb{T}}}} \leq \frac{\mathrm{Dis}_d(B)}{r_{\mathbb{T}}} \leq 1 + \frac{6}{\sqrt{r_{\mathbb{T}}}}\right) \leq 0.9444. \tag{4.28}$$

for the adjustment discrepancy ratio.

4.4.3 Maximal discrepancy

We have, for any vector h, that

$$\mathrm{E}_D(h^T B) = h^T \mathrm{E}_D(B).$$

Thus, we may equate the collection $\langle \mathrm{E}_D(B) \rangle$ with the collection

$$\{\mathrm{E}_D(F) : F \in \langle B \rangle\}.$$

It is useful to identify the element in $\langle B \rangle^+$ which leads to this maximal discrepancy.

Definition 4.9 *We call*

$$\ddot{\mathbb{Y}}_d(B) = \ddot{a}_d{}^T[B - \mathrm{E}(B)] \tag{4.29}$$

the ***discrepancy vector in*** B ***induced by the adjustment by*** D, *or more simply the* ***induced discrepancy vector****.*

$\ddot{\mathbb{Y}}_d(B)$ is the element of $\langle B \rangle^+$ with the most discrepant adjusted expectation, namely

$$\mathrm{E}_d(\ddot{\mathbb{Y}}_d(B)) = \mathrm{Dis}_d(B) = \mathrm{Dis}_d(\ddot{\mathbb{Y}}_d(B)) = \mathrm{RVar}_D(\ddot{\mathbb{Y}}_d(B)) \tag{4.30}$$

$$= \max_{F \in \langle B \rangle^+} \mathrm{Dis}_d(F). \tag{4.31}$$

Note, from (4.12), that the pair $\ddot{W}_d$, $\ddot{\mathbb{Y}}_d(B)$ satisfy, for each $F \in \langle B \rangle$,

$$\begin{aligned} \mathrm{Cov}(\ddot{W}_d, F) &= \mathrm{Cov}(\ddot{W}_d, \mathrm{E}_D(F)) = \mathrm{Cov}(\ddot{\mathbb{Y}}_d(B), \mathrm{E}_D(F)) \\ &= \mathrm{RCov}_D(\ddot{\mathbb{Y}}_d(B), F) = \mathrm{E}_d(F) - \mathrm{E}(F). \end{aligned} \tag{4.32}$$

Therefore, $\ddot{W}_d$, $\ddot{\mathbb{Y}}_d(B)$ both summarize the magnitude and direction of change in beliefs between $\mathrm{E}(\cdot)$ and $\mathrm{E}_d(\cdot)$, the former with respect to directions given by the prior variance matrix while the latter is with respect to directions determined by the resolved variance matrix.

The adjustment discrepancy vector generates the adjustment discrepancy for each element $F \in \langle B \rangle^+$ according to the relation

$$\begin{aligned}\mathrm{Dis}_d(F) &= \frac{[\mathrm{E}_d(F) - \mathrm{E}(F)]^2}{\mathrm{RVar}_D(F)} = \frac{[\mathrm{RCov}_D(F, \ddot{\mathbb{Y}}_d(B))]^2}{\mathrm{RVar}_D(F)} \\ &= [\mathrm{RCorr}_D(F, \ddot{\mathbb{Y}}_d(B))]^2 \mathrm{Dis}_d(B) \\ &= [\mathrm{RCorr}_D(F, \ddot{\mathbb{Y}}_d(B))]^2 \mathrm{E}_d(\ddot{\mathbb{Y}}_d(B)),\end{aligned} \tag{4.33}$$

where $\mathrm{RCorr}_D(F, \ddot{\mathbb{Y}}_d(B))$ is the correlation between X and Y using the resolved variance matrix,

$$\mathrm{RCorr}_D(X, Y) = \frac{\mathrm{RCov}_D(X, Y)}{\sqrt{\mathrm{RVar}_D(X)\mathrm{RVar}_D(Y)}}. \tag{4.34}$$

4.4.4 Construction over a basis

An alternative construction for the discrepancy vector is to substitute the orthonormal system derived from the canonical directions Y_i for the adjustment as calculated in (3.81) into the form (4.15), giving, for the positive canonical resolutions λ_i,

$$\ddot{W}_d = \sum_i \frac{1}{\lambda_i} \mathrm{E}_d(Y_i)\mathrm{E}_D(Y_i), \tag{4.35}$$

so that

$$\ddot{\mathbb{Y}}_d(B) = \sum_i \frac{1}{\lambda_i} \mathrm{E}_d(Y_i) Y_i, \tag{4.36}$$

$$\mathrm{Dis}_d(B) = \sum_i \mathrm{Dis}_d(Y_i) = \sum_i \frac{[\mathrm{E}_d(Y_i)]^2}{\lambda_i}. \tag{4.37}$$

It can be informative to examine the individual terms in the above sum.

Definition 4.10 *The* ***canonical standardized adjustments*** *are the values*

$$\mathrm{S}_d(Y_i) = \frac{\mathrm{E}_d(Y_i)}{\sqrt{\lambda_i}}, \tag{4.38}$$

with prior expectation zero and prior variance one.

There are two types of diagnostic information given by these values. Quantitatively, any aberrant value may require scrutiny. Qualitatively, we may look for systematic patterns. For example, a particularly revealing pattern would be a sequence of decreasing absolute values, which might suggest qualitatively a false prior classification between the more and the less informative directions of adjustment.

4.4.5 Partitioning the discrepancy

It can be informative to partition the overall data discrepancy into those aspects of the discrepancy which are of relevance to the belief adjustment of interest and those aspects of the discrepancy which affect residual portions of the data, and which may therefore suggest problems with aspects of the prior specification which do not directly influence the adjustment. We therefore partition $\langle D \rangle$ into two uncorrelated subspaces, one spanned by the elements of the adjustment $D_E = \mathrm{E}_D(B)$ and the other by the residual $D_R = \mathbb{A}_{\mathrm{E}_D(B)}(D)$. From (4.4), we have, for observed vectors $D_E = d_E$, $D_R = d_R$, the relation

$$\mathrm{Dis}(d) = \mathrm{Dis}(d_E) + \mathrm{Dis}(d_R) = \mathrm{Dis}_d(B) + \mathrm{Dis}(d_R). \qquad (4.39)$$

We term $\mathrm{Dis}(d_R) = \mathrm{Dis}(d) - \mathrm{Dis}_d(B)$ the **residual discrepancy** for the adjustment of B by d. Note, in particular, that if D_R is the zero vector, then $\mathrm{Dis}(d) = \mathrm{Dis}_d(B)$.

4.5 Examples

4.5.1 Simple one-dimensional problem

We continue the example of §4.3.1. Having obtained an observed value and assessed its consistency with beliefs specified about it, we now go ahead and evaluate the adjusted expectation. For general $X = x$, we obtain

$$\mathrm{E}_x(Y) = 0.6x - 0.2,$$

so that a value of $x = 4$ leads us to an observed adjusted expectation of $\mathrm{E}_x(Y) = 2.2$. Therefore, our prior and adjusted belief specifications are as follows.

Prior	$\mathrm{E}(Y) = 1$	$\mathrm{Var}(Y) = 1$
Adjusted	$\mathrm{E}_x(Y) = 2.2$	$\mathrm{Var}_X(Y) = 0.64$
Change	$\mathrm{E}(Y) - \mathrm{E}_x(Y) = 1.2$	$\mathrm{RVar}_X(Y) = 0.36$

Now we ask the question: was the actual change from prior expectation to adjusted expectation surprising? Equivalently, was the change in expectation consistent with the expected change in variance? One way of answering is to compute the standardized change in expectation, in the form of the **standardized adjustment** (4.21) and the **adjustment discrepancy** (4.22), which turn out in this example to be $\mathrm{S}(\mathrm{E}_x(Y)) = x - 2$ and $\mathrm{Dis}_x(Y) = (x-2)^2$ for general x, and $\mathrm{S}(\mathrm{E}_x(Y)) = 2$ and $\mathrm{Dis}_x(Y) = 4$ for $x = 4$. These match the standardized values found in §4.3.1, as they must because $\mathrm{E}_X(Y)$ is a simple linear combination of X. The comments therein concerning the magnitude of the discrepancy thus apply equally here.

4.5.2 Oral glucose tolerance test

4.5.2.1 Evaluating the adjusted expectation

When we have obtained actual observations and are happy that they are consistent with the beliefs expressed about them, we proceed to evaluating the adjusted expectations obtained in §3.8.2. We had there:

$$\mathrm{E}_D(B) = \begin{bmatrix} 0.5858D_0 - 0.0501D_2 + 2.0363 \\ 0.1904D_0 + 0.1206D_2 + 4.7047 \end{bmatrix},$$

from which we calculate

$$\begin{aligned} \mathrm{E}_d(B) = \begin{bmatrix} \mathrm{E}_d(G_0) \\ \mathrm{E}_d(G_2) \end{bmatrix} &= \begin{bmatrix} 0.5858 \times 5.4 - 0.0501 \times 9.8 + 2.0363 \\ 0.1904 \times 5.4 + 0.1206 \times 9.8 + 4.7047 \end{bmatrix} \\ &= \begin{bmatrix} 4.7085 \\ 6.9140 \end{bmatrix}. \end{aligned} \tag{4.40}$$

From the point of view of the doctor, her belief specifications and the observations $d_1 = 5.4$ and $d_2 = 9.8$ that she makes when she performs the OGT test upon herself are consistent with her revising her expectations upwards for both G_0 and G_2. In the case of the fasting blood glucose measurement, the analysis shows a revision upwards from 4.16 to 4.71; and in the case of the following 2-hour measurement, a revision upwards from 6.25 to 6.91.

Informally, as a very rough guide to the locations of G_0 and G_2, we might decide to take intervals of about two or three standard deviations in either direction from the expectation as being fairly likely to contain the relevant locations. For the prior assessments we have approximately the two and three standard deviation intervals

$$\begin{aligned} G_0: &\quad 4.16 \pm 2\sqrt{1.12} = (2.04, 6.28) \quad \text{and} \quad 4.16 \pm 3\sqrt{1.12} = (0.99, 7.33), \\ G_2: &\quad 6.25 \pm 2\sqrt{2.43} = (3.13, 9.38) \quad \text{and} \quad 6.25 \pm 3\sqrt{2.43} = (1.57, 10.93). \end{aligned}$$

For the assessments after adjusting by $[D]$ we obtain the tighter intervals

$$\begin{aligned} G_0: &\quad 4.71 \pm 2\sqrt{0.77} = (2.96, 6.46) \quad \text{and} \quad 4.71 \pm 3\sqrt{0.77} = (2.08, 7.34), \\ G_2: &\quad 6.91 \pm 2\sqrt{2.32} = (3.86, 9.96) \quad \text{and} \quad 6.91 \pm 3\sqrt{2.32} = (2.34, 11.48). \end{aligned}$$

The adjusted expectation for the 2-hour blood glucose measurement is 6.91, implying that an **average healthy** elderly patient will have a 2-hour reading on the borderline between being diagnosed as healthy and being diagnosed as having impaired glucose tolerance. Put another way, about half of the elderly will be misdiagnosed according to the thresholds set for the OGT test. Additionally, the fasting blood glucose level has adjusted expectation 4.71, suggesting that the elderly have a slightly higher fasting level than do the young. Consequently, the analysis suggests that there are static differences between the blood glucose levels for the young and the elderly; and dynamic differences between their abilities to cope with fluctuations in blood glucose levels. The three standard deviation interval for G_0 contains 7.0, suggesting that some healthy elderly patients would be classified as having impaired glucose tolerance even before ingesting the extra glucose. The three standard deviation interval for G_2 contains 10.0, suggesting that a proportion of healthy elderly patients would be wrongly classified as having diabetes.

Suppose we examine $G_h = G_2 - G_0$, the difference between the fasting and 2-hour blood glucose level for a typical elderly person, which we constructed in §3.8.2. We can adjust G_h directly by the data quantities. The prior expectation is that $\mathrm{E}(G_h) = 2.09$. The adjusted expectation is observed to be $\mathrm{E}_d(G_h) = 2.2055$, so that the data lead to a small positive change in expectation of 0.1155. Recall from §3.8.2 that the data were expected to be only weakly informative for learning about G_h. The prior three standard deviation interval for G_h is $(-2.27, 6.45)$, while the adjusted interval is $(-2.00, 6.41)$, with little difference between them.

4.5.2.2 Evaluating adjustment discrepancies

In the previous parts of this example, we have (1) obtained actual observations and checked them for consistency with the beliefs specified about them a priori, assessments which were made for the single quantities individually and collectively; (2) used the data to form the observed adjusted expectations for the two quantities of interest. It is now time to check for discrepancies amongst these adjusted expectations. By (4.21) the individual **standardized adjustments** are

$$\mathrm{S}(\mathrm{E}_d(G_0)) = \frac{\mathrm{E}_d(G_0) - \mathrm{E}(G_0)}{\sqrt{\mathrm{RVar}_D(G_0)}} = 0.93, \tag{4.41}$$

$$\mathrm{S}(\mathrm{E}_d(G_2)) = \frac{\mathrm{E}_d(G_2) - \mathrm{E}(G_2)}{\sqrt{\mathrm{RVar}_D(G_2)}} = 2.01, \tag{4.42}$$

with corresponding **adjustment discrepancies** of $\mathrm{Dis}_d(G_0) = 0.93^2 = 0.86$ and $\mathrm{Dis}_d(G_2) = 2.01^2 = 4.04$. Relative to the expected change (reduction) in variance, the change in expectation for G_0 is about 0.93 standard deviations upward, a change that does not trouble us greatly. However, the change for G_2 is fractionally over two standard deviations, and so is a little larger than we would have expected. We saw in Table 3.1 that the expected resolution of variance was only roughly 0.11 or 4.48% of prior, but we have seen a change in expectation of $6.91 - 6.25 = 0.66 \approx 2\sqrt{0.11}$.

We can use such discrepancies as important diagnostic flags. We have seen in this case a value larger than expected, and we should consider the following possibilities. First, the data may be more variable than expected, as $\mathrm{E}_d(G_2)$ is further from $\mathrm{E}_D(G_2)$ than expected. Secondly, the resolved variation $\mathrm{RVar}_D(G_2)$ may be smaller than would be consistent with such a change in expectation, implying that the prior variance $\mathrm{Var}(G_2)$ may be too tight. From our earlier check for internal data consistency, we have already noted (4.17) that the observation $d_2 = 9.8$ is perhaps suspect, and so in this particular case we would perhaps lean towards the former of these two possibilities – perhaps our doctor is not as typical a healthy elderly person as she believes.

The implication we place upon such a diagnostic depends upon the context. For example, if this had been the doctor's first try at elicitation and analysis, she might have been prepared for fairly wide discrepancies between data and belief specifications, and might thereafter reconsider her prior specification. On the other hand, the doctor may have substantial experience in quantifying her knowledge, and so feel well calibrated, in which case larger discrepancies would be a cause of concern and might lead to further investigation.

We began the continuation of this example by checking internal data consistency by evaluating the standardized observations and discrepancies for the individual quantities D_0 and D_2, and then proceeded to calculating complementary diagnostics across the collection D globally. We continue the theme of following individual quantity considerations by global considerations by calculating the **adjustment discrepancy for the collection** B taken as a whole, given by (4.23). We have already observed that the adjusted variance matrix for B is as given in (3.48), and the prior variance matrix for B is given in (3.43). Consequently, the resolved variance matrix for B is

$$\begin{aligned}\mathrm{RVar}_D(B) = \mathrm{Var}(B) - \mathrm{Var}_D(B) &= \begin{bmatrix} 1.12 & 0.72 \\ 0.72 & 2.43 \end{bmatrix} - \begin{bmatrix} 0.7718 & 0.5658 \\ 0.5658 & 2.3211 \end{bmatrix} \\ &= \begin{bmatrix} 0.3482 & 0.1542 \\ 0.1542 & 0.1089 \end{bmatrix}. \end{aligned} \tag{4.43}$$

We also have that the change in expectation is

$$\mathrm{E}_d(B) - \mathrm{E}(B) = \begin{bmatrix} 4.7085 \\ 6.9140 \end{bmatrix} - \begin{bmatrix} 4.1600 \\ 6.2500 \end{bmatrix} = \begin{bmatrix} 0.5485 \\ 0.6640 \end{bmatrix}, \tag{4.44}$$

so that we obtain the adjustment discrepancy for the collection as

$$\mathrm{Dis}_d(B) = \begin{bmatrix} 0.5485 & 0.6640 \end{bmatrix}^T \begin{bmatrix} 0.3482 & 0.1542 \\ 0.1542 & 0.1089 \end{bmatrix}^{-1} \begin{bmatrix} 0.5485 \\ 0.6640 \end{bmatrix} = 5.23 = 2.29^2. \tag{4.45}$$

This represents the largest discrepancy for any linear combination of G_0, G_2, the elements in B. That is, the standardized change in expectation for any new quantity $G_{\text{new}} = a + b_0 G_0 + b_2 G_2$, for any scalars a, b_0, b_2, is at most 2.29.

For this example, the linear combination with the most discrepant adjusted expectation turns out to be, using (4.31),

$$\begin{aligned}\ddot{\mathbb{Y}}_d(B) &= -3.01 G_0 + 10.36 G_2 - 52.20 \\ &= -3.19\,\mathrm{S}(G_0) + 16.14\,\mathrm{S}(G_2),\end{aligned}$$

where $\ddot{\mathbb{Y}}_d(B)$ is the **induced discrepancy vector for the adjustment**, and we have displayed it using the standardized and unstandardized forms of G_0 and G_2.

This quantity $\ddot{\mathbb{Y}}_d(B)$ has prior expectation zero and observed adjusted expectation $\mathrm{E}_d(\ddot{\mathbb{Y}}_d(B)) = 5.23$, which is the maximal change in adjusted expectation.

We can determine the change in expectation and the discrepancy for any constructed linear combination of interest either directly or by exploiting (4.32) and (4.33). In our example, one of the quantities of concern to the doctor is the difference between the fasting and 2-hour blood glucose level for a typical elderly person, which we constructed as $G_h = G_2 - G_0$ in §3.8.2. We can use the resolved variance matrix (4.43) directly to deduce the resolved variance matrix for G_h and the discrepancy vector $\ddot{\mathbb{Y}}_d(B)$, as both are simple linear combinations of G_0 and G_2:

$$\begin{bmatrix} \mathrm{RVar}_D(G_h) & \mathrm{RCov}_D(G_h, \ddot{\mathbb{Y}}_d(B)) \\ \mathrm{RCov}_D(G_h, \ddot{\mathbb{Y}}_d(B)) & \mathrm{RVar}_D(\ddot{\mathbb{Y}}_d(B)) \end{bmatrix} = \begin{bmatrix} 0.1487 & 0.1155 \\ 0.1155 & 5.2253 \end{bmatrix}. \tag{4.46}$$

Note that this verifies that $\mathrm{RVar}_D(\ddot{\mathbb{Y}}_d(B)) = 5.23 = \mathrm{Dis}_d(B)$. We deduce immediately from (4.32) that

$$\mathrm{E}_d(G_h) - \mathrm{E}(G_h) = \mathrm{RCov}_D(G_h, \ddot{\mathbb{Y}}_d(B)) = 0.1155, \tag{4.47}$$

which is the change in adjusted expectation that we saw at the foot of §4.5.2.1. Further, the correlation between G_h and $\ddot{\mathbb{Y}}_d(B)$ for this resolved matrix is

$$\mathrm{RCorr}_D(G_h, \ddot{\mathbb{Y}}_d(B)) = 0.131,$$

and by (4.33) the discrepancy for G_h must therefore be

$$\begin{aligned} \mathrm{Dis}_d(G_h) &= \mathrm{RCorr}_D(G_h, \ddot{\mathbb{Y}}_d(B))^2 \mathrm{Dis}_d(B) \\ &= 0.131^2 \times 5.23 = 0.09. \end{aligned}$$

We can, just as easily, deduce the discrepancy for any other linear combination of interest, through its resolved correlation with $\ddot{\mathbb{Y}}_d(B)$.

Note that, for this example, the global measure of adjustment discrepancy $\mathrm{Dis}_d(B)$ in (4.45) is equal to the global measure of data discrepancy $\mathrm{Dis}(d)$ shown in (4.18). Referring back to §4.4.5, this is because there is no part of D that is uninformative for B, there is no residual subspace D_R, and so we have straightforwardly $\mathrm{Dis}(d) = \mathrm{Dis}_d(B)$. On the other hand, for the adjustment of G_h by D there is a residual subspace D_R, and we can deduce its discrepancy via (4.39) as

$$\mathrm{Dis}(D_R) = \mathrm{Dis}(d) - \mathrm{Dis}_d(G_h) = 5.23 - 0.09 = 5.14,$$

with prior expectation unity, the rank of the residual subspace. Thus, as far as the adjustment of G_h is concerned, much of the discrepancy lies in the part of the data which is not informative for G_h.

4.6 The size of an adjustment

The discrepancy measures that we have so far discussed each compare changes in expectation to prior assessments as to the magnitudes of such changes. Thus, large discrepancies may correspond to small changes in belief, while small discrepancies may correspond to large changes in belief. However, when we adjust beliefs over B by observation of D, often it is only the beliefs over B which are of basic interest, while beliefs involving D are only of interest to assist us in modifying beliefs over B. A useful qualitative picture of the changes in our beliefs over B is based on standardizing the change between prior and adjusted expectation by the prior variance.

Definition 4.11 *The size of the adjustment of X by D is*

$$\text{Size}_d(X) = \frac{[\text{E}_d(X) - \text{E}(X)]^2}{\text{Var}(X)}. \tag{4.48}$$

Recall that we have decomposed $X - \text{E}(X)$ into two uncorrelated components, namely

$$X - \text{E}(X) = [X - \text{E}_D(X)] + [\text{E}_D(X) - \text{E}(X)].$$

$\text{Size}_d(X)$ is the ratio of the observed value of the squared magnitude of the second component, $[\text{E}_d(X) - \text{E}(X)]^2$ to the variance of the sum of the two components, as $\text{Var}(X) = \text{RVar}_D(X) + \text{Var}_D(X)$. Very large values of $\text{Size}_d(X)$ therefore might signal a diagnostic warning as there would be no variance partition consistent both with the prior variance for X and the observed change in expectation for X. Of course, the implications of such conflict will depend on the context for the assessment.

4.6.1 The size of an adjustment for a collection

Just as the discrepancy measure for an observed data vector may be expressed as the maximum discrepancy for a linear combination of the elements of the vector, by (4.10), the size of the adjustment of the collection B by $D = d$ is defined to be the maximum size of adjustment of such a linear combination, namely

$$\text{Size}_d(B) = \max_{X \in \langle B \rangle^+} \text{Size}_d(X). \tag{4.49}$$

As

$$\text{Size}_d(h^T B) = \frac{[h^T(\text{E}_d(B) - \text{E}(B))]^2}{h^T \text{Var}(B) h},$$

we deduce, using a similar argument to the derivation of (4.10) and (4.11), that the choice $\dot{h}_d$ for which $\text{Size}_d(h^T B)$ achieves the maximum is

$$\dot{h}_d = \text{Var}(B)^\dagger [\text{E}_d(B) - \text{E}(B)], \tag{4.50}$$

from which we obtain the size for a collection as follows.

Definition 4.12 *The* ***size of the adjustment of the collection*** B ***by*** $D = d$ *is*

$$\text{Size}_d(B) = [\text{E}_d(B) - \text{E}(B)]^T \text{Var}(B)^\dagger [\text{E}_d(B) - \text{E}(B)]. \tag{4.51}$$

4.7 The bearing for an adjustment

Definition 4.13 *The* ***bearing for the adjustment*** *of* B *by* $D = d$ *is*

$$\mathbb{Z}_d(B) = \dot{h}_d^T [B - \text{E}(B)] = [\text{E}_d(B) - \text{E}(B)]^T \text{Var}(B)^\dagger [B - \text{E}(B)].$$

The bearing expresses both the direction and the magnitude of the change between prior and adjusted beliefs, in this case with respect to the prior covariance specification. This is because, for any $F = u^T B \in \langle B \rangle$, we have a property analogous to (4.32), namely that, from (4.50),

$$\begin{aligned}\text{Cov}(F, \mathbb{Z}_d(B)) &= u^T \text{Var}(B) \dot{h}_d \\ &= u^T [\text{E}_d(B) - \text{E}(B)] = \text{E}_d(F) - \text{E}(F).\end{aligned} \tag{4.52}$$

Therefore, for any X which is uncorrelated with $\mathbb{Z}_d(B)$, we have

$$\text{E}_d(X) = \text{E}(X).$$

Further, if $M_d = \alpha \mathbb{Z}_d(B)$, then a bearing of M_d represents α times the change in expectation compared to a bearing of $\mathbb{Z}_d(B)$, for every element of $\langle B \rangle$, and from (4.52),

$$\text{Size}_d(B) = \text{Var}(\mathbb{Z}_d(B)). \tag{4.53}$$

Therefore, the bearing gives a simple representation of the magnitude and direction of the changes in belief with respect to the prior variance specification. This is often more straightforward to interpret than the comparison using the induced discrepancy vector, as summarized by (4.32), which is based on the adjusted variance specification, which changes with the data and so may be more complex to analyse. In particular, representation (4.52) allows us to separate out the effects of different aspects of a complex adjustment in a systematic fashion, as we shall describe in the next chapter.

Corresponding to (4.33), we may generate the size of the adjustment for any element in $\langle B \rangle^+$, from (4.52) and (4.53), by the relation

$$\begin{aligned}\text{Size}_d(F) &= \frac{[\text{E}_d(F) - \text{E}(F)]^2}{\text{Var}(F)} = \frac{[\text{Cov}(F, \mathbb{Z}_d(B))]^2}{\text{Var}(F)} \\ &= \text{Corr}(F, \mathbb{Z}_d(B))^2 \text{Size}_d(B).\end{aligned} \tag{4.54}$$

In comparison with (4.38), observe that, for any quantity U with prior mean zero and variance one, we have

$$\text{Size}_d(U) = [\text{E}_d(U)]^2. \tag{4.55}$$

4.7.1 Construction via a basis

Suppose that $(U_1, \ldots, U_{r_B})$ is *any* collection of elements of $\langle B\rangle$ which are a priori uncorrelated, with zero mean and variance one. The bearing for the adjustment of B by $D = d$, $\mathbb{Z}_d(B)$, may then be constructed as

$$\mathbb{Z}_d(B) = \sum_{i=1}^{r_B} \mathrm{E}_d(U_i)U_i. \tag{4.56}$$

$\mathbb{Z}_d(B)$ does not depend on the choice of $U_1, \ldots, U_{r_B}$. Therefore, from (4.55),

$$\mathrm{Size}_d(B) = \mathrm{Var}(\mathbb{Z}_d(B)) = \sum_{i=1}^{r_B}[\mathrm{E}_d(U_i)]^2 = \sum_{i=1}^{r_B} \mathrm{Size}_d(U_i). \tag{4.57}$$

Now consider the full collection of canonical quantities for the adjustment, $(Y_1, \ldots, Y_{r_B})$. These form an orthonormal basis for $\langle B\rangle$. Therefore, we may choose $U_i = Y_i$ in (4.56), giving

$$\begin{aligned}\mathbb{Z}_d(B) &= \sum_{i=1}^{r_B} \mathrm{E}_d(Y_i)Y_i \\ &= \sum_{i=1}^{r_{\mathbb{T}}} \mathrm{E}_d(Y_i)Y_i.\end{aligned} \tag{4.58}$$

Note that if $r_{\mathbb{T}} < r_B$, the quantities $(Y_{r_{\mathbb{T}}+1}, \ldots, Y_{r_B})$ have variance zero and so we must have $\mathrm{E}_d(Y_i) = \mathrm{E}(Y_i) = 0$ for $i = r_{\mathbb{T}} + 1, \ldots, r_B$; see §4.1.3. Comparing (4.36) and (4.58), the induced discrepancy vector and the bearing differ only in the weighting $1/\lambda_i$. Indeed, the relationship between the coefficients $\ddot{a}_d$ for the induced discrepancy vector $\ddot{\mathbb{Y}}_d(B)$ and the coefficients $\dot{h}_d$ for the bearing $\mathbb{Z}_d(B)$ is

$$\mathbb{T}_{B:D}\ddot{a}_d = \dot{h}_d. \tag{4.59}$$

4.7.2 Representing discrepancy vectors as bearings

We began this chapter by discussing discrepancy vectors for collections of observations. Notice that, for any element $F \in \langle D\rangle$, with observed value f, we must have $\mathrm{E}_d(F) = f$. Therefore the element of $\langle D\rangle$ with the largest standardized observation

$$\max_{F\in\langle D\rangle} \left[\frac{f - \mathrm{E}(F)}{\sqrt{\mathrm{Var}(F)}}\right]^2,$$

is precisely the bearing $\mathbb{Z}_d(D)$ of the adjustment of D by D. Thus, in the case where we adjust a space D by itself, $D = d$, the various discrepancy quantities coincide:

$$\begin{aligned}\mathrm{Dis}(d) &= \mathrm{Dis}_d(D) &&= \mathrm{Size}_d(D), \\ \dot{a}_d &= \ddot{a}_d &&= \dot{h}_d, \\ \dot{W}_d &= \ddot{\mathbb{Y}}_d(D) &&= \mathbb{Z}_d(D).\end{aligned}$$

4.8 Joint bearings

The relationship between the bearing for the adjustment of B by D and the adjustment of D by B follows from the relationship (3.81) between the eigenstructures of the two canonical structures as follows.

If $Y_1, Y_2, \ldots, Y_{r_{\mathbb{T}}}$ and $U_1, U_2, \ldots, U_{r_{\mathbb{T}}}$ are the canonical directions corresponding to non-zero eigenvalues for the adjustment of B by D and of D by B respectively, with corresponding canonical resolutions $\lambda_1, \lambda_2, \ldots, \lambda_{r_{\mathbb{T}}}$, where $r_{\mathbb{T}}$ is the rank of the resolution transform matrix, then the bearing for the adjustment of B by D and the adjustment of D by B may both be evaluated from the **joint bearing**, $\mathbb{Z}(B, D)$, defined as follows.

Definition 4.14

$$\mathbb{Z}(B, D) = \sqrt{\lambda_1}Y_1U_1 + \sqrt{\lambda_2}Y_2U_2 + \ldots + \sqrt{\lambda_{r_{\mathbb{T}}}}Y_{r_{\mathbb{T}}}U_{r_{\mathbb{T}}}. \tag{4.60}$$

For observed $D = d$, the bearing $\mathbb{Z}_d(B)$ is evaluated by substituting the observed values of the vector D or equivalently the observed values $U_i = u_i$, while for observed $B = b$, the bearing $\mathbb{Z}_b(D)$ is evaluated by substituting the observed values of B or equivalently the values $Y_i = y_i$, so that

$$\mathbb{Z}_d(B) = \mathbb{Z}(B, d) \quad \text{and} \quad \mathbb{Z}_b(D) = \mathbb{Z}(b, D). \tag{4.61}$$

4.9 Size diagnostics

A natural diagnostic for assessing the magnitude of an adjustment is to compare the largest standardized change in expectation that we observe to our expectation for the magnitude of the largest change, evaluated prior to observing D. We evaluate the expectation of this random quantity, $\text{Size}_D(B)$, as

$$\begin{aligned}\text{E}(\text{Size}_D(B)) &= \text{E}([\text{E}_D(B) - \text{E}(B)]^T \text{Var}(B)^\dagger [\text{E}_D(B) - \text{E}(B)]) \\ &= \textbf{tr}\{\text{E}(\text{Var}(B)^\dagger [\text{E}_D(B) - \text{E}(B)][\text{E}_D(B) - \text{E}(B)]^T)\} \\ &= \textbf{tr}\{\mathbb{T}_{B:D}\} = \sum_{i=1}^{r_{\mathbb{T}}} \lambda_i = \text{RU}_D(B).\end{aligned} \tag{4.62}$$

Thus, the expected size of the adjustment is equal to the resolved uncertainty for the structure. To compare the observed and expected values, we use the following statistic.

Definition 4.15 *The* ***size ratio for the adjustment of*** *B* ***by*** *D is*

$$\begin{aligned}\text{Sr}_d(B) &= \frac{\text{Size}_d(B)}{\text{E}(\text{Size}_D(B))} = \frac{\text{Var}(\mathbb{Z}_d(B))}{\text{RU}_D(B)} \\ &= \frac{[\text{E}_d(B) - \text{E}(B)]^T \text{Var}(B)^\dagger [\text{E}_d(B) - \text{E}(B)]}{\sum_{i=1}^{r_{\mathbb{T}}} \lambda_i}.\end{aligned} \tag{4.63}$$

We anticipate that the ratio will be near one. Large values of the size ratio suggest that we have formed new beliefs which are surprisingly discordant with our prior judgements. Values near zero might suggest that we have exaggerated our prior uncertainty. The size ratio is essentially a ratio of variances. As for our other diagnostic measures, we treat the ratio as a simple warning flag drawing our attention to possible conflicts between prior and adjusted beliefs.

Sometimes, it is useful to have simple rules of thumb to suggest warning levels for size ratios. As an example, note that were all the canonical quantities for B, D to be jointly normally distributed, then from (4.60), (4.61) it would follow that

$$\text{Var}(\text{Size}_D(B)) = \text{Var}\left(\sum_{i=1}^{r_{\mathbb{T}}} \lambda_i U_i^2\right) = 2\sum_{i=1}^{r_{\mathbb{T}}} \lambda_i^2. \tag{4.64}$$

Then, similarly to the argument leading to (4.7), a simple heuristic which can be useful in examining the size ratio for the adjustment is given by Chebychev's inequality: for any k,

$$P(-k \leq \frac{\text{Size}_D(B) - \sum_{i=1}^{r_{\mathbb{T}}} \lambda_i}{\sqrt{2\sum_{i=1}^{r_{\mathbb{T}}} \lambda_i^2}} \leq k) \leq 1 - k^{-2}.$$

The choice $k = 3\sqrt{2}$ leads to the interval

$$P\left(1 - \frac{6\sqrt{\sum_{i=1}^{r_{\mathbb{T}}} \lambda_i^2}}{\sum_{i=1}^{r_{\mathbb{T}}} \lambda_i} \leq \text{Sr}_D(B) \leq 1 + \frac{6\sqrt{\sum_{i=1}^{r_{\mathbb{T}}} \lambda_i^2}}{\sum_{i=1}^{r_{\mathbb{T}}} \lambda_i}\right) \leq 0.9444. \tag{4.65}$$

In certain circumstances, we might even find it useful to approximate the distribution of $\text{Sr}_D(B)$, for example by a gamma distribution. Matching the mean and variance suggests approximating using a two-parameter gamma distribution with shape and scale parameters

$$\text{shape} = \sum_{i=1}^{r_{\mathbb{T}}} \lambda_i, \qquad \text{scale} = \frac{2\sum_{i=1}^{r_{\mathbb{T}}} \lambda_i^2}{\sum_{i=1}^{r_{\mathbb{T}}} \lambda_i}. \tag{4.66}$$

Any such approximation is useful only as a simple heuristic for setting warning limits.

4.10 Geometric interpretation

Each of the measures that we have constructed in this chapter is based on the following geometric idea. According to the Riesz representation theorem, any bounded linear functional G on a closed inner product space I may be represented by a (unique) element $X_G \in I$ in the sense that, for any $Y \in I$,

$$G(Y) = (Y, X_G). \tag{4.67}$$

Further, the norm of G is equal to the norm of X_G, i.e.

$$\|G\|^2 = \max_{Y \in I} \frac{[G(Y)]^2}{\|Y\|^2} = (X_G, X_G). \tag{4.68}$$

The various quantities discussed in the preceding sections of this chapter may all be constructed by applying the Riesz representation to appropriate choices of functional and inner product. In each case, the magnitudes of the prior expectations for the appropriate size measures may be directly evaluated using (4.68). Thus, the discrepancy vector, $\dot{W}_d$, follows from applying the Riesz representation to the functional $G(F) = f - \mathrm{E}(F)$, with inner product being prior variance, as given by (4.12). The induced discrepancy vector for the adjustment, $\ddot{\mathbb{Y}}_d(B)$, corresponds to the functional $G(Y) = \mathrm{E}_d(Y) - \mathrm{E}(Y)$, with inner product being adjusted covariance, as given by (4.32). The bearing $\mathbb{Z}_d(B)$ corresponds to the functional $G(Y) = \mathrm{E}_d(Y) - \mathrm{E}(Y)$, with inner product being prior covariance, as given by (4.52). Notice in particular that the relation (4.59) between the bearing and the discrepancy vector follows from the property (3.89), namely that the belief transform transforms adjusted to prior covariance.

As we are mainly interested, in this book, in finite-dimensional spaces, we have preferred to give simple, direct constructions for our interpretative measures. However, the generalization of these quantities to infinite collections follows naturally by using the Riesz representation over the corresponding geometric constructions.

4.11 Linear likelihood

The bearing may be interpreted as the **linear (normalized) likelihood** by analogy with the special case of a full Bayes analysis, where the bearing vector corresponds to the normalized likelihood function. The correspondence is as follows. For simplicity, suppose that $B = \{B_1, \ldots, B_k\}$ are the indicator functions for a finite partition, so that each B_i is 1 or 0, and $\sum B_i = 1$. Now $\langle B \rangle$ is the collection of linear combinations $G = \sum_i g_i B_i$. This is equivalently the collection of all finite random variables defined on the probability space with elements $\{B_1, \ldots, B_k\}$. The bearing, given observed data d, is the random quantity $\mathbb{Z}_d(B) = \sum_i z_i B_i$ satisfying, for each $F = \sum_i f_i B_i \in \langle B \rangle$,

$$\mathrm{E}_d(F) - \mathrm{E}(F) = \mathrm{Cov}(F, \mathbb{Z}_d(B)) = \sum_i [f_i - \mathrm{E}(F)] z_i P(B_i). \tag{4.69}$$

Given a full joint probabilistic specification over B and D, so that

$$D = \{D_1, \ldots, D_r\}$$

also represents a finite partition, from (3.18) adjusted and conditional expectations are the same, so that for any observed partition member d we have

$$\mathrm{E}_d(F) - \mathrm{E}(F) = \sum_i [f_i - \mathrm{E}(F)] P(B_i|d) = \sum_i [f_i - \mathrm{E}(F)] \frac{P(d|B_i)}{P(d)} P(B_i), \tag{4.70}$$

where $P(d) = \sum_j P(d|B_j)P(B_j)$. Equating (4.69) and (4.70) gives the bearing for the adjustment as

$$\mathbb{Z}_d(B) = \sum_i \frac{P(d|B_i)}{P(d)} B_i. \tag{4.71}$$

Thus $\mathbb{Z}_d(B)$ is the normalized likelihood vector, namely the random variable which, if B_i occurs, takes value equal to the normalized likelihood of the observed data d given B_i.

In conventional likelihood analyses, interest focuses on the individual values taken, namely the coefficients $P(d|B_i)/P(d)$, while our interest, primarily, is in the composite random quantity $\mathbb{Z}_d(B)$. However, the comparison is revealing for the types of information provided by the bearing, which we may view as extending likelihood to general linear spaces. The extension of this representation to continuous probability spaces follows through the Riesz representation theorem as in §4.10.

4.12 Examples

In §4.5 we illustrated the notion of discrepancy, which relates changes in expectation to the proportion of variation explained by an adjustment. We now illustrate a related aspect, the notion of **size**, which relates changes in expectation to prior variation.

4.12.1 Algebraic example

We return now to the algebraic example considered in §3.11.2. We found there the canonical directions W_1 and W_2 for the adjustment of B by D ((3.93), (3.94)), and the canonical directions $\check{W}_1$ and $\check{W}_2$ for the reverse adjustment of D by B ((3.106), (3.107)). The canonical resolutions are the same, irrespective of direction of adjustment. In summary, the canonical quantities are:

$$\lambda_1 = \frac{4\rho^2}{(1+u)(1+v)}, \qquad W_1 = \frac{1}{\sqrt{2(1+v)}}(Y_1 + Y_2),$$

$$\check{W}_1 = \frac{1}{\sqrt{2(1+u)}}(X_1 + X_2),$$

$$\lambda_2 = 0, \qquad W_2 = \frac{1}{\sqrt{2(1-v)}}(Y_1 - Y_2),$$

$$\check{W}_2 = \frac{1}{\sqrt{2(1-u)}}(X_1 - X_2).$$

As far as an observed adjustment is concerned, we have a canonical resolution of zero for W_2 and $\check{W}_2$. As such, for any data to be consistent with these beliefs we must check that $\mathrm{E}_d(W_2) = 0$, i.e. that $\mathrm{E}_d(Y_1 - Y_2) = 0$, and similarly that $\mathrm{E}_d(X_1 - X_2) = 0$.

Using the non-degenerate canonical quantities we can calculate the bearings for the two adjustments, using (4.56), as

$$\mathbb{Z}_d(B) = \mathrm{E}_d(W_1)W_1 \quad \text{and} \quad \mathbb{Z}_b(D) = \mathrm{E}_b(\check{W}_1)\check{W}_1.$$

Before we observe d and before we observe b, these are identical because $\check{W}_1 = \frac{1}{\sqrt{\lambda_1}}\mathrm{E}_D(W_1)$ and $W_1 = \frac{1}{\sqrt{\lambda_1}}\mathrm{E}_D(\check{W}_1)$, so that

$$\mathbb{Z}(B, D) = \mathbb{Z}_D(B) = \mathbb{Z}_B(D) = \sqrt{\lambda_1}W_1\check{W}_1$$

is the **joint bearing** discussed in §4.8. Suppose we now observe $D = d$ by $X_1 = x_1$, $X_2 = x_2$, resulting in observation of $\check{W}_1$ to be

$$\check{w}_1 = \frac{1}{\sqrt{2(1+u)}}(x_1 + x_2),$$

then $\mathbb{Z}_d(B)$ is quickly obtained as $\mathbb{Z}_d(B) = \sqrt{\lambda_1}W_1\check{w}_1$. On the other hand, if $B = b$ is observed, for example by $Y_1 = y_1$, $Y_2 = y_2$, then W_1 will be observed as

$$w_1 = \frac{1}{\sqrt{2(1+v)}}(y_1 + y_2),$$

and the bearing for the reverse adjustment is $\mathbb{Z}_b(D) = \sqrt{\lambda_1}w_1\check{W}_1$.

4.12.2 Oral glucose tolerance test

By (4.48) the individual **sizes for the adjustment** are

$$\mathrm{Size}_d(G_0) = \frac{[\mathrm{E}_d(G_0) - \mathrm{E}(G_0)]^2}{\mathrm{Var}(G_0)} = 0.27 = 0.52^2,$$

$$\mathrm{Size}_d(G_2) = \frac{[\mathrm{E}_d(G_2) - \mathrm{E}(G_2)]^2}{\mathrm{Var}(G_2)} = 0.18 = 0.43^2.$$

Thus, relative to the prior variance, the change in expectation for G_0 is about 0.52 standard deviations and the change in expectation for G_2 is about 0.43 standard deviations. These changes thus appear relatively consistent with the prior specification.

Just as we obtained a global measure of discrepancy relating changes in expectation to variance resolved, we now calculate a similar global measure relating changes in expectation to prior variance, namely the **size of the adjustment for the collection** (4.51). With $\mathrm{E}_d(B) - \mathrm{E}(B)$ given by (4.44) and $\mathrm{Var}(B)$ given by (3.43), we obtain

$$\mathrm{Size}_d(B) = [\mathrm{E}_d(B) - \mathrm{E}(B)]^T \mathrm{Var}(B)^\dagger [\mathrm{E}_d(B) - \mathrm{E}(B)] \tag{4.72}$$

$$= \begin{bmatrix} 0.5485 & 0.6640 \end{bmatrix} \begin{bmatrix} 1.12 & 0.72 \\ 0.72 & 2.43 \end{bmatrix}^\dagger \begin{bmatrix} 0.5485 \\ 0.6640 \end{bmatrix}$$

$$= 0.3179, \tag{4.73}$$

which represents, relative to prior variation, the largest squared change in expectation for any linear combination of G_0, G_2, the elements in B. For this example, the linear combination with the most affected adjusted expectation turns out to be, using (4.50),

$$\mathbb{Z}_d(B) = 0.39G_0 + 0.16G_2 - 2.60 \tag{4.74}$$

$$= 0.41\,\mathrm{S}(G_0) + 0.25\,\mathrm{S}(G_2), \tag{4.75}$$

where $\mathbb{Z}_d(B)$ is the **bearing** for the adjustment. We have shown the bearing using both the standardized and unstandardized forms of G_0 and G_2. The bearing $\mathbb{Z}_d(B)$ has prior expectation zero and observed adjusted expectation

$$\mathrm{E}_d(\mathbb{Z}_d(B)) = 0.3179 = \mathrm{Var}(\mathbb{Z}_d(B)) = \mathrm{Size}_d(B).$$

This represents the maximal change in adjusted expectation relative to prior variation.

The bearing has two principal qualities: it summarizes both the magnitude and direction of the change in expectation. This means in general that for any new quantity constructed from the elements in the collection being adjusted, we can determine both the change in expectation by exploiting (4.52), and its corresponding size by exploiting (4.54). For example, we noted above the property that the changes from prior expectation to adjusted expectation for quantities in B are equivalent to the covariances of the quantities with the bearing. For example, for G_0,

$$\begin{aligned}\mathrm{E}(G_0) - \mathrm{E}_d(G_0) &= \mathrm{Cov}(\mathbb{Z}_d(B), G_0)\\ &= \mathrm{Cov}(0.39G_0 + 0.16G_2 - 2.60, G_0)\\ &= 0.39\mathrm{Var}(G_0) + 0.16\mathrm{Cov}(G_0, G_2)\\ &= 0.55\end{aligned}$$

can be seen to be the difference between its expectation $\mathrm{E}(G_0) = 4.16$ and its adjusted expectation $\mathrm{E}_d(G_0) = 4.71$. In this way, changes in expectation are expressible solely through a covariance with the bearing, and so the magnitude of a change in expectation for any quantity depends only upon the strength of correlation between the quantity and the bearing, and upon the variance of the bearing.

As a second example, suppose that we take our example of the difference between the fasting and 2-hour blood glucose level for a typical elderly person, which we constructed as $G_h = G_2 - G_0$ in §3.8.2. Both G_h and the bearing $\mathbb{Z}_d(B)$ are simple linear combinations of the original elements G_0, G_2, so it is trivial to form their prior covariance matrix: it is

$$\begin{bmatrix} \mathrm{Var}(G_h) & \mathrm{Cov}(G_h, \mathbb{Z}_d(B)) \\ \mathrm{Cov}(G_h, \mathbb{Z}_d(B)) & \mathrm{Var}(\mathbb{Z}_d(B)) \end{bmatrix} = \begin{bmatrix} 2.1100 & 0.1155 \\ 0.1155 & 0.3179 \end{bmatrix}. \tag{4.76}$$

Note that this verifies that $\text{Var}(\mathbb{Z}_d(B)) = 0.3179 = \text{Size}_d(B)$. We deduce immediately from (4.52) that

$$\text{E}_d(G_h) - \text{E}(G_h) = \text{Cov}(G_h, \mathbb{Z}_d(B)) = 0.1155,$$

which is the change in adjusted expectation that we saw at the foot of §4.5.2.1, and which we also calculated via the discrepancy vector (4.47) in §4.5.2.2. We also have that the prior correlation between G_h and $\mathbb{Z}_d(B)$ is $\text{Corr}(G_h, \mathbb{Z}_d(B)) = 0.141$, so that by (4.54) the size for G_h must be

$$\begin{aligned}\text{Size}_d(G_h) &= [\text{Corr}(G_h, \mathbb{Z}_d(B))]^2 \text{Size}_d(B)\\ &= 0.141^2 \times 0.3179 = 0.0063.\end{aligned}$$

We can, just as easily, deduce the size for any other linear combination of interest, through its resolved correlation with $\mathbb{Z}_d(B)$.

4.12.2.1 Size diagnostics

For this example, we might be a little surprised that the size for the adjustment of G_h is so small. It may be that, on reflection, the doctor feels that she has specified rather too large a prior variance for G_h, and has thus understated the value of her knowledge. In general, we would expect to find a size not too far from its prior expectation. As such, we examine the **size ratio for the adjustment**, defined in (4.63), which has expectation one.

For the adjustment of the collection $B = \{G_0, G_2\}$ by the collection $D = \{D_0, D_2\}$ we have (4.73) that the size is $\text{Size}_d(B) = 0.3179$. The **expected size of the adjustment** is given (4.62) by the sum of the canonical resolutions for the adjustment (3.60). We obtained these canonical resolutions in (3.110) as $\lambda_1 = 0.3184$ and $\lambda_2 = 0.0202$. Thus we have respectively a prior expectation for the size and a corresponding size ratio of

$$\text{E}(\text{Size}_D(B)) = 0.3184 + 0.0202 = 0.3386, \tag{4.77}$$

$$\text{Sr}_d(B) = \frac{0.3179}{0.3386} = 0.9389. \tag{4.78}$$

The value of 0.9389 for this size ratio is quite close to its expectation of unity, so we have no evidence to warn us of a conflict between data and prior specification. If we wish, we can construct a simple rule of thumb to guide us as to when size ratios are large. For example, the upper threshold given by (4.65) is

$$1 + \frac{6\sqrt{\sum_{i=1}^{2} \lambda_i^2}}{\sum_{i=1}^{2} \lambda_i} = 6.65,$$

so that a size ratio of 0.9389 is well within what we might regard as likely chance variation.

4.12.2.2 Canonical adjustment and discrepancy

It can be helpful to explore an adjustment through its canonical structure. For this example, the adjustment of B by D has two canonical directions, W_1 and W_2, which we stated in §3.11.3. These have prior expectation zero. It is simple to determine the adjusted expectation of each canonical direction as $\mathrm{E}_d(W_i)$, for example from (3.108) we have that

$$W_1 = 1.0059G_0 - 0.1136G_2 - 3.4745,$$

where the adjusted expectations for G_0, G_2 are calculated in §4.5.2.1, so that

$$\begin{aligned}\mathrm{E}_d(W_1) &= 1.0059\mathrm{E}_d(G_0) - 0.1136\mathrm{E}_d(G_2) - 3.4745\\ &= 1.0059 \times 4.7085 - 0.1136 \times 6.9140 - 3.4745\\ &= 0.4763.\end{aligned}$$

Next, each canonical direction W_i has, by (4.38) and (4.55), discrepancy $\mathrm{Dis}_d(W_i) = \frac{\mathrm{E}_d(W_i)^2}{\lambda_i}$ and size $\mathrm{Size}_d(W_i) = \mathrm{E}_d(W_i)^2$, representing diagnostics relative to the resolved variation and the prior variation, respectively. Finally, the individual canonical direction discrepancies and sizes add to give the discrepancy and size for the collection B being adjusted. The calculations are summarized in Table 4.1. The key feature is that the change in expectation for direction W_2 is quite highly discrepant (a large change in expectation compared to the variation resolved), and it is discrepancy in this direction which contributes most to the discrepancy in the collection. On the other hand, the size for W_2 is only 0.0911, so

Table 4.1 Summary of the canonical analysis.

Quantity	Resolution $\mathrm{R}_D(\cdot)$	Adj.Expect. $\mathrm{E}_d(\cdot)$	Discrepancy $\mathrm{Dis}_d(\cdot)$	Size $\mathrm{Size}_d(\cdot)$
W_i	λ_i	$\mathrm{E}_d(W_i)$	$\frac{\mathrm{E}_d(W_i)^2}{\lambda_i}$	$\mathrm{E}_d(W_i)^2$
B	$\sum \lambda_i$	$\sum \mathrm{E}_d(W_i)$	$\sum \frac{\mathrm{E}_d(W_i)^2}{\lambda_i}$	$\sum \mathrm{E}_d(W_i)^2$
Quantity	Resolution $\mathrm{R}_D(\cdot)$	Adj.Expect. $\mathrm{E}_d(\cdot)$	Discrepancy $\mathrm{Dis}_d(\cdot)$	Size $\mathrm{Size}_d(\cdot)$
W_1	0.3184	0.4763	0.7124	0.2268
W_2	0.0202	−0.3018	4.5129	0.0911
B	0.3386		5.2253	0.3179

that this discrepancy has limited impact. With regard to the row for the collection B in Table 4.1, the canonical resolutions sum to give the resolved uncertainty for the collection (3.74), which is also the prior expected value for the size of the adjustment (4.62).

5

Partial Bayes linear analysis

We have described a three-stage progression for analysing our beliefs. First, we interpret the expected adjustments, a priori. Secondly, given observations, we interpret the actual adjustments. Thirdly, we make diagnostic comparisons between actual and expected beliefs. Often, we want to explore the ways in which different aspects of the data and the prior specification combine to give the final adjustment. For example, we might be combining information of different types collected in different places by different people at different times. We therefore need to identify which aspects of the data are, a priori, most crucial to the final adjustment, in order to guide our choice of information collection. We also need ways to interpret and compare diagnostically the effects of the various portions of the observed data on our beliefs. Therefore, we now develop expressions for the partial effects of subsets of the data, as applied to the various interpretative and diagnostic measures that we have introduced.

5.1 Partial adjustment

In order to separate out the effects on our beliefs of different sub-collections of data, we evaluate partial adjustments, representing the change in adjustment resulting as we accumulate data. So, suppose that we intend to adjust our beliefs about a collection $B = \{B_1, \ldots, B_r\}$ by observation of two further collections $D = \{D_1, \ldots, D_j\}$ and $F = \{F_1, \ldots, F_k\}$ of quantities. We adjust B by the collection $(D \cup F) = \{D_1, \ldots, D_j, F_1, \ldots, F_k\}$ but separate the effects of the subsets of data. Therefore, we adjust B in stages, first by D, then adding F.

The simplest case is where the vectors D, F are uncorrelated. In this case, it is easy to check that the adjusted expectations are additive, namely that

$$D \perp F \;\Rightarrow\; \mathrm{E}_{D\cup F}(B - \mathrm{E}(B)) = \mathrm{E}_D(B - \mathrm{E}(B)) + \mathrm{E}_F(B - \mathrm{E}(B)). \tag{5.1}$$

Bayes Linear Statistics: Theory and Methods M. Goldstein and D.A. Wooff

When D, F are correlated vectors, then we obtain a modified additivity, by removing the 'common variability' between F and D, as follows. For any vectors D, F, the vectors $D, \mathbb{A}_D(F)$ are uncorrelated, and the collections of linear combinations $\langle D \cup F\rangle$, $\langle D \cup \mathbb{A}_D(F)\rangle$ are the same. From (5.1) we therefore have, for any D, F, that

$$\mathrm{E}_{D\cup F}(B - \mathrm{E}(B)) = \mathrm{E}_D(B - \mathrm{E}(B)) + \mathrm{E}_{\mathbb{A}_D(F)}(B - \mathrm{E}(B)). \tag{5.2}$$

We may assess the extra effect of adjusting B by F given that we have already adjusted by D, defined as follows.

Definition 5.1 *The* ***partial adjustment of*** B ***by*** F ***given*** D*, denoted by* $\mathrm{E}_{[F/D]}(B)$*, is*

$$\mathrm{E}_{[F/D]}(B) = \mathrm{E}_{D\cup F}(B) - \mathrm{E}_D(B). \tag{5.3}$$

Observe, from (5.2), that

$$\mathrm{E}_{[F/D]}(B) = \mathrm{E}_{\mathbb{A}_D(F)}(B - \mathrm{E}(B)).$$

Geometrically, the linearity relation

$$\mathrm{E}_{D\cup F}(B) = \mathrm{E}_D(B) + \mathrm{E}_{[F/D]}(B) \tag{5.4}$$

follows as the orthogonal projection of B into $[D \cup F]$ is equivalent to the orthogonal projection of B into $[D]$ plus the orthogonal projection of B into the space spanned by the orthogonal complement of D in F, namely $[F/D]$. Therefore, the additional adjustment of B by F, given that we have already adjusted by D, is the same as the adjustment of B by $[F/D]$. Equation (5.4) gives the sequential construction for the overall adjusted expectation, $\mathrm{E}_{D\cup F}(B)$. The corresponding adjusted quantity $\mathbb{A}_{D\cup F}(B)$ can be sequentially constructed as

$$\begin{aligned}\mathbb{A}_{D\cup F}(B) &= B - \mathrm{E}_{D\cup F}(B) = B - \mathrm{E}_D(B) - \mathrm{E}_{[F/D]}(B)\\ &= (B - \mathrm{E}_D(B)) - \mathrm{E}_{[F/D]}(B - \mathrm{E}_D(B))\\ &= \mathbb{A}_{\mathbb{A}_D(F)}(\mathbb{A}_D(B)),\end{aligned} \tag{5.5}$$

as $\mathrm{E}_{[F/D]}(\mathrm{E}_D(B)) = \mathrm{E}(\mathbb{A}_D(B)) = 0$. Therefore, to adjust B by D and F, we may first adjust both B and F by D and then adjust the adjusted form $\mathbb{A}_D(B)$ by $\mathbb{A}_D(F)$. Notice, in particular, that the 'residuals' from sequential adjustments are uncorrelated, i.e.

$$\mathrm{Cov}(\mathbb{A}_{D\cup F}(B), \mathbb{A}_D(F)) = 0. \tag{5.6}$$

This follows as $\mathbb{A}_{D\cup F}(B)$ is uncorrelated with $D \cup F$, and $\mathbb{A}_D(F) \in \langle D \cup F\rangle$.

We may represent this sequential adjustment alternatively in terms of the adjusted belief structures, so that we have

$$[B/(D \cup F)] = [[B/D]/[F/D]]. \tag{5.7}$$

Relations (5.4) and (5.7) summarize the operations that we perform to evaluate adjusted expectations and variances in stages, namely that we may adjust all beliefs by D and then further adjust all adjusted beliefs by F.

These relations correspond to the analogous properties for probabilistic conditioning, whereby we progressively condition all probabilities on information obtained from a sequence of observations. For example, for any three events A, B, C we have

$$\mathrm{P}(A|B \cap C) = \frac{\mathrm{P}(A \cap B|C)}{\mathrm{P}(B|C)}.$$

Thus, we may input pieces of information sequentially in a probabilistic analysis, updating our beliefs at each stage by simple conditioning on our current distribution. We do not need to refer explicitly to previous evidence, as its evidential content is covered by the conditioning. In the same way, we can extract a belief structure from an adjustment, simply by adjusting each of the remaining structures by this structure.

5.2 Partial variance

In §3.3, we described how the adjustment of collection B by H separated B into two uncorrelated components,

$$B = \mathrm{E}_H(B) + (B - \mathrm{E}_H(B)).$$

Dividing H as $H = D \cup F$, we further decompose $B - \mathrm{E}_H(B)$, and write

$$\begin{aligned} B &= \mathrm{E}_D(B) + [\mathrm{E}_{D\cup F}(B) - \mathrm{E}_D(B)] + [B - \mathrm{E}_{D\cup F}(B)] \\ &= \mathrm{E}_D(B) + \mathrm{E}_{[F/D]}(B) + \mathbb{A}_{D\cup F}(B). \end{aligned} \tag{5.8}$$

The vectors $\mathrm{E}_{[F/D]}(B)$, $\mathrm{E}_D(B)$, and $\mathbb{A}_{D\cup F}(B)$ are mutually uncorrelated. We may therefore partition $\mathrm{Var}_D(B)$, the 'unresolved variation' from the adjustment by D, as

$$\mathrm{Var}_D(B) = \mathrm{Var}(\mathrm{E}_{[F/D]}(B)) + \mathrm{Var}_{D\cup F}(B). \tag{5.9}$$

The second term is the adjusted variance matrix of B given $D \cup F$, and the first is the **(partial) resolved variance matrix of B by F given D,** namely

$$\mathrm{RVar}_{[F/D]}(B) = \mathrm{Var}(\mathrm{E}_{[F/D]}(B)).$$

Resolved variances are additive in the sense that

$$\mathrm{RVar}_{D\cup F}(B) = \mathrm{RVar}_D(B) + \mathrm{RVar}_{[F/D]}(B). \tag{5.10}$$

We thus have the following partitions of variation:

$$\begin{aligned} \mathrm{Var}(B) &= \mathrm{RVar}_D(B) + \rule[0.5ex]{5em}{0.4pt}\ \mathrm{Var}_D(B)\ \rule[0.5ex]{5em}{0.4pt}, \\ &= \mathrm{RVar}_D(B) + \mathrm{RVar}_{[F/D]}(B) + \mathrm{Var}_{D\cup F}(B), \\ &= \rule[0.5ex]{4em}{0.4pt}\ \mathrm{RVar}_{D\cup F}(B)\ \rule[0.5ex]{4em}{0.4pt} + \mathrm{Var}_{D\cup F}(B). \end{aligned} \tag{5.11}$$

The different ways in which we may interpret $\text{RVar}_{[F/D]}(B)$ correspond to the different variance partitions above. In particular, we may compare the value of $\text{RVar}_{[F/D]}(B)$ to the original variance of B, which we will term a **partial resolution**, or to the adjusted variance of B by D, which we will term a **relative resolution**. We now describe the different uses that we make of each comparison.

5.3 Partial resolution transforms

Definition 5.2 *For any $X \in \langle B \rangle$, we assess the further reduction in 'residual variation' from adding F, given D, as the* ***(partial) resolution****, namely*

$$\text{R}_{[F/D]}(X) = \frac{\text{RVar}_{[F/D]}(X)}{\text{Var}(X)}. \tag{5.12}$$

We analyse partial resolutions using the partial resolution transform which summarizes the effects of partial adjustments, similarly to the way in which resolution transforms summarize simple adjustments.

Definition 5.3 *The jth* ***partial canonical direction for the adjustment of*** *B* ***by*** *F* ***given*** *D is the linear combination W_j which maximizes $\text{R}_{[F/D]}(B)$ over all elements in $\langle B \rangle$ with non-zero prior variance which are uncorrelated with each W_i, $i < j$, scaled so that each $\text{Var}(W_j) = 1$. The values*

$$\zeta_i = \text{R}_{[F/D]}(W_i), \quad i = 1, \ldots, r_{\mathbb{P}}, \tag{5.13}$$

are termed the ***partial canonical resolutions****.*

The partial canonical directions for F given D are evaluated exactly as are the canonical directions for D, as described in §3.9.1, but the eigenstructure is extracted from the partial resolution transform matrix which is given as follows, and which has rank $r_{\mathbb{P}}$.

Definition 5.4 *The* ***partial resolution transform matrix*** *is*

$$\mathbb{T}_{B:[F/D]} = [\text{Var}(B)]^{\dagger}\text{RVar}_{[F/D]}(B), \tag{5.14}$$

and the ***partial adjusted belief transform matrix*** *is*

$$\mathbb{S}_{B:[F/D]} = \mathbb{I} - \mathbb{T}_{B:[F/D]}.$$

Therefore, from (5.10), we have

$$\mathbb{T}_{B:D\cup F} = \mathbb{T}_{B:D} + \mathbb{T}_{B:[F/D]}, \tag{5.15}$$

and, analogously to (3.68), we have

$$\text{RVar}_{[F/D]}(Y) = \text{Cov}(Y, \mathbb{T}_{B:[F/D]}(Y)), \tag{5.16}$$

$$\text{and} \quad \text{Var}_{[F/D]}(Y) = \text{Cov}(Y, \mathbb{S}_{B:[F/D]}(Y)). \tag{5.17}$$

The collection $(W_1, \ldots, W_{r_{\mathbb{P}}})$ forms a grid of directions over $\langle B \rangle$, summarizing the additional effects of the adjustment. Having adjusted by D, we expect to learn most additionally from F for those linear combinations of the elements of B which have large correlations with those partial canonical directions with large resolutions. The exact relation is as before, namely for any $X \in \langle B \rangle$,

$$\mathrm{R}_{[F/D]}(X) = \sum_{i=1}^{r_{\mathbb{P}}} \mathrm{c}_i(X)\zeta_i, \tag{5.18}$$

where

$$\mathrm{c}_i(X) = \mathrm{Corr}(X, W_i)^2. \tag{5.19}$$

Definition 5.5 *The* ***system partial resolution*** *is*

$$\mathrm{R}_{[F/D]}(B) = \frac{1}{r_{\mathbb{P}}} \sum_{i=1}^{r_{\mathbb{P}}} \zeta_i.$$

This compares directly to the resolution given D alone, (3.75). System resolutions are additive in the sense that

$$\mathrm{R}_D(B) + \mathrm{R}_{[F/D]}(B) = \mathrm{R}_{D \cup F}(B). \tag{5.20}$$

5.4 Relative belief adjustment

An alternative representation of the additional adjustment of B by F given D follows by assessing the adjustment ratio for F directly over the adjusted collection $\mathbb{A}_D(B)$.

Definition 5.6 *The* ***relative adjustment ratio*** *for B by F given prior adjustment by D is*

$$\mathrm{RA}_F(B/D) = \frac{\mathrm{Var}_{D \cup F}(B)}{\mathrm{Var}_D(B)}. \tag{5.21}$$

The jth ***relative canonical direction for the adjustment of B by F given D*** *is the linear combination U_j which minimizes $\mathrm{RA}_F(U/D)$ over all elements of $\langle B \rangle$ with $\mathrm{Var}_D(U) > 0$ for which $\mathrm{Cov}_D(U, U_i) = 0$, $i = 1, \ldots, j-1$. We scale each U_j so that $\mathrm{Var}_D(U_j) = 1$. The values*

$$\upsilon_j = \mathrm{RA}_F(U_j/D)$$

are termed the ***relative canonical adjustment ratios****.*

It turns out that the number of non-zero relative canonical adjustment ratios is $r_{\mathbb{P}}$. The collection $(U_1, \ldots, U_{r_{\mathbb{P}}})$ forms a grid of directions over the collection B, summarizing the additional effects of the adjustment as follows. For any X in $\langle B \rangle$,

$$\mathrm{RA}_F(X/D) = \sum_{i=1}^{r_{\mathbb{P}}} rc_i(X)\upsilon_i, \tag{5.22}$$

where

$$rc_i(X) = \mathrm{Corr}_D(X, U_i)^2, \tag{5.23}$$

and where $\mathrm{Corr}_D(X, Y)$ is the correlation between X and Y in the adjusted variance matrix $\mathrm{Var}_D(B)$. The relative canonical directions are the eigenvectors of the corresponding belief transform. We have defined $\mathbb{S}_{B:D}$ as

$$\mathbb{S}_{B:D} = \mathrm{Var}(B)^{\dagger}\mathrm{Var}_D(B),$$

and we similarly define the relative version of this transform as follows.

Definition 5.7 *The* ***relative adjusted belief transform*** *for* $[B/D]$ *given* F *is*

$$\mathbb{S}_{B:F(D)} = \mathrm{Var}_D(B)^{\dagger}\mathrm{Var}_{D\cup F}(B). \tag{5.24}$$

Just as $\mathrm{Var}_D(Y) = \mathrm{Cov}(Y, \mathbb{S}_{B:D}Y)$, we have

$$\mathrm{Var}_{D\cup F}(Y) = \mathrm{Cov}_D(Y, \mathbb{S}_{B:F(D)}Y). \tag{5.25}$$

Definition 5.8 *The* ***relative resolution transform*** *of* B *by* F *given* D *is*

$$\mathbb{T}_{B:F(D)} = I - \mathbb{S}_{B:F(D)}. \tag{5.26}$$

Relative transforms are multiplicative in the following sense:

$$\mathbb{S}_{B:D\cup F} = \mathbb{S}_{B:D}\mathbb{S}_{B:F(D)}. \tag{5.27}$$

We assess the relative transform $\mathbb{S}_{B:F(D)}$ for analyses in which we suppose that we have already adjusted all beliefs according to D, whereas we assess the partial transform $\mathbb{S}_{B:[F/D]}$ to assess the additional effects of F on B given D. We discuss how to calculate the relative resolution transform in §12.11.1.

5.5 Example: oral glucose tolerance test

For this example so far, we have adjusted one belief structure, $[B]$, representing measurements for a typically elderly person, by another belief structure, $[D]$, representing measurements our doctor makes upon herself. Have we exhausted our exploratory possibilities, or are there extra insights to be had by approaching the problem in a different way? Suppose that we consider the analogy of a traditional multiple regression, where we aim to predict a response variable Y from a collection of regressors $X_1, \ldots, X_k$. In terms of this analogy, we may be interested not only in the predictive power of the collection taken as a whole but also in whether **every** X_i is useful for the prediction; whether certain **subsets** of the X_is are more useful than others; and so forth. Some of our analyses and diagnostics so far for this example have highlighted some unusual or anomalous features. For example, we saw:

- a somewhat surprising observation d_2 (4.17), representing the 2-hour post-glucose measurement that the doctor makes upon herself;
- a rather large change in expectation for G_2 (4.42), representing the 2-hour post-glucose measurement for a typical elderly person;
- that we expect to learn very little about G_2, as we found that its resolution is only 4.48%, shown in Table 3.1.

The evidence seems to point to a surprisingly large value of d_2 being at issue. Suppose, then, that we consider D_0 and D_2 as being two distinct sources of information, separated in time, and suppose that we adjust $[B]$ first by D_0, and then by D_2. In what follows we use the notation $[D]$ and $[D_0 \cup D_2]$ synonymously.

5.5.1 Performing an initial adjustment

We begin by adjusting the collection $B = [G_0, G_2]$ by D_0 alone. For each quantity G_0, G_2, we calculate the adjusted variances and the resolutions. The adjusted variance matrix is

$$\mathrm{Var}_{D_0}(B) = \begin{bmatrix} 0.7768 & 0.5539 \\ 0.5539 & 2.3496 \end{bmatrix}. \tag{5.28}$$

The resolutions for G_0, G_2 and the system resolution are:

$$\mathrm{R}_{D_0}(G_0) = 0.3064, \quad \mathrm{R}_{D_0}(G_2) = 0.0331, \quad \mathrm{R}_{D_0}(B) = 0.1554.$$

The resolutions are only slightly smaller than those (0.3109, 0.0448, 0.1694, respectively) shown in Table 3.1 and (3.111) for the full adjustment. The single (standardized) canonical direction for the initial adjustment is

$$W_1 = 1.0507\,\mathrm{S}(G_0) - 0.1324\,\mathrm{S}(G_2), \tag{5.29}$$

with resolution

$$\mathrm{R}_{D_0}(W_1) = 0.3109.$$

The coefficients for the canonical quantity (5.29) suggest that the single piece of data D_0 will be rather more informative for the fasting measurement G_0 than for the 2-hour measurement G_2.

In summary, the adjustment of $[B]$ by D_0 alone is about in line with what was expected. Furthermore, the summaries for this adjustment were quite close to the corresponding results for the full adjustment. The implication is that D_2 is not of much value in helping us to learn about G_0 and G_2. To explore this further, we now consider the **partial** effects of adjusting by the 2-hour measurement, D_2, in addition to D_0.

5.5.2 Partial resolved variances

Every additional adjustment has the potential to reduce further the uncertainty in our quantities of interest. The extra variance reductions due to these partial adjustments are called partial resolved variances; and the extra portions of variation that are resolved relative to prior variation are called partial resolutions. When we adjust by D_2 in addition to D_0, the extra reductions in variance for G_0 and G_2, together with the partial resolutions, are as follows:

$$\mathrm{RVar}_{[D_2/D_0]}(G_0) = 0.0049, \qquad \mathrm{R}_{[D_2/D_0]}(G_0) = 0.0044, \tag{5.30}$$

$$\mathrm{RVar}_{[D_2/D_0]}(G_2) = 0.0286, \qquad \mathrm{R}_{[D_2/D_0]}(G_2) = 0.0118. \tag{5.31}$$

Thus, the partial effect of adjusting by D_2 additionally is negligible: relative to the initial uncertainty in G_2, we achieve a further reduction in uncertainty of only some 1.18%, and the relative reduction in variance for G_0 is smaller still. The decomposition of the prior variance matrix for the initial and partial adjustments (5.11) is

$$\begin{bmatrix} 1.12 & 0.72 \\ 0.72 & 2.43 \end{bmatrix} = \begin{bmatrix} 0.3432 & 0.1661 \\ 0.1661 & 0.0804 \end{bmatrix} + \begin{bmatrix} 0.0049 & -0.0119 \\ -0.0119 & 0.0286 \end{bmatrix} + \begin{bmatrix} 0.7718 & 0.5658 \\ 0.5658 & 2.3211 \end{bmatrix}. \tag{5.32}$$

5.5.3 Partial canonical directions

The solitary partial canonical direction, shown in non-standardized and standardized forms, is

$$\begin{aligned} W_{2/1} &= 0.7971G_0 - 0.6539G_2 + 0.7710 \\ &= 0.8435\mathrm{S}(G_0) - 1.0193\mathrm{S}(G_2), \end{aligned}$$

corresponding to a canonical partial resolution of

$$\mathrm{R}_{[D_2/D_0]}(W_{2/1}) = 0.0277.$$

This shows that one effect of the partial adjustment will be to reduce uncertainty in the overall belief structure $[B]$ by only $0.0277/2 = 1.39\%$, and this is the maximum partial resolution for any linear combination in $\langle B \rangle$. Note that we divide by two as this is the dimension of B. Observe from the standardized form that the extra piece of information D_2 is expected to be slightly more informative for G_2 than for G_0. The decomposition of the resolution for the initial and partial adjustments (5.20) is

$$\begin{array}{ccccc} \mathrm{R}_{D_0 \cup D_2}(B) & = & \mathrm{R}_{D_0}(B) & + & \mathrm{R}_{[D_2/D_0]}(B) \\ 0.1693 & = & 0.1554 & + & 0.0139, \end{array} \tag{5.33}$$

emphasizing that the additional partial adjustment is not expected to contribute much extra in the way of explaining variation in $\langle B \rangle$.

5.5.4 Deducing changes for other linear combinations

For the difference between a typical elderly person's fasting and 2-hour measurements, $G_h = G_2 - G_0$, we can easily deduce the implications of the initial and partial adjustments. The correlation between G_h and the initial canonical direction for the adjustment of B by D_0 alone is $\text{Corr}(G_h, W_1) = -0.3734$, whereas the correlation between G_h and the partial canonical direction for the adjustment of $[B/D_0]$ by $[D_2/D_0]$ is $\text{Corr}(G_h, W_{2/1}) = -0.9892$. Consequently, the resolutions in G_h for the initial and partial adjustments must be, by (3.71) and (5.18),

$$\text{R}_{D_0}(G_h) = 0.3734^2 \times 0.3109 = 0.0433,$$

$$\text{R}_{[D_2/D_0]}(G_h) = 0.9892^2 \times 0.0277 = 0.0272.$$

The combined resolution is thus, by (5.20),

$$\begin{array}{ccccc} \text{R}_{D_0 \cup D_2}(G_h) & = & \text{R}_{D_0}(G_h) & + & \text{R}_{[D_2/D_0]}(G_h) \\ 0.0705 & = & 0.0433 & + & 0.0272, \end{array}$$

which we calculated and showed earlier (3.114). These low resolutions show that neither measurement is expected to be of significant value for learning about G_h. Notice that the notation used here for decomposing the resolution for a single quantity is identical to that for the collection, as seen in (5.33).

5.5.5 Relative belief adjustment

We have seen that D_2 is uninformative for $[B]$ in absolute terms. However, it might still be relatively important in resolving the variation that remains having observed D_0. Where we have a partial adjustment following an initial adjustment, we can summarize the implications of the adjustment either with respect to the prior specifications (which we have concentrated on so far), or with respect to the specifications remaining after being modified for the initial adjustment. Thus, we can evaluate the partial reductions in uncertainty relative to the prior uncertainty (as considered above) or to the current adjustment variances. As an example, suppose we choose to adjust $[B]$ first by $[D_0]$ and then **relatively** by $[D_2]$. For the initial adjustment, we must adjust not only B but also D_2 by D_0. Doing so, we obtain the adjusted variance matrix

$$\text{Var}\left(\begin{bmatrix} \mathbb{A}_{D_0}(D_2) \\ \mathbb{A}_{D_0}(G_0) \\ \mathbb{A}_{D_0}(G_2) \end{bmatrix}\right) = \text{Var}_{D_0}\left(\begin{bmatrix} D_2 \\ G_0 \\ G_2 \end{bmatrix}\right) = \begin{bmatrix} 1.9671 & -0.0986 & 0.2371 \\ -0.0986 & 0.7768 & 0.5539 \\ 0.2371 & 0.5539 & 2.3496 \end{bmatrix}, \tag{5.34}$$

where $\mathbb{A}_{D_0}(X)$ is our notation for the adjusted (or residual) vector $X - \text{E}_{D_0}(X)$ when X has been adjusted by D_0. We saw part of this matrix before as (5.28), the adjusted variance matrix for G_0 and G_2 given D_0, but have been careful here to adjust D_2 by D_0 also as a necessary precursor to the relative adjustment. Equivalently, (5.34) is the variance matrix for the adjusted versions of D_2, G_0, G_2

given D_0. For the subsequent adjustment by D_2, we now treat these as the prior specifications and discard D_0 entirely. If we now adjust (the adjusted versions of) G_0, G_2 by (the adjusted version of) D_2, we find that the adjusted variance matrix is

$$\mathrm{Var}_{D_2}(\mathbb{A}_{D_0}(B)) = \begin{bmatrix} 0.7718 & 0.5658 \\ 0.5658 & 2.3211 \end{bmatrix} = \mathrm{Var}_{D_0 \cup D_2}(B), \tag{5.35}$$

which is the same as for the overall adjustment of $[B]$ by $[D]$, shown in (5.32). This illustrates that the overall adjusted variances for one collection B given another collection D are the same, whether we adjust partially (as in preceding sections) or relatively (as in this section).

However, the resolutions that we calculate depend on which specifications we prefer to regard as prior. For this, relative, adjustment we thus obtain a different set of resolutions,

$$\mathrm{R}_{\mathbb{A}_{D_0}(D_2)}(\mathbb{A}_{D_0}(G_0)) = 0.0064,$$

$$\mathrm{R}_{\mathbb{A}_{D_0}(D_2)}(\mathbb{A}_{D_0}(G_2)) = 0.0122,$$

$$\mathrm{R}_{\mathbb{A}_{D_0}(D_2)}(\mathbb{A}_{D_0}(B)) = 0.0155.$$

which are, for individual quantities such as G_0, not smaller than the corresponding partial resolutions. For example, G_0 has partial resolution 0.44% relative to the prior specification, as we saw in (5.30), but partial resolution 0.64% relative to the initial adjustment by D_0. We conclude in this example that D_2 appears uninformative for $[B]$ whether we relate the changes in variance to the initial specification or to the specification following the initial adjustment by D_0.

5.5.6 Withdrawing quantities from the adjustment

In the same way that we can introduce additional quantities into the adjustment, so too can we determine the effects of withdrawing quantities from the adjustment. We might do this for various reasons. For example, we might remove uninformative quantities, or quantities that are relatively unimportant and expensive to observe. Here, we are particularly interested in investigating the rather peculiar nature of the specifications over D_2.

When we remove D_0 from the adjustment at this stage, it as though we are left with a simple adjustment of $[B]$ by D_2. In addition, we learn about the partial adjustment of $[B]$ by $[D_0/D_2]$. For example, the resolutions for the adjustment by D_2 and for the partial adjustment removing D_0 are

$$\mathrm{R}_{D_2}(B) = 0.0448,$$

$$\mathrm{R}_{[D_0/D_2]}(B) = 0.2938.$$

This shows, as we suspected, that D_2 alone is not a good source of information (its effect is at best to reduce uncertainty by less than 5%) and that most of the information is contained wholly in D_0 (when we remove $[D_0/D_2]$ we also remove nearly all of our capability to reduce uncertainty in $[B]$).

5.6 Partial bearings

When we adjust B by the observed value $D = d$, there are a variety of interpretative and diagnostic measures that we may evaluate, as described in Chapter 4, in order to understand the ways in which our expectations have changed and to identify inconsistencies between beliefs and observations. When we make the further adjustment by $F = f$, then we may evaluate these measures for the overall adjustment by $d \cup f$. In addition, we may obtain similar qualitative insights into the changes in adjustment that follow when we add f to d, by evaluating each of the corresponding measures for the partial and relative adjustments by f given d. For example, corresponding to the canonical standardized adjustments we may evaluate the **partial canonical standardized adjustments** or the **relative canonical standardized adjustments** which are as defined by (4.38), but applied to the partial and relative adjustment by $\mathbb{A}_D(F)$. As a general principle, if we want to concentrate solely on the diagnostic implications of adding f to d, then we evaluate relative diagnostics. Partial diagnostics are appropriate if we wish to form an overall picture of diagnostic issues over the whole adjustment, using measures which separate the diagnostic effects according to the different stages of the adjustment.

In particular, it is often revealing to perform a diagnostic analysis based on the **partial bearings** for the partial adjustment, which are constructed as follows. We observe the values of $D = d$ and $F = f$. We therefore may evaluate the observed value of $\mathbb{A}_D(F)$, denoted by $\mathbb{A}_d(f)$, which we assess as

$$\mathbb{A}_d(f) = f - \mathrm{E}_d(F).$$

Definition 5.9 *The* ***size of the partial adjustment****, or* ***partial size****, is defined to be*

$$\mathrm{Size}_{[f/d]}(B) = \max_{X \in \langle B \rangle} \frac{[\mathrm{E}_{d \cup f}(X) - \mathrm{E}_d(X)]^2}{\mathrm{Var}(X)} = \max_{X \in \langle B \rangle} \frac{[\mathrm{E}_{[f/d]}(X)]^2}{\mathrm{Var}(X)}. \tag{5.36}$$

Similarly to (4.51), the value of this maximum is

$$\mathrm{Size}_{[f/d]}(B) = [\mathrm{E}_{d \cup f}(B) - \mathrm{E}_d(B)]^T \mathrm{Var}(B)^{\dagger} [\mathrm{E}_{d \cup f}(B) - \mathrm{E}_d(B)]. \tag{5.37}$$

Definition 5.10 *The random quantity which achieves this maximum is the* ***bearing for the partial adjustment****, or* ***partial bearing****, which we may construct as*

$$\mathbb{Z}_{[f/d]}(B) = \sum_{i=1}^{r_{\mathbb{P}}} \mathrm{E}_{[f/d]}(U_i) U_i, \tag{5.38}$$

for any collection $(U_1, \ldots, U_{r_{\mathbb{P}}})$ *mutually uncorrelated with unit prior variance, where* $r_{\mathbb{P}}$ *is the rank of the partial resolution transform matrix.*

From (5.38), we have that partial bearings are additive, in the sense that

$$\mathbb{Z}_{d \cup f}(B) = \mathbb{Z}_d(B) + \mathbb{Z}_{[f/d]}(B). \tag{5.39}$$

The partial bearing, $\mathbb{Z}_{[f/d]}(B)$, expresses all changes in expectation over $\langle B \rangle$ when we additionally adjust B by F given a preceding adjustment by D, through the relation

$$\mathrm{E}_{d \cup f}(X) - \mathrm{E}_d(X) = \mathrm{E}_{[f/d]}(X) = \mathrm{Cov}(X, \mathbb{Z}_{[f/d]}(B)), \quad \forall X \in \langle B \rangle. \tag{5.40}$$

We can choose to take as our basis the partial canonical directions: $U_i = W_i$, $i = 1, \ldots, r_{\mathbb{P}}$. Then, from (5.36) and (5.40), we can represent the partial size as

$$\mathrm{Size}_{[f/d]}(B) = \mathrm{Var}(\mathbb{Z}_{[f/d]}(B)) = \sum_{i=1}^{r_{\mathbb{P}}} [\mathrm{E}_{[f/d]}(W_i)]^2. \tag{5.41}$$

The corresponding random quantity is $\mathrm{Size}_{[F/D]}(B)$, with representation

$$\mathrm{Size}_{[F/D]}(B) = \mathrm{Var}(\mathbb{Z}_{[F/D]}(B)) = \sum_{i=1}^{r_{\mathbb{P}}} [\mathrm{E}_{[F/D]}(W_i)]^2,$$

which may be compared to the expected value, namely the sum of the partial canonical resolutions:

$$\mathrm{E}(\mathrm{Size}_{[F/D]}(B)) = \mathrm{RU}_{[F/D]}(B) = \sum_{i=1}^{r_{\mathbb{P}}} \zeta_i. \tag{5.42}$$

We may use the observed partial size and its prior expectation to give a partial size diagnostic corresponding to the full size ratio established in Definition 4.15.

Definition 5.11 *The **partial size ratio** is*

$$\mathrm{Sr}_{[f/d]}(B) = \frac{[\mathrm{E}_{d \cup f}(B) - \mathrm{E}_d(B)]^T \mathrm{Var}(B)^{\dagger} [\mathrm{E}_{d \cup f}(B) - \mathrm{E}_d(B)]}{\sum_{i=1}^{r_{\mathbb{P}}} \zeta_i}. \tag{5.43}$$

We may apply each of the diagnostic measures outlined for a simple adjustment directly to partial adjustments. Thus, we might evaluate simple heuristics corresponding to those suggested by (4.65), which have similar uses for examining the size ratio for the partial adjustment. For example, if all these canonical directions were to be normally distributed, then it would follow that

$$\mathrm{Var}(\mathrm{Size}_{[F/D]}(B)) = 2 \sum_{i=1}^{r_{\mathbb{P}}} \zeta_i^2, \tag{5.44}$$

suggesting the corresponding range

$$P\left(1 - \frac{6\sqrt{\sum_{i=1}^{r_{\mathbb{P}}} \zeta_i^2}}{\sum_{i=1}^{r_{\mathbb{P}}} \zeta_i} \leq \mathrm{Sr}_{[f/d]}(B) \leq 1 + \frac{6\sqrt{\sum_{i=1}^{r_{\mathbb{P}}} \zeta_i^2}}{\sum_{i=1}^{r_{\mathbb{P}}} \zeta_i}\right) \leq 0.9444. \tag{5.45}$$

Further, let $\phi_i = \zeta_i/\zeta_1$, $i = 2, \ldots, r_{\mathbb{P}}$, where $0 < \phi_i \le 1$. We may then express

$$\frac{\sqrt{\sum_{i=1}^{r_{\mathbb{P}}} \zeta_i^2}}{\sum_{i=1}^{r_{\mathbb{P}}} \zeta_i} = \frac{\sqrt{1 + \sum_{i=2}^{r_{\mathbb{P}}} \phi_i^2}}{1 + \sum_{i=2}^{r_{\mathbb{P}}} \phi_i},$$

where it is simple to show that

$$\frac{1}{r_{\mathbb{P}}} \le \frac{\sqrt{1 + \sum_{i=2}^{r_{\mathbb{P}}} \phi_i^2}}{1 + \sum_{i=2}^{r_{\mathbb{P}}} \phi_i} < 1.$$

Thus, the upper threshold in (5.45) is in the interval

$$\left[1 + 6/\sqrt{r_{\mathbb{P}}}, 7\right). \tag{5.46}$$

This provides a simple heuristic for comparing many size ratios. Similar heuristics may be developed for many of the other measures that we shall describe.

5.7 Partial data size

A particular special case of partial adjustment occurs when we adjust beliefs about a random vector, F say, by a further vector D, and subsequently we observe the value of F. We have described already how we may identify the discrepancy between the observed value of F and the prior expectation of F with the corresponding size of the adjustment by F. We may similarly quantify the portion of the discrepancy between F and $\mathrm{E}(F)$ which relates to the adjustment by D. In the definition for partial size of the adjustment of B by F given D, we replace B by F and define the corresponding **partial data size** of F given D, namely the largest change

$$\mathrm{Size}_{[f/d]}(F) = \max_{F \in \langle F \rangle} \frac{[f - \mathrm{E}_d(F)]^2}{\mathrm{Var}(F)} = \mathrm{Var}(\mathbb{Z}_{[f/d]}(F)).$$

5.8 Bearing and size for a relative adjustment

Another way in which we can assess the further changes in beliefs when we additionally adjust B by F given D is to assess directly the **bearing for the adjusted belief structure** $[B/D]$ **given** F. This quantity, denoted by $\mathbb{Z}_{f(d)}(B)$, summarizes all changes in expectation from the partial adjustment with respect to the directions within the adjusted belief structure, by the relation

$$\mathrm{Cov}_D(X, \mathbb{Z}_{f(d)}(B)) = \mathrm{E}_{d \cup f}(X) - \mathrm{E}_d(X) = \mathrm{E}_{[f/d]}(X). \tag{5.47}$$

Equating (5.40) with (5.47) gives

$$\mathrm{Cov}_D(X, \mathbb{Z}_{f(d)}(B)) = \mathrm{Cov}(X, \mathbb{Z}_{[f/d]}(B)),$$

from which we have the relationship between the two types of partial bearing,

$$\mathbb{Z}_{[f/d]}(B) = \mathbb{A}_D(\mathbb{Z}_{f(d)}(B)).$$

Corresponding to the partial data size, we have the analogous relative measure for expressing the data discrepancy given that the first stage of the adjustment has already been made.

Definition 5.12 *The* ***relative data size*** *of F given D is the largest change*

$$\text{Size}_{[f/d]}(\mathbb{A}_D(F)) = \max_{F \in \langle F \rangle} \frac{[\mathbb{A}_d(f)]^2}{\text{Var}(\mathbb{A}_D(F))}.$$

5.9 Path correlation

We now describe the path correlation. This is the correlation between the bearing for a particular data collection and the partial bearing when further data is introduced. Path correlation may be interpreted as a measure of conflict or consistency between the various sources of information based on considering whether the changes in belief that are induced by each part are similar or contradictory.

From (5.39), when we adjust beliefs in stages, the expected sizes of the respective adjustments are additive so that

$$\text{E}(\text{Size}_{D \cup F}(B)) = \text{E}(\text{Size}_D(B)) + \text{E}(\text{Size}_{[F/D]}(B)). \tag{5.48}$$

However, the observed sizes of the adjustments are not additive. The size of each adjustment is the variance of the corresponding bearing. Therefore, from (5.39),

$$\text{Var}(\mathbb{Z}_{d \cup f}(B)) = \text{Var}(\mathbb{Z}_d(B)) + \text{Var}(\mathbb{Z}_{[f/d]}(B)) + 2\text{Cov}(\mathbb{Z}_d(B), \mathbb{Z}_{[f/d]}(B)) \tag{5.49}$$

so that

$$\text{Size}_{d \cup f}(B) = \{\text{Size}_d(B) + \text{Size}_{[f/d]}(B)\} + 2\text{Cov}(\mathbb{Z}_d(B), \mathbb{Z}_{[f/d]}(B)). \tag{5.50}$$

Thus, while

$$\text{E}(\text{Cov}(\mathbb{Z}_D(B), \mathbb{Z}_{[F/D]}(B))) = 0,$$

the observed value of this covariance,

$$\text{Cov}(\mathbb{Z}_d(B), \mathbb{Z}_{[f/d]}(B)),$$

may be taken to expresses the degree of support or conflict between the two collections of evidence in determining the revision of beliefs. As a summary, we define the **path correlation** to be

$$\text{PC}(d, [f/d]) = \text{Corr}(\mathbb{Z}_d(B), \mathbb{Z}_{[f/d]}(B)). \tag{5.51}$$

The size and magnitude of the path correlation are diagnostics with interpretation as follows.

- If the path correlation is near $+1$ then the size of the adjustment of B by $D \cup F$ is much larger than the sum of the size of the adjustment by D and the size of the partial adjustment by $[F/D]$; informally, we may view the two collections of data as complementary, in that their combined effect in changing our beliefs is greater than the sum of the individual effect of each collection.
- If the path correlation is near -1 then the two collections are giving 'contradictory' messages which give smaller overall changes in belief, in combination, than we would expect from the individual adjustments with D and $[F/D]$; for example, each of the individual changes in belief might be surprisingly large but the overall change in belief might be small, masking these differences. The importance of such conflict depends on the magnitudes of the various changes in beliefs, but usually we would wish to distinguish between analyses where expectations changed but little, because all the sources of data individually implied small changes, and analyses where individual data sources suggested large changes in beliefs but these were of a contradictory nature and so cancelled each other out.

It can be helpful to visualize the bearings graphically. The three bearings $\mathbb{Z}_d(B)$, $\mathbb{Z}_{[f/d]}(B)$, $\mathbb{Z}_{d \cup f}(B)$ are three elements of $\langle B \rangle$, which, from (5.39), may be represented as three vectors which form the sides of a triangle, abc say. The squared length of the vertical side ab is equal to $\text{Size}_{d \cup f}(B)$. The squared lengths of the other two sides, ac, cb are $\text{Size}_d(B)$, $\text{Size}_{[f/d]}(B)$, respectively. The path correlation $\text{PC}(d, [f/d])$ is equal to $-\cos C$, where C is the interior angle at c. The expected value of the path covariance is always zero, so that the expected triangle is right-angled. If $C > \pi/2$ then the partial adjustment of B by F given D increases the overall change in beliefs, whereas if $C < \pi/2$ then the overall change in belief is smaller than is suggested by the individual changes in beliefs. In particular, substantial negative path correlations are of particular importance when the overall length ac is roughly as expected but the two lengths ab, bc are individually much larger than expected, so that an overall adjustment which appears to be plausible is composed of two surprising and contradictory changes in belief.

5.10 Example: oral glucose tolerance test

5.10.1 The initial observed adjustment

We began by adjusting $[B]$ solely by the doctor's fasting measurement D_0. The adjusted expectations and their observed values, given $D_0 = d_0 = 5.4$ alone, are

$$\text{E}_{D_0}(G_0) = 0.5536\, D_0 + 1.8571, \qquad \text{E}_{d_0}(G_0) = 4.8464,$$

$$\text{E}_{D_0}(G_2) = 0.2679\, D_0 + 5.1357, \qquad \text{E}_{d_0}(G_2) = 6.5821,$$

with relatively small standardized changes in adjustment of

$$\text{S}_{d_0}(G_0) = \text{S}_{d_0}(G_2) = 1.17$$

standard deviations, relative to her initial judgements. These changes are thus relatively unsurprising and, comparing them to the adjusted expectations for the full adjustment given in (4.40), namely 4.7085 and 6.9140 respectively, not much different from these. The size ratio is

$$\mathrm{Sr}_{d_0}(B) = \frac{\mathrm{Dis}_{d_0}(B)}{r_{\mathbb{T}}} = \frac{1.37}{1} = 1.37,$$

implying no conflict between (this piece of) data and the prior judgements as sources of information in the sense that this value is close to unity, its expectation.

Notice in this case that we have the same discrepancy and size values that we obtained in §4.3.3.1. That is, we have

$$\mathrm{S}(d_0) = \mathrm{S}_{d_0}(G_0) = \mathrm{S}_{d_0}(G_2) \quad \text{and} \quad \mathrm{Dis}(d_0) = \mathrm{Dis}_{d_0}(B).$$

This is a consequence of (4.39). Whenever we make a univariate adjustment by a single random quantity such as D_0, these discrepancy statistics must coincide.

We extend this example to multiple observations in Chapter 6. In particular, we address there issues which arise in examining the consistency of multiple observations, for example in §6.16.5.

5.10.2 Observed partial expectations

Suppose that we now make the additional partial adjustment of B by D_2, where D_2 is observed to be $d_2 = 9.8$. Overall, this leads to the full adjustment by D_0 and D_2 shown in (4.40). Our main interest is now in comparing the initial adjustment by D_0 alone with this full adjustment in order to identify the **partial** effects such as the change in adjusted expectation:

$$\mathrm{E}_{[D_2/D_0]}(G_0) = \mathrm{E}_d(G_0) - \mathrm{E}_{d_0}(G_0) = 4.7085 - 4.8464 = -0.1379,$$

$$\mathrm{E}_{[D_2/D_0]}(G_2) = \mathrm{E}_d(G_2) - \mathrm{E}_{d_0}(G_2) = 6.9140 - 6.5821 = +0.3319.$$

Thus, compared to the initial adjustment by D_0, the effect of the partial adjustment by D_2 on the evaluation of the adjusted expectation is to revise expectations downwards for G_0, from 4.8464 to 4.7085, and upwards for G_2, from 6.5821 to 6.9140.

Suppose that we standardize the changes with respect to the extra portion of prior variation resolved by the partial adjustment. We obtain

$$\begin{aligned}\mathrm{S}_{[d_2/d_0]}(G_0) &= \frac{\mathrm{E}_d(G_0) - \mathrm{E}_{d_0}(G_0)}{\sqrt{\mathrm{RVar}_D(G_0) - \mathrm{RVar}_{D_0}(G_0)}} = \frac{\mathrm{E}_{[D_2/D_0]}(G_0)}{\sqrt{\mathrm{RVar}_{[D_2/D_0]}(G_0)}} \\ &= \frac{-0.1379}{\sqrt{0.0049}} = -1.96. \end{aligned} \tag{5.52}$$

The standardized change for G_2 turns out to be the same, but positive:

$$\mathrm{S}_{[d_2/d_0]}(G_2) = 1.96.$$

Thus, each change in expectation has been 1.96 standard deviations relative to the partial variance resolved. That is, for such a small change in variance we saw a fairly large change in expectation by using the observed 2-hour measurement as well as the observed fasting measurement. The fact that the standardized changes are the same, except for sign, is a consequence of the partial adjustment being one-dimensional.

5.10.3 The size of the partial adjustment

In the same way that we can evaluate a size and an expected size for any general adjustment, we can also evaluate a size and an expected size for a partial adjustment, giving the partial size ratio (5.43), which gives us a useful diagnostic measure comparing actual to expected changes in behaviour specific to the partial adjustment. For our example, the sizes and size ratios for the full adjustment (shown in (4.78)), the partial adjustment, and the simple adjustment by D_0 only are shown in Table 5.1. The size ratio for the overall adjustment is roughly $\mathrm{Sr}_d(B) = 0.94$, and the size ratio for the simple adjustment by D_0 is $\mathrm{Sr}_{d_0}(B) = 1.37$, neither value being a surprise.

The partial size ratio corresponding to the partial adjustment is

$$\mathrm{Sr}_{[d_2/d_0]}(B) = \mathrm{Sr}_{[d_2/d_0]}([B/D_0]) = 3.85 = 1.96^2. \tag{5.53}$$

The interpretation is as in the previous section, a rather larger than expected squared change in expectation. Note the correspondence between size ratio (5.53) and standardized change (5.52) for a univariate adjustment. We can also show equivalence with the relative data size (5.48) in the univariate case. That is, we have

$$\mathrm{Dis}(\mathbb{A}_{d_0}(d_2)) = 1.96^2,$$

so that these diagnostics have an interpretation as the data discrepancy in the residual part of D_2, having adjusted for D_0.

We have detected a relatively surprising change in expectation. It is of interest to consider whether similar features reappear if we carry out an adjustment of B by D_2 alone. Doing so, we find that the evaluated adjusted expectations for the adjustment by D_2 solely are $\mathrm{E}_{d_2}(G_0) = 4.5983$ and $\mathrm{E}_{d_2}(G_2) = 6.8782$, representing fairly large changes ($\mathrm{S}_{d_2}(G_0) = \mathrm{S}_{d_2}(G_2) = 2.3$ standard deviations) from

Table 5.1 Sizes for the adjustments by D overall, D_0 singly, and $[D_2/D_0]$ partially.

	Adjustment		
	$D = [D_0 \cup D_2]$	$[D_2/D_0]$	D_0
Size	0.3179	0.1069	0.4268
Expected	0.3386	0.0277	0.3109
Size ratio	0.9389	3.8524	1.3729

Table 5.2 Sizes for the adjustments D_2 and $[D_0/D_2]$.

	Adjustment		
	$D = [D_0 \cup D_2]$	$[D_0/D_2]$	D_2
Size	0.3179	0.0115	0.2325
Expected	0.3386	0.2938	0.0448
Size ratio	0.9389	0.0390	5.1862

the initial values of 4.16 and 6.25, respectively. Although these are fairly large changes, they correspond to only small reductions in uncertainty about G_0 and G_2. The sizes of the adjustments are shown in Table 5.2 and display two noteworthy features. First, the size ratio for the simple adjustment by D_2 is more than five times as large as expected; and secondly, the size ratio for the partial adjustment after adjusting by D_0 additionally is very much smaller than expected. The former feature is more or less expected, given the sizes for the similar adjustment shown in Table 5.1. The latter feature may be interpreted as showing that a partial adjustment by D_0 in addition to D_2 is expected to enable changes in expectation that do not materialize.

Comparing Table 5.2 with Table 5.1, we see a size ratio of $\mathrm{Sr}_{d_2}(B) = 5.1862$ for the simple D_2 adjustment, and a size ratio of $\mathrm{Sr}_{[d_2/d_0]}(B) = 3.8524$ for the partial adjustment by $[D_2/D_0]$. Thus, the size ratio for the simple D_2 adjustment is larger than the size ratio for the adjustment where D_0 has been extracted. This might suggest that although we have identified D_2 (with its observation d_2) above as having some peculiar features, this is also true of the portion of D_0 that is common to D_2.

5.10.4 The bearing for the partial adjustment

We have seen already the bearing for the full adjustment in (4.74). We similarly calculate the bearing for the adjustment solely on D_0, which turns out to be a vector essentially in the direction of G_0. The difference between the two (5.39) is the bearing for the partial adjustment:

$$\mathbb{Z}_d(B) = 0.39G_0 + 0.16G_2 - 2.60 \tag{5.54}$$

$$\mathbb{Z}_{d_0}(B) = 0.65G_0 - 0.06G_2 - 2.35 \tag{5.55}$$

$$\begin{aligned}\mathbb{Z}_{[d_2/d_0]}(B) &= \mathbb{Z}_d(B) - \mathbb{Z}_{d_0}(B) \\ &= -0.26G_0 + 0.21G_2 - 0.25 \end{aligned} \tag{5.56}$$

(note that there is some rounding error involved for the displayed coefficients). It is solely in the direction $\mathbb{Z}_{[d_2/d_0]}(B)$ that expectations can change according to the new (i.e. not already carried by $D_0 = d_0$) information contained in $D_2 = d_2$. Note that this conclusion applies for the data observed: different data would have produced a different direction. We can see from the coefficients in (5.56) that the partial bearing is quite highly correlated with the difference $G_h = G_2 - G_0$. Actual partial changes in expectation can be calculated via (5.40), without further construction and with little extra computation. For example, the partial change in expectation for G_h is given by

$$\mathrm{E}_{[d_2/d_0]}(G_h) = \mathrm{Cov}(G_h, \mathbb{Z}_{[d_2/d_0]}(B)) = \mathrm{Cov}(G_2 - G_0, -0.26G_0 + 0.21G_2)$$

$$= \begin{bmatrix} -1 & 1 \end{bmatrix} \begin{bmatrix} 1.12 & 0.72 \\ 0.72 & 2.43 \end{bmatrix} \begin{bmatrix} -0.26 \\ 0.21 \end{bmatrix} \approx 0.47.$$

The adjusted expectation for G_h given d_0 alone is 1.7357, a standardized change of -1.17 from prior. The adjusted expectation given both d_0 and d_2 is 2.2055, with the partial change of +0.47 representing a standardized change of +1.96. As these are univariate adjustments, the standardized changes match those seen earlier. We could, of course, construct G_h directly and adjust it by D_0 and then partially by D_2 to obtain the same results.

5.10.5 The path correlation for the partial adjustment

We see in Table 5.1 a size ratio of $\mathrm{Sr}_{d_0}(B) = 1.37$ for the adjustment of B by D_0 and a rather larger partial size ratio of $\mathrm{Sr}_{[d_2/d_0]}(B) = 3.85$ for the subsequent partial adjustment by D_2. These both suggest larger changes in expectation than expected. However, the overall size ratio turns out to be $\mathrm{Sr}_{d \cup f}(B) = 0.94$ (see equation (4.78)), suggesting that the change in expectation was close to what was expected. This conundrum is explained by (5.50).

Compare (5.55) and (5.56), the bearings for the initial and partial adjustments. Clearly these two directions, which aggregate to form the overall bearing, are different, so that the overall changes in adjustment (as summarized by the overall bearing (5.54)) are the result of two somewhat contradictory changes. This aspect can be summarized by evaluating the **path correlation**: the prior correlation between the bearings for the initial and partial adjustments. In this example, the path correlation (5.51) is

$$\mathrm{PC}(d_0, [d_2/d_0]) = \mathrm{Corr}(\mathbb{Z}_{d_0}(B), \mathbb{Z}_{[d_2/d_0]}(B)) = -0.5051,$$

with

$$\mathrm{Cov}(\mathbb{Z}_{d_0}(B), \mathbb{Z}_{[d_2/d_0]}(B)) = -0.1079,$$

showing that from the point of view of revising expectations, the data are partly contradictory. Had this correlation been positive, we would have argued that the data complemented each other, with the magnitude of correlation indicating the degree of consistency. Thus, it is the negative covariance (which we see summarized as a path correlation of $\mathrm{PC}(d_0, [d_2/d_0]) = -0.5051$) between the bearings for the previous and partial adjustments which serves to diminish the size of the joint adjustment: the changes in expectation for the initial and partial adjustments are in different directions, and thus tend to cancel out each other. Numerically, we find that the overall size is obtained from (5.50) as

$$0.4268 + 0.1069 + 2 \times (-0.1079) = 0.3179.$$

5.11 Sequential adjustment

When we make a collection of sequential adjustments, the one-step changes in adjustment may be tracked in a stepwise manner giving a picture of the cumulative effects of the adjustment. Suppose that we intend to adjust B sequentially by the collections of quantities $G_1, G_2, \ldots, G_m$. We define the cumulative collection

$$G_{[i]} = \bigcup_{j=1}^{i} G_j,$$

and denote the cumulative adjustment

$$\mathrm{E}_{[i]}(B) = \mathrm{E}_{G_{[i]}}(B).$$

We may 'partial out' any stage of the adjustment as follows.

Definition 5.13 *For any $i > j$, the **partial adjustment** of B by $G_{[i]}$ given $G_{[j]}$ is*

$$\mathrm{E}_{[i/j]}(B) = \mathrm{E}_{[i]}(B) - \mathrm{E}_{[j]}(B) = \mathrm{E}_{[\bigcup_{k=j+1}^{i} G_k / \bigcup_{k=1}^{j} G_k]}(B). \tag{5.57}$$

Corresponding to the adjustment $\mathrm{E}_{[i]}(B)$ is the bearing $\mathbb{Z}_{[i]}(B)$. The bearing for the partial adjustment $\mathrm{E}_{[i/j]}(B)$ is, by (5.39),

$$\mathbb{Z}_{[i/j]}(B) = \mathbb{Z}_{[i]}(B) - \mathbb{Z}_{[j]}(B),$$

and the difference between such cumulative adjustments is, by (5.40),

$$\mathrm{E}_{[i]}(X) - \mathrm{E}_{[j]}(X) = \mathrm{Cov}(X, \mathbb{Z}_{[i/j]}(B)).$$

The bearing for the partial adjustment expresses the change, in both magnitude and direction, in beliefs between stages $[j]$ and $[i]$.

5.11.1 The data trajectory

For single-step adjustments we arrange the sequence of adjustments as follows.

Definition 5.14 *The ith **stepwise partial adjustment**, $\mathrm{E}_{[i/]}(B)$, is*

$$\mathrm{E}_{[i/]}(B) = \mathrm{E}_{[i/i-1]}(B) = \mathrm{E}_{[G_i/G_{[i-1]}]}(B), \tag{5.58}$$

with bearing

$$\mathbb{Z}_{[i/]}(B) = \mathbb{Z}_{[i]}(B) - \mathbb{Z}_{[i-1]}(B). \tag{5.59}$$

Definition 5.15 *We refer to the full sequence of stepwise adjusted bearings*

$$\mathbb{Z}_{[1]}(B),\ \mathbb{Z}_{[2/]}(B), \ldots, \mathbb{Z}_{[m/]}(B) \tag{5.60}$$

as the ***data trajectory****.*

For any accumulated adjustment, we may decompose its bearing as follows. For each j we may write

$$\mathbb{Z}_{[j]}(B) = \mathbb{Z}_{[1]}(B) + \mathbb{Z}_{[2/]}(B) + \ldots + \mathbb{Z}_{[j/]}(B), \tag{5.61}$$

from which we may express the size of the accumulated adjustment as

$$\text{Size}_{[j]}(B) = \text{Size}_{[1]}(B) + \text{Size}_{[2/]}(B) + \ldots + \text{Size}_{[m/]}(B) + 2(C_{[2]} + \ldots + C_{[j]}), \tag{5.62}$$

where

$$C_{[r]} = \text{Cov}(\mathbb{Z}_{[r-1]}(B), \mathbb{Z}_{[r/]}(B))$$

is the covariance between the bearing for the accumulated adjustment up to step $r - 1$ and the bearing for the partial adjustment at step r.

To examine the ways in which the individual terms combine to determine the overall adjustment, we must thus consider:

- the prior expectation for each change to assess which sub-collections of data are expected to be informative;
- the individual adjusted bearings $\mathbb{Z}_{[i/]}(B)$ to identify the stages at which larger than expected changes in belief occur;
- the path correlations derived from the covariances $C_{[i]}$ to see whether the evidence is internally supportive or contradictory.

5.12 The canonical trajectory

The data trajectory expresses the additional information derived at each stage from a progressive adjustment of belief, and depends, in general, on the order in which the various adjustments are made. However, there are certain cases where the order is unimportant, which we now describe.

For any collections $B,\ D$, denote by W_+ the eigenvectors of the belief transform $\mathbb{T}_{D:B}$ for the adjustment of D by B, corresponding to positive eigenvalues. Now let $M_1, \ldots, M_k$ be any partition of the elements of W_+ into k disjoint subsets. Let $\mathbb{Z}_i(B)$ be the bearing for the adjustment of B by the subset M_i, and $\mathbb{Z}_{[i/]}(B)$ the bearing for the adjustment of B by M_i given $\cup_j M_j$.

Property 5.16 *Any data trajectory created in this way has the following properties.*

5.16.1: *Bearings and adjusted bearings are the same,*

$$\mathbb{Z}_i(B) = \mathbb{Z}_{[i/]}(B),$$

for each i, as the elements of W_+, and thus of the subsets M_j, are mutually uncorrelated.

5.16.2: *From relation (4.60), as the eigenvectors of $\mathbb{T}_{B:D}$ are mutually uncorrelated, the bearings $\mathbb{Z}_1(B), \ldots, \mathbb{Z}_k(B)$ are a collection of uncorrelated random quantities. Therefore, the length of the bearing corresponding to adjustment by any sub-collection $M_{i_1} \cup \ldots \cup M_{i_j}$ is equal to the sum of the lengths of the individual bearings $\mathbb{Z}_{i_1}(B), \ldots, \mathbb{Z}_{i_j}(B)$. In particular, the expected length of $\mathbb{Z}_i(B)$ is equal to the sum of the eigenvalues of $\mathbb{T}_{B:D}$ corresponding to the eigenvectors in M_i.*

Therefore data trajectories which are built directly from partitions of the canonical directions for the belief transform will always have a simple form which is easy to interpret, as it makes no difference in which order we introduce the various portions of information, the partial bearings are always uncorrelated and the expected size of each bearing is the sum of the eigenvalues. In particular, we term the trajectory for which each M_i contains a single element of W_+ the **canonical trajectory**. Some of these ideas are illustrated in §7.6.4.3.

5.13 Detection of systematic bias

The size ratio diagnostics discussed above do not take into account the direction of discrepancy or contradiction. Sometimes it can be useful to look for such discrepancies, for example when assessing for systematic bias in a sequential adjustment for time series data. Suppose we have a series of random quantities $D_1, D_2, \ldots$ with prior beliefs $\mathrm{E}(D_i)$ and $\mathrm{Var}(D_i)$ for $i = 1, 2, \ldots$, and $\mathrm{Cov}(D_i, D_j)$ for $i = 1, 2, \ldots$ and $j > i$. Suppose that our interest is in observing $D_1 = d_1$, updating our beliefs about the remainder of the sequence, observing $D_2 = d_2$ and updating our beliefs about $D_3, D_4, \ldots$, and so forth.

For the exploration of systematic bias, consider the sequence of one-step standardized forecast errors $\epsilon_1, \epsilon_2, \ldots$, where

$$\epsilon_i = \frac{D_i - \mathrm{E}_{[i-1]}(D_i)}{\sqrt{\mathrm{Var}_{[i-1]}(D_i)}}, \qquad i = 1, 2, \ldots,$$

and where $\mathrm{E}_{[i-1]}(D_i)$ and $\mathrm{Var}_{[i-1]}(D_i)$ signify the adjusted expectation and variance for D_i given $D_1, \ldots, D_{i-1}$, with $\mathrm{E}_{[0]}(D_i)$ and $\mathrm{Var}_{[0]}(D_i)$ signifying the prior mean and variance. Key properties of the sequence of one-step forecast errors are as follows.

Property 5.17 *A sequence of such standardized adjusted expectations is a priori uncorrelated and has:*

5.17.1: $\mathrm{E}(\epsilon_i) = 0, \quad \forall i;$

5.17.2: $\mathrm{Var}(\epsilon_i) = 1, \quad \forall i;$

5.17.3: $\mathrm{E}_{\epsilon_1 \cup \epsilon_2 \cup \ldots \cup \epsilon_{i-1}}(\epsilon_i) = 0, \quad \forall i > 2.$

The first two properties are obvious. Property 5.17.3 follows directly from (5.6). If we make the extra assumption that terms in the sequence are conditionally independent of preceding terms, then it follows that the sequence $\epsilon_1, \epsilon_2, \ldots$ is a **martingale difference sequence**. Now define the standardized cumulative sum of these standardized one-step errors as

$$Q_n = \frac{1}{\sqrt{n}} \sum_{i=1}^{n} \epsilon_i, \quad n = 1, 2, \ldots. \tag{5.63}$$

Under some quite weak assumptions, for example that at least one of the higher $(r > 2)$ moments $\mathrm{E}(|\epsilon_t|^r)$ is finite, we may apply the central limit theorem for a martingale difference sequence (see, for example, Hamilton 1994), from which we have approximately that Q_n has a standard normal distribution, for large n.

For the detection of systematic bias, we calculate and plot the values $Q_1, Q_2, \ldots,$ as data arrives. The interpretation is similar to that for the cusum charts used in statistical process control. That is, the sequence should wander randomly around an expected value of zero, with rare excursions beyond 95% probability limits of around two standard deviations. If the sequence moves systematically away from zero in one direction, the implication is that there are systematic one-sided discrepancies between the prior specification and the actual observations.

5.14 Examples

5.14.1 Anscombe data sets

Anscombe (1973) discusses four fictitious data sets which have (almost) identical implications as far as linear fitting is concerned. The data are shown in Table 5.3, with rows ordered according to the value of x_1. The quantities x_1, x_2, and x_3 are the same. The four data sets have the feature that the second-order summaries (means, variances, covariances) are approximately identical for each pair. (Bassett et al. (2000) construct a similar data set with actually identical second-order summaries.) Anscombe's purpose was to demonstrate the importance of graphical analysis. We agree that such analysis is important, but also show here how the data trajectory can be used to help assess features of an adjustment and to diagnose possible conflicts.

For each data set, we have 11 pairs of values on variables Y and X. For each, we assume initially a simple linear relationship of the form

$$Y_i = \alpha + \beta x_i + \epsilon_i, \tag{5.64}$$

Table 5.3 Anscombe data sets.

	Data set 1		Data set 2		Data set 3		Data set 4	
i	y_1	x_1	y_2	x_2	y_3	x_3	y_4	x_4
1	4.26	4	3.10	4	5.39	4	7.04	8
2	5.68	5	4.74	5	5.73	5	6.89	8
3	7.24	6	6.13	6	6.08	6	5.25	8
4	4.82	7	7.26	7	6.42	7	7.91	8
5	6.95	8	8.14	8	6.77	8	5.76	8
6	8.81	9	8.77	9	7.11	9	8.84	8
7	8.04	10	9.14	10	7.46	10	6.58	8
8	8.33	11	9.26	11	7.81	11	8.47	8
9	10.84	12	9.13	12	8.15	12	5.56	8
10	7.58	13	8.74	13	12.74	13	7.71	8
11	9.96	14	8.10	14	8.84	14	12.50	19

where the quantities $\{\epsilon_i\}$ are unobserved error terms expected a priori to have mean zero and to be uncorrelated with themselves and with other quantities. Our main interest is in the collection $C = [\alpha, \beta]$. To illustrate the data trajectory, Goldstein (1988b) suggests the following initial prior specification:

$$\mathrm{E}(\alpha) = 0, \quad \mathrm{E}(\beta) = 0, \quad \mathrm{Var}(\alpha) = 2, \quad \mathrm{Var}(\beta) = 1, \quad \mathrm{Cov}(\alpha, \beta) = 0, \tag{5.65}$$

and

$$\mathrm{Var}(\epsilon_i) = 1, \quad \forall i. \tag{5.66}$$

To analyse such problems, we construct beliefs for the data quantities $\{Y_i\}$. We have, for example,

$$\begin{aligned}
\mathrm{E}(Y_i) &= \mathrm{E}(\alpha + \beta x_i + \epsilon_i) \\
&= \mathrm{E}(\alpha) + x_i \mathrm{E}(\beta) + \mathrm{E}(\epsilon_i) \\
&= 0, \\
\mathrm{Var}(Y_i) &= \mathrm{Var}(\alpha + \beta x_i + \epsilon_i) \\
&= x_i^2 + 3, \\
\mathrm{Cov}(Y_i, Y_j) &= \mathrm{Cov}(\alpha + \beta x_i + \epsilon_i, \alpha + \beta x_j + \epsilon_j) \\
&= x_i x_j + 2.
\end{aligned}$$

Data sets 1–3 will thus result in an identical specification for the 11 data quantities. We now organize these data quantities as collections; for example, let $D_1 = [Y_{1,1}, Y_{1,2}, \ldots, Y_{1,11}]$ be the collection of data quantities for the first data set. Next, we perform the adjustment of C, the collection of regression coefficients, separately by each data collection. We find, as expected, that the basic adjustment is identical for each data set D_j. That is,

$$\mathrm{E}_{d_j}(\alpha) = 2.157, \qquad \mathrm{Var}_{d_j}(\alpha) = 0.582, \qquad \mathrm{R}_{d_j}(\alpha) = 0.709,$$
$$\mathrm{E}_{d_j}(\beta) = 0.583, \qquad \mathrm{Var}_{d_j}(\beta) = 0.007, \qquad \mathrm{R}_{d_j}(\beta) = 0.993.$$

In addition, the overall data-diagnostic features are the same. That is, for each data set $D_j = d_j$ we have a bearing, size, and size ratio for the collection C of

$$\mathbb{Z}_{d_j}(C) = 1.079\,\alpha + 0.583\,\beta, \qquad \mathrm{Size}_{d_j}(C) = 1.63^2, \qquad \mathrm{Sr}_{d_j}(C) = 1.57.$$

In summary, for each data set D_j, the mean for α is adjusted from zero to 2.157; the mean for β is adjusted from zero to 0.583; about 71% of the prior variation in α has been resolved, compared to 99.3% resolution for β; the bearing shows that the largest changes in expectation relative to prior variance are roughly in the direction $2\alpha + \beta$; the largest such standardized change is 1.63 standard deviations; and the size ratio of 1.57 shows no particular conflict between the prior specification and the data.

5.14.1.1 The data trajectory

We now compute the data trajectory for each data set as described in §5.11. For each data set D_j, we order the pairs (x_i, y_i) in ascending order of the values of x_i and let $Y_{(k)}$ be the kth such constructed data quantity. We define the cumulative collection

$$G_{[i]} = \bigcup_{k=1}^{i} Y_{(k)}.$$

For example, $G_{[1]} = Y_{(1)}$ is the Y quantity corresponding to the smallest value of x; $G_{[2]} = [Y_{(1)}, Y_{(2)}]$ is the pair of Y quantities corresponding to the two smallest values of x; and

$$G_{[11]} = \bigcup_{k=1}^{11} Y_{(k)} = [Y_1, \ldots, Y_{11}]$$

is the full collection. As usual, $g_{[i]}$ represents the observed value of $G_{[i]}$.

We now carry out a sequential adjustment of the collection C by the data quantities. We begin by adjusting C by $Y_{(1)}$ alone, then partially also by $Y_{(2)}$, then partially also by $Y_{(3)}$, and so forth. At each stage, we calculate *inter alia* the following summaries, where we also indicate shorthand notation.

- $\mathrm{E}_{[i]}(\alpha) = \mathrm{E}_{g_{[i]}}(\alpha)$, the adjusted expectation for α after the ith adjustment. This shows informally how the mean for α is being sequentially updated.
- $\mathrm{E}_{[i]}(\beta) = \mathrm{E}_{g_{[i]}}(\beta)$, the adjusted expectation for β after the ith adjustment. This shows informally how the mean for β is being sequentially updated.

- $\text{Size}_{[i/]}(C) = \text{Size}_{[g_{[i]}/g_{[i-1]}]}(C)$, the size (5.41) of the partial adjustment of C by $G_{[i]} = g_{[i]}$, given $G_{[i-1]} = g_{[i-1]}$. Large values show us that the ith partial adjustment resulted in a large change in standardized expectation, relative to prior (i.e. initial) variation.

- $\text{E}(\text{Size}_{[G_{[i]}/G_{[i-1]}]}(C))$, the expected size (5.42) of the partial adjustment of C by $G_{[i]}$, given $G_{[i-1]}$.

- $\text{Sr}_{[i/]}(C) = \text{Sr}_{[g_{[i]}/g_{[i-1]}]}(C)$, the size ratio (5.43) for the partial adjustment of C by $G_{[i]} = g_{[i]}$, given $G_{[i-1]} = g_{[i-1]}$. Large values (much greater than unity) indicate unexpectedly large changes in expectation, and may indicate a conflict between data and prior specifications with respect to the partial adjustment. Small values (smaller than unity) indicate unexpectedly small changes in expectation, and may indicate perhaps that prior variability was assigned too cautiously.

- $C_{[i/]} = \text{PC}(g_{[i]}, [g_{[i]}/g_{[i-1]}])$, the path correlation (5.51) for the partial adjustment. We check to see whether the new evidence introduced at the ith adjustment is in agreement or conflict with the evidence accumulated for the preceding adjustments.

These summaries are given in Table 5.4. It is enlightening to examine them plotted against observation number. For data set 1, they are shown in Figure 5.1. The updated expectations trend from prior expectation to the full adjusted expectation given all the data. Although there are subtleties in interpreting such plots, the likely shape of such a trend, assuming that the data display no serial dependence between successive observations, is of random fluctuation around the trend from prior to adjusted expectation. There are no especially worrying features in these plots. Observations [4] and [9] appear slightly aberrant, and these are picked out in Figure 5.1(e) as having quite surprising changes in adjustment relative to prior variance. Figure 5.1(f) shows the sequence of path correlations, each multiplied by the corresponding size ratio. We do this in order to draw the eye to important features, specifically partial adjustments with large path correlations (and especially those near -1) and with partial size ratios indicating large changes in expectation.

Figure 5.2 shows the summary plots for data set 2. The relationship between Y and X is roughly quadratic. The departure from linearity is picked out on the path correlation plot by observation [6], by which time the turning point on the scatter plot begins to be noticed, and the sequence of positive path correlations becomes a sequence of negative path correlations, indicating that the new evidence is in conflict with all that has gone before. Figure 5.2(e) shows a systematically increasing size ratio, indicating large changes in expectation where small changes were expected; this is the result of the later sequential changes in expectation for α, β needing to compensate for the earlier adjustments. The latter points of Figure 5.2(f) indicate large changes in expectation together with strong consistency of direction of change.

Figure 5.3 shows the summary plots for data set 3. There is one aberrant point, observation [10], clearly picked out by the path correlation at that stage. Until then, we see in Figure 5.3(f) a sequence of increasing positive correlations (implying

Table 5.4 Diagnostic assessments for Anscombe data sets.

	Data set 1				Data set 2			
i	$\mathrm{E}_{[i]}(\alpha)$	$\mathrm{E}_{[i]}(\beta)$	$\mathrm{Sr}_{[i/]}(C)$	$C_{[i/]}$	$\mathrm{E}_{[i]}(\alpha)$	$\mathrm{E}_{[i]}(\beta)$	$\mathrm{Sr}_{[i/]}(C)$	$C_{[i/]}$
1	0.44	0.89	0.95	0.00	0.32	0.65	0.50	0.00
2	0.35	1.00	0.22	0.63	0.18	0.82	0.52	0.63
3	0.14	1.11	0.36	0.36	0.07	0.94	0.52	0.44
4	1.10	0.77	5.29	−0.36	0.28	1.02	0.26	0.49
5	1.21	0.74	0.07	0.39	0.36	1.04	0.03	0.54
6	0.92	0.81	0.50	−0.49	0.27	1.02	0.04	−0.54
7	1.23	0.74	0.67	0.36	0.02	0.96	0.43	−0.47
8	1.54	0.68	0.80	0.55	0.36	0.89	1.23	−0.29
9	1.26	0.73	0.81	−0.68	0.86	0.79	2.52	0.02
10	2.04	0.60	7.39	0.59	1.46	0.69	4.37	0.40
11	2.15	0.58	0.17	0.81	2.15	0.58	6.83	0.68

	Data set 3				Data set 4			
i	$\mathrm{E}_{[i]}(\alpha)$	$\mathrm{E}_{[i]}(\beta)$	$\mathrm{Sr}_{[i/]}(C)$	$C_{[i/]}$	$\mathrm{E}_{[i]}(\alpha)$	$\mathrm{E}_{[i]}(\beta)$	$\mathrm{Sr}_{[i/]}(C)$	$C_{[i/]}$
1	0.56	1.13	1.52	0.00	0.21	0.84	0.73	0.00
2	0.63	1.06	0.10	−0.63	0.20	0.83	0.00	−1.00
3	0.85	0.94	0.41	−0.21	0.19	0.77	1.84	−1.00
4	1.18	0.83	0.61	0.10	0.20	0.81	1.76	1.00
5	1.53	0.73	0.65	0.39	0.19	0.79	0.75	−1.00
6	1.85	0.65	0.62	0.59	0.21	0.84	4.40	1.00
7	2.13	0.59	0.53	0.72	0.20	0.83	0.00	−1.00
8	2.36	0.54	0.44	0.79	0.21	0.85	2.00	1.00
9	2.56	0.50	0.37	0.84	0.20	0.83	2.00	−1.00
10	1.70	0.65	8.94	−0.87	0.21	0.84	0.50	1.00
11	2.15	0.58	3.02	0.74	2.15	0.58	2.78	−0.01

data complementarity) corresponding to small (i.e. unsurprising) size ratios. The aberrant point has both a strong negative correlation of −0.87 and a very high size ratio of 8.94, indicating substantial discordancy.

Figure 5.4 shows the summary plots for data set 4. There is one point distant from the others, observation [11]. However, this point is not especially in conflict with the preceding evidence, and the associated size ratio is not large enough to cause alarm.

One informal means of assessing the adequacy of alternative models is via the path correlation. Suppose, for example, that we fit a quadratic regression to data set 2, modifying (5.64) to

$$Y_i = \alpha + \beta x_i + \gamma x_i^2 + \epsilon_i. \tag{5.67}$$

Following Goldstein (1988b), we specify γ to have prior mean zero, prior variance $\mathrm{Var}(\gamma) = 0.2$, and to be uncorrelated with all other quantities. We now construct

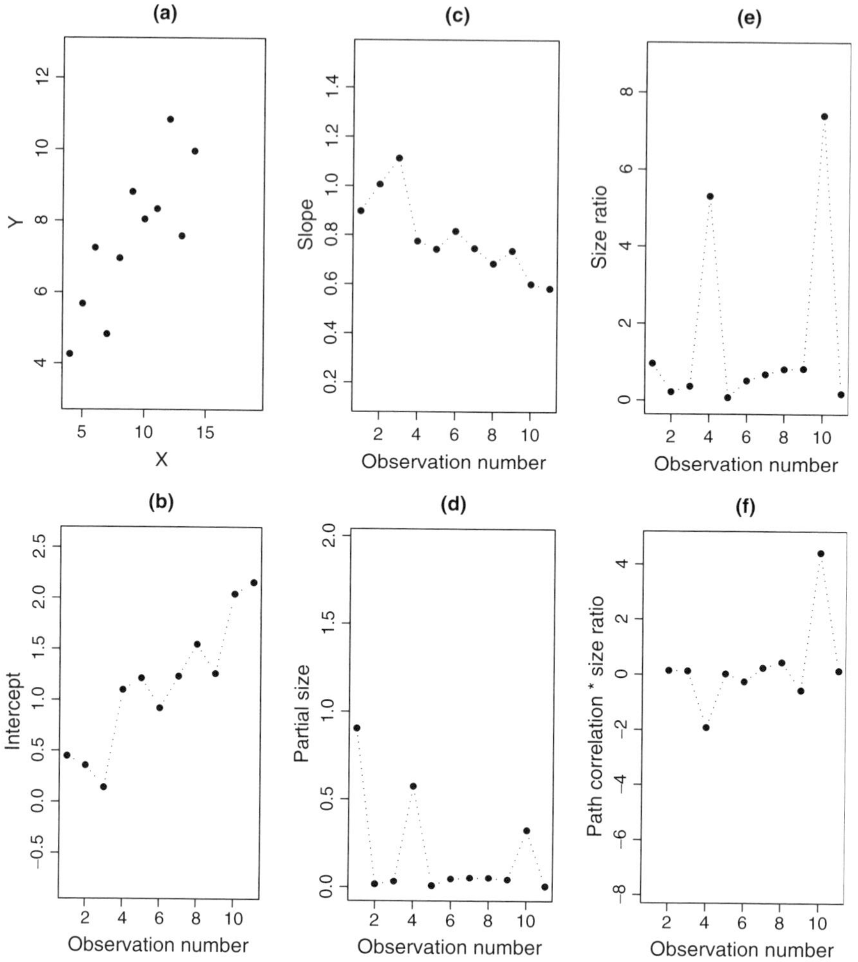

Figure 5.1 Data set 1. (a) Scatter plot of original data; (b) sequential update of expectation for α; (c) sequential update of expectation for β; (d) partial sizes for sequential adjustments; (e) partial size ratios for sequential adjustments; (f) successive path correlations multiplied by size ratios.

new data quantities for this model, and carry out a similar sequence of adjustments in order to determine the data trajectory. The results are plotted in Figure 5.5. The path correlations for the quadratic fit settle down quickly to a series of positive correlations, whilst the corresponding size ratios (which were previously increasing alarmingly) now give no reason for concern, and the maximal standardized changes in expectation are now generally quite small for each sequential adjustment. The

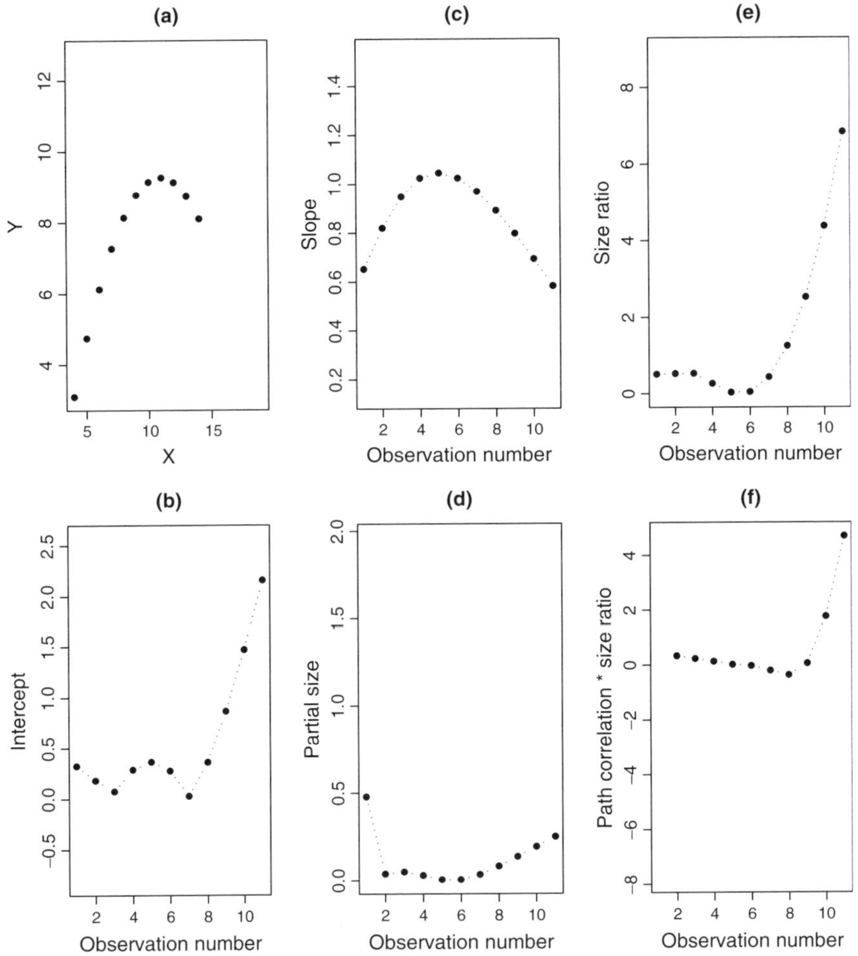

Figure 5.2 Data set 2. (a) Scatter plot of original data; (b) sequential update of expectation for α; (c) sequential update of expectation for β; (d) partial sizes for sequential adjustments; (e) partial size ratios for sequential adjustments; (f) successive path correlations multiplied by size ratios.

quadratic component γ is not plotted but can be seen in Table 5.5, which can be compared to the appropriate quadrant of Table 5.4.

5.14.2 Regression with correlated responses

The following problem is considered in Box and Tiao (1973, Chapter 8). A certain chemical process leads to a product Y and a by-product Z. The yields of both

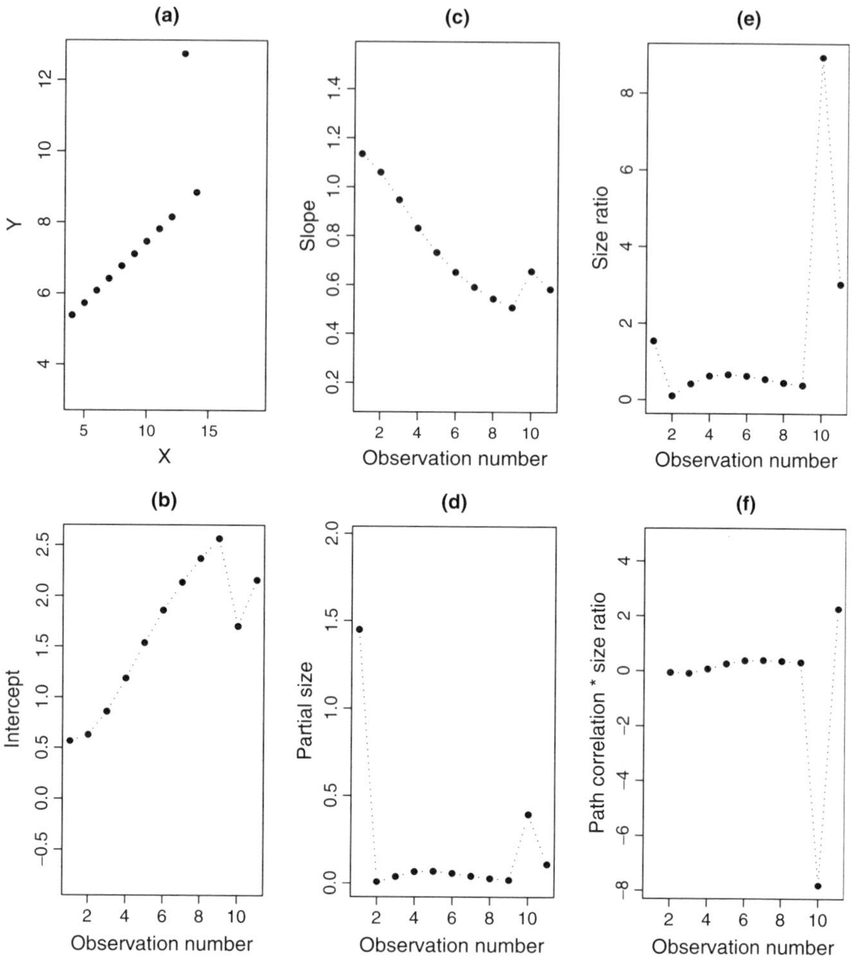

Figure 5.3 Data set 3. (a) Scatter plot of original data; (b) sequential update of expectation for α; (c) sequential update of expectation for β; (d) partial sizes for sequential adjustments; (e) partial size ratios for sequential adjustments; (f) successive path correlations multiplied by size ratios.

products are thought to be related to the temperature of the process, X^*. Twelve experiments are performed with different temperature settings (degrees Fahrenheit) to study the effect of temperature. The data are shown in Table 5.6, and plotted in panels (a) and (b) of Figure 5.6. In performing the analysis, we transform the temperature measurements to $X = (X^* - 177.86)/100$, where 177.86 is the mean temperature setting.

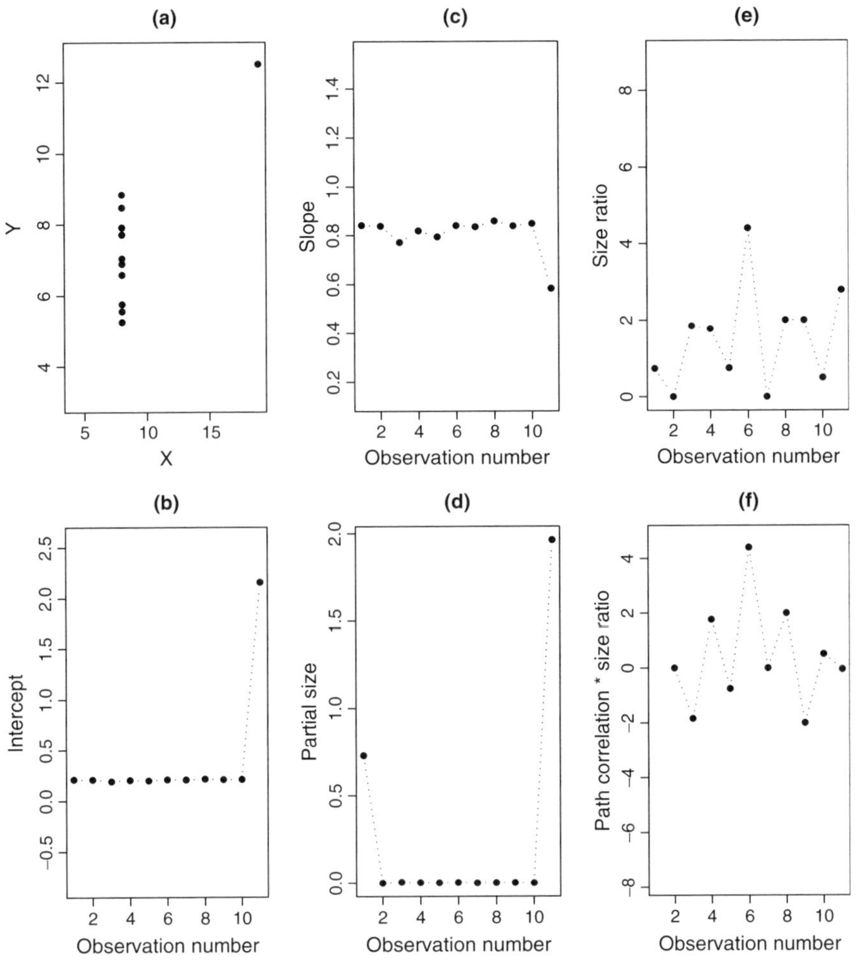

Figure 5.4 Data set 4. (a) Scatter plot of original data; (b) sequential update of expectation for α; (c) sequential update of expectation for β; (d) partial sizes for sequential adjustments; (e) partial size ratios for sequential adjustments; (f) successive path correlations multiplied by size ratios.

5.14.2.1 *The model*

The model suggested to explain relationships between the quantities is as follows:

$$Y_i = a + bx_i + e_i \tag{5.68}$$

$$Z_i = c + dx_i + f_i, \quad i = 1, \ldots, 12. \tag{5.69}$$

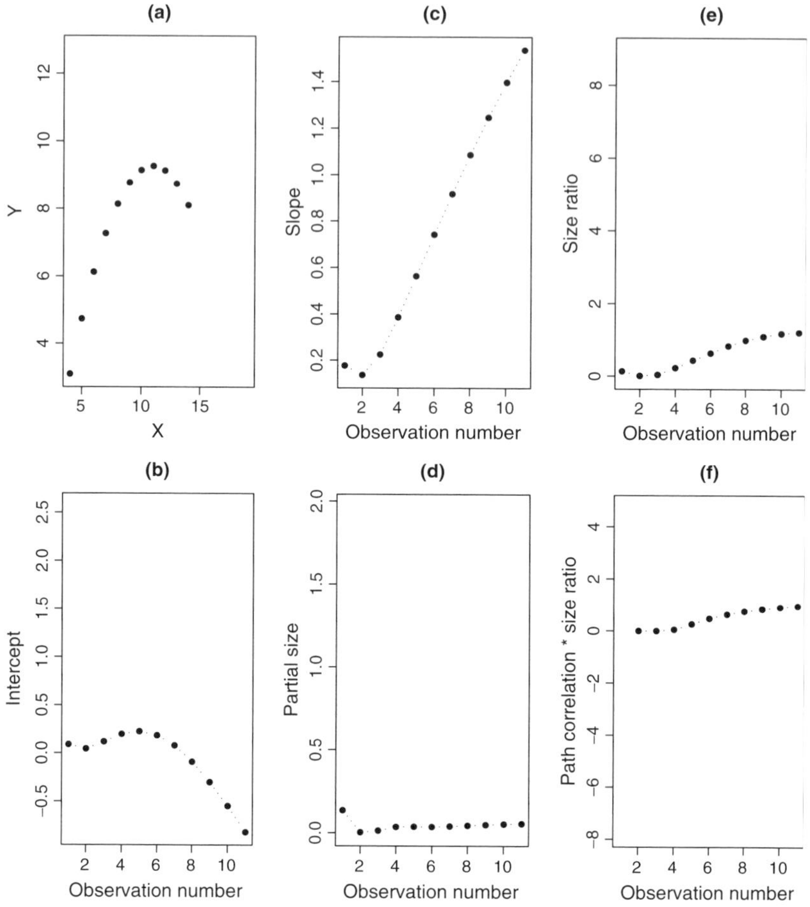

Figure 5.5 Data set 2, quadratic fit. (a) Scatter plot of original data; (b) sequential update of expectation for α; (c) sequential update of expectation for β; (d) partial sizes for sequential adjustments; (e) partial size ratios for sequential adjustments; (f) successive path correlations multiplied by size ratios.

The model reflects the beliefs that the relationships between the yields Y, Z and the temperature X are approximately linear in X over the given range of temperature values. The intercept terms a, c indicate the yields for average temperature settings, whilst the slopes of the regressions are given by b, d. The models incorporate error components e_i, f_i. Separate runs of the experiment are independent; however, in any particular run it is felt that the error components will be correlated because slight aberrations in reaction conditions or analytical procedures could simultaneously affect both product yields. We will thus suppose that $e_1, e_2, \ldots$ are an

Table 5.5 Diagnostic assessment for quadratic modelling.

	Data set 2, quadratic fit					
i	$E_{[i]}(\alpha)$	$E_{[i]}(\beta)$	$E_{[i]}(\gamma)$	$Size_{[i/]}(C)$	$Sr_{[i/]}(C)$	$C_{[i/]}$
1	0.0883	0.1766	0.1413	0.13	0.13	0.00
2	0.0438	0.1367	0.1585	0.00	0.01	0.13
3	0.1164	0.2250	0.1323	0.01	0.04	−0.15
4	0.1995	0.3862	0.0939	0.03	0.22	0.21
5	0.2272	0.5636	0.0590	0.03	0.43	0.61
6	0.1861	0.7414	0.0296	0.03	0.63	0.75
7	0.0780	0.9166	0.0048	0.03	0.82	0.77
8	−0.0883	1.0858	−0.0160	0.04	0.98	0.76
9	−0.3008	1.2465	−0.0334	0.04	1.09	0.76
10	−0.5481	1.3979	−0.0480	0.05	1.17	0.77
11	−0.8168	1.5380	−0.0603	0.05	1.20	0.78

Table 5.6 Yield of two products for a chemical process.

Temperature X^*	Main product Y	By-product Z
161.30	63.70	20.30
164.00	59.50	24.20
165.70	67.90	18.00
170.10	68.80	20.50
173.90	66.10	20.10
176.20	70.40	17.50
177.60	70.00	18.20
181.70	73.70	15.40
185.60	74.10	17.80
189.00	79.60	13.30
193.50	77.10	16.70
195.70	82.80	14.80

uncorrelated sequence of error components with expectation zero and variance σ_e^2; that $f_1, f_2, \ldots$ are an uncorrelated sequence of error components with expectation zero and variance σ_f^2; and that all pairs of error components e_i, f_j are uncorrelated except for $\mathrm{Cov}(e_i, f_i) = \sigma_{ef}$.

5.14.2.2 Prior beliefs

It is necessary to specify prior beliefs over the four quantities a, b, c, d, and for the error components. Non-informative reference prior distributions are used for all these quantities by Box and Tiao (1973). For our illustration, we attempt to portray

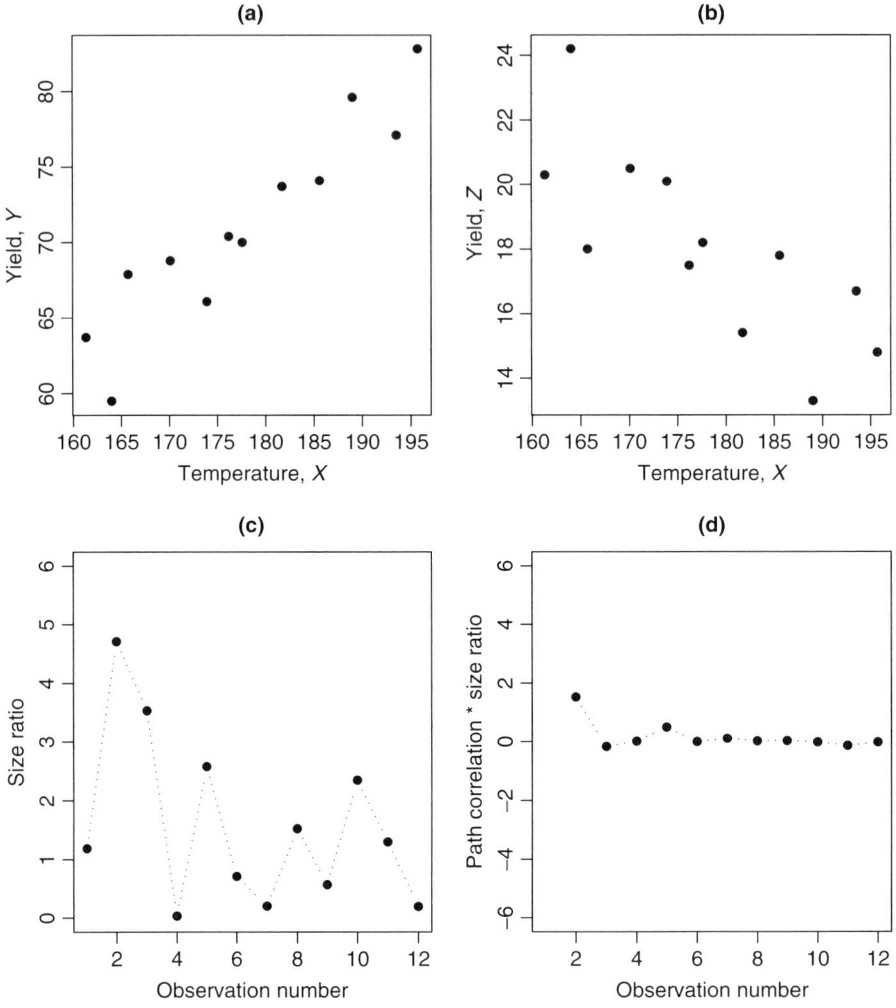

Figure 5.6 Correlated regressions: (a) scatter plot of original data, yield Y versus temperature X; (b) scatter plot of original data, yield Z versus temperature X; (c) partial size ratios for sequential adjustments; (d) successive path correlations multiplied by size ratios.

degrees of uncertainty that might plausibly be held by the process production manager. For the error quantities, we specify $\sigma_e^2 = 6.25$, $\sigma_f^2 = 4$, and $\sigma_{ef} = 2.5$, so that the correlation between the two error components for any given run is about 0.5. We chose these error variances by examining the residuals from separate least squares fits, so that the analysis would not be complicated by obvious conflicts between prior beliefs and the data; we shall discuss variance **learning** in a later chapter. We specify the following expectations and covariances between the regression coefficients:

$$\mathrm{E}\left(\begin{bmatrix} a \\ b \\ c \\ d \end{bmatrix}\right) = \begin{bmatrix} 75 \\ 40 \\ 20 \\ -30 \end{bmatrix}, \tag{5.70}$$

$$\mathrm{Var}\left(\begin{bmatrix} a \\ b \\ c \\ d \end{bmatrix}\right) = \begin{bmatrix} 4 & -6 & -1 & 0 \\ -6 & 225 & 0 & -90 \\ -1 & 0 & 1 & -2.4 \\ 0 & -90 & -2.4 & 144 \end{bmatrix}. \tag{5.71}$$

The variance–covariance specifications indicate rather more uncertainty about the first regression equation than the second. Each slope quantity is considerably more uncertain than the corresponding intercept quantity: a production manager may know roughly the yield for an average temperature, but may have only sketchy beliefs concerning the direction and magnitude of the slope. We have specified the same degree of weak negative correlation, -0.3, between slope and intercept for each regression.

If we suppose that, at any given temperature, the total amount of product yield $(Y + Z)$ will fluctuate between fairly narrow limits, then the two yields should be strongly negatively correlated. We choose to introduce this information into the model by specifying negative correlations (each -0.5) between the two intercepts and between the two slopes. We treat the slope and intercept for different regressions as being uncorrelated, as we have exhausted our intuition about the physical process. This completes the prior specification process.

We must now construct the beliefs for the data quantities. These are obtained via (5.69), (5.70), (5.71) as

$$\begin{aligned}
\mathrm{E}(Y_i) &= \mathrm{E}(a) + x_i\mathrm{E}(b), \\
\mathrm{E}(Z_i) &= \mathrm{E}(c) + x_i\mathrm{E}(d), \\
\mathrm{Var}(Y_i) &= \mathrm{Var}(a) + 2x_i\mathrm{Cov}(a, b) + x_i^2\mathrm{Var}(b) + \mathrm{Var}(e_i), \\
\mathrm{Var}(Z_i) &= \mathrm{Var}(c) + 2x_i\mathrm{Cov}(c, d) + x_i^2\mathrm{Var}(d) + \mathrm{Var}(f_i), \\
\mathrm{Cov}(Y_i, Z_j) &= \mathrm{Cov}(a, c) + x_i x_j\mathrm{Cov}(b, d), && i \neq j, \\
&= \mathrm{Cov}(a, c) + x_i^2\mathrm{Cov}(b, d) + \mathrm{Cov}(e_i, f_i), && i = j.
\end{aligned}$$

Finally we must construct the beliefs between the regression coefficients and the data quantities, for example:

$$\begin{aligned}
\mathrm{Cov}(a, Y_i) &= \mathrm{Var}(a) + x_i\mathrm{Cov}(a, b), & \mathrm{Cov}(b, Y_i) &= \mathrm{Cov}(a, b) + x_i\mathrm{Var}(b), \\
\mathrm{Cov}(c, Y_i) &= \mathrm{Cov}(c, a), & \mathrm{Cov}(d, Y_i) &= x_i\mathrm{Cov}(d, b).
\end{aligned}$$

We will arrange the regression coefficients into the collection $G = [a, b, c, d]$, and the yields as the collections $Y = (Y_1, \ldots, Y_{12})$ and $Z = (Z_1, \ldots, Z_{12})$.

5.14.2.3 *Overall adjustment*

Suppose we now calculate the adjustment of G by all the data quantities, $Y \cup Z$. The adjustment is summarized in Table 5.7. Expectations for the four regression coefficients are roughly in line with the prior specification; the intercept terms a, c have been revised downwards and the slope terms b, d revised upwards. In terms of standard deviations of variance resolved, none of the changes from prior to adjusted appear surprising. The average variance resolution across the belief structure $[G]$ is 76.4%, and the variance resolution for each of the coefficients is 80% or better. We may also calculate the canonical quantities and summarize the adjustment in terms of these, as shown in Table 5.8. This shows that there are two directions W_1, W_2 in which we expect to resolve better than 93% of the uncertainty, and two more directions for which the data are expected to be less informative. The changes in expectation for these quantities are broadly unsurprising, although the change for W_3 is quite large relative to the third canonical resolution of 0.69. The canonical quantities and the bearing vector are, in terms of the standardized versions of the regression coefficients,

$$
\begin{aligned}
W_1 &= 0.52\,\mathrm{S}(a) - 0.37\,\mathrm{S}(b) - 0.26\,\mathrm{S}(c) + 0.36\,\mathrm{S}(d),\\
W_2 &= 0.56\,\mathrm{S}(a) + 0.44\,\mathrm{S}(b) - 0.27\,\mathrm{S}(c) - 0.46\,\mathrm{S}(d),\\
W_3 &= 0.39\,\mathrm{S}(a) - 0.71\,\mathrm{S}(b) + 0.57\,\mathrm{S}(c) - 0.68\,\mathrm{S}(d),\\
W_4 &= 0.89\,\mathrm{S}(a) + 0.83\,\mathrm{S}(b) + 1.03\,\mathrm{S}(c) + 0.85\,\mathrm{S}(d),\\
\mathbb{Z}_{y\cup z}(G) &= -2.32\,\mathrm{S}(a) + 0.61\,\mathrm{S}(b) - 2.06\,\mathrm{S}(c) + 0.55\,\mathrm{S}(d),
\end{aligned}
$$

Table 5.7 Summary of adjusted expectations, standardized changes and variances for the regression coefficients a, b, c, d.

	Expectation			Uncertainty			
	Prior	Adjusted	Change	Prior	Adjusted	Resolved	Resolution
a	75	72.175	−1.48	4	0.377	3.623	0.906
b	40	51.905	0.85	225	29.794	195.206	0.868
c	20	18.993	−1.13	1	0.202	0.798	0.798
d	−30	−22.058	0.71	144	19.089	124.911	0.867
G				4	0.944	3.056	0.764

Table 5.8 Summary of adjusted expectations, standardized changes and variances for the canonical quantities $W_1, \ldots, W_4$.

	Expectation			Uncertainty			
	Prior	Adjusted	Change	Prior	Adjusted	Resolved	Resolution
W_1	0	−0.529	−0.54	1	0.053	0.947	0.947
W_2	0	−0.463	−0.48	1	0.068	0.932	0.932
W_3	0	−2.131	−2.57	1	0.310	0.690	0.690
W_4	0	−1.077	−1.54	1	0.514	0.486	0.486
G				4	0.944	3.056	0.764

with resolutions and corresponding size

$$\begin{aligned} \mathrm{R}_{Y\cup Z}(W_1) &= 0.9472, \\ \mathrm{R}_{Y\cup Z}(W_2) &= 0.9323, \\ \mathrm{R}_{Y\cup Z}(W_3) &= 0.6900, \\ \mathrm{R}_{Y\cup Z}(W_4) &= 0.4864, \\ \mathrm{Size}_{y\cup z}(G) &= 6.19, \end{aligned}$$

where $\mathrm{Var}(\mathbb{Z}_{y\cup z}(G)) = \mathrm{Size}_{y\cup z}(G)$. W_3 contrasts the intercept terms a, c with the slope terms b, d. Notice that the bearing vector $\mathbb{Z}_{y\cup z}(G)$ similarly contrasts the intercept terms a, c with the slope terms b, d, but with more emphasis on the intercept terms, implying relatively larger standardized changes in expectation for the intercept terms. The size of the adjustment turns out to be 6.19, with prior expectation 3.06, giving a size ratio of $\mathrm{Sr}_{y\cup z}(G) = 2.03$. This is well within the heuristic upper threshold of 4.09 given by expression (4.65): there appear to be no major contradictions between prior specifications and data.

5.14.2.4 *Partial adjustment: comparing two data sources*

We shall now consider two partial adjustments that may be diagnostically useful. Both concern the way in which we employ the data. The first of these is to check that the implications of the observations for the two kinds of yield, Y and Z, are consistent. Therefore, we adjust the regression coefficients G by the set of observations for the yields for the main product Y, and then adjust partially by the remaining set of observations for the yields for the by-product Z. We find that there are only two canonical quantities for the adjustment of G by Y alone:

$$\begin{aligned} W_1 &= 0.89\,\mathrm{S}(a) - 0.31\,\mathrm{S}(b), & \mathrm{R}_{Y\cup Z}(W_1) &= 0.8915, \\ W_2 &= 0.50\,\mathrm{S}(a) + 0.97\,\mathrm{S}(b), & \mathrm{R}_{Y\cup Z}(W_2) &= 0.8245, \\ \mathbb{Z}_y(G) &= -1.61\,\mathrm{S}(a) + 0.54\,\mathrm{S}(b), & \mathrm{Var}(\mathbb{Z}_y(G)) &= \mathrm{Size}_y(G) = 3.23. \end{aligned}$$

Notice that, although Y is 12-dimensional and G is four-dimensional, the canonical adjustment is two-dimensional. This because our model has Y_is constructed from coefficients a, b but not c, d. Therefore, we should find that the data set Y is essentially informative for a and b. This does not mean that we cannot use the data Y to learn about c and d, but that we do so indirectly via their covariances with a and b. Similarly, the bearing $\mathbb{Z}_y(G)$ is two-dimensional: the direction of maximal change relative to prior variation is $-1.61\mathrm{S}(a) + 0.54\mathrm{S}(b)$. Notice that this implies that we cannot expect that the data set Y will induce large changes in expectation for the coefficient d. This follows because changes in expectation arise through covariance with the bearing, by (4.52), and because d has zero prior covariance with a, the major component in this bearing. The size ratio for the adjustment by $Y = y$ is $\mathrm{Sr}_y(G) = 1.88$, showing no particular discrepancy.

We now adjust G partially by $Z = z$ in addition to $Y = y$. This partial adjustment is also two-dimensional, as the Z_is are directly informative for c, d and only indirectly informative for a, b. The partial canonical directions and the partial bearing turn out to be

$$U_1 = 0.11\,\mathrm{S}(a) - 0.29\,\mathrm{S}(b) + 0.45\,\mathrm{S}(c) - 0.92\,\mathrm{S}(d),$$

$$U_2 = 0.45\,\mathrm{S}(a) + 0.34\,\mathrm{S}(b) + 1.12\,\mathrm{S}(c) + 0.78\,\mathrm{S}(d),$$

$$\mathbb{Z}_{[z/y]}(G) = -0.71\,\mathrm{S}(a) + 0.07\,\mathrm{S}(b) - 2.06\,\mathrm{S}(c) + 0.55\,\mathrm{S}(d),$$

with corresponding resolutions and size

$$\mathrm{R}_{Y \cup Z}(U_1) = 0.7652,$$

$$\mathrm{R}_{Y \cup Z}(U_2) = 0.5748,$$

$$\mathrm{Size}_{[z/y]}(G) = 4.02,$$

where $\mathrm{Var}(\mathbb{Z}_{[z/y]}(G)) = \mathrm{Size}_{[z/y]}(G)$. The coefficients for these quantities show, as would be expected, that the main implications of adjusting partially by $Z = z$, in addition to $Y = y$, are greater resolutions of variance for c and d, and relatively larger changes in expectations for c and d. However, notice that the additional information can also indirectly resolve some of the previously unexplained variation in a, b, and can lead to changes in their expectations. The changes in variance are shown in Table 5.9. These emphasize the value of the partial adjustment for c, d, whereas the additional resolution of variance for a, b is minimal. The changes in expectation are shown in Table 5.10. None of the individual standardized changes is too alarming, the largest being a standardized change of about 2.4 standard deviations for the partial change in expectation for coefficient c. However, if we examine the initial changes in expectation for the regression coefficients (adjusting by $Y = y$), and compare these changes to the changes obtained for the partial adjustment (by $Z = z$ given $Y = y$), we see that all the partial changes are opposite in sign to the initial changes. This suggests that the two sets of observations have somewhat contradictory implications. Indeed, when we calculate the path

Table 5.9 Initial and partial changes in variance for the coefficients a, b, c, d.

	Variance			
	Prior	Resolved by Y	Resolved by $[Z/Y]$	Not resolved
a	4	3.5410	0.0821	0.3769
b	225	189.1326	6.0731	29.7942
c	1	0.2280	0.5700	0.2021
d	144	31.1530	93.7584	19.0887
	Resolution			
	Prior	Resolved by Y	Resolved by $[Z/Y]$	Not resolved
a	0	0.8853	0.0205	0.0942
b	0	0.8406	0.0270	0.1324
c	0	0.2280	0.5700	0.2021
d	0	0.2163	0.6511	0.1326

Table 5.10 Initial and partial changes in expectation for the coefficients a, b, c, d. Standardized changes are given in parentheses.

	Expectation			
	Prior	Change, given y	Partial change, given $[z/y]$	Final
a	75	−3.4378	0.6125	72.1747
		(−1.83)	(2.14)	(−1.48)
b	40	12.8964	−0.9919	51.9045
		(0.94)	(−0.40)	(0.85)
c	20	0.8057	−1.8126	18.9931
		(1.69)	(−2.40)	(−1.13)
d	−30	−3.2249	11.1674	−22.0575
		(−0.58)	(1.15)	(0.71)

correlation (5.51) we find it to be

$$\text{PC}(y, [z/y]) = \text{Corr}(\mathbb{Z}_y(G), \mathbb{Z}_{[z/y]}(G)) = -0.1468,$$

so that our global measure of data consistency shows that the two sets of observations y, z are very weakly contradictory relative to the prior specification.

5.14.2.5 Partial adjustment: sequential adjustment

For the second diagnostic partial adjustment, recall that the experiment concerns obtaining two yields at each of 12 temperatures $x_1, \ldots, x_{12}$. Thus, we are concerned with checking whether the results obtained from the sequence of measurements are consistent. To do so, we arrange the pair of measurements corresponding to temperature x_i as the collection $H_i = [Y_i, Z_i]$. We now make a series of partial

Table 5.11 Diagnostics and changes in expectation for the sequential adjustment.

Run		Coefficient estimates				Diagnostics		
i	x_i	$\mathrm{E}_{[i]}(a)$	$\mathrm{E}_{[i]}(b)$	$\mathrm{E}_{[i]}(c)$	$\mathrm{E}_{[i]}(d)$	$\mathrm{Size}_{i/}(G)$	$\mathrm{Sr}_{i/}(G)$	$C_{[i/]}$
1	161.30	74.10	44.58	19.57	−20.87	1.44	1.18	
2	164.00	72.24	53.57	20.11	−24.02	1.06	4.71	0.3234
3	165.70	73.05	52.93	19.49	−20.29	0.44	3.53	−0.0468
4	170.10	73.10	53.32	19.42	−20.24	0.00	0.03	0.4501
5	173.90	71.93	48.70	19.82	−16.80	0.54	2.57	0.1927
6	176.20	72.19	51.01	19.51	−18.16	0.13	0.70	−0.0066
7	177.60	71.99	50.36	19.46	−17.87	0.03	0.20	0.5382
8	181.70	72.35	53.64	19.08	−21.03	0.27	1.52	0.0146
9	185.60	71.99	50.43	19.21	−19.42	0.09	0.56	0.0552
10	189.00	72.44	55.10	18.85	−24.10	0.36	2.34	−0.0060
11	193.50	72.04	50.11	19.03	−21.33	0.19	1.29	−0.1086
12	195.70	72.17	51.90	18.99	−22.05	0.02	0.19	−0.0906

adjustments of G by H_1, and then partially by H_2, and so forth. At each stage, we examine the ways in which the estimates for the regression coefficients change and study the various diagnostics. These features are summarized in Table 5.11 and plotted in Figures 5.6 and 5.7. We do not observe any features to cause us to doubt the prior specifications or the data. The estimates for the regression coefficients show some fluctuation, but do not show any peculiar patterns. The estimates for the coefficient c do fall as more evidence is accumulated, but this is clearly an artefact of the configuration of the initial observations. None of the size ratios are overly large, so we do not suspect important contradictions between the data and the prior specifications. The path correlations (multiplied by the corresponding size ratio, to emphasize important features) vary between very weakly negative and quite positive, and show no systematic features. We conclude that the sequential adjustments do not reveal serious inconsistencies amongst the data.

5.14.2.6 Exploration of systematic bias

We can see whether or not the observed sequences $Y_1, \ldots, Y_{12}$ and $Z_1, \ldots, Z_{12}$ are well aligned with their prior specifications by using the ideas outlines in §5.13. We calculate and plot the standardized cumulative standardized one-step forecast errors, $Q_1, \ldots, Q_{12}$, where, for the $\{Y_i\}$ sequence,

$$Q_j = \frac{1}{\sqrt{j}} \sum_{i=1}^{j} \epsilon_i,$$

and where the one-step forecast errors are

$$\epsilon_i = \frac{y_i - \mathrm{E}_{[i-1]}(Y_i)}{\sqrt{\mathrm{Var}_{[i-1]}(Y_i)}}.$$

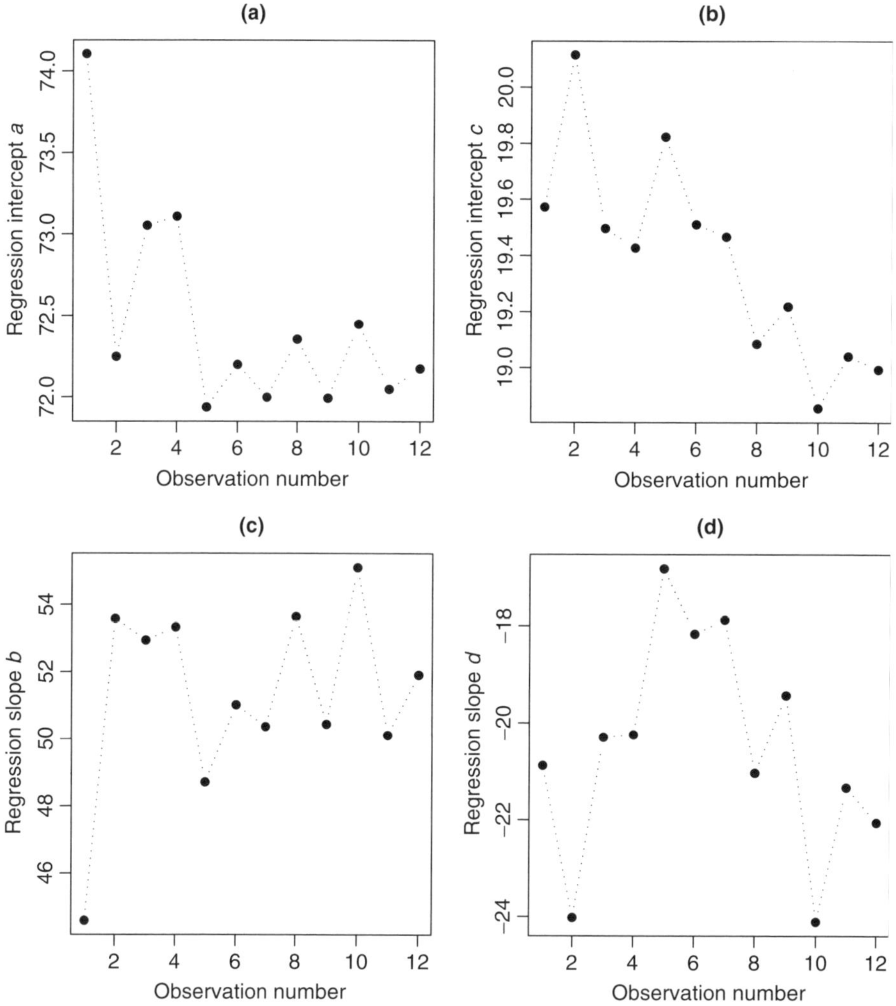

Figure 5.7 Correlated regressions: (a) sequential update of expectation for a; (b) sequential update of expectation for c; (c) sequential update of expectation for b; (d) sequential update of expectation for d.

The plotted values of Q_j are shown in the top part of Figure 5.8, with the corresponding values for the $\{Z_i\}$ sequence shown in the bottom part. Both sequences are persistently lower than the expected value of zero. We deduce that the prior expectations for both the $\{Y_i\}$ sequence and the $\{Z_i\}$ sequence are systematically higher than the values observed, taking into account all the evidence available prior to each observation. The magnitude of the differences approaches two standard deviations for much of the sequences. There is one value, $Q_2 = -2.303$, which dips below the

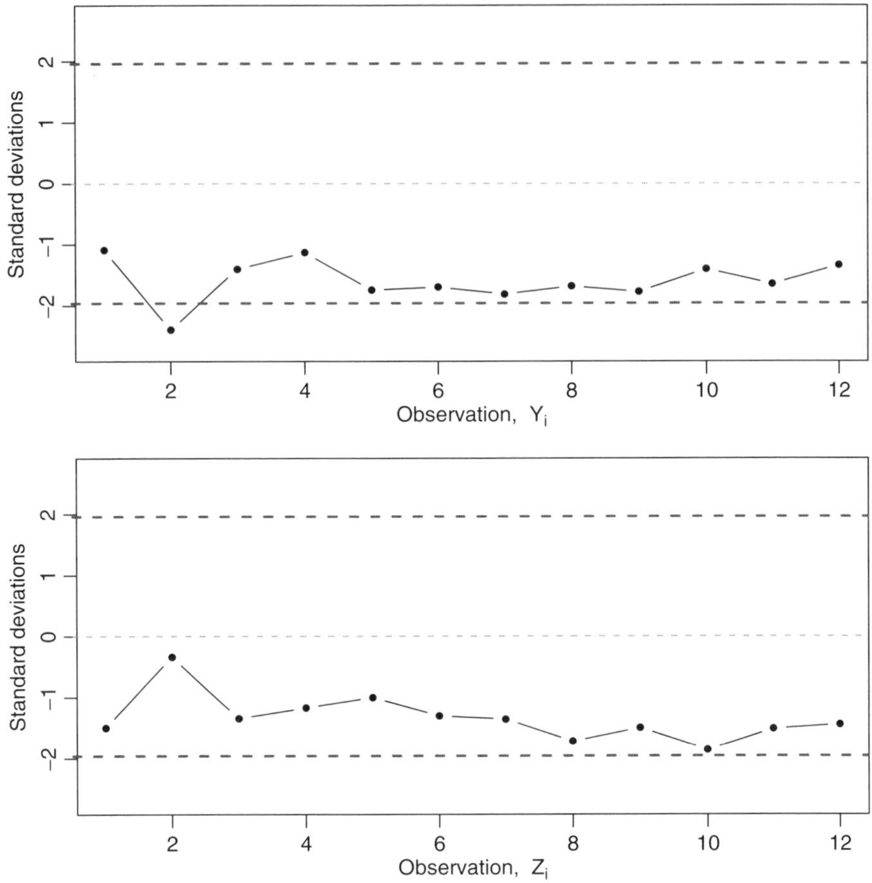

Figure 5.8 Correlated regressions: exploring systematic differences between observations and priors. The values plotted are the standardized cumulative standardized one-step forecast errors.

two-standard-deviation threshold. However, we would expect some larger values, especially at the start of such sequences where the normal approximation is not trustworthy. Taking into account the diagnostics which we checked via the sequential adjustments of the previous section, we conclude that there is some evidence of persistent overestimation of the observations, but that this has had only a marginal influence on the actual adjustment of belief for the regression coefficients.

5.15 Bayes linear sufficiency and belief separation

One of the most important relations in probabilistic modelling is that of conditional independence. Conditional independence is a sufficiency property. The random vectors A, B are conditionally independent given vector C if the conditional distribution of B given both A and C is the same as the conditional distribution of B given C alone. The corresponding Bayes linear relation is as follows.

Definition 5.18 *C is **Bayes linear sufficient** for A for adjusting B if*

$$\mathrm{E}_{C \cup A}(B) = \mathrm{E}_C(B).$$

Definition 5.19 *If A, B, C are three collections of random quantities, then C **separates** A and B, written*

$$\lfloor A \perp\!\!\!\perp B \rfloor \,/\, C,$$

*if C is Bayes linear sufficient for B for adjusting A. In this case, we say that A and B are **orthogonal given** C.*

We have various equivalent forms for the belief separation $\lfloor A \perp\!\!\!\perp B \rfloor \,/\, C$. These are based around the adjusted expectation, $\mathrm{E}_C(A)$, the adjusted belief structure $[A/C]$, and the belief transform $\mathbb{T}_{A:C}$ over A induced by the adjustment by C, as follows.

Theorem 5.20 *If A, B, C are three belief structures, then the following are all equivalent to the condition that $\lfloor A \perp\!\!\!\perp B \rfloor \,/\, C$:*

5.20.1: $\mathrm{E}_{B \cup C}(A) = \mathrm{E}_C(A)$;

5.20.2: $\mathrm{Cov}(A, B) = \mathrm{Cov}(A, C)\mathrm{Var}(C)^{\dagger}\mathrm{Cov}(C, B)$;

5.20.3: $\mathrm{E}_{\mathbb{A}_C(A)}(B) = 0$;

5.20.4: $[A/(B \cup C)] = [A/C]$;

5.20.5: $\mathbb{T}_{A:(B \cup C)} = \mathbb{T}_{A:C}$;

5.20.6: $\mathbb{A}_C(A) \perp \mathbb{A}_C(B)$;

5.20.7: $A \perp \mathbb{A}_C(B)$.

The above equivalences all follow directly from the definition of belief separation. For example, Property 5.20.4 follows as

$$[A/(B \cup C)] = [[A/C]/[B/C]] = [A/C],$$

if and only if $[A/C] \perp [B/C]$, as for any pair of collections U, V,

$$[U/V] = [U] \Leftrightarrow [U] \perp [V].$$

We may automatically generate the separations between two collections of random quantities through the analysis of the eigenstructure of the corresponding resolution transforms. From (3.83), the adjustment of B by D is performed strictly over the space of eigenvectors in $\mathbb{H}(D/B)$. It follows that, for any B, D, we have

$$\lfloor B \perp\!\!\!\perp D \rfloor \,/\, \mathbb{H}(D/B). \tag{5.72}$$

Further, if $\mathbb{H}(D/B_1) = \mathbb{H}(D/B_2) = \ldots = \mathbb{H}(D/B_k) = H$, then

$$\lfloor D \perp\!\!\!\perp (B_1 \cup \ldots \cup B_k) \rfloor \,/\, H. \tag{5.73}$$

5.16 Properties of generalized conditional independence

Belief separation is a generalized conditional independence property, which shares many of the general properties of the more familiar conditional independence relation.

Definition 5.21 *A generalized conditional independence property is a tertiary property on collections of objects which obeys the following three basic properties, for any collections B, C, D, F:*

5.21.1: $\lfloor B \perp\!\!\!\perp C \rfloor / (C \cup D)$;

5.21.2: $\lfloor B \perp\!\!\!\perp C \rfloor / D \Leftrightarrow \lfloor C \perp\!\!\!\perp B \rfloor / D$;

5.21.3: $\lfloor B \perp\!\!\!\perp (C \cup D) \rfloor / F$ *implies and is implied by the pair of conditions*

- $\lfloor B \perp\!\!\!\perp D \rfloor / F$,
- $\lfloor B \perp\!\!\!\perp C \rfloor / (D \cup F)$.

These properties reflect natural aspects of our intuitive concept of conditional independence. Informally, Property 5.21.1 requires that given C and anything else, we learn nothing about C from any other quantity B. Property 5.21.2 is a natural symmetry requirement. Property 5.21.3 expresses the notion that we cannot break the conditional independence by subdividing a collection of quantities into sub-collections, namely we require the equivalence between the statement that we learn nothing about B from a collection G given F, and the pair of statements that (i) we learn nothing about B from any sub-collection D in G, given F, and (ii) having learnt D as well as F, we still learn nothing about B from the remaining elements C in G.

It turns out that any tertiary property obeying Properties 5.21.1–5.21.3 will behave computationally as a conditional independence property; see Smith (1990). In particular, as we shall show, belief separation obeys these properties, so that the various rules for belief propagation for probabilistic structures will have direct analogues for belief adjustment. Therefore, we may build graphical models based on belief separation which will have many of the same qualitative properties as do probabilistic graphical models based on probabilistic notions of conditional independence. We construct such graphical models in Chapter 10. We have the following result.

Theorem 5.22 *Belief separation is a generalized conditional independence property.*

Proof. Properties 5.21.1 and 5.21.2 follow immediately from the definition of belief separation. Property 5.21.3 follows as

(i) if $\lfloor B \perp\!\!\!\perp (C \cup D) \rfloor / F$, then immediately we have both $\lfloor B \perp\!\!\!\perp D \rfloor / F$ and $\lfloor B \perp\!\!\!\perp C \rfloor / F$. Therefore $\mathrm{E}_{D \cup F}(B) = \mathrm{E}_F(B)$, so that $\lfloor B \perp\!\!\!\perp C \rfloor / (D \cup F)$.

(ii)

$$\lfloor B \perp\!\!\perp (C \cup D) \rfloor \; / \; F \Leftrightarrow [B/(F \cup C \cup D)] = [B/F] \tag{5.74}$$

$$\Leftrightarrow [[B/(D \cup F)]/[C/(D \cup F)]] = [B/F]. \tag{5.75}$$

The condition $\lfloor B \perp\!\!\perp C \rfloor \; / \; (D \cup F)$ establishes that

$$[[B/(D \cup F)]/[C/(D \cup F)]] = [B/(D \cup F)]$$

and the condition $\lfloor B \perp\!\!\perp D \rfloor \; / \; F$ implies that $[B/(D \cup F)] = [B/F]$. Therefore, the pair of conditions establish the right-hand side of (5.74) and so also the left-hand side of (5.74).)

■

5.17 Properties of belief separation

We may exploit belief separation to simplify the calculations required to carry out a collection of belief adjustments. The simplest case occurs when $\lfloor A \perp\!\!\perp B \rfloor \; / \; C$, where $\langle C \rangle \subseteq \langle B \rangle$, in which case it is an immediate consequence that

$$\mathrm{E}_B(A) = \mathrm{E}_C(A), \quad \mathrm{Var}_B(A) = \mathrm{Var}_C(A). \tag{5.76}$$

The general form of the above relation is as follows.

Theorem 5.23 *If* $\lfloor A \perp\!\!\perp B \rfloor \; / \; C$, *then*

5.23.1: $\mathrm{E}_B(A) = \mathrm{E}_B(\mathrm{E}_C(A))$,

5.23.2: $\mathrm{Var}_B(A) = \mathrm{Var}_C(A) + \mathrm{Var}_B(\mathrm{E}_C(A))$,

5.23.3: $\mathrm{Cov}_B(A, C) = \mathrm{Cov}_B(\mathrm{E}_C(A), C)$,

5.23.4: $\mathrm{Var}_{B \cup C}(A) = \mathrm{Var}_C(A)$,

5.23.5: $\mathrm{Cov}(A, \mathrm{E}_C(B)) = \mathrm{Cov}(A, B) = \mathrm{Cov}(\mathrm{E}_C(A), B)$.

Proof. We have

$$\begin{aligned} \mathrm{E}_B(A) &= \mathrm{E}_B(\mathrm{E}_{C \cup B}(A)) \\ &= \mathrm{E}_B(\mathrm{E}_C(A) + \mathrm{E}_{[B/C]}(A)) \\ &= \mathrm{E}_B(\mathrm{E}_C(A)) \end{aligned}$$

as $A \perp [B/C]$. Further

$$\begin{aligned} \mathrm{Var}_B(A) &= \mathrm{Var}_B(A - \mathrm{E}_C(A) + \mathrm{E}_C(A)) \\ &= \mathrm{Var}_B(A - \mathrm{E}_C(A)) + \mathrm{Var}_B(\mathrm{E}_C(A)) \\ &= \mathrm{Var}_C(A) + \mathrm{Var}_B(\mathrm{E}_C(A)). \end{aligned}$$

Property 5.23.3 follows as $(A - \mathrm{E}_C(A))$ is uncorrelated with $C \cup B$. The remaining properties follow similarly. ■

Theorem 5.23 is an example of the role of belief separation in local computation. Bayes linear local computation is similar in purpose to probabilistic local computation. We simplify the task of evaluating a large collection of belief adjustments by breaking the overall adjustment into a collection of smaller adjustments. Thus, the above theorem shows that if $\lfloor A \perp\!\!\!\perp B \rfloor / C$ then the adjustment of (A, C) by B is completely determined by the adjustment of A by C and the adjustment of C by B. In some cases, such staged adjustment is most efficiently carried out through the corresponding resolution transforms, as outlined in the following theorem.

Theorem 5.24 *Write the normalized eigenvectors of* $\mathbb{T}_{C:B}$ *as* $Z_1, Z_2, \ldots,$ *and the normalized eigenvectors of* $\mathbb{T}_{B:C}$ *as* $W_1, W_2, \ldots,$ *respectively, with corresponding eigenvalues* $\lambda_1, \lambda_2, \ldots.$ *If* $\lfloor A \perp\!\!\!\perp B \rfloor / C$, *then we have the following properties.*

5.24.1: *For any* $X \in \langle A \rangle$,

$$\mathrm{E}_B(X) - \mathrm{E}(X) = \sum_i \sqrt{\lambda_i}\mathrm{Cov}(X, Z_i)W_i = \sum_i \sqrt{\lambda_i}\mathrm{Cov}(\mathrm{E}_C(X), Z_i)W_i, \tag{5.77}$$

$$\mathrm{Var}_B(X) = \mathrm{Var}(X) - \sum_i \lambda_i[\mathrm{Cov}(X, Z_i)]^2. \tag{5.78}$$

5.24.2: *For any* $X \in \langle A \rangle$,

$$\mathbb{T}_{A:B}(X) = \mathrm{E}_A(\mathbb{T}_{C:B}(\mathrm{E}_C(X))).$$

5.24.3: W *is an eigenvector of* $\mathbb{T}_{A:B}$, *with eigenvalue* λ *if and only if* $\mathrm{E}_C(W)$ *is an eigenvector of* $\mathbb{T}_{C:A}(\mathbb{T}_{C:B}(\cdot))$, *with eigenvalue* λ.

5.24.4: $\mathbf{tr}\{\mathbb{T}_{A:B}\} = \mathbf{tr}\{\mathbb{T}_{C:B}\mathbb{T}_{C:A}\}$.

5.24.5: *If* $\mathbb{T}_{C:B}$ *and* $\mathbb{T}_{C:A}$ *have the same eigenvectors* $Y_1, Y_2, \ldots,$ *with eigenvalues* $\theta_1, \theta_2, \ldots$ *for* $\mathbb{T}_{C:B}$, *and* $\phi_1, \phi_2, \ldots$ *for* $\mathbb{T}_{C:A}$, *then the quantities* $\mathrm{E}_A(Y_1), \mathrm{E}_A(Y_2), \ldots$ *are the eigenvectors of* $\mathbb{T}_{A:B}$ *and the quantities* $\mathrm{E}_B(Y_1), \mathrm{E}_B(Y_2), \ldots$ *are the eigenvectors of* $\mathbb{T}_{B:A}$, *and with eigenvalues* $\lambda_i = \theta_i\phi_i$, *in each case.*

Proof. For Property 5.24.1, the canonical directions Z_i for the adjustment of C by B, and the canonical directions W_i for the adjustment of B by C, are related (see §3.9.4) by $\mathrm{E}_B(Z_i) = \sqrt{\lambda_i}W_i$. For any $X \in A$ we have $\mathrm{E}_C(X) \in \langle C \rangle$ and $Z_1, Z_2, \ldots$ form a basis for $\langle C \rangle$ (see §3.9.2). Thus, we may write

$$\mathrm{E}_C(X) - \mathrm{E}(X) = \sum_i \mathrm{Cov}(\mathrm{E}_C(X), Z_i)Z_i,$$

so that

$$\mathrm{E}_B(\mathrm{E}_C(X) - \mathrm{E}(X)) = \mathrm{E}_B\left(\sum_i \mathrm{Cov}(\mathrm{E}_C(X), Z_i)Z_i\right),$$

so that equation (5.77) follows from Property 5.23.1, as

$$\text{Cov}(X - \text{E}_C(X), Z_i) = 0,$$

from Property 5.20.7.

The variance property (5.78) follows from Property 5.23.2, as

$$\begin{aligned}
\text{Var}_B(X) &= \text{Var}_C(X) + \text{Var}_B(\text{E}_C(X)) \\
&= \text{Var}(X) - \text{RVar}_C(X) + \text{Var}(\text{E}_C(X)) - \text{RVar}_B(\text{E}_C(X)) \\
&= \text{Var}(X) - \text{RVar}_B(\text{E}_C(X)) \\
&= \text{Var}(X) - \sum_i \lambda_i \text{Cov}(\text{E}_C(X), Z_i)^2 \quad \text{by (3.70)}.
\end{aligned}$$

For any $X \in \langle A \rangle$, we have, from Property 5.23.1 that

$$\text{E}_B(X) = \text{E}_B(\text{E}_C(X))$$

so that

$$\begin{aligned}
\mathbb{T}_{A:B}(X) &= \text{E}_A(\text{E}_B(X)) \\
&= \text{E}_A(\text{E}_C(\text{E}_B(\text{E}_C(X)))) \quad \text{(from Property 5.23.1)} \\
&= \text{E}_A(\mathbb{T}_{C:B}(\text{E}_C(X))),
\end{aligned}$$

giving Property 5.24.2

W is an eigenvector of $\mathbb{T}_{A:B}$, with eigenvalue λ if

$$\mathbb{T}_{A:B}(W) = \lambda W,$$

so that, from Property 5.24.2,

$$\text{E}_C(\text{E}_A(\mathbb{T}_{C:B}(\text{E}_C(W)))) = \mathbb{T}_{C:A}(\mathbb{T}_{C:B}(\text{E}_C(W))) = \lambda \text{E}_C(W),$$

so that $\text{E}_C(W)$ is an eigenvector of $\mathbb{T}_{C:A}(\mathbb{T}_{C:B})$, with eigenvalue λ. Property 5.24.4 follows as

$$\begin{aligned}
\mathbf{tr}\{\mathbb{T}_{A:B}\} &= \mathbf{tr}\{\text{E}_A(\mathbb{T}_{C:B}(\text{E}_C(.)))\} \\
&= \mathbf{tr}\{\mathbb{T}_{C:B}(\text{E}_C(\text{E}_A(.)))\} = \mathbf{tr}\{\mathbb{T}_{C:B}(\mathbb{T}_{C:A})\}.
\end{aligned}$$

Finally, suppose that the transforms $\mathbb{T}_{C:B}$ and $\mathbb{T}_{C:A}$ have the same eigenvectors $Y_1, Y_2, \ldots,$ with eigenvalues $\theta_1, \theta_2, \ldots,$ for $\mathbb{T}_{C:B}$, and $\phi_1, \phi_2, \ldots,$ for $\mathbb{T}_{C:A}$. Let

$$W_i = \frac{1}{\sqrt{\theta_i}} \text{E}_B(Y_i), \quad U_i = \frac{1}{\sqrt{\phi_i}} \text{E}_A(Y_i).$$

From (3.81), U_i, W_i are the normalized eigenvectors of $\mathbb{T}_{A:C}$, $\mathbb{T}_{B:C}$ respectively. Therefore, $\text{Cov}(U_i, Y_j) = 0$, for all $i \neq j$, so that from (5.77) we have

$$\begin{aligned}\text{E}_B(U_i) &= \frac{1}{\sqrt{\phi_i}}\text{Cov}(\text{E}_C(\text{E}_A(Y_i)), W_i)W_i \\ &= \sqrt{\phi_i}\text{Cov}(Y_i, W_i)W_i = \sqrt{\phi_i\theta_i}W_i,\end{aligned}$$

from (3.82). Similarly,

$$\text{E}_A(W_i) = \sqrt{\phi_i\theta_i}U_i$$

and Property 5.24.5 follows. ■

In subsequent sections, we shall construct models based on belief separation. It will be important in such models to consider which separations are preserved as we adjust beliefs. We have the following result, giving the conditions under which belief separation is preserved by partial belief adjustment.

Theorem 5.25 *Suppose that* A, B, C *are three belief structures for which we have* $\lfloor A \perp\!\!\!\perp B \rfloor / C$. *For any further belief structure* D *we have that*

$$\lfloor [A/D] \perp\!\!\!\perp [B/D] \rfloor / [C/D]$$

if and only if $\lfloor A \perp\!\!\!\perp B \rfloor / (C \cup D)$. *In particular, a sufficient condition for*

$$\lfloor [A/D] \perp\!\!\!\perp [B/D] \rfloor / [C/D]$$

is that $\lfloor A \perp\!\!\!\perp (B \cup D) \rfloor / C$.

Proof. The condition $\lfloor [A/D] \perp\!\!\!\perp [B/D] \rfloor / [C/D]$ is equivalent to the condition

$$(A - \text{E}_D(A)) - \text{E}_{\mathbb{A}_D(C)}(A - \text{E}_D(A)) \perp (B - \text{E}_D(B)) - \text{E}_{\mathbb{A}_D(C)}(B - \text{E}_D(B))$$

which reduces to the condition $\lfloor A \perp\!\!\!\perp B \rfloor / (C \cup D)$. A sufficient condition to ensure that $\lfloor A \perp\!\!\!\perp B \rfloor / (C \cup D)$ is that $\lfloor A \perp\!\!\!\perp B \cup D \rfloor / C$, from Property 5.21.3. ■

5.18 Example: regression with correlated responses

5.18.1 Exploiting separation

We continue the example from §5.14.2. Organize the quantities as follows. Let $G = [a, b, c, d]$ be the collection of uncertain quantities. Let

$$H = Y_1, \ldots, Y_{12}, Z_1, \ldots, Z_{12}$$

represent the quantities which we observed. Suppose that we are interested in predicting the responses

$$Y_r = a + bx_r + e_r, \quad Z_r = c + dx_r + f_r, \qquad r > 12,$$

for some, as yet unspecified, design point x_r; and collect these responses as the structure H_r.

First, we have that Y_r, Z_r are linear combinations of quantities in G, excepting the error terms e_r, f_r, which are uncorrelated with all other quantities. It follows that $\mathbb{A}_G(H_r) \perp H$. Thus, by Property 5.20.7 we have $\lfloor H_r \perp\!\!\!\perp H \rfloor / G$. That is, a future observation is separated from past observations by the set of uncertain quantities. Theorem 5.23 now shows how we may adjust the mean and variance of H_r by H via the adjustment of G by H. We have $Y_r \in \langle H_r \rangle$, so that, for example, by Property 5.23.1,

$$\begin{aligned} \mathrm{E}_H(Y_r) &= \mathrm{E}_H(\mathrm{E}_G(Y_r)) = \mathrm{E}_H(\mathrm{E}_G(a + bx_r + e_r)) \\ &= \mathrm{E}_H(a + bx_r), \quad \text{as } a + bx_r \in \langle G \rangle \text{ and } \mathrm{E}_G(e_r) = \mathrm{E}(e_r) = 0, \\ &= \mathrm{E}_H(a) + x_r \mathrm{E}_H(b), \end{aligned}$$

which depends only on the chosen value x_r and the quantities a, b adjusted by the data already available. By Property 5.23.2,

$$\begin{aligned} \mathrm{Var}_H(Y_r) &= \mathrm{Var}_G(Y_r) + \mathrm{Var}_H(\mathrm{E}_G(Y_r)) \\ &= \mathrm{Var}_G(a + bx_r) + \mathrm{Var}_G(e_r) + \mathrm{Var}_H(a + bx_r) \\ &= 0 + \mathrm{Var}(e_r) + \mathrm{Var}_H(a + bx_r), \\ &= \mathrm{Var}(e_r) + \mathrm{Var}_H(a) + 2x_r \mathrm{Cov}_H(a, b) + x_r^2 \mathrm{Var}_H(b), \end{aligned}$$

which depends only on an irreducible error variance and on the chosen value x_r and the adjusted variances and covariance for the quantities a, b given the initial data. The required adjusted expectations and variances are given in Table 5.7, excepting $\mathrm{Cov}_H(a, b) = -0.3372$ and $\mathrm{Var}(e_r) = 6.25$. Figure 5.9 shows the observations, the prediction as x_r varies, $\mathrm{E}_H(a) + x_r \mathrm{E}_H(b)$, a two-standard-deviation envelope for the prediction, and a two-standard-deviation envelope for the underlying mean component, $a + bx$. The envelopes are narrowest at

$$x_r = \frac{-\mathrm{Cov}_H(a, b)}{\mathrm{Var}_H(b)} = 178.9918,$$

after rescaling, and wider (more uncertain) as we move away from the observed region. Note that the second observation, (164, 59.50), appears relatively most distant from the predicted value for such a temperature: it is this observation which has the highest diagnostic value in Table 5.11.

5.18.2 Heart of the transform

The heart of the transform for this example is obtained by calculating the canonical directions and resolutions for the adjustment $\mathbb{T}_{H:G}$ of the data quantities H by the set of uncertain quantities G. As the former is 24-dimensional and the latter is

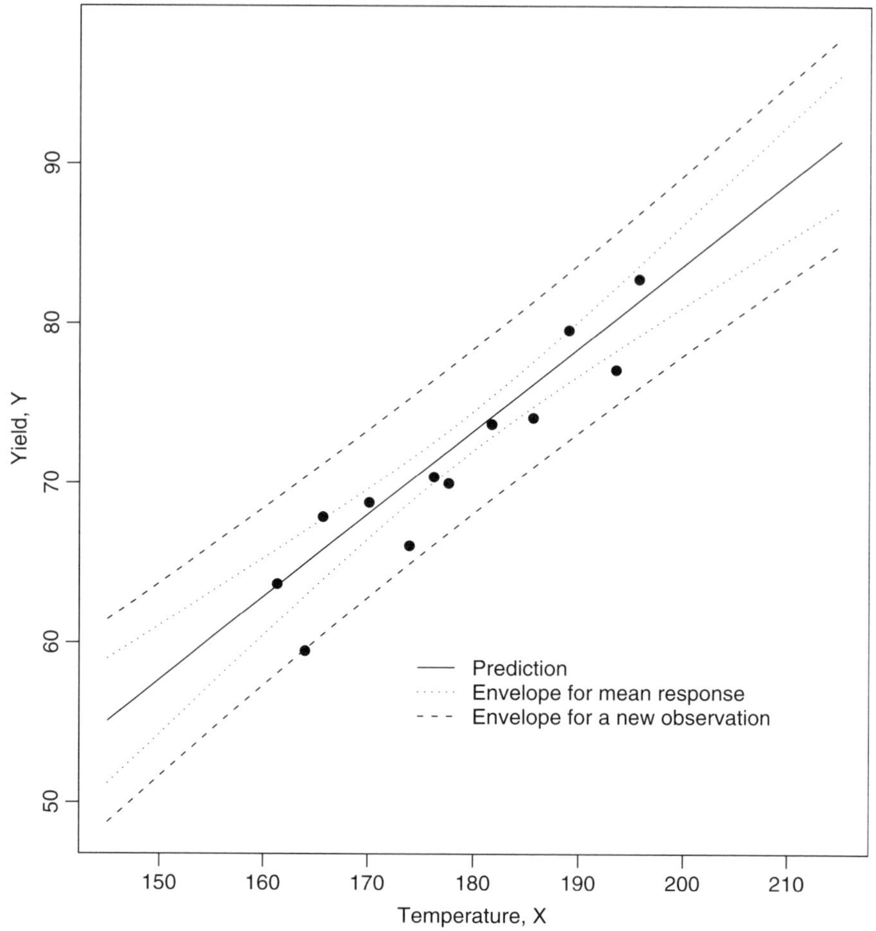

Figure 5.9 The original observations, the prediction as temperature varies, a two-standard-deviation envelope for the prediction, and a two-standard-deviation envelope for the underlying mean component.

four-dimensional, the heart of the transform is at most four-dimensional, possibly less, depending on the covariances between G and H. As it turns out, the heart is four-dimensional, with canonical resolutions

$$\lambda_1 = 0.9472, \quad \lambda_2 = 0.9323, \quad \lambda_3 = 0.6900, \quad \lambda_4 = 0.4864,$$

and corresponding directions $W_1, \ldots, W_4$, shown in Table 5.12 in the form

$$W_1 = 0.0487\, Y_1 + \ldots + 0.0147\, Z_{12} - 12.0698$$

Table 5.12 The four directions in H comprising the heart of the transform for the adjustment of G by H.

	W_1	W_2	W_3	W_4
Y_1	0.0487	0.0101	0.0580	0.0181
Y_2	0.0442	0.0048	0.0508	0.0109
Y_3	0.0414	0.0014	0.0462	0.0064
Y_4	0.0341	−0.0072	0.0343	−0.0053
Y_5	0.0277	−0.0146	0.0241	−0.0154
Y_6	0.0239	−0.0191	0.0178	−0.0216
Y_7	0.0216	−0.0219	0.0141	−0.0253
Y_8	0.0147	−0.0299	0.0030	−0.0362
Y_9	0.0082	−0.0376	−0.0075	−0.0466
Y_{10}	0.0026	−0.0443	−0.0167	−0.0556
Y_{11}	−0.0049	−0.0531	−0.0288	−0.0676
Y_{12}	−0.0086	−0.0574	−0.0347	−0.0734
Z_1	−0.0547	−0.0200	0.0930	−0.0036
Z_2	−0.0493	−0.0132	0.0842	−0.0127
Z_3	−0.0459	−0.0088	0.0787	−0.0185
Z_4	−0.0370	0.0023	0.0645	−0.0335
Z_5	−0.0293	0.0120	0.0522	−0.0464
Z_6	−0.0246	0.0179	0.0447	−0.0543
Z_7	−0.0218	0.0214	0.0402	−0.0590
Z_8	−0.0135	0.0318	0.0269	−0.0730
Z_9	−0.0057	0.0418	0.0143	−0.0862
Z_{10}	0.0012	0.0504	0.0033	−0.0978
Z_{11}	0.0103	0.0618	−0.0113	−0.1131
Z_{12}	0.0147	0.0674	−0.0184	−0.1206
Constant	−12.0698	17.1141	−21.3308	37.8097

and so forth. We arrange these directions as the collection W^+. These directions identify the four 'sufficient statistics' for the 24 observed data quantities. They divide the data into two spaces. One, $\mathbb{H}(H/G)$, is spanned by W^+ and is used for the belief adjustment. The second space, $\mathbb{H}^{\perp}(H/G)$, which we discuss below, is not relevant for adjusting G, but instead plays a diagnostic role for our specification. The observed values of these four directions are −0.54, 0.48, −2.57, 1.54, corresponding to prior expectation zero and prior variance one, so that the third 'sufficient statistic' is slightly at odds with the belief specification. Note the correspondence with the canonical directions summarized in Table 5.8.

5.18.2.1 Diagnostics in the complementary observed space

There remain 20 directions in the data space $[H]$ which are not informative for G, but which have been observed and thus can be compared diagnostically to prior beliefs specified about them. The simplest construction is to form the partial

Table 5.13 Squared observed adjustments for the 20 directions in H uninformative for adjustment of G.

14.0126	0.7170	11.6216	0.0602	0.8846
0.0018	0.4855	0.4776	0.6094	0.0167
0.1690	2.3612	0.5421	1.7612	0.3511
0.1414	0.2387	0.0567	1.0093	0.1033

adjustment of H by H given W^+: that is, we adjust H first by the heart W^+ and then by itself. This yields partial canonical directions (Definition 5.3) which we will label $W_5, \ldots, W_{24}$ and arrange as the collection W^0. These directions span $\mathbb{H}^{\perp}(H/G)$, the orthogonal complement of $\mathbb{H}(H/G)$ in H. The squared observed adjustments for these quantities, $\mathrm{E}_{[h/w^+]}(W_i)$, are shown in Table 5.13. By (5.41), the partial size of the adjustment can be obtained by summing these squares, giving $\mathrm{Size}_{[h/w^+]}(H) = 35.621$. The corresponding size ratio is thus $35.621/20 = 1.78$. As this is relatively close to its expectation, we judge that overall there is no strong evidence that the data are inconsistent with their prior specifications. There are, it is true, two quite large squared standardized adjustments in Table 5.13. Inspection of the first and third partial canonical directions (not shown here), corresponding to these largest discrepancies, reveals no obvious problems. For such inspection, we look for any patterns amongst the coefficients, taking into account the nature of the model. Here, the model encompasses a time sequence, $t = 1, 2, \ldots, 12$, and pairing of observations, (Y_t, Z_t), so that it is natural to look for patterns across time and for pairs.

We should also examine the partial bearing, which provides the direction in W^0 with maximal squared partial change in adjustment, equal to 35.621. This direction (not shown here) also has no obvious pattern amongst its coefficients.

5.19 Further reading

The construction and analysis of adjusted belief structures are discussed in Goldstein (1988a). The bearing and the data trajectory were introduced and illustrated in Goldstein (1988b). Our suggested diagnostic measures lend themselves to natural graphical representations; see Farrow and Goldstein (1996) and Williams and Goldstein (1999). For a more general discussion of Bayes linear sufficiency, with application to the types of problem discussed in O'Hagan et al. (1992), see Goldstein and O'Hagan (1996). Belief separation, as a generalized conditional independence property, is discussed in Goldstein (1990).

6

Exchangeable beliefs

So far we have been concerned with features which are general to any adjustment of beliefs. Now, we introduce additional features which are characteristic of statistical applications, and in particular we exploit the notion of exchangeability as a fundamental subjective judgement underlying many statistical models. Because we only require consideration of second-order beliefs, it turns out to be both feasible and desirable to use restricted exchangeability to build our second-order models strictly from prior specifications over observable quantities. We now describe this restricted notion of exchangeability, and derive the appropriate representation theorem for second-order exchangeable sequences. We introduce these ideas in the simple context of coin tossing, and then describe the general construction. In the latter part of the chapter, we show how exchangeable beliefs may be adjusted by observation, and in particular we show that the canonical structure for the adjustment may be constructed in a simple and intuitive form.

6.1 Exchangeability

To introduce our approach to statistical modelling, compare the following two situations.

(A) We want to estimate the proportion, p, of people in some large, finite population who possess some characteristic, say the number who smoke. Therefore, we take a random sample, of size n, with replacement. We count how many in our sample smoke, r say, and we use the ratio r/n, or some appropriate modification if n is small, as an estimate of p.

(B) We want to estimate the probability, p, that a spun coin lands heads. Therefore, we spin the coin n times, and count the number of heads, r, that we obtain. We use the ratio r/n, or some appropriate modification if n is small, as an estimate of p.

Bayes Linear Statistics: Theory and Methods M. Goldstein and D.A. Wooff

Traditional statistical methodology emphasizes the similarity between these two problems. In each case, we view the observed value r as having a binomial distribution, with parameters n, p. Depending on our approach to statistics, we may create some estimate for p or revise a prior probability distribution representing our beliefs concerning p, performing essentially the same analysis for (A) and (B).

However, there is a fundamental difference between the two cases. For (A), the quantity p of interest is a well-defined and, in principle, observable quantity, with a definite physical meaning, while for (B) p is not observable, even in principle, is not clearly defined, and has no immediate physical meaning. Therefore, in the second problem, before we can carry out a meaningful analysis we must first be clear just what it is that we are learning about.

In a sense, this difficulty arises from the problem of giving a precise statement of the relative frequency definition of probability. But, as the example suggests, even if we start from a subjectivist position, we often find that we need to create quantities such as 'the probability that a coin lands heads' in order to learn from the experience of spinning the coin.

One way that we may appear to avoid this problem is to argue that we are really interested in certain observable outcomes, such as the event that the coin will land heads on the next spin, and therefore to use the previous observations simply to change our predictive probabilities for such future observables. If we carry out such an analysis purely by specifying beliefs linking different observable outcomes, then our procedure does have a clear meaning, but we must pay a very high price for this, as the probabilistic specification that we require over the observables is usually very complex. Therefore, even if we are only interested in predictive statements about future coin spins, usually we will derive these predictive statements by supposing that there is an underlying probability that the coin will land heads, updating our beliefs about this probability by observing various spins of the coin, and then using our updated beliefs about the underlying probability for heads to generate predictive probabilities for future tosses.

In this way, we move the probability of heads from being a physical quantity which is of intrinsic interest to a mental construct which we introduce in order to simplify an analysis which otherwise would prove very complex. How can we justify the use of this mental construct?

The most careful interpretation of relative frequency probability as a mental construct comes from the notion of exchangeability. Suppose that we have an in principle infinite collection of coin spins $S = (S_1, S_2, \ldots)$, where each S_i takes value 1, if heads, or 0, if tails. We say that the collection S is **exchangeable** if our beliefs over S are unaffected by any permutation of the subscripts, or equivalently if, for each k, our probability for obtaining precisely k heads in n spins is the same irrespective of which subset of n spins we pick from S. In this case, we may apply the exchangeability representation theorem of de Finetti (see de Finetti 1937), which tells us that we may construct a probability measure Q on the interval

[0,1] such that our probability of k heads in n spins is given by

$$\int_0^1 \binom{n}{k} p^k (1-p)^{n-k} dQ(p). \tag{6.1}$$

Our beliefs therefore must be exactly as though we consider that there is a 'true but unknown' probability p that a spin will land heads, so that given p the sequence of tosses is independent and identically distributed with probability p of heads on each toss, and we specify a prior probability distribution Q for p. The distribution Q is uniquely specified by the prior specification over outcomes for the collection of coin spins.

In principle, this representation resolves our problem. We may construct beliefs about the unobservable 'probability' of spinning a head, by expressing beliefs over the collection of observables. We may then treat p as if it were an actual random quantity. For example, if we observe r heads in m spins, our conditional probability of observing k heads in the next n spins is given by

$$\int_0^1 \binom{n}{k} p^k (1-p)^{n-k} dQ(p|r,m), \tag{6.2}$$

where $Q(.|r,m)$ is the conditional probability measure that we obtain by updating beliefs about p using Bayes' theorem. Thus, we obtain exactly the same answer if (i) we restrict the prior specification and subsequent conditioning entirely to statements of belief about observables, i.e. the joint distribution of the first $n+m$ tosses, or (ii) we treat p as though it were an actual random quantity, and derive conditional predictive beliefs about future observables from (6.2). Usually (ii) is technically easier than (i), so that the introduction of unobservable parameters can be viewed as a natural way of organizing and simplifying both the task of coherent prior specification and the calculations which are involved in a predictive analysis. Versions of the exchangeability representation theorem have been developed for a great many statistical problems, in which there is some form of invariance of beliefs over permutations of the (possibly transformed) observables.

We can therefore make a strong case for the view that exchangeability is the fundamental judgement which gives meaning to the kinds of assumptions and modelling which characterize the usual types of statistical analysis. However, there is a basic problem with the view that we start with exchangeable beliefs over observables, and construct probabilistic models directly from these judgements. Consider the coin spinning example. In order to construct the prior probability measure p, we must specify, for every k, n, our prior probability for k heads in n spins. It is extremely difficult to make such prior specifications, even for quite moderate values of k, n, and it is therefore completely impractical to actually make all of the prior specifications over observables which are required in order to apply the exchangeability representation. Therefore, in practice, we cannot construct the prior measure Q from our stated beliefs over observables, nor can we view the construction as a pragmatic simplification of beliefs which we could, with a certain amount of effort, express. However, logically, we cannot construct these beliefs in any other way, as the quantity p has no operational meaning outside of the representation theorem. Therefore, we appear to have no choice but to treat p as a real but unobservable random quantity which exists separately from our ability to specify all of the prior beliefs over observables which are necessary in the representation theorem.

If we cannot apply the representation theorem constructively in the simplest possible example, namely spinning coins, then in any genuine problem of interest we will obtain little practical help from the exchangeability representation theorem, so that a Bayes analysis will usually be conducted entirely in terms of some parametric model. Thus, while in principle the exchangeability representation is fundamental to statistical modelling, in practice it is much less important, and it is unusual for exchangeability to play a constructive role in belief specification and analysis. The difficulty that we face in using the representation is one which we have already identified as a frequent problem for the Bayesian approach, namely that we need to specify prior beliefs to an extreme level of detail before we may carry out any analysis of aspects of these beliefs.

Further, the full exchangeability specification is not only overly detailed but also often misleading. In most sampling situations, there are certain sequences which we might observe which would lead us to question the exchangeability formulation; for example, long alternating sequences of heads and tails might suggest that the outcome of the previous spin influenced the current spin, perhaps by the way the coin is facing at start of each spin. In a fully exchangeable specification, such systematic patterns are ignored, and prior exchangeability always leads to conditional exchangeability for all sequences. Thus, full exchangeability rarely reflects our actual beliefs, but it would usually be very difficult to anticipate and specify beliefs over all of the departures from exchangeability that we might observe and, in any case, this would lose the essential simplicity of the exchangeable analysis.

The problems that we have discussed in applying the exchangeability representation show the price that we pay for the requirement of full probabilistic prior specification. In the Bayes linear approach, however, we work with far more modest belief specifications. For this reason, we will find that we are able to exploit exchangeability in practice, as well as in principle, so that the statistical models that we shall construct will be built directly from beliefs over observables. We shall first introduce our approach for the example of spinning coins.

6.2 Coin tossing

Suppose that we have an in principle infinite collection of coin spins, given by $S = (S_1, S_2, \ldots)$, where each S_i takes value 1, if heads, or 0, if tails. We want to create a representation for p, the 'true but unknown probability' that the coin will land heads. The relative frequency interpretation of such a probability is based on a hypothetical physical limit for the proportion of heads in n tosses. The usual Bayesian representation for this probability under exchangeability replaces the hypothetical physical limit with a hypothetical infinite collection of probability assessments, which, in practice, we would never be able to make. We now describe informally an alternative approach to the representation theorem in which p is constructed strictly on the basis of a small number of actual probability assessments.

Let us therefore suppose that certain actual beliefs that we may express about the sequence are not sensitive to the ordering of the sequence. If we only wish to make a minimal prior specification we may consider the following two numbers.

(i) We consider the probability that an individual spin, S_i, lands heads. Suppose that we judge this probability to be the same for each spin. Call the common value q_1.

(ii) We consider the probability that two individual spins S_i and S_j both land heads. Suppose that we judge this probability to be the same for each pair of spins $i \neq j$. Call the common value q_2.

If the sequence of spins satisfies these conditions, we say that the sequence is **second-order exchangeable**. Second-order exchangeability is the simplest possible belief specification for the sequence of spins. Because the requirements that such a specification imposes are so weak, they will apply in a great many statistical problems. Indeed, we can hardly avoid evaluating the values q_1, q_2 if we want to analyse the sequence of coin tosses. However, these two values alone will be sufficient to generate the representation that we require for the underlying probability p of heads. We proceed as follows. Call P_n the proportion of heads in the first n spins, so that

$$P_n = \frac{1}{n}(S_1 + \ldots + S_n).$$

Conditions (i), (ii) are equivalent to the prior specifications

$$\mathrm{E}(S_i) = q_1, \quad \mathrm{E}(S_i S_j) = q_2, \qquad \forall i \neq j. \tag{6.3}$$

Given these specifications, we have, for any $n < m$, that

$$\mathrm{E}((P_n - P_m)^2) = \left(\frac{1}{n} - \frac{1}{m}\right)(q_1 - q_2). \tag{6.4}$$

Therefore, P_n is a Cauchy sequence in mean square (i.e. for any $\epsilon > 0$, there is an integer N_ϵ for which $\mathrm{E}((P_n - P_m)^2) < \epsilon$, $\forall n, m > N_\epsilon$). This condition is sufficient to ensure that our beliefs are consistent with the existence of a further random quantity P for which

$$\lim_{n \to \infty} \mathrm{E}([P_n - P]^2) = 0. \tag{6.5}$$

The quantity P exists in the closure of the inner product space containing the sequence of quantities P_n. We will discuss the geometric details of this convergence in the context of the full second-order representation theorem in §6.4.

Property 6.1 *From* (6.5), *we may deduce the following properties of* P.

6.1.1: *Beliefs about* P *follow from the prior specification* (6.3) *as*

$$\mathrm{E}(P) = q_1, \quad \mathrm{Var}(P) = q_2 - q_1^2. \tag{6.6}$$

6.1.2: *If* $R_i = S_i - P$, *then, for each* i,

$$\mathrm{E}(R_i) = 0, \quad \mathrm{Var}(R_i) = q_1 - q_2. \tag{6.7}$$

6.1.3: *The sequence* $R_1, R_2, \ldots$ *is uncorrelated and each* R_i *is uncorrelated with* P.

We have therefore constructed the following exchangeability representation for the coin spins. For each i,

$$S_i = P + R_i, \tag{6.8}$$

so that our actual exchangeability specifications lead directly to a further quantity, P, which plays an analogous role to the frequency probability p which emerges from the full exchangeability representation. The quantity p in the full representation may be viewed as the relative frequency limit of the sequence P_n, and, given the value of p, the sequence of spins is independent, with probability p. Similarly, the second-order beliefs that we have stated must be consistent with the condition that the relative frequency P_n will converge to P, in the sense of (6.5). Given the value of P, the sequence of spins is uncorrelated, each with the same variance and with expectation P.

The difference between the two representations is that, unlike the full representation theorem, beliefs about P genuinely are constructed from specifications about observable random quantities. The only beliefs about P that are specified are the probabilities for the outcomes for one and for two spins, but these specifications are precisely what we need in order to carry out the Bayes linear analysis.

We now describe informally the basis for such an analysis. Consider first the framework which is provided by the usual exchangeability representation for analysing the coin spins. We observe k heads in n spins. How may we update our beliefs? As we have observed, either we may ignore the representation theorem and derive all further probability statements concerning the sequence directly from conditioning on the joint probability distribution that we have specified over the observables, or we may treat the probability p in the representation theorem as a real unknown quantity, update beliefs for p using Bayes' theorem and derive all statements of belief about the sequence from the revised mixture distribution, using relations such as (6.2). Each approach gives the same answer for all probability assessments for future observables, supporting the view that 'belief' in p is an organizing principle which provides an efficient computational algorithm for updating beliefs over the sequence.

In our development, we have a similar equivalence between the two approaches. This equivalence is as follows. We have developed an exchangeability representation (6.8) in which each spin, S_i, may be written as $S_i = P + R_i$. We have two alternatives approaches. First, we can ignore the representation and directly evaluate adjusted means and variances for future observations given the observed value of P_n. Secondly, we can evaluate an adjusted mean and variance for P given P_n and then derive all adjusted beliefs about future spins from representation (6.8), using the adjusted beliefs for P. We may see that these approaches are equivalent as follows.

To adjust the further spin S_i by P_n, where $i > n$, is equivalent to writing

$$S_i = \alpha_i P_n + \beta_i + Q_i, \tag{6.9}$$

where Q_i has expectation zero and is uncorrelated with P_n. The adjusted expectation of S_i is given by the observed value of $\alpha_i P_n + \beta_i$, while the adjusted variance is the variance of Q_i.

We may similarly write

$$P = \alpha P_n + \beta + Q, \tag{6.10}$$

where Q has expectation zero and is uncorrelated with P_n. The adjusted mean and variance of P are the observed value of $\alpha P_n + \beta$ and the variance of Q, respectively.

Now, $S_i = P + R_i$, and R_i is uncorrelated with P_n, as R_i is uncorrelated with P and with each R_j, for $j < n$. Therefore, if we substitute for P in (6.8) from (6.10), we will obtain (6.9), i.e. comparing (6.10) with (6.9), we have

$$\alpha_i = \alpha, \quad \beta_i = \beta, \quad Q_i = Q + R_i.$$

Therefore, the two approaches to adjusting beliefs about future spins are equivalent. Formally, we have, for $i > n$, that $(\lfloor P_n \perp\!\!\!\perp S_i \rfloor \,/\, P)$, so that equivalence between the two updates is as described in Theorem 5.23.

Note from (6.5) that observing a sufficiently large number of spins of the coin reduces your uncertainty about P to an arbitrarily small value (i.e. $\mathrm{Var}(Q)$ goes to zero with n). Thus, as $n \to \infty$, the adjusted beliefs over future spins reduce to a collection of uncorrelated quantities $P^* + R_i$, where P^* is the observed large-sample relative frequency of heads.

In the next section we turn to the general version of the second-order exchangeability representation theorem.

6.3 Exchangeable belief structures

We now describe the general approach to modelling and analysing beliefs over second-order exchangeable collections. We begin by explaining how we represent such exchangeable objects and discuss the prior judgements that we must make.

To motivate the discussion, let us suppose that a doctor is examining a collection of patients. There is a certain collection of r measurements that she makes on each patient. We collect these measurements as the **measurement vector**

$$M = (M_1, \ \ldots, \ M_r).$$

For example, suppose that M_1 is the blood pressure, M_2 is the weight and M_3 is the age for each patient. We have a list of individuals for whom these measurements are to be evaluated. The value of measurement M_i for patient j in the list is M_{ij}. In our example, M_{35} would therefore be the age of patient 5 in the list. Therefore, for patient j, we have a measurement vector

$$M_j = (M_{1j}, M_{2j}, \ldots, M_{rj}).$$

In common with other writers, we face difficulties in finding a satisfactory notation able to distinguish between (a) M_j meaning element j of the conceptual measurement vector M, and (b) M_j meaning the vector of measurements for patient j. In this book, we will leave it to context as to which is intended, clarifying when necessary.

We shall suppose that the number of individuals is, at least in principle, infinite. This may be interpreted in two ways: first, as a pragmatic approximation to a very large finite collection of individuals; or secondly, as for coin tossing, we may consider our beliefs about hypothetical infinite collections to guide us in our belief specifications for actual finite collections. We will discuss below the modifications that we must make when the collection of individuals is necessarily finite.

Now, we want to learn about certain aspects of each element M_i of the measurement vector. Thus, we specify a collection $X = (X_1, \ldots, X_r)$ of functions of the elements of M, over which we intend to express and adjust second-order beliefs. For example, we might choose

$$X_1 = M_1, \quad X_2 = M_1^2, \quad X_3 = M_1 M_2, \quad X_4 = M_2, \quad X_5 = M_2^2,$$

and so forth, and we might include indicator functions for the ranges of some of the quantities. Corresponding to each patient j there is a measurement vector M_j and a corresponding vector $X_j = (X_{1j}, \ldots, X_{rj})$ representing the measurements and functions of those measurements for that patient. Our intention is to take a sample of individuals, measure vector M_j and thus vector X_j for each patient j and therefore adjust beliefs about further vectors X_k. We term X the **observation vector**, as all of the analysis will be carried out in terms of observations of the value of X. Thus, we do not often need to refer to the original measurement vector, M, but sometimes it is conceptually helpful to distinguish between the measurements that we make and the aspects of our beliefs about the measurements over which we specify and adjust beliefs.

Definition 6.2 *The collection of vectors $X_1, X_2, X_3, \ldots$ is* ***second-order exchangeable*** *if the first- and second-order belief specification for the sequence of vectors is unaffected by any permutation of the order of the vectors, so that*

(i) the mean vector and variance matrix is the same for each individual,

$$\mathrm{E}(X_i) = \mu, \quad \mathrm{Var}(X_i) = \Sigma, \qquad \forall i; \tag{6.11}$$

(ii) *the covariance matrix between any two different individuals is the same,*

$$\mathrm{Cov}(X_i, X_j) = \Gamma, \qquad \forall i \neq j. \tag{6.12}$$

Here, μ is a vector with r elements $(\mu_1, \ldots, \mu_r)$; Σ is an $r \times r$ non-negative definite matrix with elements $\{\sigma_{ij}\}$; and Γ is also an $r \times r$ non-negative definite matrix with elements $\{\gamma_{ij}\}$. Thus, we make the required prior specification by first considering our mean and variance specification, μ and Σ, for a single future case. We then consider our covariance matrix Γ between two such cases. We may specify such covariance directly. However, it will often be more natural to consider the difference between two individuals, and therefore to assess our variance for the difference $X_i - X_j$. If we intend to use observations on X for certain individuals to adjust beliefs about the value of X for further individuals, then it is unavoidable that such variance and covariance specifications must be made, over all pairs of individuals under consideration. The second-order exchangeable specification is the simplest specification that we can make over such a collection of individuals. As we will show, this specification over observables leads directly to the representation theorem for second-order exchangeable vectors, so that our statistical models may be constructed directly from simple belief specifications over observable quantities.

6.4 The representation theorem

De Finetti's representation theorem for an infinite exchangeable sequence of random vectors summarizes all of the probabilistic relationships which have been asserted between the infinite collection as follows. These beliefs are shown to be consistent with the existence of a further 'random entity', namely a random joint probability distribution function, F say, such that beliefs concerning the values of the sequence, conditional on the value of F, are that the sequence is independent and identically distributed from F. We have a similar purpose, namely to summarize all of the relationships that we have expressed between the members of the sequence of vectors $X_1, X_2, \ldots,$ as represented by our second-order exchangeable specification of means, variances, and covariances over the sequence. Within our formulation, this means that we must construct a further random vector, $\mathcal{M}(X)$, consistent with the beliefs that we have expressed such that adjusting the sequence by $\mathcal{M}(X)$ results in a sequence of uncorrelated quantities, each with mean zero and the same variance matrix. Given such a vector $\mathcal{M}(X)$, all of the covariances that we have expressed between the vectors $X_1, X_2, \ldots$ may be represented by the common relationship between each X_i and $\mathcal{M}(X)$. We may therefore update beliefs about the sequence simply by updating beliefs about $\mathcal{M}(X)$.

We now show how the vector $\mathcal{M}(X)$ is constructed. Essentially the construction is as for the coin spinning example. We form the sequence of vectors $\bar{X}_n = (\bar{X}_{1n}, \ldots, \bar{X}_{rn})$ of averages, namely

$$\bar{X}_n = \frac{1}{n} \sum_{j=1}^{n} X_j. \tag{6.13}$$

This sequence turns out to be Cauchy in expected mean square, and so converges to a limiting vector, namely $\mathcal{M}(X)$, which will have the properties that we require. We have the following representation theorem.

Theorem 6.3 (Representation theorem for an infinite sequence of second-order exchangeable random vectors)
If $X_1, X_2, \ldots$ is an infinite second-order exchangeable sequence of random vectors, with mean and variance structure given by (6.11), (6.12), *then we may introduce the further random vector $\mathcal{M}(X)$, termed the* ***population mean vector****, and also the infinite sequence*

$$\mathcal{R}_1(X),\ \mathcal{R}_2(X),\ \ldots,$$

termed the ***individual residual vectors****, which satisfy the following properties.*

6.3.1: *For each individual j,*

$$X_j = \mathcal{M}(X) + \mathcal{R}_j(X). \tag{6.14}$$

6.3.2: *The mean and variance for $\mathcal{M}(X)$ are*

$$\mathrm{E}(\mathcal{M}(X)) = \mu, \quad \mathrm{Var}(\mathcal{M}(X)) = \Gamma. \tag{6.15}$$

6.3.3: *The collection $\mathcal{R}_1(X), \mathcal{R}_2(X), \ldots$ is second-order exchangeable, with, for each individual j,*

$$\mathrm{E}(\mathcal{R}_j(X)) = 0, \quad \mathrm{Var}(\mathcal{R}_j(X)) = \Sigma - \Gamma, \tag{6.16}$$

and the vectors $\mathcal{R}_1(X), \mathcal{R}_2(X), \ldots$ are mutually uncorrelated.

6.3.4: *Each $\mathcal{R}_j(X)$ is uncorrelated with $\mathcal{M}(X)$.*

Proof. Denote by $\langle X\rangle^+$ the inner product space on the collection of elements X_{ij}, constructed as in §3.10 under the inner product $(Y, Z) = \mathrm{Cov}(Y, Z)$, where we identify all quantities which differ by a constant.

For each j and $n < m$, we have

$$\|\bar{X}_{jn} - \bar{X}_{jm}\|^2 = \mathrm{Var}(\bar{X}_{jn} - \bar{X}_{jm}) = \left(\frac{1}{n} - \frac{1}{m}\right)(\sigma_{jj} - \gamma_{jj}). \tag{6.17}$$

Any inner product space S may be embedded in the minimal closure of the space, S^* say, by adding, for each Cauchy sequence $s_1, s_2, \ldots$ of elements of S for which the limit point of the sequence does not exist in S, a new element s^* whose inner product with each element $s \in S$ is $(s, s^*) = \lim_n (s, s_n)$.

Denote the minimal closure of $\langle X\rangle^+$ by $[X]$. From (6.17), each sequence $\bar{X}_{jn}$ is a Cauchy sequence in $\langle X\rangle^+$. Therefore, $\mathcal{M}(X_j) = \lim_n \bar{X}_{jn}$ exists in $[X]$, and for each $Z \in [X]$ we have

$$(\mathcal{M}(X_j), Z) = \lim_n (\bar{X}_{jn}, Z) = \lim_n \frac{1}{n}\sum_{r=1}^{n} \mathrm{Cov}(X_{jr}, Z). \tag{6.18}$$

Therefore, for each r, n, j, we have

$$\text{Cov}(\mathcal{M}(X_j), X_{rn}) = \gamma_{jr} = \text{Cov}(\mathcal{M}(X_j), \mathcal{M}(X_r)),$$

so that

$$\text{Cov}(\mathcal{M}(X), X_n) = \Gamma = \text{Var}(\mathcal{M}(X)).$$

Therefore, for each $r \neq s$, we have

$$\text{Cov}(X_s - \mathcal{M}(X), \mathcal{M}(X)) = \text{Cov}(X_r - \mathcal{M}(X), X_s - \mathcal{M}(X)) = 0,$$

and the theorem follows. ■

We have thus shown that specification of second-order beliefs over any two vectors X_i, X_j, for $i \neq j$, and the symmetric extension of such beliefs to all pairs of vectors must correspond to beliefs which are consistent with the existence of a further vector, $\mathcal{M}(X)$, which is the limit of the sample mean vectors. For each X_j we have

$$\text{E}_{\mathcal{M}(X)}(X_j) = \mathcal{M}(X), \tag{6.19}$$

so that, having adjusted the sequence by $\mathcal{M}(X)$, we have a residual sequence

$$\mathbb{A}_{\mathcal{M}(X)}(X_j) = \mathcal{R}_j(X) \tag{6.20}$$

whose elements are uncorrelated with zero mean and the same variance. Therefore, the representation theorem serves as a basis for the subjectivist modelling of statistical problems.

As the residual vectors are uncorrelated over individuals, the population mean $\mathcal{M}(X)$ induces the separation of beliefs over individuals. That is, if $D_n = (X_1, \ldots, X_n)$ and $D_{n,r} = (X_{n+1}, \ldots, X_{n+r})$ are collections of measurements for different groups of individuals, then

$$\lfloor D_n \perp\!\!\!\perp D_{n,r} \rfloor \,/\, \mathcal{M}(X). \tag{6.21}$$

Bayes linear analysis is usually concerned with belief adjustments over collections of linear combinations of quantities. Thus, we denote by $\langle X \rangle$ the collection of linear combinations of the form

$$Y = \sum_{i=1}^{r} \alpha_i X_i, \tag{6.22}$$

where $(X_1, \ \ldots, X_r)$ are the elements of the conceptual vector X of observations. Similarly, for measurements on individual j, $\langle X_j \rangle$ denotes the collection of linear combinations of the elements of measurement vector X_j, so that the value of quantity Y in (6.22) for individual j is given by $Y_j \in \langle X_j \rangle$ determined as

$$Y_j = \sum_{i=1}^{r} \alpha_i X_{ij}. \tag{6.23}$$

Second-order exchangeability for X implies second-order exchangeability for the collection $\langle X \rangle$. We extend the exchangeability representation over the linear spaces $\langle X_j \rangle$ as follows. It is straightforward to check that, for each Y as defined by (6.22), the exchangeability representation for the corresponding sequence of scalars Y_j is

$$Y_j = \sum_i^r \alpha_i X_{ij} \in \langle X_j \rangle \quad \Rightarrow \quad Y_j = \mathcal{M}(Y) + \mathcal{R}_j(Y),$$

where

$$\mathcal{M}(Y) = \sum_{i=1}^r \alpha_i \mathcal{M}(X_i),$$

$$\mathcal{R}_j(Y) = \sum_{i=1}^r \alpha_i \mathcal{R}_j(X_i).$$

6.5 Finite exchangeability

The infinite sequences that form the basis for the second-order exchangeability representation theorem are either the members of a very large finite population, for example in survey sampling, or correspond to hypothetical repetitions of some random experiment, such as coin tossing. In certain circumstances, however, we may apply the representation theorem for intrinsically finite sequences. For example, we might have a fixed number, r say, of possible treatments which we could apply to some unit, so that $X_1, \ldots, X_r$ would be the response of the unit under each of the r treatments. For such cases, there may be no sensible interpretation, even as a mental construct, to an infinite extension over additional hypothetical treatments. Therefore, if $X_1, \ldots, X_r$ is a finite second-order exchangeable sequence, then we must construct the representation based on the finite average $\bar{X}_r = (1/r)\sum_i X_i$. In this case, we may still write the same representation, namely, for each i,

$$X_i = \bar{X}_r + \mathcal{R}_i(X),$$

where each $\mathcal{R}_i(X)$ is uncorrelated with $\bar{X}_r$, but the residual vectors are correlated, as

$$\text{Var}(\mathcal{R}_i(X)) = \frac{r-1}{r}(\Sigma - \Gamma),$$

and

$$\text{Cov}(\mathcal{R}_i(X), \mathcal{R}_j(X)) = -\frac{1}{r-1}\text{Var}(\mathcal{R}_i(X)), \quad \forall i \neq j.$$

Therefore, the exchangeability representation, which is based on strict orthogonality for infinite sequences, must be modified to correspond to orthogonality of order $1/r$. Large finite collections of random quantities can therefore be treated as effectively infinite, provided that we do not sample a substantial proportion of the whole collection, but small collections must be analysed with the above covariance specification.

6.6 Example: oral glucose tolerance test

In §3.8.3 we returned to our example concerning the OGT test and examined the implications, for reducing variation, of observing tests on more than one healthy elderly individual. We now go back to this example to see how the notion of exchangeability provides a structure for the features of interest. To do so, we must return to Chapter 2, in which we obtained qualitative representations for the components of variation comprising our measurements. There, we considered the fasting and 2-hour measurements on blood glucose level for a particular elderly person (G_0 and G_2) and a similar pair of measurements for our doctor (D_0 and D_2). In §2.6 we showed how relationships between these four quantities could be established through intermediary quantities. In summary, we established

$$G_0 = R_0 + C_0 + Y_0, \qquad D_0 = Z_0 + C_0 + Y_0,$$

$$G_2 = R_2 + C_2 + Y_2, \qquad D_2 = Z_2 + C_2 + Y_2.$$

Here, Y_0 and Y_2 represent the fasting and 2-hour measurements for an average young person, and C_0 and C_2 represent differences between the fasting and 2-hour measurements between typical young and typical healthy elderly persons. The difference between G_0 and D_0 is that the measurement D_0 includes a term Z_0 representing individual variation between the doctor's fasting measurement and the average such measurement, whilst the measurement G_0 includes another term R_0 representing the individual variation between that individual's fasting measurement and the average such measurement. Further, the quantities R_0 and Z_0 are judged to be uncorrelated, uncorrelated with all other quantities, to have expectation zero, and to have the same prior variance. These judgements arose because we treated our doctor and another typical elderly person as **exchangeable**. We specified for each an underlying mean component $C_0 + Y_0$ common to all such typical elderly persons, and individual variation for each, uncorrelated with any other quantity.

Let us now extend our example. Suppose that we consider taking measurements for a sample of typical elderly patients, and that we make the judgement that these measurements are second-order exchangeable across individuals. All the ingredients that we need for the full specification over these measurements are already in place as we have already considered relationships for and between two individuals, and these extend to any number of individuals. Using the notation of §3.8.3, let the fasting and 2-hour measurements be D_{i0} and D_{i2} for the ith person measured, and collect this pair of measurements into the vector D_i. Following the arguments in §2.6, we write these as

$$D_i = \begin{bmatrix} D_{i0} \\ D_{i2} \end{bmatrix} = \begin{bmatrix} R_{i0} + C_0 + Y_0 \\ R_{i2} + C_2 + Y_2 \end{bmatrix} = \begin{bmatrix} \mathcal{R}_i(D_0) + \mathcal{M}(D_0) \\ \mathcal{R}_i(D_2) + \mathcal{M}(D_2) \end{bmatrix} \tag{6.24}$$

where

$$\mathcal{M}(D_0) = C_0 + Y_0, \qquad \mathcal{M}(D_2) = C_2 + Y_2,$$

$$\mathcal{R}_i(D_0) = R_{i0}, \qquad \mathcal{R}_i(D_2) = R_{i2}$$

are respectively the mean and residual components for the two measurements. Expectation and variance–covariance specifications for the measurements for a pair of individuals are given in (3.54) and (3.56). However, it is as meaningful to show these specifications through their exchangeability representation as:

$$\mathrm{E}(\mathcal{M}(D)) = \mathrm{E}\left(\begin{bmatrix}\mathcal{M}(D_0)\\ \mathcal{M}(D_2)\end{bmatrix}\right) = \begin{bmatrix}4.16\\ 6.25\end{bmatrix}, \tag{6.25}$$

$$\mathrm{E}(\mathcal{R}_i(D)) = \mathrm{E}\left(\begin{bmatrix}\mathcal{R}_i(D_0)\\ \mathcal{R}_i(D_2)\end{bmatrix}\right) = \begin{bmatrix}0\\ 0\end{bmatrix}, \quad \forall i, \tag{6.26}$$

$$\mathrm{Var}(\mathcal{M}(D)) = \mathrm{Var}\left(\begin{bmatrix}\mathcal{M}(D_0)\\ \mathcal{M}(D_2)\end{bmatrix}\right)$$

$$= \begin{bmatrix}\mathrm{Var}(C_0 + Y_0) & \mathrm{Cov}(C_0 + Y_0, C_2 + Y_2)\\ \mathrm{Cov}(C_0 + Y_0, C_2 + Y_2) & \mathrm{Var}(C_2 + Y_2)\end{bmatrix} \tag{6.27}$$

$$= \begin{bmatrix}0.62 & 0.30\\ 0.30 & 0.43\end{bmatrix}, \tag{6.28}$$

$$\mathrm{Var}(\mathcal{R}_i(D)) = \mathrm{Var}\left(\begin{bmatrix}\mathcal{R}_i(D_0)\\ \mathcal{R}_i(D_2)\end{bmatrix}\right)$$

$$= \begin{bmatrix}\mathrm{Var}(R_{i0}) & \mathrm{Cov}(R_{i0}, R_{i2})\\ \mathrm{Cov}(R_{i0}, R_{i2}) & \mathrm{Var}(R_{i0})\end{bmatrix}$$

$$= \begin{bmatrix}0.50 & 0.42\\ 0.42 & 2.00\end{bmatrix}, \quad \forall i, \tag{6.29}$$

$$\mathrm{Cov}(\mathcal{M}(D), \mathcal{R}_i(D)) = 0, \quad \forall i, \tag{6.30}$$

$$\mathrm{Cov}(\mathcal{R}_j(D), \mathcal{R}_i(D)) = 0, \quad \forall i \neq j. \tag{6.31}$$

Variances and covariances for and between $C_0, C_2, Y_0, Y_2, R_0, R_2$ are given in §2.7. With respect to variance–covariance specifications for an individual and any pair of individuals, the representation has two important consequences. The first is that we have decomposed the variance matrix for the pair of measurements for an individual into

$$\mathrm{Var}(D_i) = \mathrm{Var}(\mathcal{M}(D) + \mathcal{R}_i(D)) = \mathrm{Var}(\mathcal{M}(D)) + \mathrm{Var}(\mathcal{R}_i(D))$$

$$= \begin{bmatrix}0.62 & 0.30\\ 0.30 & 0.43\end{bmatrix} + \begin{bmatrix}0.50 & 0.42\\ 0.42 & 2.00\end{bmatrix},$$

and this is the same for all individuals. The second is that we can express the covariance matrix between the measurements for two different individuals as

$$\text{Cov}(D_i, D_j) = \text{Cov}(\mathcal{M}(D) + \mathcal{R}_i(D), \mathcal{M}(D) + \mathcal{R}_j(D)) = \text{Var}(\mathcal{M}(D))$$
$$= \begin{bmatrix} 0.62 & 0.30 \\ 0.30 & 0.43 \end{bmatrix}, \tag{6.32}$$

as for (6.27), and this is the same for all pairs of individuals. In the notation of §6.3, we thus have

$$\mu = \begin{bmatrix} 4.16 \\ 6.25 \end{bmatrix}, \tag{6.33}$$

$$\Gamma = \begin{bmatrix} 0.62 & 0.30 \\ 0.30 & 0.43 \end{bmatrix}, \tag{6.34}$$

$$\Sigma = \begin{bmatrix} 1.12 & 0.72 \\ 0.72 & 2.43 \end{bmatrix}, \tag{6.35}$$

$$\Sigma - \Gamma = \begin{bmatrix} 0.50 & 0.42 \\ 0.42 & 2.00 \end{bmatrix}. \tag{6.36}$$

In §3.8.3 we constructed the difference between the fasting and 2-hour blood glucose measurement as $G_h = G_2 - G_0$. It is simple to demonstrate exchangeability over linear spaces for such quantities, (6.23), as follows. Define $D_{ih} = D_{i0} - D_{i2}$ to be the 2-hour difference for individual i. From (6.24) we have

$$\begin{aligned} D_{ih} &= D_{i0} - D_{i2} \\ &= [\mathcal{M}(D_2) + \mathcal{R}_i(D_2)] - [\mathcal{M}(D_0) + \mathcal{R}_i(D_0)] \\ &= [\mathcal{M}(D_2 - D_0)] + [\mathcal{R}_i(D_2 - D_0)] \\ &= \mathcal{M}(D_h) + \mathcal{R}_i(D_h), \end{aligned}$$

so that each such 2-hour difference has the representation as an underlying mean component plus a residual component which has zero expectation and is uncorrelated with all other quantities. In the notation of §6.3, it is straightforward to show that, for the exchangeable sequence of 2-hour differences,

$$\mu = \begin{bmatrix} 2.09 \end{bmatrix}, \quad \Gamma = \begin{bmatrix} 0.45 \end{bmatrix}, \quad \Sigma = \begin{bmatrix} 2.11 \end{bmatrix}, \quad \Sigma - \Gamma = \begin{bmatrix} 1.66 \end{bmatrix}.$$

6.7 Example: analysing exchangeable regressions

6.7.1 Introduction

In an industrial process for extracting aluminium by electrolysis from a solution of alumina, experiments are run, under similar operating conditions, to measure the percentage concentration of alumina in solution every 10 minutes, terminating when the concentration falls to a pre-specified level. These experiments are expensive in terms of both time and money. The measurements are the responses

Y_{rt} representing, for run r, the concentrations of alumina remaining in solution at time t from the end of the run; the alumina level is essentially fixed at the end of the run. In all, there turn out to be 13 equally spaced time points. We considered this example in Goldstein (1991) and Goldstein and Wooff (1998). We are grateful to our colleague Dr Malcolm Farrow for providing these data and judgements concerning the process.

We want to learn both about the uncertain quantities underlying the process and about future runs of the experiment. One way to organize this problem is (a) to make judgements about the evolution of the process over time, and (b) to make judgements about the similarity of measurements at the same time point but for different runs. Suppose that we consider the evolution of the process over time. Although there are complicating operating features, Faraday's law and practical experience suggest that aluminium is extracted at a constant rate over time. This implies a regression model of the form:

$$Y_{rt} = a_r + tb_r + \epsilon_{rt}, \tag{6.37}$$

where a_r and b_r are regression coefficients specific to run r, and the terms $\{\epsilon_{rt}\}$ express departures from the linear model.

Now let us consider how runs might differ for the same time point. Suppose we decide that the regression coefficients a_r and b_r are second-order exchangeable over runs. This corresponds to the judgement that the underlying evolution across time is the same, but disturbed by variation specific to a run. It is as though we consider a very large number of similar experiments, in each of which we model the amount of alumina being extracted at time t, as t times a slope term b plus an intercept term a. Thus, on a particular run we treat the intercept for that run as comprising an underlying mean intercept, plus a discrepancy specific to that run. Similarly, we treat the slope for a particular run as comprising an underlying mean slope, plus a discrepancy specific to that run. Thus, following Theorem 6.3, we can decompose each a_r and b_r into uncorrelated mean and residual components,

$$a_r = \mathcal{M}(a) + \mathcal{R}_r(a), \tag{6.38}$$

$$b_r = \mathcal{M}(b) + \mathcal{R}_r(b), \tag{6.39}$$

so that, for example, $\mathcal{M}(a)$ is the underlying mean intercept term for all runs, and $\mathcal{R}_r(a)$ is the discrepancy from the intercept obtained for the rth run.

6.7.2 Error structure and specifications

The model becomes fully specified when we have made second-order prior judgements over it. First, we deal with the error terms ϵ_{rt}. These are constructed from uncorrelated components as follows:

$$\epsilon_{rt} = E_{rt} + V_{rt} + H_{rt},$$

$$V_{rt} = V_{r,t-1} + F_{rt},$$

$$H_{rt} = \phi H_{r,t-1} + U_{rt}, \quad t \geq 2.$$

These terms express discrepancies from the linear trend as the sum of:

- a pure measurement error for each reading, E_{rt};
- a stochastic development of the discrepancy as a random walk with drift, V_{rt};
- an autoregressive term expressing the measurement of the suspended particles in the chemical analysis, H_{rt}.

The error quantities are uncorrelated with all other quantities and among themselves, except as follows from these belief specifications:

$$\mathrm{Var}(U_{rt}) = 0.0204, \qquad \mathrm{Var}(H_{r1}) = 0.04, \qquad \mathrm{Var}(E_{rt}) = 0.01,$$
$$\phi = 0.7, \qquad \mathrm{Var}(F_{rt}) = 0.01.$$

Further details concerning the reasoning underlying the specification of the error structure are given in Goldstein (1991).

6.7.3 Regression coefficient specifications

Next we consider specifications for the regression coefficients. These can be judged via the representation (6.38) or, as follows, directly: we judge each a_r to be uncorrelated with each b_r, with the remaining prior specifications being

$$\mathrm{E}(a_r) = 1.4, \qquad \mathrm{E}(b_r) = 0.1,$$
$$\mathrm{Var}(a_r) = 0.058, \qquad \mathrm{Var}(b_r) = 0.0017,$$
$$\mathrm{Cov}(a_r, a_s) = 0.038, \qquad \mathrm{Cov}(b_r, b_s) = 0.0016,$$

for all r and $s \neq r$. In terms of the representation (6.38), these judgements amount to

$$\mathrm{E}(\mathcal{M}(a)) = 1.4, \qquad \mathrm{E}(\mathcal{M}(b)) = 0.1,$$
$$\mathrm{Var}(\mathcal{M}(a)) = 0.038, \qquad \mathrm{Var}(\mathcal{M}(b)) = 0.0016,$$
$$\mathrm{Var}(\mathcal{R}_r(a)) = 0.020, \qquad \mathrm{Var}(\mathcal{R}_r(b)) = 0.0001,$$
$$\mathrm{Cov}(\mathcal{M}(a), \mathcal{M}(b)) = 0,$$

with additionally $\mathcal{R}_1(a), \ldots$ and $\mathcal{R}_1(b), \ldots$ being sequences which are uncorrelated amongst themselves and uncorrelated with $\mathcal{M}(a)$ and $\mathcal{M}(b)$, and which have prior expectation zero. In the notation of §6.3, for the exchangeable sequence of pairs (a_i, b_i) of regression coefficients, we have

$$\mu = \begin{bmatrix} 1.4 \\ 0.1 \end{bmatrix},$$

$$\Gamma = \begin{bmatrix} 0.038 & 0 \\ 0 & 0.0016 \end{bmatrix},$$

$$\Sigma = \begin{bmatrix} 0.058 & 0 \\ 0 & 0.0017 \end{bmatrix},$$

$$\Sigma - \Gamma = \begin{bmatrix} 0.020 & 0 \\ 0 & 0.0001 \end{bmatrix}.$$

6.7.4 Structural implications

Our explicit modelling has a number of implications in this example. First, notice that it is the evolution of the process over time that is exchangeable, as we can write

$$a_r + tb_r = \mathcal{M}(a + tb) + \mathcal{R}_r(a + tb).$$

Indeed, by (6.37) and (6.38), each set of runs for a specified time point is exchangeable and has the representation

$$Y_{rt} = \mathcal{M}(Y_t) \qquad + \mathcal{R}_r(Y_t) \qquad + \epsilon_{rt}, \tag{6.40}$$

$$= \mathcal{M}(a) + t\mathcal{M}(b) \qquad + \mathcal{R}_r(a) + t\mathcal{R}_r(b) \qquad + \epsilon_{rt}, \tag{6.41}$$

where $\mathcal{M}(Y_t)$ is the underlying mean for the measurement at time t across all runs, whilst $\mathcal{R}_r(Y_t)$ is a residual component measuring the discrepancy at time t between the measurement and the mean component for run r. We gather the mean components $\{\mathcal{M}(Y_1), \ldots, \mathcal{M}(Y_{13})\}$ into the collection $\mathcal{M}(Y)$, and the mean components for the regression coefficients into the collection $\mathcal{M}(Q) = \{\mathcal{M}(a), \mathcal{M}(b)\}$. $\langle \mathcal{M}(Y) \rangle$ can be constructed entirely from the underlying slope and intercept quantities in $\mathcal{M}(Q)$, so that

$$\langle \mathcal{M}(Y) \rangle \equiv \langle \mathcal{M}(Q) \rangle. \tag{6.42}$$

As such, $\langle \mathcal{M}(Y) \rangle$ is two-dimensional; as we shall see later, this has consequences for the belief revision when we adjust by data.

6.8 Adjusting exchangeable beliefs

We now describe the use of exchangeable data for adjusting beliefs over underlying population quantities and future observables. In the notation of §6.3, we have an infinite sequence $X_1, X_2, \ldots$ of second-order exchangeable random vectors, where, for each individual i, the prior mean vector and variance matrix are $\mathrm{E}(X_i) = \mu$, $\mathrm{Var}(X_i) = \Sigma$ and the covariance matrix for any two different individuals $i \neq j$ is $\mathrm{Cov}(X_i, X_j) = \Gamma$. We construct the exchangeability representation

$$X_j = \mathcal{M}(X) + \mathcal{R}_j(X) \tag{6.43}$$

for each j, according to Theorem 6.3, where each $\mathcal{R}_j(X)$ has prior mean zero and variance matrix $\Sigma - \Gamma$, and $\mathcal{M}(X), \mathcal{R}_1(X), \mathcal{R}_2(X), \ldots$ are mutually uncorrelated.

We now consider how beliefs for $\mathcal{M}(X)$ and for values of X_j, $j > n$, are adjusted when we observe a sample of values $D_n = (X_1, X_2, \ldots, X_n)$. We begin by discussing some basic sufficiency conditions which simplify the analysis of exchangeable models.

6.9 Predictive sufficiency for exchangeable models

In principle, the usual Bayes analysis of an exchangeable sample proceeds as follows. We specify prior beliefs over all combinations of possible samples. From these beliefs and the representation theorem for infinite exchangeable sequences, we may proceed as though the observed sample and all future samples form an independent and identically distributed sequence drawn from an unknown probability distribution, F, for which we have a prior distribution P. Having observed the sample, we may either (i) update predictive beliefs about future observations directly from the joint distribution over the observable quantities, or (ii) update the prior distribution P over possible values of F and derive all predictive statements over future observations directly from the posterior distribution for F. Of course, (i) and (ii) will give identical results, but usually (ii) will be much simpler than (i). Further, as the size of the observed sample increases, the posterior distribution over F usually tends to a point mass on a single probability distribution (roughly, the empirical distribution of the sample), so that the limiting predictive distribution is that future observations are independent with common distribution given as the limiting sample distribution.

The equivalence between the two forms of belief adjustment is similar for second-order exchangeable sequences. This follows from the belief separation between D_n and X_j, $j > n$, induced by $\mathcal{M}(X)$, namely, from (6.21), we have

$$\lfloor D_n \perp\!\!\perp X_j \rfloor \,/\, \mathcal{M}(X). \tag{6.44}$$

Therefore, to adjust beliefs over X_j given D_n, we may adjust beliefs about the separating collection, in this case $\mathcal{M}(X)$, and from these adjusted beliefs derive the adjusted beliefs about the further observation X_j. To simplify notation, we write, for any random quantity U, the adjustment by the sample of n observations D_n as

$$\mathrm{E}_n(U) = \mathrm{E}_{D_n}(U), \quad \mathrm{Var}_n(U) = \mathrm{Var}_{D_n}(U).$$

Let X_j, $j > n$, be any further observation. From (6.44) and Property 5.23.1, we have

$$\mathrm{E}_n(X_j) = \mathrm{E}_n(\mathrm{E}_{\mathcal{M}(X)}(X_j)),$$

so that, from (6.19), we have

$$\mathrm{E}_n(X_j) = \mathrm{E}_n(\mathcal{M}(X)). \tag{6.45}$$

Further, from (6.44) and Property 5.23.2, we have

$$\mathrm{Var}_n(X_j) = \mathrm{Var}_n(\mathrm{E}_{\mathcal{M}(X)}(X_j)) + \mathrm{Var}_{\mathcal{M}(X)}(X_j)$$

so that

$$\mathrm{Var}_n(X_j) = \mathrm{Var}_n(\mathcal{M}(X)) + \mathrm{Var}(\mathcal{R}_j(X)). \tag{6.46}$$

Equations (6.45) and (6.46) demonstrate the equivalence of the two forms of adjustment. If we observe a sample $D_n = (X_1, X_2, \ldots, X_n)$, then we may either,

(i) evaluate adjusted beliefs, $\mathrm{E}_n(X_j)$, $\mathrm{Var}_n(X_j)$, about future vectors, X_j, directly from the joint covariance structure over the observable quantities, or (ii) evaluate adjusted beliefs, $\mathrm{E}_n(\mathcal{M}(X))$, $\mathrm{Var}_n(\mathcal{M}(X))$, over $\mathcal{M}(X)$, and derive all predictive statements over future observations directly from the revised beliefs for $\mathcal{M}(X)$ in (6.43). As we have shown, (i) and (ii) will give identical results, but usually (ii) will be simpler than (i). As the size of the observed sample increases, the adjusted variance of $\mathcal{M}(X)$ tends to zero, so that in the limit, future observations are uncorrelated with variance equal to the common variance of each $\mathcal{R}_j(X)$.

6.10 Bayes linear sufficiency for sample means

Given the second-order exchangeable sequence $X_1, X_2, \ldots$, we denote the mean of the first n vectors as $\bar{X}_n = (1/n)\sum_{j=1}^n X_j$. We now show that the sample mean vector is Bayes linear sufficient for the sample D_n for adjusting beliefs both for the population mean vector $\mathcal{M}(X)$ and for future observations X_r, $r > n$.

First observe that, for any $i \leq n$, we have

$$\mathrm{Cov}(\bar{X}_n, \bar{X}_n) = \frac{1}{n}\sum_{j=1}^{n} \mathrm{Cov}(X_j, \bar{X}_n) = \mathrm{Cov}(X_i, \bar{X}_n),$$

so that

$$\mathrm{Cov}(X_i - \bar{X}_n, \bar{X}_n) = 0. \tag{6.47}$$

Therefore, for each $i \leq n$,

$$\mathrm{E}_{\bar{X}_n}(X_i) = \bar{X}_n. \tag{6.48}$$

Therefore, as $\bar{X}_n$ is of the form

$$\bar{X}_n = \mathcal{M}(X) + \frac{1}{n}\sum_{j=1}^{n} \mathcal{R}_j(X), \tag{6.49}$$

we have

$$\mathbb{A}_{\bar{X}_n}(X_i) = X_i - \bar{X}_n = \mathcal{R}_i(X) - \frac{1}{n}\sum_{j=1}^{n} \mathcal{R}_j(X). \tag{6.50}$$

Therefore, as $\mathcal{R}_i(X) \perp \mathcal{M}(X)$, for each i, we have that

$$\lfloor D_n \perp\!\!\!\perp \mathcal{M}(X) \rfloor \,/\, \bar{X}_n. \tag{6.51}$$

Therefore, the sample mean is Bayes linear sufficient for adjusting beliefs for the population mean, and hence is also Bayes linear sufficient for adjusting beliefs over future observations X_i, for $i > n$. From (5.76), (6.45), (6.46), we have therefore derived the following result.

Theorem 6.4 *Let $X_1, X_2, \ldots$ be an infinite second-order exchangeable sequence of vectors. Then the sample mean vector $\bar{X}_n$ from a sample*

$$D_n = (X_1, \ldots, X_n)$$

is Bayes linear sufficient for D_n for adjusting both $\mathcal{M}(X)$ and any values X_i, $i > n$, namely

$$\mathrm{E}_n(\mathcal{M}(X)) = \mathrm{E}_{\bar{X}_n}(\mathcal{M}(X)), \quad \mathrm{Var}_n(\mathcal{M}(X)) = \mathrm{Var}_{\bar{X}_n}(\mathcal{M}(X)) \tag{6.52}$$

and, for any $i > n$,

$$\mathrm{E}_n(X_i) = \mathrm{E}_{\bar{X}_n}(X_i) = \mathrm{E}_{\bar{X}_n}(\mathcal{M}(X)), \tag{6.53}$$

$$\mathrm{Var}_n(X_i) = \mathrm{Var}_{\bar{X}_n}(X_i) = \mathrm{Var}_{\bar{X}_n}(\mathcal{M}(X)) + \mathrm{Var}(\mathcal{R}_i(X)). \tag{6.54}$$

Therefore, in order to adjust beliefs over the mean vector and future observations, it is sufficient to adjust $\mathcal{M}(X)$ by the sample mean vector.

6.11 Belief adjustment for scalar exchangeable quantities

As an illustration, we now derive the Bayes linear adjustment for a scalar sequence $X_1, X_2, \ldots$ of second-order exchangeable random quantities, with

$$\mathrm{E}(X_i) = \mu, \qquad \mathrm{Var}(X_i) = \sigma^2, \qquad \mathrm{Cov}(X_i, X_j) = \gamma, \quad i \neq j.$$

We have, from the representation theorem, for each i, that X_i may be written as the uncorrelated sum

$$X_i = \mathcal{M}(X) + \mathcal{R}_i(X)$$

where

$$\mathrm{E}(\mathcal{M}(X)) = \mu, \quad \mathrm{Var}(\mathcal{M}(X)) = \gamma,$$
$$\mathrm{E}(\mathcal{R}_i(X)) = 0, \quad \mathrm{Var}(\mathcal{R}_i(X)) = \sigma^2 - \gamma = \psi.$$

Therefore, we can write the sample mean, from a sample of size n, as

$$\bar{X}_n = \mathcal{M}(X) + \bar{R}_n(X),$$

where $\bar{R}_n(X)$ is the mean of the n residuals, so that

$$\mathrm{E}(\bar{X}_n) = \mu,$$
$$\mathrm{Var}(\bar{X}_n) = \gamma + \frac{1}{n}\psi,$$
$$\mathrm{Cov}(\bar{X}_n, \mathcal{M}(X)) = \gamma.$$

As $\bar{X}_n$ is Bayes linear sufficient for $\mathcal{M}(X)$, we may evaluate the adjusted expectation for $\mathcal{M}(X)$ given a sample of n as

$$\begin{aligned}\mathrm{E}_n(\mathcal{M}(X)) &= \mathrm{E}(\mathcal{M}(X)) + \mathrm{Cov}(\mathcal{M}(X), \bar{X}_n)(\mathrm{Var}(\bar{X}_n))^{-1}(\bar{X}_n - \mathrm{E}(\bar{X}_n)) \\ &= \frac{\gamma \bar{X}_n + \frac{1}{n}\psi\mu}{\gamma + \frac{1}{n}\psi},\end{aligned} \tag{6.55}$$

with corresponding adjusted variance

$$\begin{aligned}\mathrm{Var}_n(\mathcal{M}(X)) &= \mathrm{Var}(\mathcal{M}(X)) \\ &\quad - \mathrm{Cov}(\mathcal{M}(X), \bar{X}_n)(\mathrm{Var}(\bar{X}_n))^{-1}\mathrm{Cov}(\bar{X}_n, \mathcal{M}(X)) \\ &= \frac{\frac{1}{n}\psi\gamma}{\frac{1}{n}\psi + \gamma}.\end{aligned} \tag{6.56}$$

Therefore, we see that the adjusted expectation weighs the prior expectation and the data mean in inverse proportion to the expected squared difference between each quantity and $\mathcal{M}(X)$. Note also that (6.56) can be written equivalently as

$$\frac{1}{\mathrm{Var}_n(\mathcal{M}(X))} = \frac{1}{\gamma} + \frac{n}{\psi}$$

which is the familiar form whereby precision (i.e inverse variance) is additive.

6.12 Canonical structure for an exchangeable adjustment

We now derive the general relationship between exchangeable adjustments for a vector X based on samples of different sizes. This relationship is based on the eigenstructure of the resolution transform for the adjustment which changes with sample size in a very simple way. We denote the resolution transform for the adjustment of $\mathcal{M}(X)$ by D_n as

$$\mathbb{T}_n = \mathbb{T}_{\mathcal{M}(X):D_n}.$$

Theorem 6.5 *The eigenvectors of $\mathbb{T}_n$ are the same for each n. Further, if eigenvector W has eigenvalue λ for $\mathbb{T}_1$, then the corresponding eigenvalue $\lambda_{(n)}$ for W as an eigenvector of $\mathbb{T}_n$ is*

$$\lambda_{(n)} = \frac{n\lambda}{(n-1)\lambda + 1}. \tag{6.57}$$

Proof. From Theorem 6.4, the sample mean $\bar{X}_n$ is Bayes linear sufficient for D_n for the adjustment of $\mathcal{M}(X)$. We therefore evaluate the matrix representation of $\mathbb{T}_n$ as

$$\mathbb{T}_n = \mathrm{Var}(\mathcal{M}(X))^{-1}\mathrm{Cov}(\mathcal{M}(X), \bar{X}_n)\mathrm{Var}(\bar{X}_n)^{-1}\mathrm{Cov}(\bar{X}_n, \mathcal{M}(X)).$$

As $\text{Var}(\mathcal{M}(X)) = \Gamma$ and, for each i, $\text{Var}(\mathcal{R}_i(X)) = \Sigma - \Gamma$ we have, from (6.49), that

$$\text{Var}(\bar{X}_n) = \Gamma + \frac{1}{n}(\Sigma - \Gamma), \quad \text{Cov}(\mathcal{M}(X), \bar{X}_n) = \Gamma, \tag{6.58}$$

so that

$$\mathbb{T}_n = \left(\Gamma + \frac{1}{n}(\Sigma - \Gamma)\right)^{-1} \Gamma. \tag{6.59}$$

Therefore, v is an eigenvector of $\mathbb{T}_n$, with eigenvalue λ, so that $\mathbb{T}_n v = \lambda v$, if and only if

$$\Gamma v = \lambda \left(\Gamma + \frac{1}{n}(\Sigma - \Gamma)\right) v. \tag{6.60}$$

Rearranging (6.60), we have equivalently that v is an eigenvector of $\mathbb{T}_n$, with eigenvalue λ if and only if

$$\Gamma v = \frac{\lambda}{n - (n-1)\lambda} \Sigma v. \tag{6.61}$$

In particular, when $n = 1$, relation (6.61) reduces to

$$\Gamma v = \lambda \Sigma v. \tag{6.62}$$

Equating the conditions for λ and v to satisfy relation (6.62) and condition (6.61), we have that v is an eigenvector of $\mathbb{T}_n$ with eigenvalue λ if and only if, for each n, v is an eigenvector of $\mathbb{T}_n$ with eigenvalue $\lambda_{(n)}$, where

$$\lambda = \frac{\lambda_{(n)}}{n - (n-1)\lambda_{(n)}},$$

or equivalently, where

$$\lambda_{(n)} = \frac{n\lambda}{(n-1)\lambda + 1},$$

which gives the result. ∎

We have, in this proof, assumed positive definite variance matrices for simplicity of exposition. The results also hold when the variance matrices are non-negative definite; we provide full details in §12.12.

The canonical directions for the adjustment of $\mathcal{M}(X)$ by D_n are therefore the same for each sample size n. We term these directions the **canonical directions induced by exchangeability**. As the canonical directions remain the same for all n, the qualitative features of the adjustment will remain the same for all sample sizes. This is important both computationally and qualitatively. Qualitatively, as the canonical directions remain the same, no matter what the sample size, the underlying features of the adjustment will remain the same for all sample sizes. Thus, it is simple and natural to compare possible choices of sample size based on the effects on the underlying eigenstructure.

Computationally, we may exploit (6.57) to simplify any design problem for which we must choose the sample size to achieve variance reductions over elements of $\langle \mathcal{M}(X) \rangle$. For example, we have the following corollary.

Corollary 6.6 *Suppose that W is an eigenvector of $\mathbb{T}_1$ with eigenvalue $\lambda > 0$. Then the sample size n required to achieve a proportionate variance reduction of α for W, $0 < \alpha < 1$, i.e. so that* $\text{Var}_n(W) \leq (1-\alpha)\text{Var}(W)$, *is*

$$n \geq \frac{\alpha}{1-\alpha}\frac{1-\lambda}{\lambda}.$$

Further, if the minimal positive eigenvalue of $\mathbb{T}_1$ is $\lambda_{\min}$, then a sample size of

$$n \geq \frac{\alpha}{1-\alpha}\frac{1-\lambda_{\min}}{\lambda_{\min}}$$

is the minimum sample which is sufficient to achieve a proportionate variance reduction of α for every element of $\langle \mathcal{M}(X) \rangle$.

For each non-zero eigenvalue λ for $\mathbb{T}_1$, we have $\lambda_{(n)} \to 1$, $n \to \infty$. Therefore, as $n \to \infty$, we reduce variance about each such component to zero.

6.12.1 Standard form for the adjustment

If $\text{Var}(\mathcal{M}(X))$ is of lower rank than $\text{Var}(X_1)$, then there will be many alternative forms for the eigenvectors of each $\mathbb{T}_n$, as there will be many linear combinations of the elements of $\langle \mathcal{M}(X) \rangle$ with zero variance. In such cases, there is a natural choice for the form of the eigenvectors, termed the standard form, that we now define. Consider the eigenvectors of the corresponding transform $\mathbb{T}_1^* = \mathbb{T}_{X_1:\mathcal{M}(X)}$. Any eigenvector $\alpha^T X_1$, of $\mathbb{T}_1^*$ with eigenvalue λ satisfies

$$\text{E}_1(\text{E}_{\mathcal{M}(X)}(\alpha^T X_1)) = \lambda \alpha^T X_1.$$

However, from (6.19), we have

$$\text{E}_{\mathcal{M}(X)}(\alpha^T X_1) = \alpha^T \mathcal{M}(X),$$

so that

$$\text{E}_1(\alpha^T \mathcal{M}(X)) = \lambda \alpha^T X_1.$$

From §3.9.4, the eigenvalues of $\mathbb{T}_1^*$ and $\mathbb{T}_1$ are the same, and $\alpha^T X_1$ is an eigenvector of $\mathbb{T}_1^*$ if and only if $\text{E}_{\mathcal{M}(X)}(\alpha^T X_1) = \alpha^T \mathcal{M}(X)$ is an eigenvector of $\mathbb{T}_1$. For each eigenvector $\alpha^T X_1$ of $\mathbb{T}_1^*$, we term $\alpha^T \mathcal{M}(X)$ the **standard form** for the corresponding eigenvector for $\mathbb{T}_1$. Conversely, if

$$\text{E}_1(\beta^T \mathcal{M}(X)) = \lambda \beta^T X_1$$

for some value λ and choice of vector β, then it follows that $\beta^T \mathcal{M}(X)$ is an eigenvector of $\mathbb{T}_1$ with eigenvalue λ, so that this property uniquely characterises the collection of eigenvectors of $\mathbb{T}_1$. The argument is the same for each sample size, so that we have the following corollary.

Corollary 6.7 *If the positive eigenvalues of* $\mathbb{T}_1$ *are distinct, then the canonical directions induced by exchangeability, expressed in standard form, are the unique collection of elements of* $\langle\mathcal{M}(X)\rangle$ *with the property that, for each n, the adjusted expectation of eigenvector* $\alpha^T\mathcal{M}(X)$*, corresponding to eigenvalue* $\lambda_{(n)}$*, is given by*

$$\mathrm{E}_n(\alpha^T\mathcal{M}(X)) = \lambda_{(n)}\alpha^T\bar{X}_n. \tag{6.63}$$

If an eigenvalue of $\mathbb{T}_1$ *is repeated k times, then we may identify a unique corresponding linear subspace* $\langle\mathcal{M}(X)\rangle_\lambda$*, of dimension k, such that each element of* $\langle\mathcal{M}(X)\rangle_\lambda$ *satisfies* (6.63) *for each n for this value of* λ.

We write the standard form of the eigenvectors for $\mathbb{T}_1$ as

$$W_i = \alpha_i^T\mathcal{M}(X), \quad i = 1, \ldots, r,$$

where each W_i is normalized to prior variance one, with corresponding sample means $\bar{W}_{jn} = \alpha^T\bar{X}_n$. We have shown that

$$\mathrm{E}_n(W_j) = \lambda_{j(n)}\bar{W}_{jn}, \qquad \text{for each } j, n.$$

Note that, in this parameterization, we have

$$\lambda_{j(n)} = \frac{1}{\mathrm{Var}(\bar{W}_{jn})}.$$

For any element $Y \in \langle\mathcal{M}(X)\rangle$, we therefore have

$$\mathrm{E}_n(Y) = \sum_i \frac{n\lambda_{i(1)}}{(n-1)\lambda_{i(1)}+1}\mathrm{Cov}(Y, W_i)\bar{W}_{in}, \tag{6.64}$$

$$\mathrm{Var}_n(Y) = \sum_i \frac{1-\lambda_{i(1)}}{(n-1)\lambda_{i(1)}+1}[\mathrm{Cov}(Y, W_i)]^2, \tag{6.65}$$

$$\mathcal{R}_n(Y) = \sum_i \frac{n\lambda_{i(1)}}{(n-1)\lambda_{i(1)}+1}[\mathrm{Corr}(Y, W_i)]^2, \tag{6.66}$$

from (3.71) and (6.57).

6.12.2 Further properties of exchangeable adjustments

In general, for two elements Y, U of $\langle X\rangle$, if $\mathrm{Var}_1(\mathcal{M}(Y)) < \mathrm{Var}_1(\mathcal{M}(U))$, then it need not follow that $\mathrm{Var}_n(\mathcal{M}(Y)) < \mathrm{Var}_n(\mathcal{M}(U))$, for each n, and indeed the inequality may be reversed several times as n increases. However, we have the following result for the relationship between two different canonical resolutions as n increases.

Corollary 6.8 *Suppose that* λ, μ *are two eigenvalues of* $\mathbb{T}_1$ *such that* $\lambda < \mu$. *Then* $\lambda_{(n)} < \mu_{(n)}$ *for each n. Thus, for every pair of canonical quantities* W_i, W_j, $\mathrm{Var}_1(W_i) < \mathrm{Var}_1(W_j)$ *implies* $\mathrm{Var}_n(W_i) < \mathrm{Var}_n(W_j)$, *for each n.*

Finally, note that, in many applications, it is natural to express the basis for the population in a more convenient form. As the belief transform reveals the changes in information over the linear space, we may apply Theorem 6.5 to any such re-expression, by the following corollary.

Corollary 6.9 *Under the conditions of Theorem 6.5, let F be any collection of quantities for which $\langle F \rangle = \langle \mathcal{M}(X) \rangle$. The eigenvalues of resolution transform $\mathbb{T}_F = \mathrm{E}_F(\mathrm{E}_n(.))$ are the same as for $\mathbb{T}_n$ and the eigenvectors of $\mathbb{T}_F$ are the elements of $\langle F \rangle$ corresponding to the eigenvectors of $\mathbb{T}_n$.*

6.13 Algebraic example

For this algebraic example, suppose we wish to learn about the mean components, $\mathcal{M}(X)$, for the data quantities in the example discussed in §3.8.1 and §3.11.2. The data objects there were X_1, X_2. Suppose that we can instead observe the exchangeable sequence

$$X_{11}, X_{21}; X_{12}, X_{22}; \ldots, X_{1n}, X_{2n}; \ldots,$$

where each pair X_{11}, X_{21} has the same variance matrix as X_1, X_2. We collect the ith pair of observables in the sequence as $D_i = [X_{1i}, X_{2i}]$, and we let D be the ordered sequence of quantities

$$D = [D_1, D_2, \ldots, D_n].$$

To begin with, we will ignore our results on Bayes linear sufficiency and proceed as though we had to collect all data quantities into one large vector.

Compared to §3.8.1, we need one further set of belief specifications to describe the relationships between the exchangeable sequence of observables. For this example, we will suppose that

$$\mathrm{Cov}(X_{ij}, X_{kl}) = \begin{cases} \gamma, & i = k, \quad j \neq l, \\ 0, & i \neq k, \quad j \neq l. \end{cases} \tag{6.67}$$

For example, for $n = 2$, belief specifications for this problem are:

$$\mathrm{E}\left(\begin{bmatrix} D_1 \\ D_2 \end{bmatrix}\right) = \mathrm{E}\left(\begin{bmatrix} X_{11} \\ X_{21} \\ X_{12} \\ X_{22} \end{bmatrix}\right) = \begin{bmatrix} 0 \\ 0 \\ 0 \\ 0 \end{bmatrix}, \tag{6.68}$$

$$\mathrm{Var}\left(\begin{bmatrix} D_1 \\ D_2 \end{bmatrix}\right) = \begin{bmatrix} \mathrm{Var}(D_1) & \mathrm{Cov}(D_1, D_2) \\ \mathrm{Cov}(D_2, D_1) & \mathrm{Var}(D_2) \end{bmatrix}$$

$$= \mathrm{Var}\left(\begin{bmatrix} X_{11} \\ X_{21} \\ X_{12} \\ X_{22} \end{bmatrix}\right) = \begin{bmatrix} 1 & u & \gamma & 0 \\ u & 1 & 0 & \gamma \\ \gamma & 0 & 1 & u \\ 0 & \gamma & u & 1 \end{bmatrix}. \tag{6.69}$$

For general n, the variance specifications may be written more elegantly using direct product notation (§11.12.2):

$$\begin{aligned}\mathrm{Var}(D) &= \mathbf{I}_n \otimes \left(\begin{bmatrix} 1 & u \\ u & 1 \end{bmatrix} - \begin{bmatrix} \gamma & 0 \\ 0 & \gamma \end{bmatrix}\right) + \mathbf{J}_n \otimes \gamma \mathbf{I}_2, \qquad (6.70)\\ &= \mathbf{I}_n \otimes (\Sigma - \Gamma) + \mathbf{J}_n \otimes \Gamma,\end{aligned}$$

with $\Gamma = \gamma \mathbf{I}_2$. Suppose also that we construct the data averages for a sample of size n as

$$\bar{D}_n = \frac{1}{n}[D_1 + D_2 + \ldots + D_n].$$

It is simple to show that $\bar{D}_n$ has

$$\mathrm{Var}(\bar{D}_n) = \Gamma + \frac{1}{n}(\Sigma - \Gamma), \qquad (6.71)$$

$$\mathrm{Cov}(\mathcal{M}(X), \bar{D}_n) = \Gamma. \qquad (6.72)$$

6.13.1 Representation

From (6.67), we deduce the representation

$$\begin{aligned} X_{1i} &= \mathcal{M}(X_1) + \mathcal{R}_i(X_1),\\ X_{2j} &= \mathcal{M}(X_2) + \mathcal{R}_j(X_2),\end{aligned}$$

with belief specifications

$$\begin{aligned}
\mathrm{Var}(\mathcal{M}(X)) &= \Gamma, & \\
\mathrm{Var}(\mathcal{R}_i(X)) &= \Sigma - \Gamma, & \forall i,\\
\mathrm{Cov}(\mathcal{R}_i(X), \mathcal{M}(X)) &= 0, & \forall i,\\
\mathrm{Cov}(\mathcal{R}_i(X), \mathcal{R}_j(X)) &= 0, & \forall i \neq j,\\
\mathrm{E}(\mathcal{M}(X)) &= 0, & \\
\mathrm{E}(\mathcal{R}_i(X)) &= 0, & \forall i.
\end{aligned}$$

6.13.2 Coherence

In §3.8.1 we specified $|u| < 1$ for this example, in order that the variance matrix Σ be invertible. For the further specifications to be coherent we need the matrices $\Sigma - \Gamma$ and Γ to be non-negative definite (for full details of coherence conditions for exchangeable adjustments, see §12.12.2). Γ is non-negative definite for $\gamma \geq 0$. For $\Sigma - \Gamma$ to be non-negative definite, we require $\gamma \leq 1 - |u|$. However, for this example we will need to ensure that both matrices are invertible. Thus we shall impose the condition

$$0 < \gamma < 1 - |u| < 1.$$

6.13.3 Bayes linear sufficiency

We begin by ignoring the simplifications provided by Bayes linear sufficiency, and obtain the resolution transform by brute force. By (3.65), this is

$$\mathbb{T}_n = \text{Var}(\mathcal{M}(X))^{-1}\text{Cov}(\mathcal{M}(X), D)\text{Var}(D)^{-1}\text{Cov}(D, \mathcal{M}(X)). \tag{6.73}$$

First, notice that

$$\begin{aligned}
&\text{Cov}(\mathcal{M}(X), D)\text{Var}(D)^{-1}\text{Cov}(D, \mathcal{M}(X))\\
=&[\mathbf{1}_n^T \otimes \Gamma][\mathbf{I}_n \otimes (\Sigma - \Gamma) + \mathbf{J}_n \otimes \Gamma]^{-1}[\mathbf{1}_n \otimes \Gamma]\\
=&[\mathbf{1}_n^T \otimes \Gamma][\mathbf{I}_n \otimes (\Sigma - \Gamma)^{-1} - \mathbf{J}_n \otimes (\Sigma + (n-1)\Gamma)^{-1}\Gamma(\Sigma - \Gamma)^{-1}][\mathbf{1}_n \otimes \Gamma]\\
&\text{(by Lemma 11.61; see §11.12.2)}\\
=&\Gamma\left(\Gamma + \frac{1}{n}(\Sigma - \Gamma)\right)^{-1}\Gamma\\
=&\text{Cov}(\mathcal{M}(X), \bar{D}_n)\text{Var}(\bar{D}_n)^{-1}\text{Cov}(\bar{D}_n, \mathcal{M}(X)).
\end{aligned}$$

It follows that the resolution transforms for the sample averages and for the full set of observations are identical, and this provides an informal demonstration of Bayes linear sufficiency.

We can obtain the resolution transform for a sample of size n directly as

$$\begin{aligned}
\mathbb{T}_n &= \left(\Gamma + \frac{1}{n}(\Sigma - \Gamma)\right)^{-1}\Gamma\\
&= \frac{n\gamma}{[1+(n-1)\gamma]^2 - u^2}\begin{bmatrix} 1+(n-1)\gamma & -u \\ -u & 1+(n-1)\gamma \end{bmatrix}.
\end{aligned}$$

From the resolution matrix we compute the canonical resolutions as follows:

$$\lambda_{1(n)} = \frac{n\gamma}{1+u+(n-1)\gamma}, \tag{6.74}$$

$$\lambda_{2(n)} = \frac{n\gamma}{1-u+(n-1)\gamma}, \tag{6.75}$$

where the ordering of the canonical resolutions depends on the sign of u. $\mathbb{T}_n$ has corresponding algebraic eigenvectors proportional to $[1\ \ 1]^T$ and $[1\ \ -1]^T$, so that the canonical quantities are

$$W_1 = \alpha_1 \begin{bmatrix}1 & 1\end{bmatrix}\begin{bmatrix}\mathcal{M}(X_1)\\ \mathcal{M}(X_2)\end{bmatrix}, \quad W_2 = \alpha_2 \begin{bmatrix}1 & -1\end{bmatrix}\begin{bmatrix}\mathcal{M}(X_1)\\ \mathcal{M}(X_2)\end{bmatrix},$$

with α_1 and α_2 chosen to ensure that $\text{Var}(W_1) = \text{Var}(W_2) = 1$. The canonical quantities are thus

$$W_1 = \frac{1}{\sqrt{2\gamma}}(\mathcal{M}(X_1) + \mathcal{M}(X_2)), \tag{6.76}$$

$$W_2 = \frac{1}{\sqrt{2\gamma}}(\mathcal{M}(X_1) - \mathcal{M}(X_2)). \tag{6.77}$$

For a sample of size $n = 1$ we have $\lambda_{1(1)} = \gamma/(1+u)$ and $\lambda_{2(1)} = \gamma/(1-u)$, and it is straightforward to verify the relationship (6.57). It is obvious that the canonical directions do not depend on the sample size. Indeed, we could have calculated the canonical structure via (6.62) by solving the eigenvalue problem $\Sigma^{-1}\Gamma v = \lambda v$, without reference to the sample size, and then evaluating the canonical resolutions for whatever sample size we deem relevant, by (6.57).

The implication of the canonical structure for this example is straightforward to understand, but, in general, such implications are far from obvious without recourse to such canonical analysis. Examining the canonical resolutions, we observe that large values of u imply that one such resolution is large and the other small, so that there will be one strongly informative direction and one weakly informative direction. Here, u is partly a measure, in the residual structure, of the correlation between X_1 and X_2. The system resolution (3.75) for this example, for $n = 1$, is

$$\begin{aligned} \text{R}_{D_1}(\mathcal{M}(X)) &= \frac{\mathbf{tr}\{\mathbb{T}_n\}}{\mathbf{rk}\{\text{Var}(\mathcal{M}(X))\}} = \frac{\lambda_{1(1)} + \lambda_{2(1)}}{2} = \frac{\frac{\gamma}{1+u} + \frac{\gamma}{1-u}}{2} \\ &= \frac{\gamma}{1-u^2}, && (6.78) \\ &< \frac{1}{1+|u|}. && (6.79) \end{aligned}$$

Note that in (6.78) we should interpret the appearance of γ, which is the underlying mean component prior variance, as representing the natural scaling. The final result (6.79) indicates that the larger the value of $|u|$, the smaller the resolution possible; in other words, the more highly correlated the information, the less predictive value it has, all other things being equal.

6.14 Example: adjusting exchangeable regressions

6.14.1 Bayes linear sufficiency

To continue the example of §6.7, suppose that we contemplate n full runs of the experiment. That is, we obtain n sets of observations, each of which is a vector. We arrange the n measurements corresponding to a given time point t as the vector

$$\tilde{Y}_t = \begin{bmatrix} Y_{1t} & Y_{2t} & \ldots & Y_{nt} \end{bmatrix}^T,$$

and we collect the vectors at the 13 time points as the collection

$$C(n) = (Y_1, \ \ldots, \ Y_{13}).$$

Suppose we let

$$\mathcal{S}_n(\tilde{Y}_t) = \frac{1}{n}(Y_{1t} + Y_{2t} + \ldots + Y_{nt}), \quad t = 1, \ldots, 13,$$

be the average, across all runs, of the observations at time point t, and collect these averages into the collection

$$\mathcal{S}(n) = [\mathcal{S}_n(\tilde{Y}_1), \mathcal{S}_n(\tilde{Y}_2), \ldots, \mathcal{S}_n(\tilde{Y}_{13})]. \tag{6.80}$$

By Theorem 6.4, this collection of averages, $\mathcal{S}(n)$, is Bayes linear sufficient for the collection of original observations $C(n)$ for adjusting the collection of mean components $\mathcal{M}(Y)$. Thus, we do not have to construct explicitly variance and expectations for the averages $\mathcal{S}_n(\cdot)$, but can instead exploit exchangeability. The computational implementation for such cases is described in full detail in Chapter 12. Here, for example, we would employ Theorem 12.65 in §12.12.5 to evaluate the adjusted expectation, adjusted variance, and resolution transform for the quantities being adjusted.

6.14.2 Adjustment

In practice, three full runs of the experiment were carried out, so that $n = 3$. The data are shown in Table 6.1 and plotted in Figure 6.1. We now construct beliefs over the averaged quantities, using the specifications and relationships given in §6.7, and use them to adjust $\mathcal{M}(Y)$. A summary of the adjustment is shown in Table 6.2, together with a column showing the observed averages. For example,

Table 6.1 Data for three runs of the exchangeable regressions experiment.

Time	Run1	Run 2	Run 3	Average
1	1.79	1.93	1.54	1.75
2	2.14	1.76	1.48	1.79
3	2.13	1.61	1.57	1.77
4	2.07	2.32	1.28	1.89
5	2.08	1.87	1.50	1.82
6	1.88	1.80	1.79	1.82
7	1.94	2.21	1.88	2.01
8	2.01	2.23	2.11	2.12
9	2.35	2.42	2.48	2.42
10	2.23	2.58	2.28	2.36
11	2.58	2.60	3.39	2.86
12	2.48	2.65	3.44	2.86
13	2.82	2.70	2.80	2.77

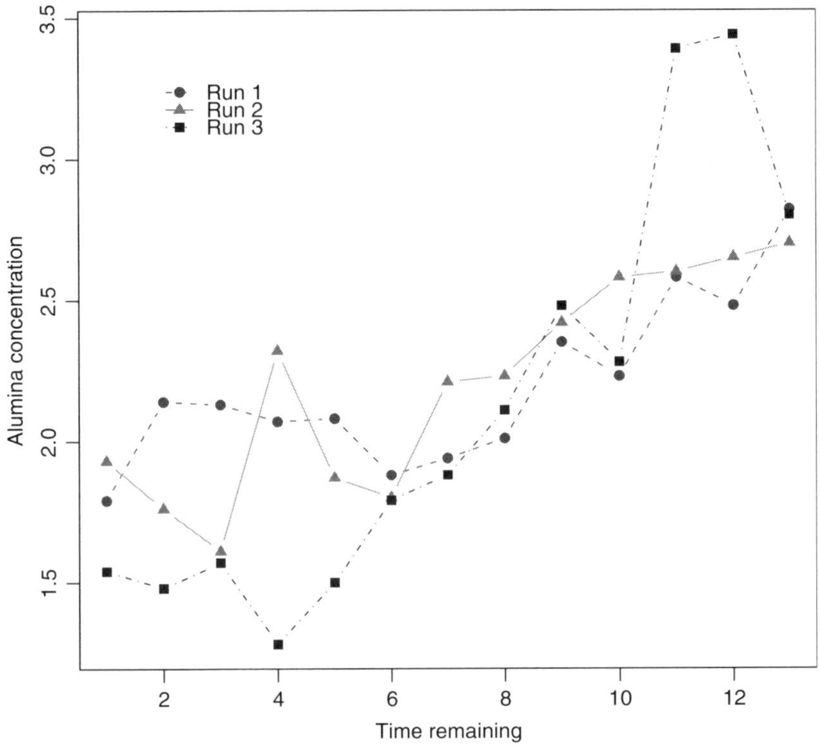

Figure 6.1 The concentrations of alumina remaining in solution at time t from the end of the experiment: results from three independent experiments.

the underlying mean for a run at time $t = 1$, $\mathcal{M}(Y_1)$, has respectively prior and adjusted expectation

$$\mathrm{E}(\mathcal{M}(Y_1)) = 1.5, \quad \mathrm{E}_{\mathcal{S}(3)}(\mathcal{M}(Y_1)) = 1.6031,$$

and the change in expectation is about 0.64 standard deviations relative to the resolved variance in $\mathcal{M}(Y_1)$; that is,

$$\mathrm{S}(\mathrm{E}_{\mathcal{S}(3)}(\mathcal{M}(Y_1))) = \frac{\mathrm{E}_{\mathcal{S}(3)}(\mathcal{M}(Y_1)) - \mathrm{E}(\mathcal{M}(Y_1))}{\sqrt{\mathrm{RVar}_{\mathcal{S}(3)}(\mathcal{M}(Y_1))}} = 0.64.$$

The change in variance for $\mathcal{M}(Y_1)$ from prior to adjusted is calculated as $0.0396 - 0.0258 = 0.0138$, so that about 65% of prior variance is resolved; that is,

$$\mathrm{R}_{\mathcal{S}(3)}(\mathcal{M}(Y_1)) = 0.65.$$

The prior and adjusted means are plotted in Figure 6.2, together with prior and adjusted three-standard-deviation bounds in each case. It can be seen that expectations for the mean components are revised upwards in accordance with generally

Table 6.2 Adjusted expectations, standardized changes, variances, and variance resolutions for the mean components.

	Mean of	Expectations			Variances		
	3 runs	Prior	Adjusted	Change	Prior	Adjusted	Res.
$\mathcal{M}(Y_1)$	1.7533	1.5	1.6031	0.6414	0.0396	0.0138	0.6528
$\mathcal{M}(Y_2)$	1.7933	1.6	1.7013	0.5741	0.0444	0.0133	0.7009
$\mathcal{M}(Y_3)$	1.7700	1.7	1.7994	0.5044	0.0524	0.0135	0.7418
$\mathcal{M}(Y_4)$	1.8900	1.8	1.8976	0.4405	0.0636	0.0145	0.7719
$\mathcal{M}(Y_5)$	1.8167	1.9	1.9958	0.3853	0.0780	0.0162	0.7921
$\mathcal{M}(Y_6)$	1.8233	2.0	2.0939	0.3386	0.0956	0.0186	0.8050
$\mathcal{M}(Y_7)$	2.0100	2.1	2.1921	0.2994	0.1164	0.0218	0.8127
$\mathcal{M}(Y_8)$	2.1167	2.2	2.2902	0.2665	0.1404	0.0257	0.8171
$\mathcal{M}(Y_9)$	2.4167	2.3	2.3884	0.2386	0.1676	0.0303	0.8193
$\mathcal{M}(Y_{10})$	2.3633	2.4	2.4866	0.2148	0.1980	0.0356	0.8201
$\mathcal{M}(Y_{11})$	2.8567	2.5	2.5847	0.1944	0.2316	0.0417	0.8201
$\mathcal{M}(Y_{12})$	2.8567	2.6	2.6829	0.1767	0.2684	0.0485	0.8195
$\mathcal{M}(Y_{13})$	2.7733	2.7	2.7811	0.1613	0.3084	0.0560	0.8185

larger than expected data averages. A substantial portion of the variance in each mean component is resolved, and we learn proportionately more about the end of the series than about its beginning.

Theorem 6.4 also explains that adjusted expectations for future runs, Y_{rt}, $r > n$, $t = 1, \ldots, 13$, are the same as for $\mathcal{M}(Y_t)$, whilst adjusted variances are increased by adding $\text{Var}(\mathcal{R}_r(Y_t))$. Thus the observation at $t = 1$ for a future run, Y_{r1}, also has prior expectation 1.5, adjusted expectation 1.6031, and standardized change 0.64. The change in variance for Y_{r1} from prior to adjusted is $(0.0396 + 0.0801) - 0.0258 = (0.0138 + 0.0801)$, where $\text{Var}(\mathcal{R}_r(Y_1)) = 0.0801$ is the residual variance component.

6.14.3 Resolution transforms

We now calculate the resolution transform for the exchangeable system of §6.7. Suppose we abbreviate the transform for the adjustment of the collection of mean components, $\mathcal{M}(Y)$, by the collection of averages $\mathcal{S}(n)$ as

$$\mathbb{T}_n = \mathbb{T}_{\mathcal{M}(Y):\mathcal{S}(n)}.$$

We first evaluate the transform $\mathbb{T}_1$ for a sample of size $n = 1$. This resolution transform for adjusting $\mathcal{M}(Y)$ by $\mathcal{S}(1)$ has two positive canonical resolutions,

$$\lambda_{1(1)} = 0.6032,$$

$$\lambda_{2(1)} = 0.2976,$$

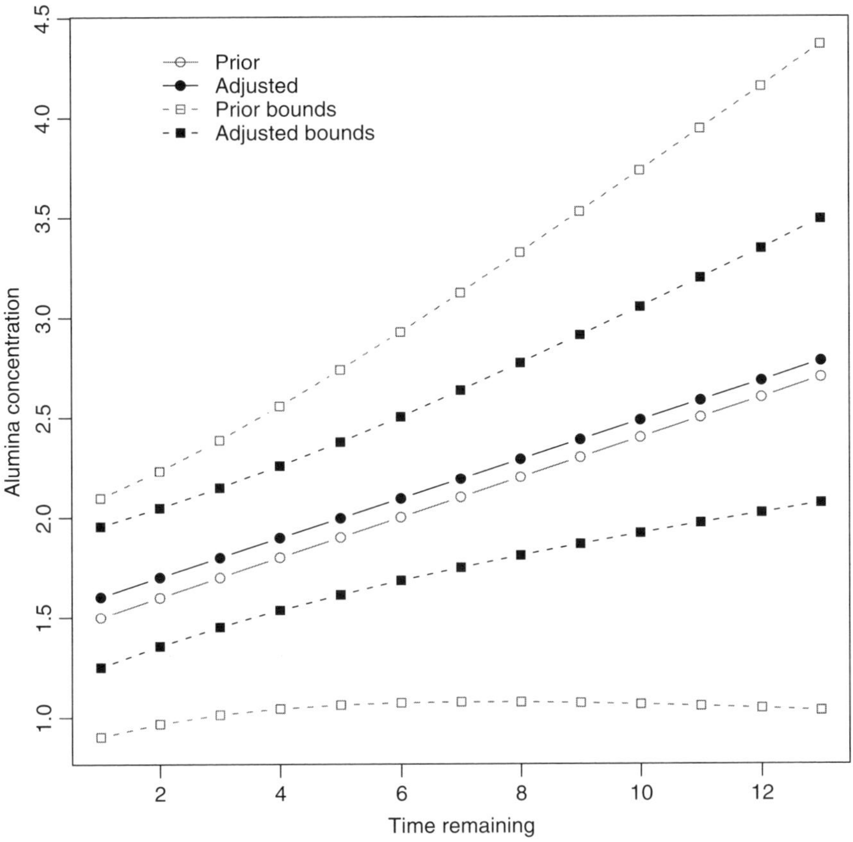

Figure 6.2 Prior and adjusted means for the concentrations of alumina remaining in solution at time t from the end of the experiment, with three-standard-deviation bounds.

corresponding to the canonical directions shown in Table 6.3, i.e.

$$Z_1 = 0.10\mathcal{M}(Y_1) + 0.05\mathcal{M}(Y_2) + \ldots + 0.98\mathcal{M}(Y_{13}) - 5.31, \tag{6.81}$$

$$Z_2 = 3.72\mathcal{M}(Y_1) + 1.12\mathcal{M}(Y_2) + \ldots - 1.42\mathcal{M}(Y_{13}) - 5.44. \tag{6.82}$$

There are only two non-zero canonical resolutions as, from (6.41), we see that all 13 mean components are formed from linear combinations of the two underlying regression coefficient components $\mathcal{M}(a)$ and $\mathcal{M}(b)$. Therefore, the prior variance matrix $\text{Var}(\mathcal{M}(Y))$ has rank two, so that there can be (at most) only two canonical quantities about which any data source can be informative for $\mathcal{M}(Y)$.

A rough interpretation of the coefficients for the first canonical quantity, Z_1, indicates that the data are expected to be most informative for the tail-weighted average of the mean components, whereas the coefficients for the second canonical quantity, Z_2, show that the data are expected next, separately, to be most informative about the difference between the end and the beginning of a run.

6.14.4 Resolution partition for exchangeable cases

Table 6.3 also shows the **resolution partition** for each mean component. Following (3.71), we partition the overall reduction in variance for any quantity into additive contributions from each canonical direction. For a sample of size $n = 1$ the overall resolution for $\mathcal{M}(Y_1)$ is $\mathrm{R}_{\mathcal{S}(1)}(\mathcal{M}(Y_1)) = 0.4068$, attributable to roughly equal contributions, 0.2155 and 0.1913, in the two informative directions. In contrast, the contribution to resolution for $\mathcal{M}(Y_{13})$ from the second direction is negligible compared with that from the first.

The final two columns in Table 6.3 display the resolution partition for the mean components when $n = 3$, and demonstrate how we decompose, and thereby trace the source of, the variance resolutions reported in the last column of Table 6.2. By Theorem 6.5, the canonical directions for the adjustment of $\mathcal{M}(Y)$ by $\mathcal{S}(3)$, are the same combinations, Z_1, Z_2, that we evaluated for the adjustment by a sample with $n = 1$. From (6.57) we deduce that the corresponding canonical resolutions are

$$\lambda_{1(3)} = \frac{3\lambda_{1(1)}}{(3-1)\lambda_{1(1)}+1} = \frac{3 \times 0.6032}{(3-1)0.6032+1} = 0.8202, \tag{6.83}$$

$$\lambda_{2(3)} = \frac{3\lambda_{2(1)}}{(3-1)\lambda_{2(1)}+1} = \frac{3 \times 0.2976}{(3-1)0.2976+1} = 0.5597. \tag{6.84}$$

Table 6.3 Canonical directions and resolution contributions for the mean components.

	Coefficient in		Contribution from		Contribution from	
Component	Z_1	Z_2	$Z_1, n=1$	$Z_2, n=1$	$Z_1, n=3$	$Z_2, n=3$
$\mathcal{M}(Y_1)$	0.10	3.72	0.2155	0.1913	0.2930	0.3597
$\mathcal{M}(Y_2)$	0.05	1.12	0.3271	0.1362	0.4447	0.2562
$\mathcal{M}(Y_3)$	0.04	0.58	0.4217	0.0896	0.5733	0.1685
$\mathcal{M}(Y_4)$	0.05	0.40	0.4913	0.0552	0.6680	0.1038
$\mathcal{M}(Y_5)$	0.05	0.30	0.5383	0.0320	0.7319	0.0603
$\mathcal{M}(Y_6)$	0.06	0.22	0.5680	0.0174	0.7723	0.0326
$\mathcal{M}(Y_7)$	0.07	0.15	0.5860	0.0085	0.7967	0.0160
$\mathcal{M}(Y_8)$	0.08	0.09	0.5961	0.0035	0.8105	0.0066
$\mathcal{M}(Y_9)$	0.10	0.02	0.6012	0.0010	0.8174	0.0019
$\mathcal{M}(Y_{10})$	0.12	−0.04	0.6031	0.0001	0.8200	0.0001
$\mathcal{M}(Y_{11})$	0.16	−0.13	0.6030	0.0001	0.8198	0.0003
$\mathcal{M}(Y_{12})$	0.30	−0.37	0.6016	0.0008	0.8179	0.0015
$\mathcal{M}(Y_{13})$	0.98	−1.42	0.5994	0.0019	0.8150	0.0035
Constant	−5.31	−5.44				

Thus, the implication of a sample size of three rather than a sample size of one is to increase the explanation of variation from 63% to 82% in the major direction of interest, and to increase the explanation of variation from 30% to 56% in the minor direction of interest. The explanation of variation in any other linear combination of the mean components will lie somewhere between these extremes.

6.14.5 Data diagnostics

The bearing for the adjustment (§4.6) is shown in Table 6.4, i.e. the linear combination of mean components for which we obtain the largest change in expectation, relative to its prior variance, is the linear combination

$$\mathbb{Z}_{\mathcal{S}(3)}(\mathcal{M}(Y)) = 0.89\mathcal{M}(Y_1) + 0.78\mathcal{M}(Y_2) + \ldots - 0.46\mathcal{M}(Y_{13}) - 3.75. \quad (6.85)$$

The standardized change is

$$\text{Size}_{\mathcal{S}(3)}(\mathcal{M}(Y)) = 0.2920, \quad (6.86)$$

corresponding to a prior expectation, before observing the three experiments, of 1.3799. The size ratio for the adjustment (4.63) is thus

$$\text{Sr}_{\mathcal{S}(3)}(\mathcal{M}(Y)) = \frac{0.2920}{1.3799} = 0.2116,$$

with the implication that the data induced smaller changes in expectation than were expected at the outset.

We also examine the changes in expectation relative to the variation resolved. The induced discrepancy vector for the adjustment (Definition 4.9) is shown in

Table 6.4 The bearing and discrepancy vector for the adjustment.

	Coefficient in Bearing vector	Coefficient in Discrepancy vector
$\mathcal{M}(Y_1)$	0.89	3.39
$\mathcal{M}(Y_2)$	0.78	1.02
$\mathcal{M}(Y_3)$	0.66	0.53
$\mathcal{M}(Y_4)$	0.55	0.38
$\mathcal{M}(Y_5)$	0.44	0.29
$\mathcal{M}(Y_6)$	0.33	0.22
$\mathcal{M}(Y_7)$	0.21	0.16
$\mathcal{M}(Y_8)$	0.10	0.10
$\mathcal{M}(Y_9)$	−0.01	0.04
$\mathcal{M}(Y_{10})$	−0.13	−0.01
$\mathcal{M}(Y_{11})$	−0.24	−0.08
$\mathcal{M}(Y_{12})$	−0.35	−0.27
$\mathcal{M}(Y_{13})$	−0.46	−1.06
Constant	−3.75	−6.14

the final column of Table 6.4, i.e. the linear combination of mean components for which we obtain the largest change in expectation, relative to its resolved variance, is the linear combination

$$\ddot{\mathbb{Y}}_{\mathcal{S}(3)}(\mathcal{M}(Y)) = 3.39\mathcal{M}(Y_1) + 1.02\mathcal{M}(Y_2) + \ldots - 1.06\mathcal{M}(Y_{13}) - 6.14. \quad (6.87)$$

The standardized change is

$$\text{Dis}_{\mathcal{S}(3)}(\mathcal{M}(Y)) = 0.5020, \quad (6.88)$$

corresponding to a prior expectation, before observing the three experiments, of 2.0. The discrepancy ratio for the adjustment is thus

$$\text{Dr}(\mathcal{S}(3))\mathcal{M}(Y) = \frac{0.5020}{2} = 0.2510,$$

about the same as the size ratio. We comment on these, and examine further diagnostics, in §6.14.8.

6.14.6 Sample size choice

The values of the canonical resolutions also determine the **rates** at which fresh information contributes to reducing variation in each canonical direction as we increase the sample size, n. We quickly resolve almost all the variance in quantities which are strongly correlated with initially informative canonical directions, but we need very large sample sizes to learn about quantities that are strongly correlated with directions that are only weakly informative at the outset. To illustrate this point, Figure 6.3 plots the proportion of variance resolved in Z_1, Z_2, and $\mathcal{M}(Y_1)$ as we increase the sample size, n. The gain in information is fastest in the direction of Z_1, and slowest in the direction of Z_2, and these bound the information gains for any other quantities correlated with them, such as $\mathcal{M}(Y_1)$. The bounds follow as the ordering of the canonical resolutions is preserved when n increases, by Corollary 6.8, and by Corollary 6.6, as the two canonical resolutions in this case are the maximal and minimal positive values. Figure 6.3 also illustrates that, for any positive canonical resolution λ, $\lambda_{(n)}$ is a monotonic increasing function of n with limit unity: we can reduce the variance in the corresponding canonical direction to an arbitrarily small level by taking sufficiently large n.

It is straightforward, using Corollary 6.6, to calculate a sample size to achieve a desired level of variance resolution across every possible linear combination of the mean components. For example, if we wish to explain at least $\alpha = 90\%$ of the variance in every such linear combination, we must take a sample size of

$$n > \frac{\alpha}{1-\alpha}\frac{1-\lambda_{\min}}{\lambda_{\min}} = \frac{0.9}{1-0.9}\frac{1-0.2976}{0.2976} = 21.24, \quad (6.89)$$

i.e. a sample of size $n = 22$. If, instead, we want to achieve a minimum variance reduction in a specified linear combination, we can exploit (6.66) numerically. For

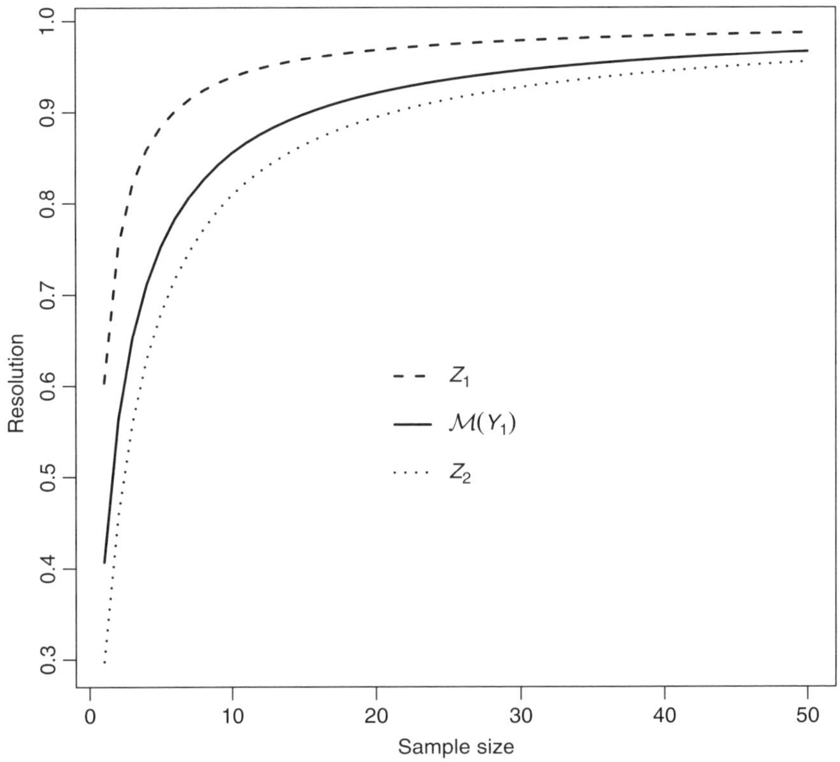

Figure 6.3 Variance resolutions, in the two canonical directions Z_1, Z_2 and the mean component $\mathcal{M}(Y_1)$, for larger sample sizes.

example, suppose that we want to achieve a variance reduction of at least 75% in the difference between $\mathcal{M}(Y_1)$ and $\mathcal{M}(Y_2)$. It is straightforward to show, using (6.81) and (6.82), that

$$\text{Corr}(\mathcal{M}(Y_2) - \mathcal{M}(Y_1), Z_1) = 0.9055,$$

$$\text{Corr}(\mathcal{M}(Y_2) - \mathcal{M}(Y_1), Z_2) = -0.4244.$$

Thus, by (6.66), we have that the resolution for a sample size of n for the adjustment of $\mathcal{M}(Y_2) - \mathcal{M}(Y_1)$ is

$$\begin{aligned} \text{R}_n(\mathcal{M}(Y_2) - \mathcal{M}(Y_1)) &= \sum_i \frac{n\lambda_{i(1)}}{(n-1)\lambda_{i(1)} + 1}[\text{Corr}(\mathcal{M}(Y_2) - \mathcal{M}(Y_1), Z_i)]^2 \\ &= \frac{n \times 0.6032}{(n-1)0.6032 + 1} 0.9055^2 \\ &\quad + \frac{n \times 0.2976}{(n-1)0.2976 + 1}(-0.4244)^2. \end{aligned}$$

It is simple numerically to establish that $\mathrm{R}_2(\mathcal{M}(Y_2) - \mathcal{M}(Y_1)) = 0.6996$ and that $\mathrm{R}_3(\mathcal{M}(Y_2) - \mathcal{M}(Y_1)) = 0.7733$, so that a sample size of $n = 3$ leads to the required resolution of at least 75%.

6.14.7 Adjustment for an equivalent linear space

We now consider the implications of the data for learning about the underlying slope and intercept quantities, $\mathcal{M}(a)$ and $\mathcal{M}(b)$, gathered into the collection $\mathcal{M}(Q)$. By (6.42), adjustment of this collection is equivalent to the adjustment of the collection $\mathcal{M}(Y)$, and we can apply Corollary 6.9. Thus, the resolution transform $\mathbb{T}_{\mathcal{M}(Q):\mathcal{S}(1)}$ for the adjustment of $\mathcal{M}(Q)$ by $\mathcal{S}(1)$ has two canonical directions,

$$W_1 = 2.18\mathcal{M}(a) + 22.64\mathcal{M}(b) - 5.31,$$

$$W_2 = 4.65\mathcal{M}(a) - 10.61\mathcal{M}(b) - 5.44,$$

corresponding to canonical resolutions $\psi_1 = 0.6032$ and $\psi_2 = 0.2976$, respectively. This is structurally identical to the canonical structure obtained for the adjustment of $\mathcal{M}(Y)$ by $\mathcal{S}(1)$ in that $\lambda_i = \psi_i$ and $Z_i = W_i$: for example, by (6.41), the coefficient of $\mathcal{M}(a)$ in W_1 must equate to the sum of the coefficients in Z_1 displayed in Table 6.3. Clearly, the value of new information for larger n will be the same for both $\mathcal{M}(Q)$ and $\mathcal{M}(Y)$. We could add the resolutions for $\mathcal{M}(a)$ and $\mathcal{M}(b)$ for increasing sample size to Figure 6.3: we must and do find that they are bounded by the resolutions for $Z_1 = W_1$ and $Z_2 = W_2$. This equivalence illustrates the way in which the resolution transform fully captures the geometric structure underlying our beliefs over the observables.

6.14.8 Data diagnostics for an equivalent linear space

The adjustment diagnostics for two equivalent linear spaces must be identical. We have that $\langle\mathcal{M}(Q)\rangle$ and $\langle\mathcal{M}(Y)\rangle$ are equivalent. Consequently, comparing to (6.86) and (6.88), we must have

$$\mathrm{Size}_{\mathcal{S}(3)}(\mathcal{M}(Q)) = \mathrm{Size}_{\mathcal{S}(3)}(\mathcal{M}(Y)) = 0.2920,$$

$$\mathrm{Dis}_{\mathcal{S}(3)}(\mathcal{M}(Q)) = \mathrm{Dis}_{\mathcal{S}(3)}(\mathcal{M}(Y)) = 0.5020.$$

The bearing and induced discrepancy vector (Definition 4.9) must similarly be identical. Direct calculation shows that, for the adjustment of the collection $\mathcal{M}(Q)$, comprising the underlying mean slope and intercept, these are respectively

$$\mathbb{Z}_{\mathcal{S}(3)}(\mathcal{M}(Q)) = 2.76\mathcal{M}(a) - 1.15\mathcal{M}(b) - 3.75, \tag{6.90}$$

$$\ddot{\mathbb{Y}}_{\mathcal{S}(3)}(\mathcal{M}(Q)) = 4.70\mathcal{M}(a) - 4.45\mathcal{M}(b) - 6.14. \tag{6.91}$$

Recalling that we have, for each $\mathcal{M}(Y_t)$, $\mathcal{M}(Y_t) = \mathcal{M}(a) + t\mathcal{M}(b)$, we may check that (6.90) matches (6.85) and that (6.91) matches (6.87).

6.14.9 Compatibility of data sources

Where we have exchangeable data, one obvious question is whether the data arising from different samples appear to be compatible. To investigate, we adjust the underlying mean regression components, $\mathcal{M}(Q)$, sequentially by the data available from the first, second, and third experiments, and calculate diagnostics for these adjustments as described in §5.11. For this we must compute the bearings (the directions of maximal standardized change in expectation) for (1) the adjustment by data from the first experiment alone; (2) the partial adjustment by data from the second experiment, over and above data from the first experiment; (3) the partial adjustment by data from the third experiment, over and above data from the first two experiments. Suppose we represent these three adjustments as D_1, D_2, D_3, observed to be d_1, d_2, d_3, respectively. Using our standard notation (§5.6), these bearings and partial bearings, with their corresponding size ratios, turn out to be:

$$\mathbb{Z}_{d_1}(\mathcal{M}(Q)) = +2.80\mathcal{M}(a) - 4.10\mathcal{M}(b) - 3.51, \tag{6.92}$$

$$\mathrm{Sr}_{d_1}(\mathcal{M}(Q)) = 0.3612, \tag{6.93}$$

$$\mathbb{Z}_{[d_2/d_1]}(\mathcal{M}(Q)) = +1.43\mathcal{M}(a) - 2.74\mathcal{M}(b) - 1.72, \tag{6.94}$$

$$\mathrm{Sr}_{[d_2/d_1]}(\mathcal{M}(Q)) = 0.2877, \tag{6.95}$$

$$\mathbb{Z}_{[d_3/d_1 \cup d_2]}(\mathcal{M}(Q)) = -1.44\mathcal{M}(a) + 6.71\mathcal{M}(b) + 1.35, \tag{6.96}$$

$$\mathrm{Sr}_{[d_3/d_1 \cup d_2]}(\mathcal{M}(Q)) = 0.4871. \tag{6.97}$$

Inspecting the size ratios, none of the standardized changes in expectation were large relative to prior variation. Indeed, there may be a suspicion of systematically smaller than expected changes in expectation, indicating perhaps that the prior variances assigned to these quantities overstated the assessor's uncertainties about them. We need next to compute the path correlations (§5.9) between these bearings. Inspecting the vectors in (6.92), (6.94), and (6.96), it seems obvious that the first two are quite similar, whilst the third appears quite different. The path correlation for data from the first two experiments turns out to be, using (5.51),

$$\mathrm{PC}(d_1, [d_2/d_1]) = \mathrm{Corr}(\mathbb{Z}_{d_1}(\mathcal{M}), \mathbb{Z}_{[d_2/d_1]}(\mathcal{M})) = 0.9964,$$

where $\mathcal{M} = \mathcal{M}(Q)$ for convenience, whilst the path correlation between data from the first two experiments and data from the third experiment turns out to be

$$\begin{aligned}\mathrm{PC}(d_1 \cup d_2, [d_3/d_1 \cup d_2]) &= \mathrm{Corr}(\mathbb{Z}_{d_1 \cup d_2}(\mathcal{M}), \mathbb{Z}_{[d_3/d_1 \cup d_2]}(\mathcal{M})) \\ &= \mathrm{Corr}(\mathbb{Z}_{d_1}(\mathcal{M}) + \mathbb{Z}_{d_1/d_2}(\mathcal{M}), \mathbb{Z}_{[d_3/d_1 \cup d_2]}(\mathcal{M})) \\ &= -0.9386.\end{aligned}$$

Thus, there is substantial agreement between the data from the first two experiments, but there is a portion of the data from the third experiment that is in substantial disagreement. In this particular example, however, the relevance of the disagreement is minor, as the corresponding size ratios are all small. Had the size ratios been larger (i.e. larger than expected changes in expectation), this finding would perhaps have motivated an investigation of the third experiment in an attempt to discover the root of the discordancy.

One further check we can make is to assess the consistency of the observations made at different time points. To do so, suppose that we let

$$G_t = [Y_{1t}, Y_{2t}, Y_{3t}]$$

be the collection of observations to be made at time t, and let g_t represent the actual observations. We will again calculate diagnostics pertaining to the adjustment of $\mathcal{M}(Q)$, as these will be the same as the diagnostics for the adjustment of $\mathcal{M}(Y)$ as the two corresponding linear spaces are identical. We will form diagnostics for the sequence of exchangeable adjustments where we adjust $\mathcal{M}(Q)$ by G_1, and then partially by G_2, and so forth, until finally we adjust partially by the three observations G_{13}. At each time point t, the average $\bar{G}_t = Y_{1t} + Y_{2t} + Y_{3t}$ is Bayes linear sufficient for the collection G_t for any adjustment, including partial adjustments, of $\mathcal{M}(Q)$.

Table 6.5 summarizes the sequential diagnostics for this problem, and these are also plotted in Figure 6.4. Only one of the partial adjustments stands out. At time $t = 11$, the data are discordant with the data at preceding time points (path correlation −0.73), and there is simultaneously a surprisingly large change in expectation: the size ratio is 8.66. This combination of features is most evident

Table 6.5 Diagnostic assessment for sequential exchangeable adjustments.

i	$\mathrm{E}_{[i]}(\mathcal{M}(a))$	$\mathrm{E}_{[i]}(\mathcal{M}(b))$	$\mathrm{Size}_{[i/]}(\mathcal{M}(Q))$	$\mathrm{Sr}_{[i/]}(\mathcal{M}(Q))$	$C_{[i/]}$
1	1.5452	0.1061	0.5782	0.99	
2	1.5387	0.1026	0.0086	0.08	−0.54
3	1.5310	0.0902	0.0976	0.81	−0.22
4	1.5310	0.0907	0.0001	0.00	−0.32
5	1.5341	0.0763	0.1298	1.43	0.37
6	1.5385	0.0675	0.0487	0.64	0.73
7	1.5340	0.0736	0.0238	0.38	−0.84
8	1.5313	0.0764	0.0052	0.10	−0.82
9	1.5155	0.0902	0.1256	2.88	−0.81
10	1.5217	0.0855	0.0151	0.41	0.61
11	1.4925	0.1054	0.2694	8.66	−0.73
12	1.4926	0.1053	0.0000	0.00	0.04
13	1.5050	0.0982	0.0357	1.57	0.07

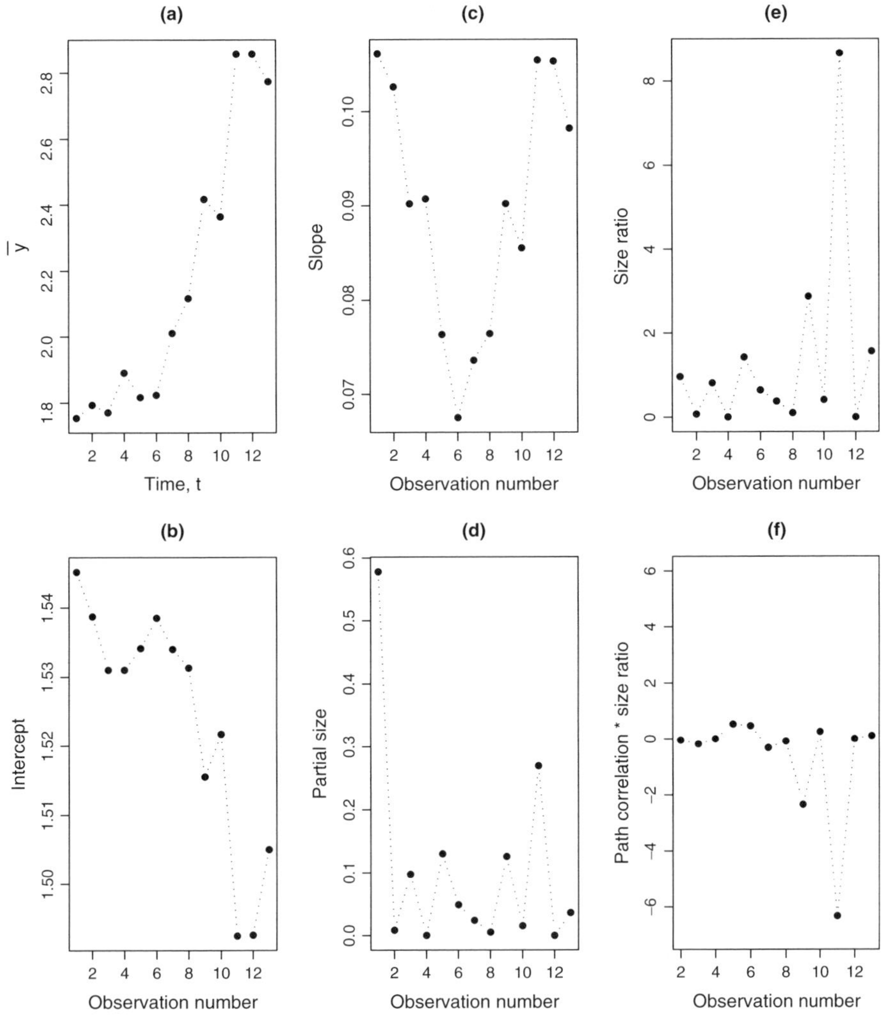

Figure 6.4 (a) Scatter plot of observed means at each time point; (b) sequential update of expectation for $\mathcal{M}(a)$ (c) sequential update of expectation for $\mathcal{M}(b)$; (d) partial sizes for sequential adjustments; (e) partial size ratios for sequential adjustments; (f) successive size-weighted path correlations.

in Figure 6.5(f). As a further check, the observations are divided into the set of all observations at time points up to and including $t = 10$, D_{1-10}, and all other observations D_{11-13}, and a diagnostic assessment carried out. The path correlation turns out to be -0.73, confirming some differences between the beginning and the end of the process that are not captured by our model. Referring back to Figure 6.1,

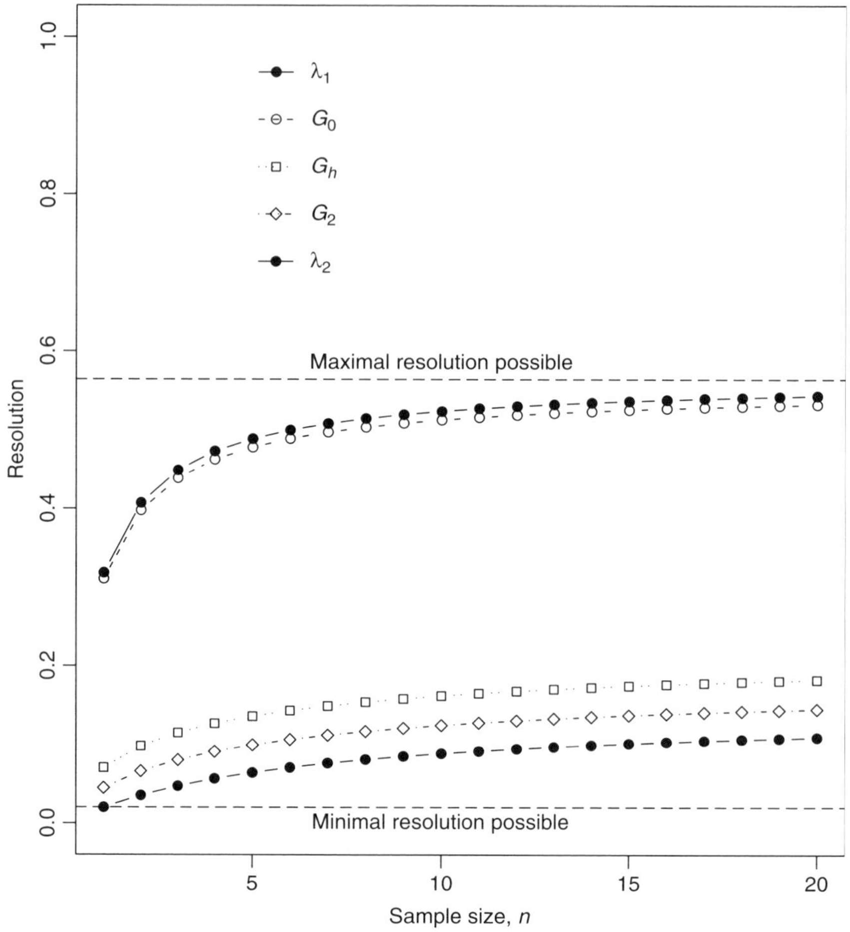

Figure 6.5 Variance resolutions in the two canonical quantities and the three quantities G_0, G_2, and G_h for sample sizes up to $n = 20$.

visually a large change occurs between $t = 10$ and $t = 11$ for experiment $r = 3$. We will see later (§9.13.2) that the third experiment appears somewhat aberrant, compared to the other two experiments.

6.15 Predictive adjustment

We now extend the results of §6.12 to the case of predictive adjustment. Thus, suppose that we observe $D_n = (X_1, \ldots, X_n)$ and we intend to adjust beliefs over the further collection of r observations $D_{n,r} = (X_{n+1}, \ldots, X_{n+r})$.

We may adjust beliefs over $D_{n,r}$ by first adjusting beliefs over $\mathcal{M}(X)$ and then using (6.45), (6.46) to adjust beliefs over $D_{n,r}$. From Theorem 6.4, the sample mean $\bar{X}_n$ is Bayes linear sufficient for D_n for adjusting $\mathcal{M}(X)$ and so is also Bayes linear sufficient for D_n for adjusting $D_{n,r}$. If we denote the sample mean vector for $D_{n,r}$ as

$$\bar{X}_{n,r} = \frac{1}{r} \sum_{j=n+1}^{n+r} X_j,$$

then, by a similar argument to (6.51), we have

$$\lfloor D_{n,r} \perp\!\!\!\perp \mathcal{M}(X) \rfloor \ / \ \bar{X}_{n,r}.$$

Therefore, the adjustment of $D_{n,r}$ by D_n is precisely equivalent to the adjustment of $\bar{X}_{n,r}$ by $\bar{X}_n$, as (i) adjustment of $D_{n,r}$ by D_n or $\bar{X}_n$ gives precisely the same results, and (ii) $\mathbb{A}_{\bar{X}_{n,r}}(D_{n,r})$ is orthogonal to D_n, so that any linear combination of the elements of $D_{n,r}$ which is orthogonal to the mean vector $\bar{X}_{n,r}$ has adjusted mean equal to the prior mean.

We now consider the resolution transform for the adjustment of $D_{n,r}$ by D_n. By the above argument, this transform may be equivalently assessed as the resolution transform for the adjustment of $\bar{X}_{n,r}$ by $\bar{X}_n$, which we denote by $\mathbb{T}_{n,r}$. We show that the eigenvectors of $\mathbb{T}_{n,r}$ are essentially the same as for $\mathbb{T}_n$, the resolution transform for the adjustment of $\mathcal{M}(X)$ given D_n, and derive the corresponding eigenvalues. We have the following result.

Theorem 6.10 *Suppose that W is an eigenvector of $\mathbb{T}_s$ for each s, with corresponding eigenvalue $\lambda_{(s)}$, and standard form $W = \alpha^T \mathcal{M}(X)$. Then*

$$W_{n,r} = \alpha^T \bar{X}_{n,r}$$

is an eigenvector of $\mathbb{T}_{n,r}$ with eigenvalue

$$\lambda_{(n,r)} = \lambda_{(n)}\lambda_{(r)}. \tag{6.98}$$

Proof. We have $\lfloor D_n \perp\!\!\!\perp D_{n,r} \rfloor \ / \ \mathcal{M}(X)$. Thus, from Theorem 5.24, $W^*_{(n,r)}$ is an eigenvector of $\mathbb{T}_{n,r}$, with eigenvalue λ if and only if $\mathrm{E}_{\mathcal{M}(X)}(W^*_{(n,r)}) = W$ is an eigenvector of $\mathbb{T}_r\mathbb{T}_n$, with eigenvalue λ. From Theorem 6.5, the transforms $\mathbb{T}_n$, $\mathbb{T}_r$ have the same eigenvectors for each value of n and r, and the result follows. ■

Thus, we obtain the eigenvectors of $\mathbb{T}_{n,r}$ from the standard form for the eigenvectors of $\mathbb{T}_1$ by replacing each linear combination of $\mathcal{M}(X)$ by the corresponding linear combination of $\bar{X}_{n,r}$ and evaluating the eigenvalues by (6.98). Thus, predictive adjustment shares the same qualitative features as does adjustment over the population structure, with similar implications for design and interpretation.

We may alternatively derive the result of Theorem 6.10 directly by constructing the matrix representation of $\mathbb{T}_{n,r}$, which may be written as

$$\begin{aligned}\mathbb{T}_{n,r} &= (\mathrm{Var}(\bar{X}_{n,r}))^{-1}\mathrm{Cov}(\bar{X}_{n,r}, \bar{X}_n)(\mathrm{Var}(\bar{X}_n))^{-1}\mathrm{Cov}(\bar{X}_n, \bar{X}_{n,r}) \\ &= \left(\Gamma + \frac{1}{r}(\Sigma - \Gamma)\right)^{-1}\Gamma\left(\Gamma + \frac{1}{n}(\Sigma - \Gamma)\right)^{-1}\Gamma \\ &= \mathbb{T}_r\mathbb{T}_n. \end{aligned} \tag{6.99}$$

As before, for simplicity of exposition we assume that the variance matrices here are positive definite, deferring consideration of the non-negative definite case to §12.12. For sample size choice for a predictive adjustment, we have the following corollary.

Corollary 6.11 *Suppose that W is an eigenvector of $\mathbb{T}_1$ with corresponding eigenvalue $\lambda > 0$. Then the minimal sample size n required to achieve a proportionate variance reduction α for the canonical direction $W^*_{(n,r)}$ of $\mathbb{T}_{(n,r)}$ is, for $0 < \alpha < \lambda_{(r)}$,*

$$n \geq \frac{\alpha(1-\lambda)}{\lambda(\lambda_{(r)} - \alpha)}. \tag{6.100}$$

The maximum possible resolution of variance for $W^*_{(n,r)}$ is $\lambda_{(r)}$. The maximum possible proportionate reduction in variance for any element of $\langle D_{n,r}\rangle$ is thus

$$\frac{r\lambda_{\max}}{1 + (r-1)\lambda_{\max}}, \qquad \text{as } n \to \infty,$$

where $\lambda_{\max}$ is the maximum eigenvalue of $\mathbb{T}_1$.

6.16 Example: oral glucose tolerance test

6.16.1 Context of exchangeability

We now continue the example of §6.6 in which we considered an exchangeable sample for the OGT test problem. In earlier chapters, our focus has been on using our doctor's own measurements to learn about the effect of the test on a typical elderly person, with whom our doctor is exchangeable. In the context of exchangeability, we can now see that this involved a predictive adjustment. Consequently, we first explore the implications of exchangeability for an underlying mean component for this example.

6.16.2 Mean component adjustment

The underlying quantity that connects individuals is given by $\mathcal{M}(D)$, with prior variance matrix given by (6.32). For an exchangeable sample of size n of measurements $D_1, \ldots, D_n$, the adjustment of these mean components has a basic resolution transform which is calculated from (6.59) with Σ, Γ defined in (6.34) and (6.35). In practice, we can calculate the resolution transform (6.59) with $n = 1$ and deduce the canonical structure for an adjustment by any sample size. The canonical quantities and canonical directions (in standardized and unstandardized forms) for the adjustment of the mean component by the first observation from an exchangeable sequence are

$$\lambda_1 = 0.5643, \qquad W_1 = 1.3390\mathcal{M}(G_0) - 0.1512\mathcal{M}(G_2) - 4.6253, \tag{6.101}$$

$$W_1^* = 1.0543\mathcal{M}(G_0) - 0.0992\mathcal{M}(G_2), \tag{6.102}$$

$$\lambda_2 = 0.1420, \qquad W_2 = 0.8012\mathcal{M}(G_0) - 1.8676\mathcal{M}(G_2) - 8.3395, \tag{6.103}$$

$$W_2^* = 0.6309\mathcal{M}(G_0) - 1.2247\mathcal{M}(G_2). \tag{6.104}$$

One of the great advantages of the Bayes linear approach is that, with respect to the belief specifications, these quantities completely characterize the adjustment:

- for the underlying mean components, given any sample size;
- for predicting a single other individual, given any sample size;
- for predicting the mean of a further collection of such individuals, given any sample size;
- for any linear combination of these mean components and/or further such individuals, given any sample size.

Indeed, with respect to the belief specifications, we can deduce all required results straightforwardly from (6.101) and (6.103).

6.16.3 Variance reduction for a predictive adjustment

In §3.8.3 we made brute-force calculations to show the implications of a larger sample size for learning about a typical elderly person's fasting and 2-hour blood glucose levels, and the difference between them. We saw in Figure 3.1 that we appeared to learn relatively little about these quantities, even for quite large sample sizes. Let us now use the underlying canonical structure to investigate and explain what is going on here. Theorem 6.10 shows us that the ith canonical resolution for a predictive adjustment of an individual is

$$\lambda_{i(n,1)} = \lambda_{i(n)}\lambda_{i(1)}. \tag{6.105}$$

Here, $\lambda_{i(n,1)}$ is the resolved variance in W_i, the ith canonical direction for the adjustment of the mean component. As $n \to \infty \Rightarrow \lambda_{i(n)} \to 1$, we can resolve all of the prior variance in every canonical direction W_i by taking a sufficiently large sample size. However, for the predictive adjustment, there is an unresolvable portion of prior variation, and from (6.105) it follows that the unresolved portion for the ith canonical direction is $1 - \lambda_{i(1)}$. For this predictive adjustment for a single

Table 6.6 Canonical and maximal canonical resolutions.

Quantity	Prior variance	Variance resolved $n = 1$	Maximal variance resolved $n \to \infty$
	1	$\lambda_{i(1,1)} = \lambda^2_{i(1)}$	$\lambda_{i(1)}$
W_1	1	0.3184	0.5643
W_2	1	0.0202	0.1420
G_0	1	0.3109	0.5536
G_2	1	0.0448	0.1770
G_h	1	0.0705	0.2133

individual, the variance resolutions for the canonical quantities are summarized in the first part of Table 6.6. Notice that the variance resolutions for $n = 1$ agree with those reported in (3.110) in §3.11.3. Also notice that the implication of these belief specifications is that we do not expect to learn very much, particularly in the second canonical quantity, even by taking very large sample sizes. Even in the first canonical direction, W_1, for which we will learn most, we expect to resolve only 31.84% of its variance using one observation, up to a maximum of 56.43% using a very large number of observations.

The minimal and maximal resolutions (6.105) provide bounds on the proportion of variance resolved for any linear combination of the quantities being adjusted. It follows that for G_0, G_2, and $G_h = G_2 - G_0$, the proportion of variance we expect to resolve lies somewhere between about 2% and 31% if we take a sample size $n = 1$, and at most somewhere between about 14% and 56% if we take a huge sample size. It is for this reason that Figure 3.1 shows very little explanation of variance for these three quantities, for the range of sample sizes shown. It is simple to calculate what the variance resolutions are, via (6.65), in that for any linear combination Y in this predictive space we have

$$\mathrm{R}_n(Y) = \frac{\sum_i \lambda_{i(n)}\lambda_{i(1)}(\mathrm{Cov}(Y, W_i))^2}{\sum_i (\mathrm{Cov}(Y, W_i))^2}. \tag{6.106}$$

The resolutions for G_0, G_2, and G_h for $n = 1$ and maximal n are shown in the second part of Table 6.6, with their prior variances rescaled to unity for convenience. We commented at length on the relationships between these three quantities and the canonical quantities in §3.11.3, and those comments apply equally here, but now underpinned by the insights which the exchangeability representation provides.

Figure 3.1 suggests that most of the resolvable variance is resolved by a relatively small sample size. Figure 6.5 shows variance resolutions in the two canonical quantities and the three quantities G_0, G_2, and G_h for sample sizes up to $n = 20$. Notice how the variance resolutions are bounded by the canonical resolutions. The resolutions for G_0 (which is highly correlated with the primary canonical quantity) quite quickly approach the maximum, suggesting that a fairly small sample size would be quite efficient for learning about G_0. On the other hand, the information gains are much slower for G_2 and G_h, which are more highly correlated with the secondary canonical quantity. If we felt that it was important to remove a large proportion of the explainable variation in G_h, it is clear that we would have to pay the penalty of taking a very large sample size.

6.16.4 Observed exchangeable adjustments

For this example, a sample of size $n = 15$ was obtained by taking 15 elderly healthy individuals, and administering to them the OGT test. These data are a part of the data set discussed in Farrow and Leyland (1991). We are very grateful to Malcolm Farrow for providing the data and belief judgements for them. The data are shown in Table 6.7 and plotted in Figure 6.6. Our illustrations using this example in Chapter 4 were performed using the first of these observations. Inspecting the scatter plot, we observe that the 2-hour measurement G_2 tends to rise with fasting measurement, G_0. This is consistent with the prior correlation specified between them, $\text{Corr}(G_0, G_2) = 0.4364$. There are two pairs of observations, (4.8, 2.3) and (4.6, 3.7), which look unusual. We would not normally expect the 2-hour measurement to exceed the fasting measurement,

Observed exchangeable adjustments proceed with the observed sample means summarizing the observed quantities. Before this stage is reached, however, the data must be checked to ensure that they are consistent with beliefs specified about them. This is the case whenever $\text{Var}(D)$ is full rank, as in this example. Notice that when $\text{Var}(D)$ is not full rank, every vector observation d_i must be checked for consistency (using Definition 12.61 of §12.12.3), as the fact that the sample means, $\bar{d}$, are consistent does not imply that the individual observations, d_i, are consistent.

Next, the sample means are calculated and the adjusted expectations and variances obtained for $n = 15$. The calculations are made using the results from earlier

Table 6.7 Blood glucose levels, in mmol/litre, for 15 healthy elderly individuals, measured before and 2 hours after administration of the oral glucose tolerance test.

Observation	1	2	3	4	5
Blood glucose level, fasting	5.4	4.8	4.0	4.0	5.2
Blood glucose level, 2-hour	9.8	2.3	5.0	8.1	8.9
Observation	6	7	8	9	10
Blood glucose level, fasting	4.5	4.9	4.7	4.6	3.9
Blood glucose level, 2-hour	6.8	7.0	7.7	3.7	4.8
Observation	11	12	13	14	15
Blood glucose level, fasting	4.4	4.6	4.9	6.3	4.1
Blood glucose level, 2-hour	7.9	6.0	9.8	8.4	5.1

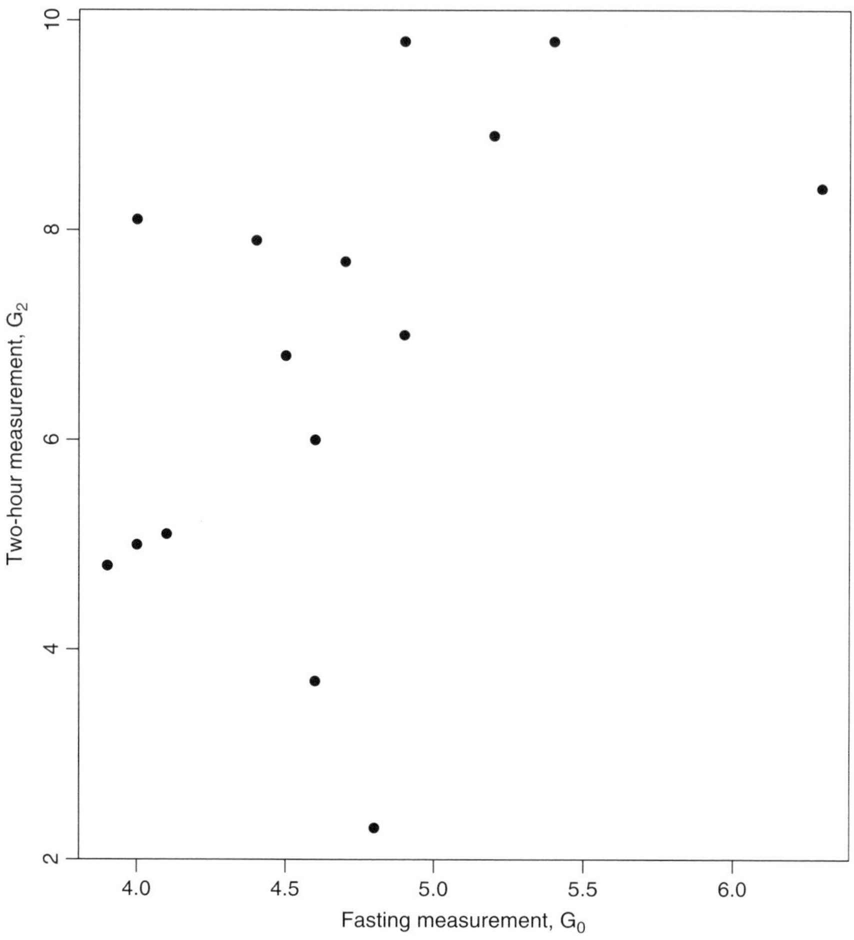

Figure 6.6 Scatter plot of fasting and 2-hour blood glucose levels following the oral glucose tolerance test on 15 healthy individuals.

in this chapter or, more generally, using those of §12.12.5. The main features are summarized in Tables 6.8 and 6.9 and are as follows.

- The data for the basic measurements G_0 and G_2 are generally higher on average than expected, and by about the same amount, so that the prior judgement that the 2-hour measurement exceeds the fasting measurement by about 2 mmol/l seems justified.

- The adjusted expectations for the underlying population means corresponding to G_0 and G_2 are roughly 4.65 and 6.66 mmol/l, and so these are the predictions for a new individual's two readings.

Table 6.8 Summary of the adjustment by 15 observations: means and standard deviations of the observations; and prior and adjusted expectations with standardized change in adjustment, relative to variance resolved.

	Data		Expectation		
	Mean	SD	Prior	Adjusted	Change
G_0	4.6867	0.6065	4.16	4.6530	0.64
G_2	6.7533	2.1497	6.25	6.6593	0.71
G_h	2.0667	1.9492	2.09	2.0062	−0.14

Table 6.9 Summary of the adjustment by 15 observations: prior and adjusted variances and variance resolutions for the mean and predictive components.

	Mean component variation			Predictive component variation		
	Prior	Adjusted	Resolution	Prior	Adjusted	Resolution
G_0	0.62	0.0313	0.9495	1.12	0.5313	0.5256
G_2	0.43	0.0964	0.7757	2.43	2.0964	0.1373
G_h	0.45	0.0814	0.8192	2.11	1.7414	0.1747

- Relative to variance resolved, neither of the changes in expectation is particularly surprising: the larger standardized change is $\mathrm{S}_d(G_2) = 0.71$ standard deviations (see §4.4.1). Note that the adjusted expectation and standardized change in adjustment are identical for the mean and predictive components.

- Variance resolutions for the mean components are quite substantial: 95% and 78% of the variation in $\mathcal{M}(G_0)$, $\mathcal{M}(G_2)$ respectively, is resolved by this sample. However, as suggested by the canonical analysis summarized in Table 6.6, variance resolutions for the predictive components are much smaller. Given a sample size of $n = 15$, the percentage of variance explained for another individual's G_2 measurement is only 13.7%. Table 6.6 implies that, whilst this is small, we cannot do much better even by taking a huge sample size. The reason is straightforward. The mean component for G_2 has prior variance 0.43, all of which can be resolved by taking a sufficiently large sample. The predictive component for G_2 has prior variance $0.43 + 2.00$ of which the first part, corresponding to mean component variation, can similarly be resolved fully by taking a sufficiently large sample size; whereas the second part, corresponding to variation specific to an individual, can never be resolved through measuring other individuals.

As in §4.5.2.1, we take intervals of about two or three standard deviations in either direction from the expectation as being fairly likely to contain the relevant locations. For the prior assessments we have (restating these from §4.5.2.1) approximately the three standard deviation intervals

$$G_0 : 4.16 \pm 3\sqrt{1.12} = (0.99, 7.33),$$

$$G_2 : 6.25 \pm 3\sqrt{2.43} = (1.57, 10.93).$$

For the assessments after adjusting by $[D]$ we obtain the tighter intervals

$$G_0 : 4.65 \pm 3\sqrt{0.53} = (2.46, 6.84),$$

$$G_2 : 6.66 \pm 3\sqrt{2.10} = (2.32, 11.00).$$

The interpretation is as in §4.5.2.1, though somewhat less dramatic. The evidence does seem to support the notion that the fasting measurement for healthy elderly individuals exceeds that for younger people. It also seems plausible that the effect of the test is to raise the blood glucose level of individuals (young and old) by about 2 mmol/l. The consequence for the elderly is that their normal 2-hour measurement, which we assess at present as being around 6.66 mmol/l, is only just below the level of 7.0 mmol/l deemed to be the threshold for impaired glucose tolerance. The caveat is that we have resolved very little of the variation in the 2-hour measurement G_2, and in the derived quantity G_h, so the evidence for the locations of these two quantities remains inconclusive.

Globally, the induced discrepancy vector (Definition 4.9) and the bearing (§4.6) turn out to be

$$\ddot{\mathbb{Y}}_d(B) = 0.4273G_0 + 0.8724G_2 - 7.2302,$$

$$\mathbb{Z}_d(B) = 0.4100G_0 + 0.0469G_2 - 1.9991.$$

These indicate the maximal changes in expectation with respect to resolved and prior uncertainty, respectively. The diagnostics for these two quantities reveal nothing untoward. For example, the adjustment discrepancy (4.22) is $\text{Dis}_d(B) = 0.5677$, with prior expectation 2 and discrepancy ratio 0.2839; and the size of the adjustment (4.51) is $\text{Size}_d(B) = 0.2214$, with prior expectation $\lambda_{1(15)} + \lambda_{2(15)} = 0.6379$ and size ratio $\text{Sr}_d(B) = 0.3470$. These indicate smaller than expected changes in expectation.

6.16.5 Path diagnostics

Finally, we inspect the data for the concordance of individual observations. One way to do this is to sort the data according to the fasting measurement (this seems appropriate as we expect these measurements to be better understood than the 2-hour measurements), and then to examine the one-step sequential adjustments by each fresh data pair for discordancy. The results are shown in Figure 6.7. This shows, in panel (b), that the arrival of the new evidence leads consistently to higher and higher adjusted expectations for G_0, but that the evidence is more ambiguous for the 2-hour measurement G_2 shown in panel (c). Inspection of panel (f), in which

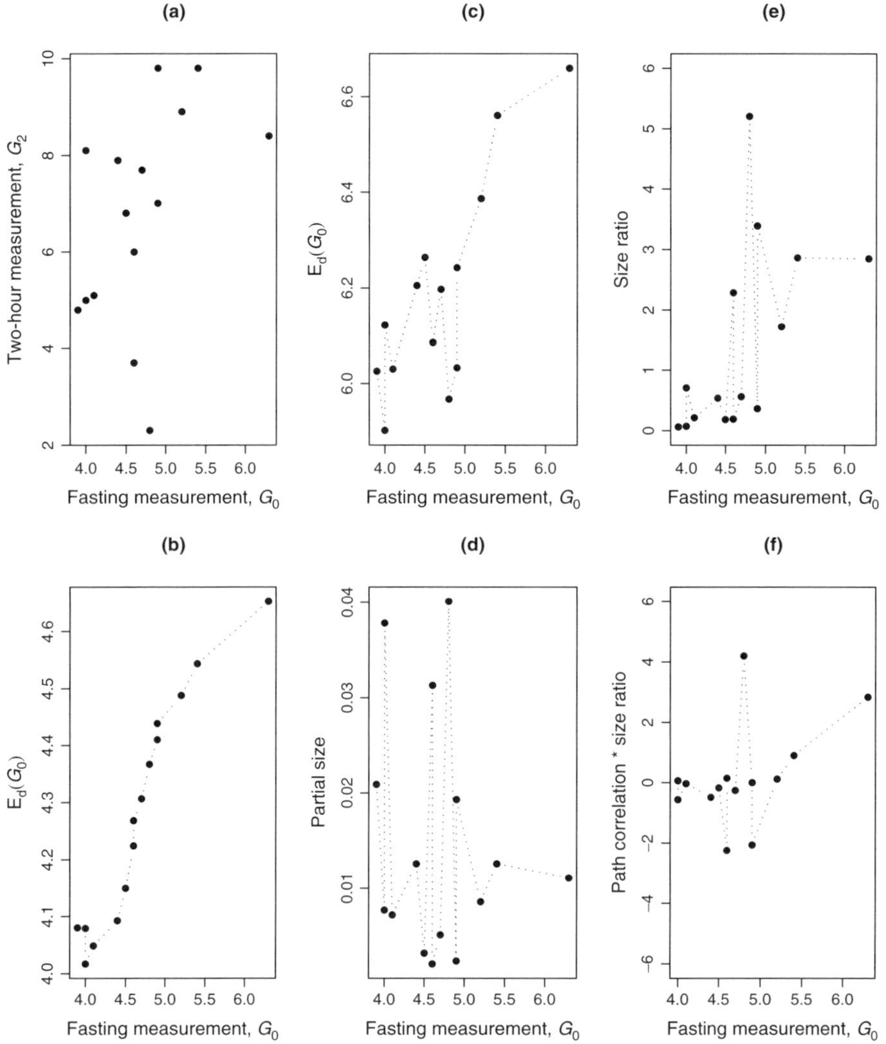

Figure 6.7 Path diagnostics, full data set: (a) scatter plot of original data, fasting measurement versus 2-hour measurement; (b) sequential update of expectation for G_0; (c) sequential update of expectation for G_2; (d) partial sizes for sequential adjustments; (e) partial size ratios for sequential adjustments; (f) successive path correlations multiplied by size ratios.

are plotted the successive path correlations, weighted by the corresponding size ratios, shows no particularly disturbing contradictions: there is one combination of quite highly positive path correlation with quite high size ratio when we partially adjust by the measurement (4.8, 2.3); however, recall that we would be mostly concerned if we had found large **negative** combinations. Examining panel (e), we find that the two largest size ratios correspond to the observations (4.8, 2.3) and (4.6, 3.7), implying that these are quite unusual observations in relation to the preceding belief adjustment. These are the two unusual observations noted earlier. In summary, these two pairs of observations are not strongly discordant with the other 13 observations, but they are discrepant in the context of the full belief and data analysis.

As such, one option, for the purpose of comparison, is to exclude this discrepant pair of observations and recalculate the observed adjustment. The diagnostic analysis is summarized in Figure 6.8, in which we see no strongly discrepant features. The recalculated adjustment is summarized in Table 6.10. The adjusted expectation for G_2 is markedly higher than before, and exceeds the threshold for impaired glucose tolerance. The adjusted expectation for G_h is also higher than before, $\mathrm{E}_d(G_h) = 2.4066$, suggesting that the normal healthy elderly individuals do react differently to younger individuals. We might reasonably conclude that the doctor's suspicions about the validity of the OGT test for the elderly appear well founded. We should also be concerned with the two individuals excluded from the final analysis because their patterns of reaction to the OGT test appear atypical. Exclusion may seem reasonable on the grounds that we desire our sample of observations to be representative of elderly healthy individuals with a normal pattern of response. On the other hand, we run the risk of ignoring genuine features relevant to a minority of the population.

6.17 Example: predictive analysis for exchangeable regressions

For the exchangeable regressions example of §6.7 and §6.14, suppose that we form the collection $C_f = Y_{1,f}, \ldots, Y_{13,f}$ of values of one future experiment based on a sample of size $n < f$, and that we want to assess the implication of using that sample to learn about the elements of the collection C_f. We can do so by exploiting Theorem 6.10. In particular, for our actual sample size of $n = 3$, the canonical directions of $\mathbb{T}_{(3;1)}$ are the projections of the canonical directions for $\mathbb{T}_1$, and there are two positive canonical resolutions which, by (6.98), are equal to

$$\lambda_{1(3,1)} = \lambda_{1(3)}\lambda_{1(1)} = \frac{3\lambda_{1(1)}}{(3-1)\lambda_{1(1)}+1}\lambda_{1(1)} = 0.4947,$$

$$\lambda_{2(3,1)} = \lambda_{2(3)}\lambda_{2(1)} = \frac{3\lambda_{2(1)}}{(3-1)\lambda_{2(1)}+1}\lambda_{2(1)} = 0.1666.$$

The canonical directions are as shown in Table 6.3, except that the each $\mathcal{M}(Y_i)$ is replaced by the corresponding predictive component, $Y_{i,f}$, so that the two canonical directions are

$$\tilde{Z}_1 = 0.10Y_{1,f} + 0.05Y_{2,f} + \ldots + 0.98Y_{13,f} - 5.31, \tag{6.107}$$

$$\tilde{Z}_2 = 3.72Y_{1,f} + 1.12Y_{2,f} + \ldots - 1.42Y_{13,f} - 5.44, \tag{6.108}$$

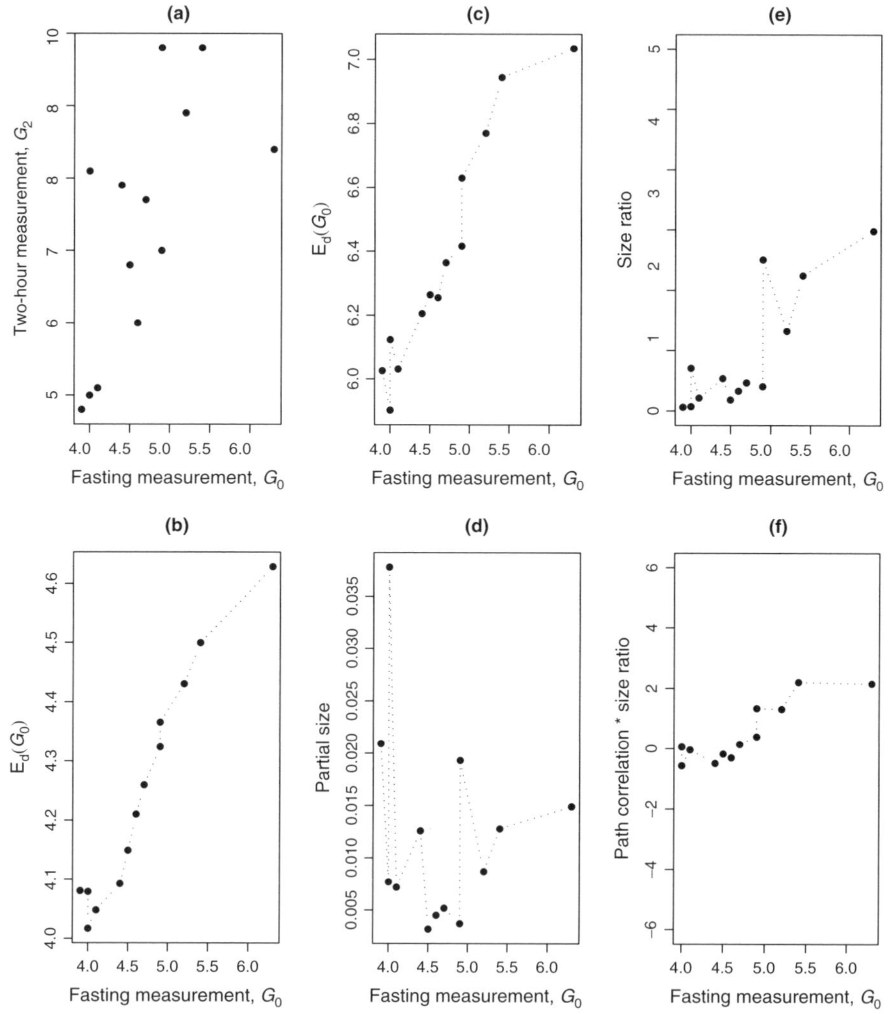

Figure 6.8 Path diagnostics, discrepant observations excluded. (a) Scatter plot of original data, fasting measurement versus 2-hour measurement; (b) Sequential update of expectation for G_0; (c) Sequential update of expectation for G_2; (d) partial sizes for sequential adjustments; (e) partial size ratios for sequential adjustments; (f) successive path correlations multiplied by size ratios.

where (6.107) corresponds to (6.81) and (6.108) corresponds to (6.82). Table 6.11 summarizes the implications of data from the initial three experiments for predicting the values of a future experiment. Prior and adjusted variances for each quantity are given together with the resolution partition and total and maximal resolutions. The data from three initial runs are expected to remove about half of our uncertainty

Table 6.10 Summary of the adjustment by 13 observations. Shown are the means and standard deviations of the observations; prior and adjusted expectations with standardized change in adjustment, relative to variance resolved; and prior and adjusted variances with resolutions.

	Data		Expectation			Variation		
	Mean	SD	Prior	Adjusted	Change	Prior	Adjusted	Resol.
G_0	4.685	0.650	4.16	4.629	0.61	1.12	0.536	0.522
G_2	7.331	1.660	6.25	7.035	1.38	2.43	2.107	0.133
G_h	2.646	1.329	2.09	2.407	0.53	2.11	1.750	0.171

Table 6.11 Variances for predicting future observables $Y_{1,F}, \ldots, Y_{13,F}$.

	Variances		Variance resolutions			
Quantity	Prior	Adjusted	in Z_1	in Z_2	Total	Max
$Y_{1,F}$	0.1197	0.0939	0.0969	0.1190	0.2160	0.3308
$Y_{2,F}$	0.1348	0.1037	0.1465	0.0844	0.2309	0.3294
$Y_{3,F}$	0.1533	0.1144	0.1960	0.0576	0.2535	0.3418
$Y_{4,F}$	0.1752	0.1261	0.2425	0.0377	0.2802	0.3630
$Y_{5,F}$	0.2005	0.1387	0.2847	0.0234	0.3082	0.3890
$Y_{6,F}$	0.2292	0.1522	0.3221	0.0136	0.3358	0.4171
$Y_{7,F}$	0.2613	0.1667	0.3549	0.0071	0.3620	0.4455
$Y_{8,F}$	0.2968	0.1821	0.3834	0.0031	0.3865	0.4730
$Y_{9,F}$	0.3357	0.1984	0.4081	0.0009	0.4090	0.4993
$Y_{10,F}$	0.3780	0.2156	0.4295	0.0001	0.4296	0.5238
$Y_{11,F}$	0.4237	0.2338	0.4481	0.0001	0.4483	0.5466
$Y_{12,F}$	0.4728	0.2529	0.4643	0.0009	0.4652	0.5677
$Y_{13,F}$	0.5253	0.2729	0.4785	0.0021	0.4806	0.5871

in the first predictive canonical quantity, but only one-sixth of the uncertainty in the second predictive canonical quantity. Comparing the total resolutions for the mean and predictive adjustments summarized in Table 6.2 and 6.11, we observe how much smaller are the latter. The maximal resolutions displayed in the final column show how much variance would be resolved by explaining all of the mean component variation.

6.17.1 Choice of canonical directions

This is an example where the resolution transform has a rank ($r_{\mathbb{T}} = 2$) smaller than its dimension (13), and there is thus some arbitrariness in the canonical quantities. Standard form for such situations is discussed in §6.12.1. As we are here interested in analysing both mean components and predictive components, we need to ensure that the canonical quantities corresponding to adjustments for equivalent linear spaces match algebraically. This issue is discussed in detail in Chapter 12 (see §12.12.4 in particular). With regard to the results there, in making the computations for this example, we used (12.60) to obtain the resolution transform for the mean component adjustment, and (12.52) to obtain the resolution transform for the predictive component adjustment. By Theorem 12.63 and Theorem 12.64, these transforms then share the same algebraic directions. Note that there is no arbitrariness in the canonical resolutions.

6.18 Further reading

The fundamental role of exchangeability within the subjectivist approach is developed in de Finetti (1937), available in Kyburg and Smokler (1964); see also de Finetti (1974, 1975). The second-order representation theorem was given in Goldstein (1986a). The basic properties of the adjustment of such exchangeable structures are described in Goldstein and Wooff (1998). Adjusting exchangeable beliefs raises certain foundational questions that we do not pursue here; for example, how do we interpret the analysis of an exchangeable structure which we expect no longer to be exchangeable by the end of the analysis, so that the mean quantities in the representation will cease to have meaning? A careful discussion of such issues is given in Goldstein (1994b).

In Goldstein and Wooff (1997), the properties of the canonical structure for the adjustment are exploited to choose appropriate sample sizes for balanced experimental design. Extensions of these ideas to handle many variables cross-classified in many ways are discussed in Shaw and Goldstein (1999). An overview of various interpretative and diagnostic tools appropriate for the analysis of exchangeable structures is given in Farrow and Goldstein (1993) in the context of grouped multivariate repeated measurement studies, and illustrated by analysis of a crossover trial. Exchangeability modelling as a basis for partition testing is described in Coolen et al. (2001); the approach of the example is pursued with a more careful mixture of Bayes and Bayes linear modelling, exploiting Bayes linear kinematics, in Goldstein and Shaw (2004). The role of exchangeable belief analysis in identifying experimental designs which balance gains in information against relevant financial and ethical costs is outlined in Farrow and Goldstein (1992) and extended in detail to allow for imprecise utility trade-offs in Farrow and Goldstein (2006).

7

Co-exchangeable beliefs

We now extend our exchangeability representation to partially exchangeable collections of groups of individuals, for which individuals within a group are judged exchangeable, and the relationship between individuals in different groups obeys certain natural invariance properties. When such relationships are expressed only over means, variances and covariances, we term such collections co-exchangeable. We develop the representation in two stages, by first considering the relation between an exchangeable group and a single further quantity, and then generalizing this relation to co-exchangeable collections.

7.1 Respecting exchangeability

Suppose that we have an infinite second-order exchangeable sequence of vectors $X = (X_1, X_2, \ldots)$, and a further random vector, F.

Definition 7.1 *We say that* F ***respects exchangeability*** *over* X *if*

$$\mathrm{Cov}(F, X_i) = \Sigma_F, \tag{7.1}$$

a constant for all i.

In this case, all the relationships between F and X may be expressed via the single relationship between F and the mean vector $\mathcal{M}(X)$ as, for each j,

$$\mathrm{Cov}(F, \mathcal{R}_j(X)) = 0. \tag{7.2}$$

This follows as, for each j,

$$\mathrm{Cov}(F, \mathcal{M}(X)) = \lim_n \frac{1}{n} \sum_{r=1}^{n} \mathrm{Cov}(F, X_r) = \mathrm{Cov}(F, X_j) = \Sigma_F, \quad \forall j.$$

Bayes Linear Statistics: Theory and Methods M. Goldstein and D.A. Wooff

7.2 Adjustments respecting exchangeability

We now consider the adjustment of F by samples from X. Thus, we have observations on a second-order exchangeable collection $D_n = (X_1, \dots, X_n)$ using which we want to adjust beliefs over F. We suppose, for notational simplicity, that $\mathcal{M}(X)$ has been transformed to vector W which is in the standard form (§6.12.1) for the corresponding exchangeable adjustment. We construct the diagonal matrix of eigenvalues

$$\Lambda_n = \mathbf{diag}\left\{\frac{n\lambda_1}{1+(n-1)\lambda_1}, \dots, \frac{n\lambda_r}{1+(n-1)\lambda_r}\right\}, \tag{7.3}$$

where λ_i are the eigenvalues for the adjustment of W by a sample of size one on W. We write the adjusted expectation of F by D_n as $\mathrm{E}_n(F)$ and the resolution transform over F induced by D_n as $\mathbb{T}_{(n,F)}(\cdot)$. We have the following form for the belief adjustment.

Theorem 7.2 *If F respects exchangeability over X then, for each n, the collection of sample means $\bar{W}_n$ is Bayes linear sufficient for the adjustment of F by D_n. The adjusted expectation and resolution transforms are:*

$$\mathrm{E}_n(F) = \mathrm{E}(F) + \mathrm{Cov}(F, W)\Lambda_n \bar{W}_n, \tag{7.4}$$

$$\mathbb{T}_{(n,F)}(\cdot) = \mathrm{E}_F(\mathbb{T}_n(\mathrm{E}_{\mathcal{M}(X)}(\cdot))), \tag{7.5}$$

with matrix representation

$$\mathbb{T}_{(n,F)}(\cdot) = \mathrm{Var}(F)^{\dagger}\mathrm{Cov}(F, W)\Lambda_n\mathrm{Cov}(W, F). \tag{7.6}$$

Proof. We have, from (7.2), that $\lfloor D_n \perp\!\!\!\perp F \rfloor \,/\, W$. As $\lfloor D_n \perp\!\!\!\perp W \rfloor \,/\, \bar{W}_n$, the adjustment of F by D_n is equivalent to the adjustment of F by $\bar{W}_n$. As $\mathrm{Var}(W)$ is the identity matrix and $\mathrm{Var}(W_1)$, is diagonal, we may write

$$\mathrm{E}_n(F) = \mathrm{E}_{\bar{W}_n}(F) = \mathrm{E}_{\bar{W}_n}(\mathrm{E}_W(F)) \tag{7.7}$$

$$= \mathrm{E}(F) + \mathrm{Cov}(F, W)\Lambda_n \bar{W}_n. \tag{7.8}$$

The representation for $\mathbb{T}_{(n,F)}$ follows as

$$\mathbb{T}_{(n,F)}(\cdot) = \mathrm{E}_F(\mathrm{E}_n(\cdot)) = \mathrm{E}_F(\mathrm{E}_{\bar{W}_n}(\cdot)). \tag{7.9}$$

■

We may therefore assess $\mathbb{T}_{(n,F)}$ by evaluating $\mathbb{T}_n$ and then pre- and post-multiplying by the projections $\mathrm{E}_F(\cdot)$, $\mathrm{E}_{\mathcal{M}(X)}(\cdot)$, respectively. This type of representation has two principal advantages. First, we need only evaluate $\mathbb{T}_n$ once, and then apply the representation over whatever collections F we require. Secondly, $\mathbb{T}_n$ has a particularly simple representation, in terms of the natural basis of eigenvectors, as, for each eigenvector Z, $\mathbb{T}_n(Z) = \lambda_{(n)}Z$. As the eigenvectors do not change with the sample size n, and the eigenvalues are given for each n by (6.57), we can easily assess the effect of increasing sample size on the adjustment. In particular, we may deduce the following corollary describing adjusted expectations and variances over $[F]$.

Corollary 7.3 *If F respects exchangeability over X and $\mathbb{T}_1$ has a full set of orthonormal standard eigenvectors $\mathcal{M}(W_1), \mathcal{M}(W_2), \ldots$, with corresponding eigenvalues $\lambda_1, \lambda_2, \ldots$, then, for each n and each $Z \in [F]$, we have*

$$\mathrm{E}_n(Z) = \mathrm{E}(Z) + \sum_i \frac{n\lambda_i}{1 + (n-1)\lambda_i} \mathrm{Cov}(Z, \mathcal{M}(W_i))\bar{W}_{in}, \tag{7.10}$$

$$\mathrm{Var}_n(Z) = \mathrm{Var}(Z) - \sum_i \frac{n\lambda_i}{1 + (n-1)\lambda_i} [\mathrm{Cov}(Z, \mathcal{M}(W_i))]^2. \tag{7.11}$$

In particular, the maximal resolved variance for any sample size, for each Z is

$$\sum_i [\mathrm{Cov}(Z, \mathcal{M}(W_i))]^2. \tag{7.12}$$

7.3 Example: simple algebraic problem

For a simple algebraic demonstration, we return to the example of §3.8.1, §3.11.2 and §6.13. We began in §3.8.1 with two pairs of quantities, Y_1, Y_2 and X_1, X_2, and explored the adjustment of the Y pair by the X pair. In §6.13 we extended the X pair to an exchangeable sequence of observables

$$X_{11}, X_{21}; X_{12}, X_{22}; \ldots, X_{1n}, X_{2n}; \ldots,$$

with belief specifications summarized in (6.70), and we examined these specifications for learning about the underlying mean components for the exchangeable sequence, $\mathcal{M}(X)$. We did this by obtaining the resolution transform $\mathbb{T}_n = \mathbb{T}_{\mathcal{M}(X):D_n}$ and its canonical structure for this, pure exchangeable, adjustment. We now complete the example by adjusting the Y pair by the full exchangeable sequence of X quantities. To do so we carry out a general exchangeable adjustment of the Y pair by the X quantities, via the pure exchangeable adjustment carried out in §6.13.

We have specified all the covariances between these quantities in earlier chapters. In summary, we form the collections

$$B = [Y_1, Y_2], \qquad D_i = [X_{1i}, X_{2i}], \quad i = 1, 2, \ldots, n.$$

The variance matrix over any pair D_i, D_j, $i \neq j$, is given in (6.69), whilst the variance matrix over B, D_i is given in (3.33). In terms of the notation of §7.1, B respects exchangeability over D. For example, when $n = 2$ we have the variance matrix

$$\mathrm{Var}\left(\begin{bmatrix} D_1 \\ D_2 \\ B \end{bmatrix}\right) = \begin{bmatrix} \mathrm{Var}(D_1) & \mathrm{Cov}(D_1, D_2) & \mathrm{Cov}(D_1, B) \\ \mathrm{Cov}(D_2, D_1) & \mathrm{Var}(D_2) & \mathrm{Cov}(D_2, B) \\ \mathrm{Cov}(B, D_1) & \mathrm{Cov}(B, D_2) & \mathrm{Var}(B) \end{bmatrix}$$

$$= \mathrm{Var}\left(\begin{bmatrix} X_{11} \\ X_{21} \\ X_{12} \\ X_{22} \\ Y_1 \\ Y_2 \end{bmatrix}\right) = \begin{bmatrix} 1 & u & \gamma & 0 & \rho & \rho \\ u & 1 & 0 & \gamma & \rho & \rho \\ \gamma & 0 & 1 & u & \rho & \rho \\ 0 & \gamma & u & 1 & \rho & \rho \\ \rho & \rho & \rho & \rho & 1 & v \\ \rho & \rho & \rho & \rho & v & 1 \end{bmatrix}. \tag{7.13}$$

For general n, the variance specifications may be written more elegantly using direct product notation: $\mathrm{Var}(D)$ is given in (6.70), and

$$\mathrm{Cov}(B, D) = \mathbf{1}_n^T \otimes \begin{bmatrix} \rho & \rho \\ \rho & \rho \end{bmatrix}. \tag{7.14}$$

7.3.1 Coherence

For these specifications to be coherent we need some conditions additional to the requirements listed in §3.8.1 and §6.13, as follows. Following the coherence requirements which we detail in §12.12.2 and Definition 12.60, we require the variance matrix

$$\begin{bmatrix} \mathrm{Var}(B) & \mathrm{Cov}(B, D) \\ \mathrm{Cov}(D, B) & \mathrm{Var}(\mathcal{M}(X)) \end{bmatrix} = \begin{bmatrix} 1 & v & \rho & \rho \\ v & 1 & \rho & \rho \\ \rho & \rho & \gamma & 0 \\ \rho & \rho & 0 & \gamma \end{bmatrix} \tag{7.15}$$

to be non-negative definite. By Lemma 12.3, this is so when $|v| < 1$ and when $\gamma \geq 0$ and when $\mathrm{Var}(B) - \mathrm{Cov}(B, D)\mathrm{Var}(\mathcal{M}(X))^{-1}\mathrm{Cov}(D, B)$ is non-negative definite. It is straightforward to show that the last condition is satisfied when

$$|\rho| \leq \frac{1}{2}\sqrt{\gamma(1 + v)}. \tag{7.16}$$

This is a more stringent condition than (3.35).

7.3.2 Resolution transform

By Theorem 7.2, and as B respects exchangeability with D, the data averages

$$\bar{D}_n = \frac{1}{n}[D_1 + D_2 + \ldots + D_n]$$

are Bayes linear sufficient for the adjustment of B by D. We can arrive at the resolution transform for this adjustment, $\mathbb{T}_{n,B}$, using (7.6) (or, algebraically, (12.68)). We need the standard form (§6.12.1) for the corresponding exchangeable adjustment. This was obtained in §6.13 as canonical quantities (6.76) and (6.77), with canonical resolutions (6.74) and (6.75). To find the resolution transform we need to form the covariances of elements in B with the canonical quantities for the corresponding exchangeable adjustment:

$$\begin{aligned}\mathrm{Cov}(Y_1, W_1) &= \mathrm{Cov}\left(Y_1, \frac{1}{\sqrt{2\gamma}}[\mathcal{M}(X_1) + \mathcal{M}(X_2)]\right) = \rho\sqrt{\frac{2}{\gamma}} \\ &= \mathrm{Cov}(Y_2, W_1), \\ \mathrm{Cov}(Y_1, W_2) &= \mathrm{Cov}\left(Y_1, \frac{1}{\sqrt{2\gamma}}[\mathcal{M}(X_1) - \mathcal{M}(X_2)]\right) = 0 \\ &= \mathrm{Cov}(Y_2, W_2).\end{aligned}$$

Consequently, we obtain the resolution transform $\mathbb{T}_{n,B}$, using (7.6), as

$$\begin{aligned}\mathbb{T}_{n,B} &= \mathrm{Var}(B)^{-1}\mathrm{Cov}(B, W)\Lambda_{(n)}\mathrm{Cov}(W, B) \\ &= \begin{bmatrix}1 & v \\ v & 1\end{bmatrix}^{-1}\begin{bmatrix}\rho\sqrt{\frac{2}{\gamma}} & 0 \\ \rho\sqrt{\frac{2}{\gamma}} & 0\end{bmatrix}\begin{bmatrix}\frac{n\gamma}{1+u+(n-1)\gamma} & 0 \\ 0 & \frac{n\gamma}{1-u+(n-1)\gamma}\end{bmatrix}\begin{bmatrix}\rho\sqrt{\frac{2}{\gamma}} & \rho\sqrt{\frac{2}{\gamma}} \\ 0 & 0\end{bmatrix} \\ &= \frac{2n\rho^2}{(1+v)(1+u+[n-1]\gamma)}\begin{bmatrix}1 & 1 \\ 1 & 1\end{bmatrix}. \qquad (7.17)\end{aligned}$$

The eigenvalues of $\mathbb{T}_{n,B}$ are

$$\lambda_{1(n,B)} = \frac{4n\rho^2}{(1+v)(1+u+[n-1]\gamma)}, \qquad (7.18)$$

$$\lambda_{2(n,B)} = 0, \qquad (7.19)$$

with corresponding eigenvectors proportional to $[1 \quad 1]^T$ and $[1 \quad -1]^T$, so that the canonical quantities are

$$W_{1(n,B)} = \alpha_1 \begin{bmatrix}1 & 1\end{bmatrix}\begin{bmatrix}Y_1 \\ Y_2\end{bmatrix}, \quad W_{2(n,B)} = \alpha_2 \begin{bmatrix}1 & -1\end{bmatrix}\begin{bmatrix}Y_1 \\ Y_2\end{bmatrix},$$

with α_1 and α_2 chosen to ensure that $\mathrm{Var}(W_{1(n,B)}) = \mathrm{Var}(W_{2(n,B)}) = 1$. The canonical quantities are thus:

$$W_{1(n,B)} = \frac{1}{\sqrt{2(1+v)}}(Y_1 + Y_2), \qquad (7.20)$$

$$W_{2(n,B)} = \frac{1}{\sqrt{2(1-v)}}(Y_1 - Y_2). \qquad (7.21)$$

For the special case $n = 1$, these results should and do match the results of §3.11.2, where (7.17) corresponds to (3.91) and where the canonical structure (7.18)–(7.21) corresponds to (3.93), (3.94).

In general, the canonical quantities for a transform $\mathbb{T}_{n,F}$ for a general exchangeable adjustment are not usually the same for different values of n. However, the symmetric structure for this example leads to canonical quantities which are not functions of n. Thus, as for pure exchangeable adjustments, they have the advantage that changes in variance resolution for varying n can be assessed entirely by inspection of the canonical quantities.

7.4 Co-exchangeable adjustments

We now consider collections of partially exchangeable observations. For example, we may consider patients undergoing a variety of treatments. Patients under each particular treatment may be judged to be exchangeable, but we may not judge patients under different treatments to be exchangeable. However, we will often be prepared to judge the covariances between responses for patients on different treatments to be unaffected by reordering within individual patient groups. As such, suppose that we have a collection of vectors Y_{in}, where, for example, Y_{in} might be the response vector for the nth patient on treatment i.

Definition 7.4 *The sequences $Y_1, Y_2, \ldots$ are* ***co-exchangeable*** *if they satisfy the following properties.*

> **7.4.1:** *For fixed i, the sequence $Y_i = (Y_{i1}, Y_{i2}, \ldots)$ is infinite second-order exchangeable.*
>
> **7.4.2:** *For any pair $i \neq j$, $\text{Cov}(Y_{im}, Y_{jn}) = \Sigma_{ij}, \forall m, n$.*

For each $i \neq j$, and each n, the vector Y_{jn} respects exchangeability with the sequence Y_i, so that, from (7.2), $\text{Cov}(Y_{jn}, \mathcal{R}_m(Y_i)) = 0$ for each m. Therefore, for each $i \neq j$ and each m,

$$\text{Cov}(\mathcal{M}(Y_i), \mathcal{R}_m(Y_j)) = 0,$$

so that

$$\text{Cov}(Y_{im}, Y_{jn}) = \text{Cov}(\mathcal{M}(Y_i), \mathcal{M}(Y_j)). \tag{7.22}$$

Therefore, all the relationships between the various vectors Y_{in} may be expressed in terms of the relationships between the mean vectors for the series.

We may therefore generalize the analysis of §7.1 to cover adjustments of co-exchangeable data structures, by reducing the sample from each collection to the corresponding sample mean, and then deducing all of the adjustments that we require from the corresponding adjustments of the population mean vectors by the corresponding sample means.

For example, suppose that we wish to predict future responses for one group from current observations on a sample from a different but related group. Thus, we have observations on a second-order exchangeable collection $D_n = (X_1, \ldots, X_n)$ and we want to predict outcomes for a further collection $F_m = (Y_1, \ldots, Y_m)$, where $Y_1, \ldots, Y_m$ are a further second-order exchangeable collection which is co-exchangeable with X, so that $\mathrm{Cov}(X_i, Y_j)$ has the same value for each i, j.

We suppose, for notational simplicity, that $\mathcal{M}(X), \mathcal{M}(Y)$ have been transformed to vectors W, U which are in the standard form for the corresponding exchangeable adjustment. We construct the diagonal matrices of eigenvalues

$$\Lambda_n = \mathbf{diag}\left\{\frac{n\lambda_1}{1+(n-1)\lambda_1}, \ldots, \frac{n\lambda_r}{1+(n-1)\lambda_r}\right\},$$

$$\Lambda_m^* = \mathbf{diag}\left\{\frac{n\lambda_1^*}{1+(n-1)\lambda_1^*}, \ldots, \frac{n\lambda_s^*}{1+(n-1)\lambda_s^*}\right\},$$

where λ_i, λ_j^* are the eigenvalues for the adjustment of W, U by a sample of size one on W, U, respectively.

Theorem 7.5

$$\mathrm{E}_n(\bar{U}_m) = \mathrm{Cov}(U, W)\Lambda_n \bar{W}_n, \tag{7.23}$$

$$\mathbb{T}_{\bar{U}_m:\bar{W}_n} = \mathrm{Cov}(U, W)\Lambda_n \mathrm{Cov}(W, U)\Lambda_m^*, \tag{7.24}$$

$$\mathbb{T}_{U:\bar{W}_n} = \mathrm{Cov}(U, W)\Lambda_n \mathrm{Cov}(W, U). \tag{7.25}$$

Proof. From (7.22), we have

$$\lfloor D_n \perp\!\!\!\perp F_m \rfloor \,/\, (W, U).$$

As $\lfloor D_n \perp\!\!\!\perp W \rfloor \,/\, \bar{W}_n$ and $\lfloor F_m \perp\!\!\!\perp U \rfloor \,/\, \bar{U}_m$, the adjustment of F_m by D_n is equivalent to the adjustment of $\bar{U}_m$ by $\bar{W}_n$. As $\mathrm{Var}(W)$ and $\mathrm{Var}(U)$ are equal to the identity matrix and $\mathrm{Var}(W_1)$, $\mathrm{Var}(U_1)$ are both diagonal, we may write

$$\begin{aligned}\mathrm{E}_n(\bar{U}_m) &= \mathrm{E}_{\bar{W}_n}(\bar{U}_m) = \mathrm{E}_{\bar{W}_n}(\mathrm{E}_U(\bar{U}_m)) \\ &= \mathrm{E}_{\bar{W}_n}(U) = \mathrm{E}_{\bar{W}_n}(\mathrm{E}_W(U)) \\ &= \mathrm{Cov}(U, W)\Lambda_n \bar{W}_n.\end{aligned}$$

Similarly,

$$\mathrm{E}_{\bar{U}_m}(\bar{W}_n) = \mathrm{E}_{\bar{U}_m}(\mathrm{E}_U(W)),$$

so that the operator $\mathbb{T}_{\bar{U}_m:\bar{W}_n} = \mathrm{E}_{\bar{U}_m}(\bar{W}_n)$ may be written as (7.24). Equation (7.25) follows similarly. ■

Replacing $\mathrm{Cov}(W, U)$ by the identity matrix gives Theorem 6.10 as a special case of (7.24). In general, from (7.24), the elements of $\bar{U}_m$ will be eigenvectors of $\mathbb{T}_{\bar{U}_m:\bar{W}_n}$ if and only if $\mathrm{Cov}(U, W)$ is diagonal, or equivalently if and only if the elements of U are eigenvectors of the transform $\mathbb{T}_{U:W}$.

7.5 Example: analysing further exchangeable regressions

To illustrate co-exchangeable structures, we continue the exchangeable regressions example analysed in §6.7 and §6.14. That example concerned the amount of alumina extracted over time in an industrial smelter, using a particular solvent. The experiment can be, and was, run a number of times to gain more precise information about the rate of extraction for that solvent.

An alternative, but similar, solvent may be used to extract the alumina. Thus, we envisage a parallel set of experimental runs to determine the rate of extraction for the new solvent. One question is how informative are the first set of experiments for a parallel set of experiments using a different solvent. (In practical terms, it may be feasible only to run experiments using a single kind of solvent.) We can address this question as follows.

We judge that the two solvents are sufficiently similar in anticipated performance for the model for the new solvent to be generally the same as the model for the old solvent. Thus, we will 'copy' the model for the old solvent set out in §6.7, and use an asterisk to denote quantities which relate to the new solvent. Thus, for the rth run of the set of experiments for the new solvent, we let Y^*_{rt} represent the concentration of alumina in solution at time t, and we model

$$Y^*_{rt} = a^*_r + tb^*_r + \epsilon^*_{rt} \tag{7.26}$$

similarly to (6.37), and with prior beliefs specified over the further quantities $\{a^*_r, b^*_r, \epsilon^*_{rt}\}$ identically to the priors for $\{a_r, b_r, \epsilon_{rt}\}$. Summaries and structural implications under the new model are the same as those for the old model. For example, the prior variance matrix expressed over the Y^*s is identical to that for the Ys, and the structure noted in (6.41) and (6.41) applies similarly. Also, the underlying slope and intercept quantities have the same second-order structure; for example $\mathrm{E}(\mathcal{M}(a^*)) = \mathrm{E}(\mathcal{M}(a))$ and $\mathrm{E}(\mathcal{M}(b^*)) = \mathrm{E}(\mathcal{M}(b))$, so that

$$\mathrm{E}(Y^*_{rt}) = \mathrm{E}(\mathcal{M}(a)) + t\mathrm{E}(\mathcal{M}(b)) = \mathrm{E}(Y_{lt}), \quad \forall r, l. \tag{7.27}$$

If we are to use the first set of experiments to help us revise our beliefs about the second set of experiments, we need to make additional judgements about the relationships between them. As shown in §6.7.4, the model for the old solvent is completely characterized by judgements about the underlying intercept and slope quantities, $\mathcal{M}(a)$ and $\mathcal{M}(b)$, and the error quantities, $\{\epsilon_{rt}\}$. Similarly, the model for the new solvent is completely characterized by judgements about the corresponding intercept and slope quantities for the new model, $\mathcal{M}(a^*)$ and $\mathcal{M}(b^*)$, and the error quantities, $\{\epsilon^*_{rt}\}$. Consequently, the relationships between the two models are captured by relationships between these two sets of quantities.

Suppose that we judge the two sets of error components $\epsilon_{rt}, \epsilon^*_{lk}$ as being uncorrelated with all other quantities, and that we view the pair of underlying slope quantities as uncorrelated with the pair of underlying intercept quantities. The relationship between the pairs (Y^*_{rt}, Y_{lk}) therefore depends solely on the values

$\mathrm{Cov}(\mathcal{M}(a), \mathcal{M}(a^*))$ and $\mathrm{Cov}(\mathcal{M}(b), \mathcal{M}(b^*))$. Suppose that

$$\mathrm{Corr}(\mathcal{M}(a), \mathcal{M}(a^*)) = \rho_a \geq 0 \quad \text{and} \quad \mathrm{Corr}(\mathcal{M}(b), \mathcal{M}(b^*)) = \rho_b \geq 0.$$

Then, for example,

$$\mathrm{Cov}(Y^*_{rt}, Y_{lk}) = \rho_a \mathrm{Var}(\mathcal{M}(a)) + tk\rho_b \mathrm{Var}(\mathcal{M}(b))), \quad \forall r, l. \tag{7.28}$$

The observations for the different runs of the experiments using the old solvent at time t, $\{Y_{1t}, Y_{2t}, \ldots\}$, are exchangeable; the observations for the different runs of the experiments using the new solvent at time k, $\{Y^*_{1k}, Y^*_{2k}, \ldots\}$, are also exchangeable; and the two sets of these quantities are second-order co-exchangeable, for all t, k, by (7.27) and (7.28).

We now consider the implications of the data $C(n)$ (the observations from the first set of experiments, analysed in §6.7 and §6.14) for learning about the mean components for the quantities for the new solvent,

$$\mathcal{M}(Y^*) = \{\mathcal{M}(Y^*_1), \ldots, \mathcal{M}(Y^*_{13})\}.$$

We could perform the adjustment by brute force. However, an alternative is to calculate the adjustment via the standard forms for the separate exchangeable adjustments, as described in §7.4. To do this, we must obtain the adjustment of the underlying mean components $\mathcal{M}(Y)$ by data $C(n)$, and for the new solvent we must obtain the adjustment of the underlying mean components $\mathcal{M}(Y^*)$ by notional data $C^*(m)$ from a sample of m runs of the experiment using the new solvent.

We found in §6.14.3 that the adjustment of $\mathcal{M}(Y)$ by the corresponding exchangeable sample of size n has two canonical quantities, (6.81) and (6.82), corresponding to positive canonical resolutions, $\lambda_{11} = 0.6032$, $\lambda_{21} = 0.2976$. This provides the first standard form, which we shall write as

$$\begin{aligned} W_1 &= 0.10\mathcal{M}(Y_1) + 0.05\mathcal{M}(Y_2) + \ldots + 0.98\mathcal{M}(Y_{13}) - 5.31 \\ &= g_1\mathcal{M}(a) + h_1\mathcal{M}(b) - 5.31, \end{aligned} \tag{7.29}$$

$$\begin{aligned} W_2 &= 3.72\mathcal{M}(Y_1) + 1.12\mathcal{M}(Y_2) + \ldots - 1.42\mathcal{M}(Y_{13}) - 5.44 \\ &= g_2\mathcal{M}(a) + h_2\mathcal{M}(b) - 5.44, \end{aligned} \tag{7.30}$$

where $g_1 = 2.18$, $g_2 = 4.65$, $h_1 = 22.64$, $h_2 = -10.61$ are the coefficients reported in §6.14.7.

Next, we consider the corresponding exchangeable adjustment for the new solvent. As beliefs for the model for the new solvent are identical to those for the old solvent, the canonical structure is the same as for the old solvent. This provides the following standard form for the adjustment: canonical resolutions $\lambda^*_{11} = 0.6032$, $\lambda^*_{21} = 0.2976$ and canonical quantities

$$U_1 = g_1\mathcal{M}(a^*) + h_1\mathcal{M}(b^*) - 5.31, \tag{7.31}$$

$$U_2 = g_2\mathcal{M}(a^*) + h_2\mathcal{M}(b^*) - 5.44. \tag{7.32}$$

We may now use (7.25) to obtain the resolution transform for the adjustment of the mean components for the new solvent, $\mathcal{M}(Y^*)$, by the data for the old solvent $C(n)$. The covariances between the standard forms for the two initial adjustments are

$$\mathrm{Cov}(U_i, W_i) = \rho_a \mathrm{Var}(\mathcal{M}(a))g_i^2 + \rho_b \mathrm{Var}(\mathcal{M}(b))h_i^2, \quad i = 1, 2,$$

$$\mathrm{Cov}(U_i, W_j) = \rho_a \mathrm{Var}(\mathcal{M}(a))g_1 g_2 + \rho_b \mathrm{Var}(\mathcal{M}(b))h_1 h_2, \quad i = 1, 2 \neq j = 1, 2,$$

recalling that $\mathrm{Cov}(\mathcal{M}(a), \mathcal{M}(b)) = 0$ and $\mathrm{Cov}(\mathcal{M}(a^*), \mathcal{M}(b^*)) = 0$ by specification. As U_1, U_2 are uncorrelated and have variance one, we may thus write the covariance matrix between $\bar{U} = [U_1, U_2]$ and $\bar{W} = [W_1, W_2]$ as

$$\mathrm{Cov}(\bar{U}, \bar{W}) = \rho_a \mathbf{I}_2 + (\rho_b - \rho_a)\mathrm{Var}(\mathcal{M}(b))hh^T, \tag{7.33}$$

where $h = [h_1\ h_2]^T$. We can now write down the resolution transform for this co-exchangeable adjustment via (7.25), $\mathbb{T}_{\mathcal{M}(Y^*):C(n)}$, as

$$[\rho_a \mathbf{I}_2 + (\rho_b - \rho_a)\mathrm{Var}(\mathcal{M}(b))hh^T]\Lambda_{(n)}[\rho_a \mathbf{I}_2 + (\rho_b - \rho_a)\mathrm{Var}(\mathcal{M}(b))hh^T]. \tag{7.34}$$

One immediate consequence is that if $\rho_a = \rho_b = \rho > 0$ then we have

$$\mathbb{T}_{\mathcal{M}(Y^*):C(n)} = \rho^2 \mathbb{T}_{\mathcal{M}(Y):C(n)},$$

so that the canonical directions for the adjustment of $\mathcal{M}(Y)$ by $C(n)$ and the canonical directions for the adjustment of $\mathcal{M}(Y^*)$ by $C(n)$ are of the same form, but with the corresponding canonical resolutions multiplied by ρ^2, i.e. $0.6032\rho^2$ and $0.2976\rho^2$. Hence the data quantities $C(n)$ have similar structural implications for learning about $\mathcal{M}(Y)$ and $\mathcal{M}(Y^*)$, but are less informative for the latter, depending on the magnitude of ρ. For $\rho = 1$, $[\mathcal{M}(Y^*)] = [\mathcal{M}(Y)]$ and the implications of the data for $\mathcal{M}(Y^*)$ are exactly as calculated for $\mathcal{M}(Y)$ in §6.14, in particular §6.14.3. When $\rho = 0$, $\mathcal{M}(Y^*)$ is uncorrelated with $\mathcal{M}(Y)$, and the data are valueless for learning about $\mathcal{M}(Y^*)$. As $\lim_{n\to\infty} \Lambda_{(n)} = \mathbf{I}_2$, we also see in this case that the most we can learn about any quantity in $[\mathcal{M}(Y^*)]$ (standardized to have variance one) is ρ^2.

If we wish to predict the mean, $\bar{Y}_m^*$, of a future sample of m runs of the experiment, then we can use (7.24) instead of (7.25). For the case with $\rho_a = \rho_b = \rho > 0$, this results in canonical resolutions for the predicted average of

$$\rho^2 \lambda_{i(n)} \lambda_{i(m)}, \quad i = 1, 2.$$

These three components represent (1) the basic level of information that $C(\cdot)$ has for predicting $[\mathcal{M}(Y^*)]$; (2) the extra information available from observing a sample of size n rather than a sample of size 1; (3) extra precision inherent in predicting the average of m observations rather than one observation.

7.5.1 The resolution envelope

For general $\rho_a, \rho_b > 0$, the canonical directions for the resolution transform (7.34) change as n varies. However, it remains simple to compute the maximal and minimal resolutions for each potential sample size, given the adjustment for $n = 1$. For example, Figure 7.1 plots the canonical resolutions for the adjustment of $\mathcal{M}(Y^*)$ by $C(n)$ for $n = 1$, $n = 3$ and $n = 10$; and for $\rho_a = 0.7$ and non-negative ρ_b. This is plausible when we believe that the intercept in one series of experiments is fairly informative for the intercept in a parallel series using a different solvent, as there is only so much aluminium to extract, but we are much less confident about the relationship between the rates of extraction. Two canonical resolutions, λ_1 and λ_2, are plotted, as $\mathcal{M}(Y^*)$ is two-dimensional, except when

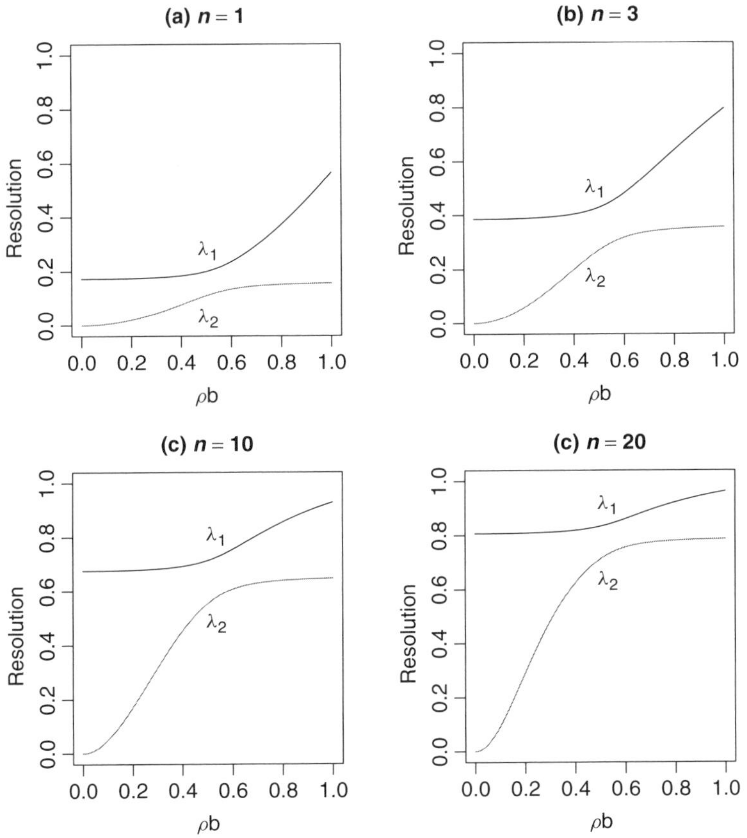

Figure 7.1 The effect of sample size on maximal and minimal variance resolutions for adjusting mean components $\mathcal{M}(Y^*)$ by co-exchangeable data $C(n)$ for $\rho_a = 0.7$ and $0 < \rho_b < 1$.

$\rho_b = 0$, when $\lambda_2 = 0$. These form a **resolution envelope** bounding the resolution in any linear combination of $\mathcal{M}(a^*)$ and $\mathcal{M}(b^*)$, including all $\mathcal{M}(Y_i^*)$. Figure 7.1 shows that the minimal canonical resolution is very sensitive to changes in ρ_b for about $\rho_b < 0.6$ but not for larger values, whilst the reverse is true for the maximal canonical resolution. Increasing the sample size roughly maintains the shape of the envelope, but, in lifting and stretching it, exacerbates the sensitivity problem.

7.6 Example: exchangeability in a population dynamics experiment

The following is a summary of a population dynamics experiment considered in Arthur and Farrow (1987), and for subsequent experiments Mitchell et al. (1992). There are two species of fruit fly, *Drosophila melanogaster* (DM) and *Drosophila hydei* (DH), which are believed to compete. Flies are put into a cage and the numbers of each species are counted every fortnight for one year. The counts at time point $t = 1$ are fixed as follows. In three cages are placed 20 flies of species DM and 80 flies of species DH. In three further cages are placed 80 flies of species DM and 20 flies of species DH. Subsequent counts are made at $t = 2, \ldots, 26$. Each such count is materially expensive and extremely time-consuming. The counts are transformed by taking logarithms, as this is a common practice among biologists. The data are shown, as raw counts, in Tables 7.1 and 7.2. The transformed data are plotted in Figures 7.2 and 7.3. There is a fair amount of cage variation. The trend appears to be that the DM species has higher counts at the end of the year, irrespective of starting count. Figure 7.4, which shows the averages of the log counts across cages, more clearly suggests this feature.

7.6.1 Model

The following model, which should be treated as exploratory, was suggested by a statistician, M. Farrow, in consultation with the experimenter, W. Arthur, and later slightly modified by us: we consider the original specifications in §9.14. The counts are modelled as follows. Let Y_{psct} be the natural log of the number in species s and cage c at time t, given starting point p. There are two starting points, corresponding to the different starting numbers of flies of each species in cages. We write Y_{psct} as the sum of two uncorrelated components,

$$Y_{psct} = M_{pst} + R_{psct}, \quad t = 2, 3, \ldots, 26,$$

where M represents a local mean and R is a residual component representing individual cage variation. The residuals are modelled as

$$R_{psct} = \theta_s R_{psct-1} + \psi_s (R_{psct-1} - R_{psct-2}) + \phi_s R_{ps'ct-1} + H_{psct},$$

Table 7.1 Cages with 20 *D. melanogaster* and 80 *D. hydei*.

	Cage 1		Cage 2		Cage 3	
t	DM	DH	DM	DH	DM	DH
1	20	80	20	80	20	80
2	14	80	18	94	19	85
3	7	156	20	45	31	349
4	29	271	39	141	53	291
5	36	345	99	193	19	162
6	42	233	92	146	28	337
7	43	341	93	58	45	527
8	43	450	151	125	16	177
9	63	284	98	91	21	139
10	57	321	242	245	84	347
11	60	279	168	340	58	295
12	109	351	163	245	55	179
13	73	236	92	330	46	242
14	177	877	71	226	97	730
15	321	170	129	276	186	469
16	499	388	227	525	156	341
17	358	405	263	416	159	722
18	112	887	109	147	393	1121
19	42	691	293	652	441	940
20	41	1610	282	554	382	739
21	31	935	84	226	64	173
22	71	1055	38	527	27	715
23	108	1546	11	520	19	1231
24	196	1301	9	1142	41	1346
25	264	1091	25	1964	32	836
26	551	564	30	1113	84	1333

where θ, ψ, ϕ are constants which depend on species. This model is a second-order autoregression with an additional cross-species term and a slightly unusual parameterization in the second term, to aid elicitation. The H_{psct} quantities are noise terms with mean 0, variance ν_s, and are uncorrelated with each other and all other quantities. The process is initiated with

$$R_{psc0} = R_{psc1} = 0, \quad \forall p.$$

The local mean is modelled as

$$M_{pst} = L_s + D_{pst}, \quad t = 2, 3, \ldots, 26,$$

where L_s is the **equilibrium level** for species s in the presence of the other species s'. The deviation of the local mean from the equilibrium is

$$D_{pst} = \alpha_s D_{pst-1} + \beta_s(D_{pst-1} - D_{pst-2}) + \gamma_s D_{ps't-1} + G_{pst},$$

Table 7.2 Cages with 80 *D. melanogaster* and 20 *D. hydei*.

	Cage 1		Cage 2		Cage 3	
t	DM	DH	DM	DH	DM	DH
1	80	20	80	20	80	20
2	186	43	96	18	105	28
3	321	202	123	50	309	198
4	255	151	246	70	404	219
5	96	198	208	96	317	220
6	42	262	231	206	132	360
7	55	310	270	201	42	364
8	63	233	292	292	113	868
9	45	170	216	96	77	925
10	75	265	238	200	62	881
11	38	84	165	16	128	1309
12	40	195	196	148	195	834
13	44	23	86	10	363	1776
14	97	159	251	139	391	483
15	152	233	621	135	408	866
16	114	230	431	211	479	875
17	120	396	334	294	477	847
18	166	301	126	245	242	532
19	231	767	77	726	250	703
20	205	403	96	1439	307	1416
21	170	220	75	409	164	295
22	72	370	107	818	253	733
23	58	244	85	413	233	1117
24	41	174	121	359	136	453
25	129	305	60	227	121	851
26	337	106	90	832	82	1093

where G_{pst} is a noise term with mean 0 and variance

$$\text{Var}(G_{pst}) = \lambda^{t-1}\omega_s, \quad t \geq 1,$$

in which $0 < \lambda < 1$ is chosen to ensure that, in the long term, the local means tend to L_s. The G_{pst} quantities are uncorrelated with each other and all other quantities. α, β, γ are constants which depend on species. We initiate the process with $D_{ps0} = D_{ps1}$ chosen to ensure that the local mean M_{ps1} is equal to the number of flies put into the cage at $t = 1$ for starting point p, i.e.

$$D_{p11} = D_{p10} = \frac{1}{\alpha_1\alpha_2 - \gamma_1\gamma_2}(\alpha_2[\upsilon_{p1} - \text{E}(L_1)] - \gamma_1[\upsilon_{p2} - \text{E}(L_2)]),$$

$$D_{p21} = D_{p20} = \frac{1}{\alpha_1\alpha_2 - \gamma_1\gamma_2}(\alpha_1[\upsilon_{p2} - \text{E}(L_2)] - \gamma_2[\upsilon_{p1} - \text{E}(L_1)]),$$

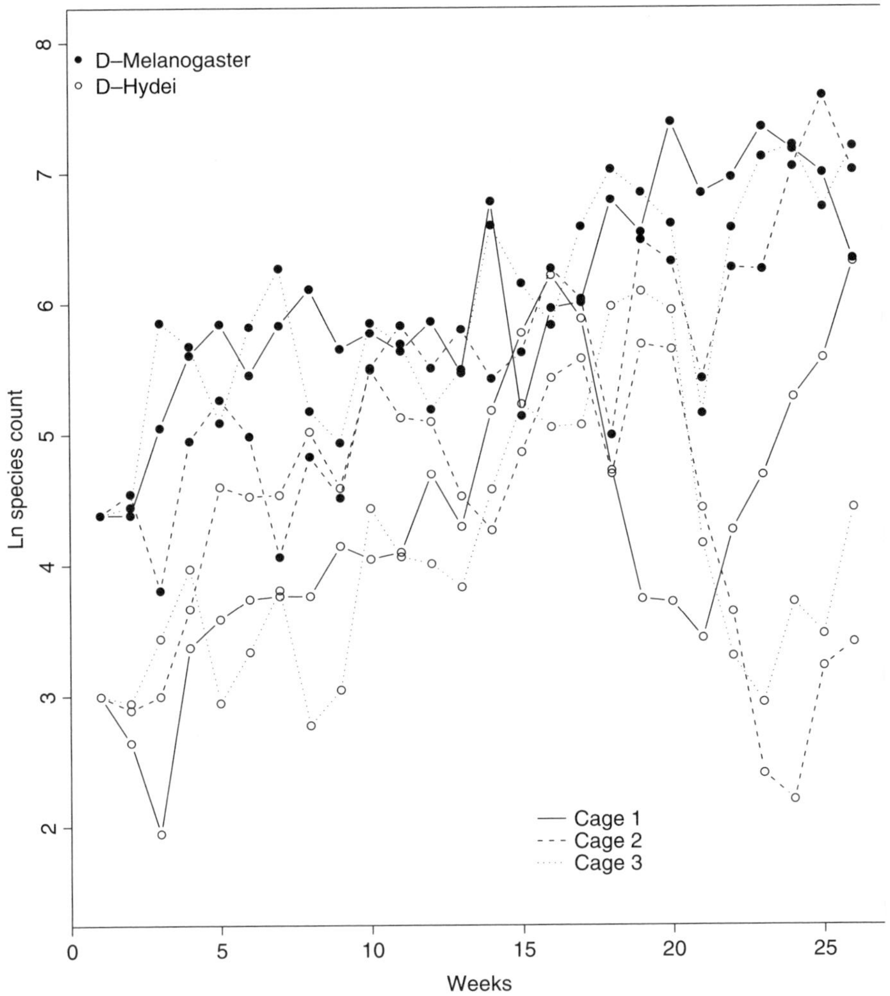

Figure 7.2 Counts of two species of fly in six cages, $p = 1$: cages with starting counts of 80 DM flies and 20 DH flies. Counts are shown on a log scale.

where $\upsilon_{p1}, \upsilon_{p2}$ are the starting counts for the species:

$$p = 1 : \upsilon_{11} = \ln(20),\ \ \upsilon_{12} = \ln(80),$$

$$p = 2 : \upsilon_{21} = \ln(80),\ \ \upsilon_{22} = \ln(20).$$

The model induces oscillations in the mean profile, M_{st}, but these are expected to die out, partly because the oscillations in the individual series die out, but more

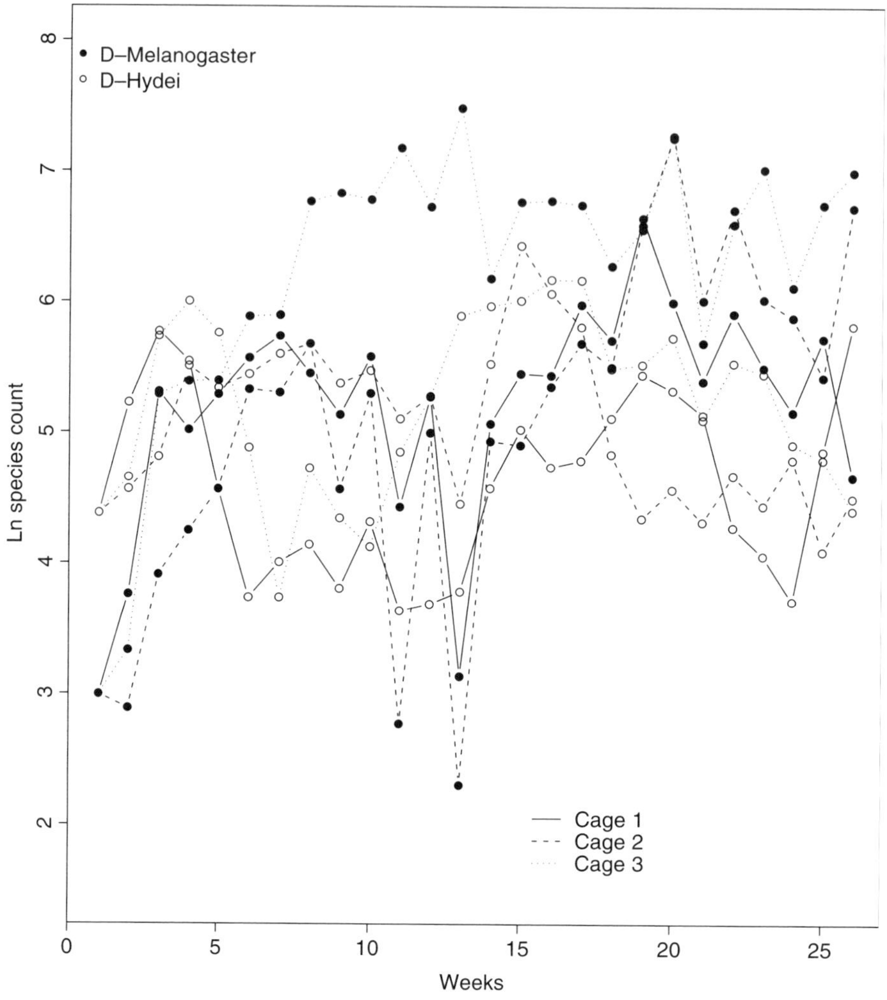

Figure 7.3 Counts of two species of fly in six cages, $p = 2$: cages with starting counts of 20 DM flies and 80 DH flies. Counts are shown on a log scale.

because these oscillations are expected to become out of phase for individual series (Farrow and Goldstein 1996).

7.6.2 Specifications

Constants required for the model were specified as shown in Table 7.3, following some basic elicitation. We choose $\lambda = 0.9$. The equilibrium levels L_s are specified each to have prior expectation $E(L_s) = 6.0$, prior variance $Var(L_s) = 0.49$, and

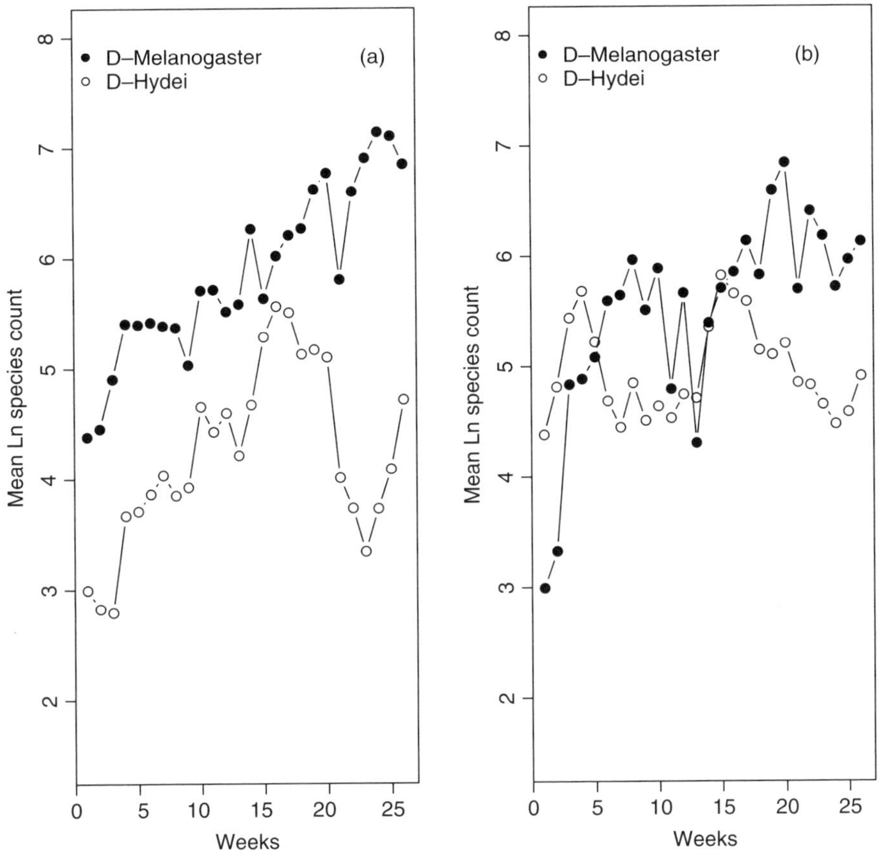

Figure 7.4 Mean counts of two species of fly in six cages: (a) cages starting with 80 DM flies and 20 DH flies; (b) cages starting with 20 DM flies and 80 DH flies. Counts are shown on a log scale.

Table 7.3 Constants required for the model.

	DM	DH		DM	DH
θ_s	0.45	0.40	β_s	0.1	0.1
ψ_s	0.26	0.10	γ_s	−0.1	−0.1
ϕ_s	−0.34	−0.18	ν_s	0.04	0.02
α_s	0.85	0.85	ω_s	0.2	0.2

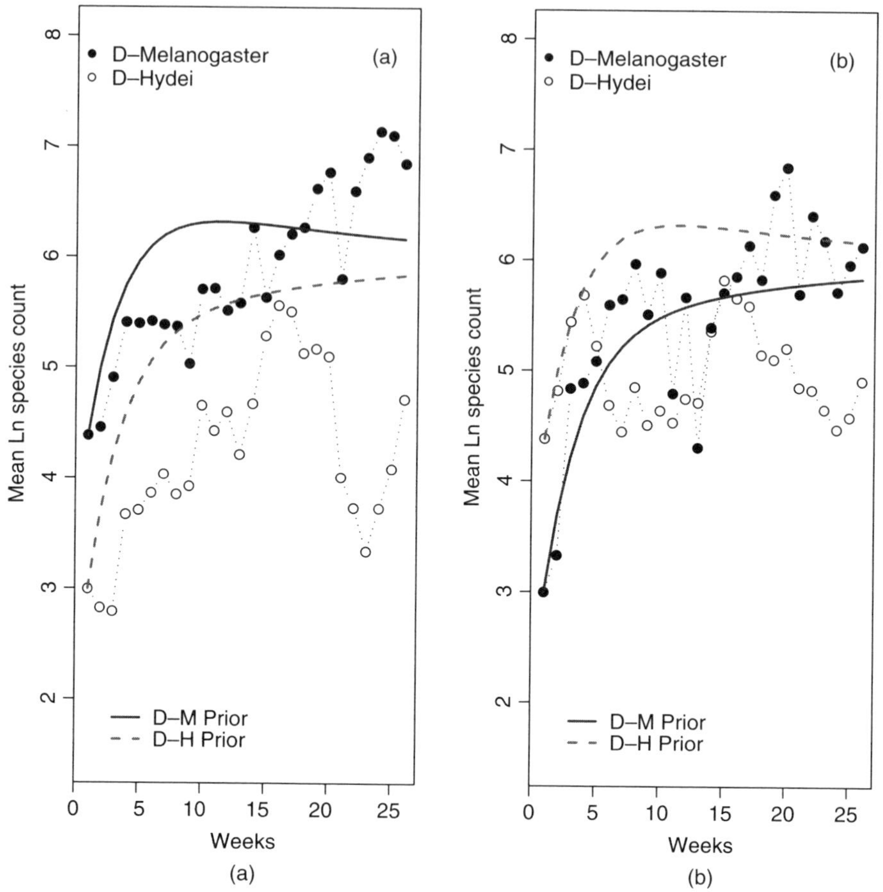

Figure 7.5 Mean counts and prior local means for two species of fly: (a) cages starting with 80 DM flies and 20 DH flies; (b) cages starting with 20 DM flies and 80 DH flies. Counts are shown on a log scale.

prior covariance $\text{Cov}(L_s, L_{s'}) = -0.2$. These specifications and starting points are such that the same second-order specification is made over M_{11} and M_{22} and over M_{12} and M_{21}. The prior expectations for the local means are shown in Figure 7.5, and exhibit slow convergence to the same equilibrium point of 6.0. Compared to the data, the priors for $\{M_{st}\}$ look reasonable for cages with starting point $p = 2$, but a little high for cages with starting point $p = 1$. In general, the data do not appear to track the priors well, except perhaps for *D. melanogaster* for starting point $p = 2$. In particular, the priors for *D. hydei* look to be far from the observed means.

7.6.3 Issues

There are several questions of interest raised in performing and analysing this experiment, which should be considered as exploratory. We focus briefly on a few of the issues.

- The equilibrium levels are unknown and of much interest. We consider how the adjusted expectations for these equilibria change as the experiment progresses by using all the data up to a certain time point.
- The experiments are costly to perform. As an aspect of design, we exploit exchangeability to consider sample size implications for increasing the number of cages to be assessed at each time point, and we examine the behaviour of the canonical structure to provide information about the duration of the experiment.
- We consider whether the prior specification appears appropriate, partly through basic exploration and partly through data diagnostics.

7.6.4 Analysis

We construct the model described above. The counts of flies of the same species and in cages with the same starting point are second-order exchangeable for each given time point. That is, for given p, s, t, the sequence

$$Y_{ps1t}, Y_{ps2t}, \ldots$$

is an exchangeable sequence. Further, the collection $L = [L_1, L_2]$ respects exchangeability (§7.1) with this sequence. In addition, the mean

$$\mathcal{S}_3(Y_{pst}) = \frac{1}{3}\sum_{c=1}^{3} Y_{psct}$$

is Bayes linear sufficient for this sequence for adjusting a collection respecting exchangeability with it. Thus, we may take advantage of Corollary 7.3 and, from the point of view of practical implementation, Theorem 12.65, when forming the adjustment of L by the data.

We could simply use all the data in one go to do this, but we are mainly interested in various model diagnostics at this stage. Therefore we perform the sequential adjustment of L by the collections $G_1, \ldots, G_{26}$, where

$$G_t = \{\mathcal{S}_3(Y_{11t}), \mathcal{S}_3(Y_{12t}), \mathcal{S}_3(Y_{21t}), \mathcal{S}_3(Y_{22t})\}$$

is the collection of averages of all the data available at time t. At each stage we obtain the adjustment by all the data up to time t inclusive, $G_{[t]}$, and diagnostically compare changes in adjustment between the full adjustment at time $t-1$ and the full adjustment at time t, to provide the data trajectory described in §5.11.

Table 7.4 Sequential adjustment: adjusted expectations and variances, partial size ratio $\mathrm{Sr}(\cdot) = \mathrm{Sr}_{[G_t/G_{[t]}]}(L)$, path correlation $C = \mathrm{PC}(G_{[t-1]}, G_t)$ and canonical resolutions. Prior means and variances are given as the first row.

t	$\mathrm{E}_{G_{[t]}}(\cdot)$		$\mathrm{Var}_{G_{[t]}}(\cdot)$					
	L_1	L_2	L_1	L_2	$\mathrm{Sr}(\cdot)$	C	λ_1	λ_2
–	6.000	6.000	0.490	0.490				
2	6.093	6.158	0.079	0.076			0.879	0.753
3	6.106	6.144	0.078	0.076	0.086	−0.269	0.879	0.759
4	6.105	6.123	0.075	0.073	0.080	−0.912	0.880	0.773
5	6.000	6.047	0.073	0.071	4.161	−0.988	0.880	0.786
6	5.924	6.005	0.071	0.069	1.859	−0.726	0.881	0.799
7	5.850	5.937	0.069	0.068	2.679	0.819	0.881	0.812
8	5.812	5.908	0.067	0.066	0.622	0.986	0.882	0.824
9	5.691	5.772	0.065	0.064	9.163	0.967	0.882	0.836
10	5.745	5.843	0.063	0.062	2.270	−0.983	0.883	0.847
11	5.615	5.680	0.061	0.060	13.016	0.975	0.884	0.858
12	5.627	5.717	0.060	0.058	0.421	−0.931	0.884	0.868
13	5.486	5.525	0.058	0.057	18.157	0.982	0.885	0.878
14	5.609	5.704	0.056	0.055	15.647	−0.989	0.888	0.885
15	5.676	5.671	0.055	0.054	0.957	−0.552	0.896	0.887
16	5.695	5.736	0.053	0.052	1.471	−0.944	0.905	0.888
17	5.702	5.791	0.052	0.051	0.970	−0.876	0.912	0.889
18	5.612	5.750	0.050	0.049	3.789	0.993	0.919	0.890
19	5.633	5.893	0.049	0.048	7.216	−0.833	0.926	0.891
20	5.642	5.953	0.047	0.046	1.340	−0.700	0.932	0.893
21	5.439	5.670	0.046	0.045	57.403	0.842	0.938	0.894
22	5.375	5.865	0.045	0.044	11.418	−0.474	0.943	0.896
23	5.262	5.851	0.043	0.043	5.339	0.997	0.948	0.897
24	5.250	5.813	0.042	0.041	0.748	0.747	0.953	0.899
25	5.260	5.845	0.041	0.040	0.549	−0.760	0.957	0.901
26	5.358	5.833	0.040	0.039	3.819	−0.958	0.960	0.903

The adjusted expectations and adjusted variances for the equilibrium points L_1, L_2 are shown in Table 7.4 and graphed in Figure 7.6. Note that the scale chosen is narrower than for the raw data graphed in Figures 7.2 and 7.3, in order to help reveal detail. However, this does tend to overemphasize changes over time. The principal features are as follows. First, there is close agreement between the equilibrium points for the first two species for the first few weeks, diverging thereafter. The accumulation of evidence points to species *D. melanogaster* having a higher equilibrium point than *D. hydei* when both species are present. Secondly, the adjusted expectations fall in time until about week $t = 13$, despite the fact that the species counts tend on average to be larger as time progresses. This tends to confirm what we saw in Figure 7.5: that the prior values $\mathrm{E}(L_s)$ were on the high side. Thirdly, the three-standard-deviation envelopes do get narrower over

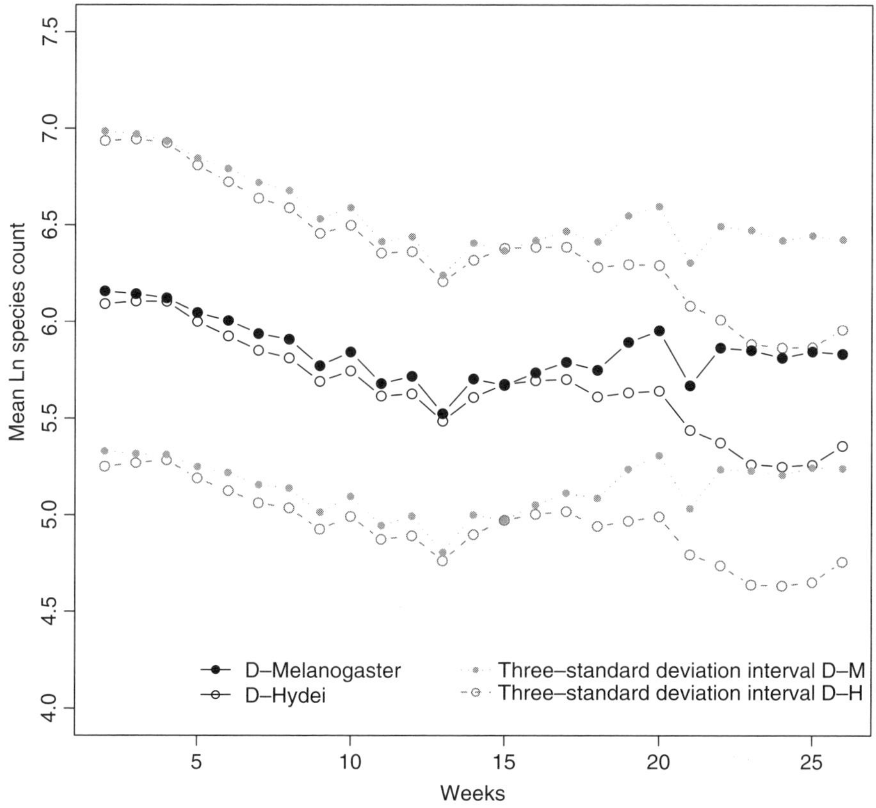

Figure 7.6 Adjusted expectations and three-standard-deviation intervals for the equilibrium points, adjusting sequentially by all the data available to time t inclusive.

time, but only marginally so: the prior variances for L_s and the adjusted variances for L_s given (1) data from week $t = 2$ and (2) all data up to week $t = 26$ are:

$$\text{Var}(L_1) = 0.490, \quad \text{Var}_{G_2}(L_1) = 0.079, \quad \text{Var}_{G_{[26]}}(L_1) = 0.040,$$

$$\text{Var}(L_2) = 0.490, \quad \text{Var}_{G_2}(L_1) = 0.076, \quad \text{Var}_{G_{[26]}}(L_2) = 0.039.$$

This shows that the first batch of information, data at week $t = 2$, resolves a large amount of the prior variation: the percentage of variation explained is around 84% of prior. However, by week $t = 26$, we have resolved only about half the variation remaining at week $t = 2$, so that the gain in information is rather slow once we get past the initial adjustment. Fourthly, while the prior three-standard-deviation interval for L_1 does contain all the adjusted values $\text{E}_{G_{[t]}}(L_1)$, the adjusted values for L_2 finally fall below the corresponding prior three-standard-deviation interval, suggesting either that the prior variance was specified too confidently or that the value specified for $\mathrm{E}(L_2)$ was too high, or both.

7.6.4.1 Diagnostics

The sequential diagnostic assessment suggests various inconsistencies and incompatibilities. The data trajectory is summarized in Table 7.4 and Figure 7.7. The diagnostic plot reveals several large partial size ratios (§5.43), the largest occurring at $t = 21$ for which

$$\mathrm{Sr}_{[G_{21}/G_{[20]}]}(L) = 57.4. \tag{7.35}$$

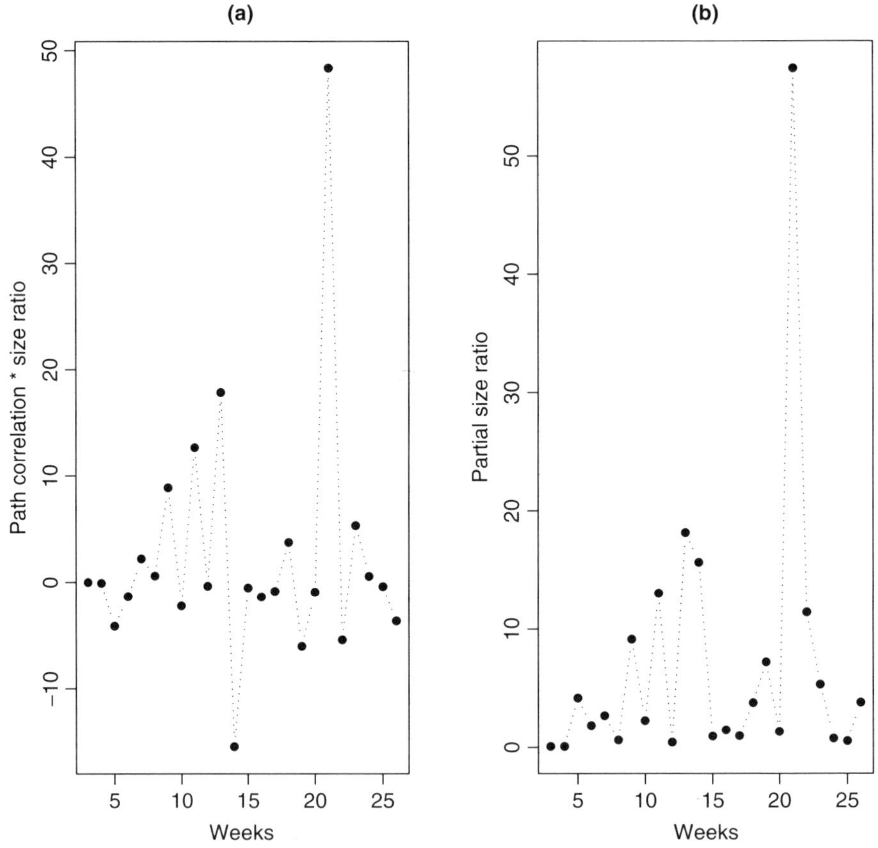

Figure 7.7 The data trajectory for the sequential adjustment of equilibrium points as data accumulate. (a) Successive path correlations multiplied by size ratios; (b) partial size ratios for sequential adjustments.

The diagnostic threshold suggested in (5.46) is around [5.24, 7], so the diagnostics at weeks $t = 9, 11, 13, 14, 19, 21, 22$, and perhaps week $t = 23$, all suggest unusually large changes in expectation relative to prior variance. The path correlations between the adjustments up to time $t - 1$ and the partial adjustments at time t are also shown in Table 7.4 and plotted, weighted by values of the corresponding size ratio to emphasize important contradictions, in Figure 7.7. We note two particular features: that the partial adjustment at week $t = 21$ was highly unusual, but in the same direction as the aggregated adjustment up to week $t = 20$ inclusive; and that the partial adjustment at week $t = 14$ was highly unusual, and in an opposite direction to the aggregated adjustment up to week $t = 13$ inclusive. Returning to the plots of the data in Figures 7.2 and 7.3, we see that, from week $t = 20$ to week $t = 21$, counts fell for both species substantially in all six cages. This suggests that there were outside factors involved, perhaps environmental, which caused drops in counts during that week. For the partial adjustment at week $t = 14$, the size ratio is

$$\mathrm{Sr}_{[G_{14}/G_{[13]}]}(L) = 15.6,$$

with corresponding path correlation

$$\mathrm{PC}(G_{[13]}, G_{14}) = -0.9893.$$

Examining the data plots and the counts in Tables 7.1 and 7.2, we notice that many of the counts were rather higher in week $t = 14$ than in week $t = 13$. This led to relatively large positive changes in adjusted expectation from $t = 13$ to $t = 14$, bucking the trend of generally falling adjusted expectations from the start of the experiment. Again, we surmise that there may have been outside factors influencing the experiment during this week.

There are other useful diagnostic assessments we might carry out. For example, the aggregated adjustment uses all the data, generated from two different starting points. Are the data from these two starting points telling the same story? To help find out, we collect all the counts for starting points $p = 1, 2$ into

$$F_p = \{Y_{psct}\}, \qquad s = 1, 2, \quad c = 1, 2, 3, \quad t = 2, 3, \ldots, 26,$$

and adjust L by F_1 and then partially by F_2. We find nothing much surprising: the bearing for the first adjustment is

$$\mathbb{Z}_{F_1}(L) = -1.78L_1 - 0.90L_2 + 16.07,$$

with size ratio $\mathrm{Sr}_{F_1}(L) = 0.7510$, and the partial bearing for the additional adjustment by F_2 is

$$\mathbb{Z}_{[F_2/F_1]}(L) = 0.05L_1 - 0.15L_2 + 0.66,$$

with partial size ratio $\mathrm{Sr}_{[F_2/F_1]}(L) = 0.13$. The path correlation between these two directions is $\mathrm{PC}(F_1, F_2) = -0.13$. We conclude that the data from the two starting points are more or less compatible taken as a whole.

7.6.4.2 Sample size choice

Next, we consider whether the sample size of $n = 3$ is adequate. For the accumulated adjustment at each week we evaluate the canonical resolutions λ_1 and λ_2 for the adjustment for our sample size of $n = 3$. These resolutions are plotted in Figure 7.8. Also shown are the **maximal** canonical resolutions μ_1 and μ_2, the resolutions obtained as we let the sample size $n \to \infty$, via (7.12). The canonical directions change over time, as the adjustments are not exchangeable with respect to time t. There are two key features. The first, and most important, is that the maximal resolutions for $n \to \infty$ are hardly larger than the canonical resolutions for $n = 3$, implying that there is virtually no value in taking a larger sample size. This is contrary to our instincts for the process under study: we have almost certainly underspecified the amount of residual variation, $\text{Var}(R_{psct})$, compared to variation for the local means, $\text{Var}(M_{psct})$.

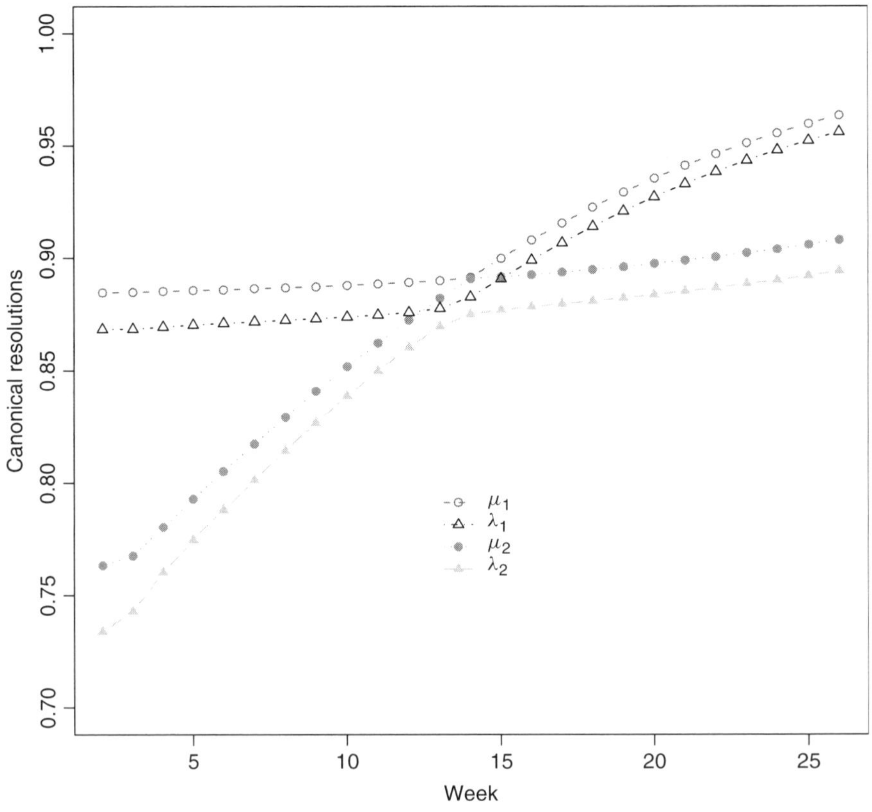

Figure 7.8 Canonical resolutions, λ_1, λ_2, and maximal canonical resolutions, μ_1, μ_2, for the sequential adjustment of equilibrium points.

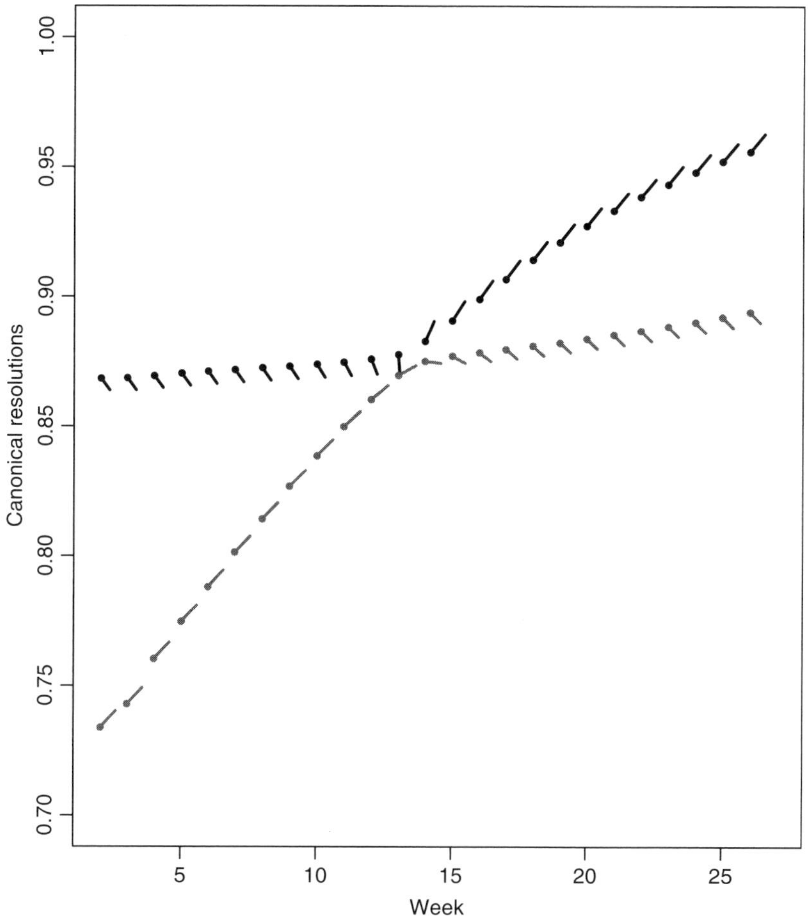

Figure 7.9 Canonical resolutions, λ_1, λ_2, and corresponding canonical directions, W_{t1}, W_{t2}, at each time point t for the sequential adjustment of equilibrium points.

The other interesting feature is the behaviour of the resolution envelope. The first canonical resolution rises only very slowly up to about week 14, and quickly thereafter, while the second canonical resolution rises very quickly up to week 14 and then slowly thereafter. This can be explained by examining Figure 7.9. Each line represents a canonical direction. A line is drawn from each canonical resolution point, considered as (0, 0), to the coordinates given by the standardized coefficients of (L_1, L_2) for the corresponding direction (this representation does not preserve graphically the orthogonality between the directions). We observe that the canonical direction at week $t = 2$ stays pretty much the same over time, and that all the learning over time takes place in an orthogonal direction. At

Table 7.5 The canonical directions, W_{t1}, W_{t2}, represented as coefficients of L_1, L_2, with L_1, L_2 standardized to have variance one.

Week	W_{t1}		W_{t2}	
	L_1	L_2	L_1	L_2
2	0.5463	−0.6438	0.9495	0.8863
3	0.5434	−0.6465	0.9511	0.8843
4	0.5451	−0.6449	0.9501	0.8855
5	0.5444	−0.6456	0.9506	0.8849
6	0.5405	−0.6492	0.9528	0.8823
7	0.5340	−0.6551	0.9564	0.8779
8	0.5248	−0.6635	0.9615	0.8716
9	0.5113	−0.6757	0.9687	0.8622
10	0.4898	−0.6946	0.9798	0.8470
11	0.4509	−0.7273	0.9983	0.8191
12	0.3640	−0.7946	1.0331	0.7540
13	0.0859	−0.9619	1.0920	0.5241
14	0.5347	1.0910	0.9561	−0.0979
15	0.7600	1.0303	0.7888	−0.3719
16	0.8239	0.9953	0.7218	−0.4576
17	0.8514	0.9767	0.6892	−0.4960
18	0.8663	0.9656	0.6704	−0.5172
19	0.8754	0.9584	0.6585	−0.5304
20	0.8814	0.9535	0.6504	−0.5391
21	0.8856	0.9500	0.6447	−0.5453
22	0.8885	0.9475	0.6406	−0.5497
23	0.8907	0.9456	0.6376	−0.5529
24	0.8923	0.9443	0.6354	−0.5552
25	0.8934	0.9433	0.6338	−0.5569
26	0.8942	0.9426	0.6327	−0.5581

about week 14, this takes over as the primary canonical direction. The coefficients for the two canonical directions, displayed in Table 7.5, reveal that we know relatively little about the sum of the equilibria at the start of the experiment, but that as time passes it is this sum about which we continue to learn, whereas we learn only very slowly about the difference between them. In the context of the original motivation for these experiments, which was to explore the inhibitory influence of one species in competition with another, this is unwelcome news.

Farrow and Goldstein (1996) also considered a model and the data for this experiment. They used the same model structure, but with slightly different specifications for the parameters, and examined some Bayes linear data diagnostics for the series of observations for starting point $p = 1$. They similarly found a large number of diagnostic warnings, and concluded that the belief specification demonstrates faster than actual convergence to equilibrium, leading to too much weight being placed on early observations for learning about the equilibria. This is evident in our modified model in Figure 7.5. Farrow and Goldstein (1996) also present a graphical diagnostic, the **diagnostic triangle**, to facilitate diagnostic monitoring for changes in expectation for multivariate time series. We compare the specifications we have used in this chapter to those used in Farrow and Goldstein (1996) in §9.14.

7.6.4.3 A canonical trajectory

Next, we illustrate a canonical trajectory (§5.12) for this problem. There are several such trajectories we might examine, for example for the equilibrium quantities L_1, L_2. We choose to examine the canonical trajectory for the 50 unknown local means for starting point $p = 1$, $\{M_{1st}\}$, $s = 1, 2$, $t = 2, \ldots, 26$, adjusting by all the information available up to and including week $t = 26$, namely three observations on each $\{Y_{1st}\}$. We then repeat to form the canonical trajectory for starting point $p = 2$. In both cases, this involves a pure exchangeable adjustment, in that – in the notation of Theorem 6.3 – we can form a 50-dimensional variance matrix Γ expressing variances across the local means and a 50-dimensional variance matrix $\Sigma - \Gamma$ expressing variances across the corresponding residual terms $\{R_{pst}\}$, $p = 1, 2$.

For exchangeable adjustments, there are a variety of ways of organizing the data into parts so that the partial adjustments are orthogonal: for example, we could separate out influences from observations from different cages at the same time point and starting point. One simple arrangement is to construct the canonical quantities for the exchangeable adjustment, also constructing their observed values as the corresponding linear combinations of the $\{Y_{pst}\}$. For a pure exchangeable adjustment, the constructed quantities are orthogonal both in the mean space and in the residual space, by design. Therefore, we can simply carry out the sequential adjustment of the original quantities by the canonical quantities. These give necessarily orthogonal partial adjustments, and in each case the length of the bearing is equal to the size of the canonical adjustment and the expected length of the corresponding bearing is the resolution for that canonical quantity.

Suppose we write M_1 to represent the collection of the 50 local means $\{M_{1st}\}$, $s = 1, 2$, $t = 2, \ldots, 25$, for starting point $p = 1$. Suppose also that we write M_2 for the corresponding collection of means for starting point $p = 2$. The belief specifications over M_1 and M_2 are identical except for expectations. Consequently, excepting differences in expectation, the canonical quantities and canonical resolutions are the same for each adjustment. Suppose that we write these canonical quantities as $Z_1, \ldots, Z_{50}$, with corresponding canonical resolutions $\lambda_1, \ldots, \lambda_{50}$, and suppose that we define the cumulative collection

$$Z_{[i]} = \bigcup_{k=1}^{i} Z_i.$$

We write $z_{[i]}$ for the observed value of $Z_{[i]}$, and z_i for the observed value of Z_i. In fact, for this exchangeable adjustment we have $c = 3$ cages for each starting point and so there are three observations, $z_{ip1}, z_{ip2}, z_{ip3}$, for each canonical quantity for each starting point.

We could now carry out the sequential adjustment of M_1 by each of the $Z_1, \ldots, Z_{50}$ in turn, and so obtain

1. $\text{Size}_{[i/]}(M_1) = \text{Size}_{[z_{[i]}/z_{[i-1]}]}(M_1)$, the size (5.41) of the partial adjustment of M_1 by $Z_{[i]} = z_{[i]}$, given $Z_{[i]} = z_{[i-1]}$;

2. $\text{E}(\text{Size}_{[Z_{[i]}/Z_{[i-1]}]}(M_1))$, the expected size (5.42) of the partial adjustment of M_1 by $G_{[i]}$, given $G_{[i-1]}$.

There is, however, a short-cut which we may use here. It follows because the successive Z_i are uncorrelated, so that $[Z_{[i]}/Z_{[i-1]}] = [Z_i]$, and have prior expectation zero and variance unity by design. Therefore, we have

$$\text{Size}_{[z_{[i]}/z_{[i-1]}]}(M_1) = \text{Size}_{z_i}(M_1) = \bar{z}_{i1}^2,$$

by (4.48), where

$$\text{E}_{z_i}(Z_i) = \bar{z}_{i1} = \frac{1}{3}\sum_{j=1}^{3} z_{i1j}$$

is the average of the observations for starting point $p = 1$; and

$$\text{E}(\text{Size}_{[Z_{[i]}/Z_{[i-1]}]}(M_1)) = \text{E}(\text{Size}_{Z_i}(M_1)) = \lambda_i.$$

Thus, this canonical trajectory may be deduced simply from the canonical structure provided by adjustment of the local means by the data. The results are identical for starting point $p = 2$, in that the canonical quantities and resolutions are the same, except that the observed values of the canonical quantities differ. For $p = 2$, we have instead

$$\text{Size}_{[z_{[i]}/z_{[i-1]}]}(M_2) = \text{Size}_{z_i}(M_2) = \bar{z}_{i2}^2,$$

where

$$\bar{z}_{i2} = \frac{1}{3}\sum_{j=1}^{3} z_{i2j}.$$

Table 7.6 Canonical trajectories for the adjustments of the collections of local means M_1 and M_2, $i = 1, \ldots, 25$. The expected sizes are the same for both adjustments.

i	$\mathrm{E}(\mathrm{Size}_{[Z_{[i]}/Z_{[i-1]}]}(\cdot)) = \lambda_i$	$\mathrm{Size}_{[z_{[i]}/z_{[i-1]}]}(M_1)$	$\mathrm{Size}_{[z_{[i]}/z_{[i-1]}]}(M_2)$
1	0.9988	0.6298	0.0523
2	0.9963	1.7265	1.3205
3	0.9881	0.4946	1.9425
4	0.9826	3.8163	2.8733
5	0.9704	0.7145	3.8555
6	0.9652	0.2375	1.3774
7	0.9548	0.3254	0.0782
8	0.9503	0.0014	0.9322
9	0.9387	0.0620	10.0195
10	0.9325	0.0152	0.1931
11	0.9307	0.0201	0.3143
12	0.9233	0.0033	0.3585
13	0.9147	0.0037	1.7982
14	0.9055	1.3944	0.0387
15	0.9021	4.4850	0.2802
16	0.8942	1.3640	0.1130
17	0.8892	0.1545	2.0725
18	0.8821	2.4387	4.2239
19	0.8769	0.7469	2.9171
20	0.8690	2.1301	6.1409
21	0.8641	0.9739	3.0525
22	0.8544	0.1188	16.6066
23	0.8499	1.2758	0.5850
24	0.8384	0.1610	19.2568
25	0.8346	0.4865	0.0004

The calculations are summarized in Tables 7.6 and 7.7. Figure 7.10 plots the sizes for the partial adjustments and their expected sizes. We accumulate the partial sizes as their sums provide the sizes for the overall adjustments. Similarly, we accumulate the expected sizes to provide the expected size for the overall adjustment.

We observe that the overall sizes of the adjustments are rather larger than the expected sizes, and particularly so for starting point $p = 2$. For starting point $p = 1$, the major contributions to size are in the directions Z_{26}, Z_{37}, and Z_{46}. For starting point $p = 2$, the major contributions to size are in the directions Z_{22}, Z_{24}, and Z_{37}. We may also plot the size ratios,

$$\mathrm{Sr}_{[z_{[i]}/z_{[i-1]}]}(M) = \frac{\mathrm{Size}_{[z_{[i]}/z_{[i-1]}]}(M)}{\mathrm{E}(\mathrm{Size}_{[Z_{[i]}/Z_{[i-1]}]}(M))} = \frac{\bar{z}_{ip}^2}{\lambda_i}, \quad p = 1, 2.$$

Table 7.7 Canonical trajectories for the adjustments of the collections of local means M_1 and M_2, $i = 26, \ldots, 50$. The expected sizes are the same for both adjustments. The bottom row shows the expected size and size for the full adjustments of M_1 and M_2.

i	$\mathrm{E}(\mathrm{Size}_{[Z_{[i]}/Z_{[i-1]}]}(\cdot)) = \lambda_i$	$\mathrm{Size}_{[z_{[i]}/z_{[i-1]}]}(M_1)$	$\mathrm{Size}_{[z_{[i]}/z_{[i-1]}]}(M_2)$
26	0.8208	21.7085	2.2741
27	0.8182	0.0000	3.3876
28	0.8025	0.9686	0.1373
29	0.7996	0.1505	0.2382
30	0.7836	2.5627	6.5975
31	0.7784	0.1683	0.7310
32	0.7637	0.1961	1.0748
33	0.7547	1.0964	9.5587
34	0.7424	5.4600	0.5720
35	0.7289	0.1981	0.0092
36	0.7198	0.0193	0.0000
37	0.7009	17.6308	23.5500
38	0.6956	0.8440	0.0619
39	0.6725	0.1750	3.5056
40	0.6681	0.0486	3.9743
41	0.6454	1.0229	0.1429
42	0.6353	0.0259	5.3114
43	0.6178	1.1943	0.5423
44	0.6001	0.1495	0.1491
45	0.5880	6.5359	6.1695
46	0.5595	13.7163	2.0969
47	0.5293	2.6541	0.2180
48	0.5003	4.5518	0.2206
49	0.4611	1.0565	0.4378
50	0.4391	0.5669	1.0281
M	39.7324	106.4809	152.3919

These are shown in Figure 7.11 and emphasize both the number of canonical quantities with unusually large changes in expectation, and that there are more such unusually large changes for starting point $p = 2$. We are free to explore such features in greater detail as we desire, through examining the canonical quantities with large changes in expectation, and exploring their relationships with the original quantities. One interesting feature arising here is that there are large discrepancies for both starting points for canonical quantity Z_{37}. This canonical quantity is approximately

$$Z_{37} \approx (2.7M_{1,20} - 2.1M_{1,21}) + (4.6M_{2,20} - 6.1M_{2,21}),$$

identically for both starting points, where the M quantities are in standardized form (mean zero, variance unity), and where we have ignored coefficients smaller

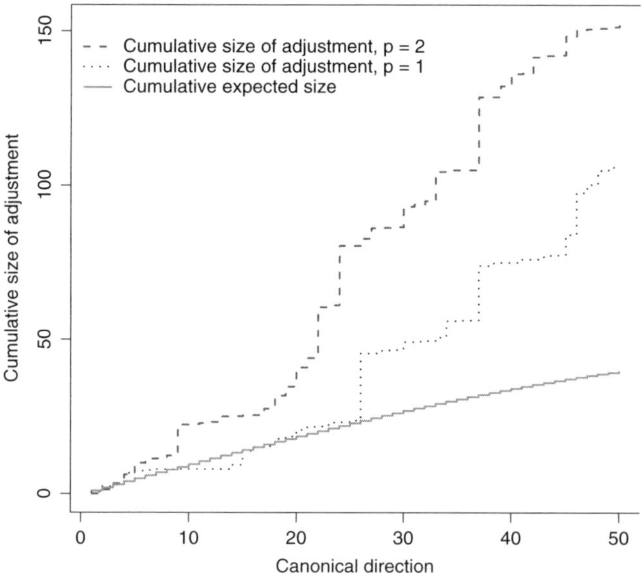

Figure 7.10 Canonical trajectory: cumulative sizes and expected sizes.

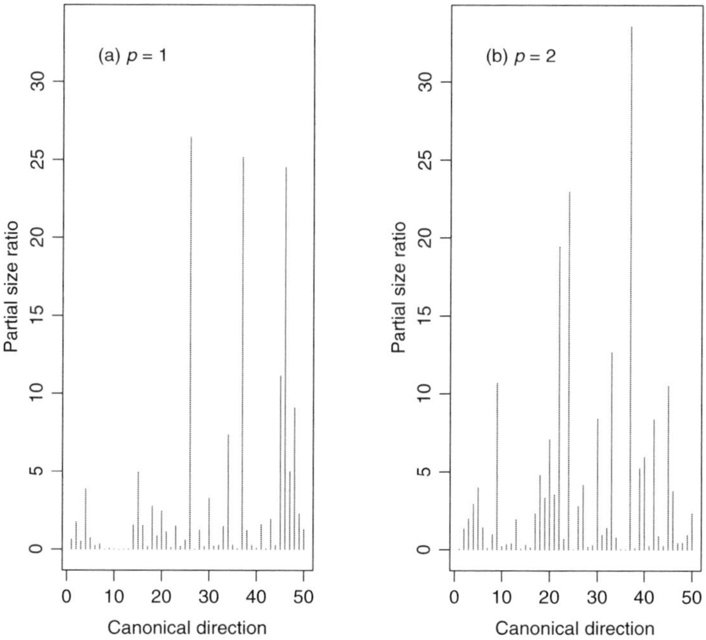

Figure 7.11 Canonical trajectory: partial size ratios (a) for starting point $p = 1$, (b) for starting point $p = 2$.

than 1.7. Thus, this canonical quantity represents approximately changes in local mean between weeks $t = 20$ and $t = 21$, for both species, and one conclusion is that the change in adjusted expectation for the differences between these local means were far larger than expected, for both species and for both starting points. This corresponds to the feature we observed earlier in (7.35). However, that earlier finding was clouded by the sequential adjustments not being orthogonal. The canonical trajectory analysis allows us firmly to pinpoint the very unusual behaviour between weeks $t = 20$ and $t = 21$, separated out from earlier behaviour.

8

Learning about population variances

In the preceding chapters, we showed how second-order exchangeability judgements could be used to adjust beliefs about collections of population means. We now consider how we may carry out similar analyses to learn about collections of population variances and covariances. This analysis raises several new features. First, our uncertainty about variances must be expressed through fourth-order moments. Secondly, the quantities which we use to adjust our beliefs, for example sample variances, often have a more complicated structure than do the sample means. Thirdly, there are additional coherence constraints involved in constructing collections of adjusted variances and covariances. Finally, the approach raises interesting questions about the relationship between the analysis of beliefs about population variances and the corresponding analysis of beliefs about population means. We begin by considering the simplest case, that of learning about an individual population variance when the population mean is known.

8.1 Assessing a population variance with known population mean

Suppose that $X = \{X_1, X_2, \ldots\}$ is an infinite exchangeable sequence of scalar random quantities, where $\mathrm{E}(X_k) = \mu$, $\mathrm{Var}(X_k) = \sigma^2$, and $\mathrm{Cov}(X_k, X_j) = \gamma$. As such, we have the exchangeability representation

$$X_k = \mathcal{M}(X) + \mathcal{R}_k(X), \quad k = 1, 2, \ldots, \tag{8.1}$$

where the sequence $\mathcal{R}_1(X), \mathcal{R}_2(X), \ldots$ is uncorrelated and has expectation $\mathrm{E}(\mathcal{R}_k(X)) = 0$ and variance

$$\mathrm{Var}(\mathcal{R}_k(X)) = \sigma^2 - \gamma = V_R, \tag{8.2}$$

Bayes Linear Statistics: Theory and Methods M. Goldstein and D.A. Wooff

say. Now suppose that the population mean $\mathcal{M}(X)$ is known, so that

$$\gamma = \text{Var}(\mathcal{M}(X)) = \text{Cov}(X_k, X_j) = 0.$$

To learn about the population variance, we must construct a representation for the corresponding quantity. Thus, let $[\mathcal{R}_k(X)]^2 = (X_k - \mu)^2 = V_k$, and suppose that we judge that the sequence $V_1, V_2, \ldots$ is also second-order exchangeable. We therefore have the representation

$$[\mathcal{R}_k(X)]^2 = V_k = \mathcal{M}(V) + \mathcal{R}_k(V) \tag{8.3}$$

where $\text{E}(\mathcal{M}(V)) = V_R$, and the sequence $\mathcal{R}_1(V), \mathcal{R}_2(V), \ldots$ is uncorrelated with zero mean and constant variance $V_{R(V)}$ and each element $\mathcal{R}_k(V)$ is uncorrelated with $\mathcal{M}(V)$. $\mathcal{M}(V)$ represents the population variance. We denote the variance of $\mathcal{M}(V)$ by V_M.

We may specify V_M and $V_{R(V)}$ directly. V_M expresses our judgement as to how much our beliefs about the population variance $\mathcal{M}(V)$ might change were we able to observe a large sample, while $V_{R(V)}$ reflects our judgements as to the shape of the population distribution. In §8.3, we will consider the specification of these quantities in more detail.

Just as for the sample mean, $\bar{X}_n^{(2)} = \frac{1}{n}\sum_{i=1}^n (X_i - \mu)^2$ is Bayes linear sufficient for all of the individual $(X_i - \mu)^2$ quantities for adjusting $\mathcal{M}(V)$. We may evaluate the adjusted expectation for $\mathcal{M}(V)$ given a sample of n from (6.55) and (6.56) as

$$\text{E}_n(\mathcal{M}(V)) = \frac{V_M \bar{X}_n^{(2)} + \frac{1}{n} V_{R(V)} V_R}{V_M + \frac{1}{n} V_{R(V)}}, \tag{8.4}$$

with corresponding adjusted variance

$$\text{Var}_n(\mathcal{M}(V)) = \frac{\frac{1}{n} V_{R(V)} V_M}{\frac{1}{n} V_{R(V)} + V_M}. \tag{8.5}$$

8.2 Assessing a population variance with unknown population mean

Now suppose, as before, that $X = \{X_1, X_2, \ldots\}$ is judged to be an infinite exchangeable sequence of scalar random quantities, where $\text{E}(X_k) = \mu$, $\text{Var}(X_k) = \sigma^2$, and $\text{Cov}(X_k, X_j) = \gamma$. We have the exchangeability representation (8.1) where the sequence $\mathcal{R}_1(X)$,$\mathcal{R}_2(X), \ldots$ is uncorrelated and has expectation $\text{E}(\mathcal{R}_k(X)) = 0$ and variance, V_R, given by (8.2), where $\gamma \geq 0$.

As above, we construct a representation for the corresponding population variance. Thus, let $[\mathcal{R}_k(X)]^2 = V_k$ and suppose that the sequence $V_1, V_2, \ldots$ is also second-order exchangeable. We have the representation (8.3) as above, and, again, $\mathcal{M}(V)$ represents the population variance.

The difference between the present case and our previous representation is that, as $\mathcal{M}(X)$ is unknown, the quantities V_k are not observable. Therefore, we must construct various combinations of the observables which are informative for $\mathcal{M}(V)$. Suppose that we have a sample, $(X_1, \ldots, X_n)$, of size $n \geq 2$. A simple construction is to introduce the squared residuals

$$(X_k - \bar{X}_n)^2 = (\mathcal{R}_k(X) - \bar{R}_n)^2,$$

where

$$\bar{R}_n = \frac{1}{n}\sum_{k=1}^{n} \mathcal{R}_k(X).$$

By symmetry, the adjusted mean for $\mathcal{M}(V)$ given the squared residuals is a function of the sum of the squared residuals, which we standardize in the usual way to give the sample variance

$$s_n^2 = \frac{1}{n-1}\sum_{k=1}^{n}(X_k - \bar{X}_n)^2. \tag{8.6}$$

We can create the following representation for s_n^2:

$$\begin{aligned} s_n^2 &= \frac{1}{n-1}\sum_k(\mathcal{R}_k(X) - \bar{R}_n)^2 = \frac{1}{n-1}\left\{\sum_k \mathcal{R}_k(X)^2 - \frac{1}{n}\left[\sum_k \mathcal{R}_k(X)\right]^2\right\} \\ &= \frac{1}{n}\sum_k \mathcal{R}_k(X)^2 - \frac{2}{n(n-1)}\sum_{k<j}\mathcal{R}_k(X)\mathcal{R}_j(X) \\ &= \mathcal{M}(V) + T, \end{aligned} \tag{8.7}$$

where, from (8.3),

$$T = \frac{1}{n}\sum_k \mathcal{R}_k(V) - \frac{2}{n(n-1)}\sum_{k<j}\mathcal{R}_k(X)\mathcal{R}_j(X).$$

Suppose that we consider that the residuals $\mathcal{R}_j(X)$ satisfy certain natural fourth-order uncorrelated properties, as follows. Suppose that, for $k \neq j \neq i$, the product $\mathcal{R}_k(X)\mathcal{R}_j(X)$ is uncorrelated with $\mathcal{M}(V)$ and $\mathcal{R}_i(V)$, i.e.

$$\text{Cov}(\mathcal{M}(V), \mathcal{R}_k(X)\mathcal{R}_j(X)) = \text{Cov}(\mathcal{R}_i(V), \mathcal{R}_k(X)\mathcal{R}_j(X)) = 0; \tag{8.8}$$

and if $k > j$, $w > u$, then

$$\text{Cov}(\mathcal{R}_k(X)\mathcal{R}_j(X), \mathcal{R}_w(X)\mathcal{R}_u(X)) = 0, \quad \text{unless } k = w,\ j = u. \tag{8.9}$$

It follows that

$$\text{E}(T) = 0, \tag{8.10}$$

$$V_T = \text{Var}(T) = \frac{1}{n}V_{R(V)} + \frac{2}{n(n-1)}[V_M + V_R^2], \tag{8.11}$$

$$\text{Cov}(\mathcal{M}(V), T) = 0, \tag{8.12}$$

so that

$$\mathrm{E}(s_n^2) = V_R, \quad \mathrm{Var}(s_n^2) = V_M + V_T, \quad \mathrm{Cov}(s_n^2, \mathcal{M}(V)) = V_M. \tag{8.13}$$

With these specifications, the adjusted mean and variance for $\mathcal{M}(V)$ given s_n^2 are

$$\mathrm{E}_{s_n^2}(\mathcal{M}(V)) = \frac{V_M s_n^2 + V_T V_R}{V_M + V_T}, \tag{8.14}$$

$$\mathrm{Var}_{s_n^2}(\mathcal{M}(V)) = \frac{V_M V_T}{V_M + V_T}. \tag{8.15}$$

8.3 Choice of prior values

The quantity $V_{R(V)}$ reflects our judgements as to the shape of the population distribution. Suppose we consider the population variance to act as a scale parameter, so that

$$\mathcal{R}_i(X) = \sqrt{\mathcal{M}(V)}Z_i, \tag{8.16}$$

where $\mathrm{E}(Z_i) = 0$, $\mathrm{Var}(Z_i) = 1$, Z_i is independent of the value of $\mathcal{M}(V)$, and the $Z_1, Z_2, \ldots$ are independent. Then, from representation (8.3),

$$\mathcal{R}_i(V) = \mathcal{M}(V)(Z_i^2 - 1)$$

so that

$$\begin{aligned} V_{R(V)} = \mathrm{Var}(\mathcal{R}_i(V)) &= (\mathrm{Var}(\mathcal{M}(V)) + [\mathrm{E}(\mathcal{M}(V))]^2)\mathrm{Var}(Z_i^2) \\ &= (V_M + V_R^2)\mathrm{Var}(Z_i^2). \end{aligned} \tag{8.17}$$

This is determined by the kurtosis of Z_i, given by

$$\mathrm{Kur}(Z_i) = \frac{\mathrm{E}(Z_i^4)}{\mathrm{E}(Z_i^2)^2} = \mathrm{E}(Z_i^4), \quad \text{as } \mathrm{Var}(Z_i^2) = \mathrm{Kur}(Z_i) - 1.$$

We may judge the Z_i to be approximately Gaussian, for which $\mathrm{Kur}(Z_i) = 3$ and $\mathrm{Var}(Z_i^2) = 2$. Otherwise, we might employ a non-Gaussian distribution, for example one with fatter tails. A possibility is to use a t distribution, but scaled to have variance 1. That is, we take

$$Z_i = \sqrt{\frac{\nu}{\nu - 2}}T_\nu,$$

where T_ν has a t distribution with $\nu > 4$ degrees of freedom. Such a distribution has kurtosis

$$\mathrm{Kur}(Z_i) = \frac{3(\nu - 2)}{\nu - 4},$$

leading to the choice

$$\mathrm{Var}(Z_i^2) = \frac{2(\nu - 1)}{\nu - 4}. \tag{8.18}$$

Small values for ν lead to higher kurtosis, and thereby to a higher variance for the residuals of the squares, $V_{R(V)}$, and thence to a higher variance for V_T, with the implication that the observed value of s_n^2 receives less weight in the update formula (8.14). The smallest practicable value for ν is $\nu = 5$, leading to the choice $\text{Var}(Z_i^2) = 8$. As regards smaller values for $\text{Var}(Z_i^2)$, it is straightforward to show that the kurtosis for a uniform distribution centred on zero is 1.8. Indeed, any regular unimodal symmetric distribution has kurtosis no smaller than 1.8 (Stuart and Ord 1994). Thus, it is often appropriate to choose $\text{Var}(Z_i^2) \geq 0.8$.

Suppose that we are prepared to use a representation of the form (8.16), and to specify values for $\text{Var}(Z_i^2)$ and for V_R. From (8.17), our specification will be completed by specifying V_M. Two suggestions are as follows.

First, it is convenient to write $V_M = cV_R^2$ for some $c > 0$, and so instead to choose c. For convenience, write

$$\kappa = \frac{1}{n}[(n-1)\text{Var}(Z_i^2) + 2]. \tag{8.19}$$

Then we may write the proportion of (8.15) resolved, relative to prior, as

$$\frac{\text{Var}_{s_n^2}(\mathcal{M}(V))}{\text{Var}(\mathcal{M}(V))} = \frac{1}{1 + \frac{n-1}{\kappa}\frac{c}{c+1}}, \tag{8.20}$$

which decreases monotonically as a function of c between one and $[1 + \frac{n-1}{\kappa}]^{-1}$. One way of choosing V_M is now to explore our attitudes to the implications of various sample sizes, given κ. If we feel that sample information will quite quickly reduce remaining variance as a proportion of prior, then we should choose a small value of c. For $\kappa = 2$, a nomogram showing the relationship between sample size n, scaling choice c, and proportion of the variation in $\mathcal{M}(V)$ explained by that sample size and that choice of c, is graphed in Figure 8.1. For $\kappa \neq 2$, simply replace n in Figure 8.1 by

$$n' = (n-1)\kappa/2 + 1.$$

Notice that the construction (8.20) makes plain that higher kurtosis values, as evidenced via $\kappa > 2$, have the same effect as reducing the sample size, and so weakening the impact of observations on the updated variance.

An alternative method is to make a direct judgement as to the value of our prior information through the notion of equivalent sample size. We can write (8.14) in the form

$$\text{E}_{s_n^2}(\mathcal{M}(V)) = \alpha s_n^2 + (1-\alpha)\text{E}(\mathcal{M}(V)), \tag{8.21}$$

with

$$\alpha = \frac{V_M}{V_M + V_T}.$$

Suppose that we consider our prior information to be worth a notional sample size of m. In combination with a sample size n, it is then reasonable to form an adjusted

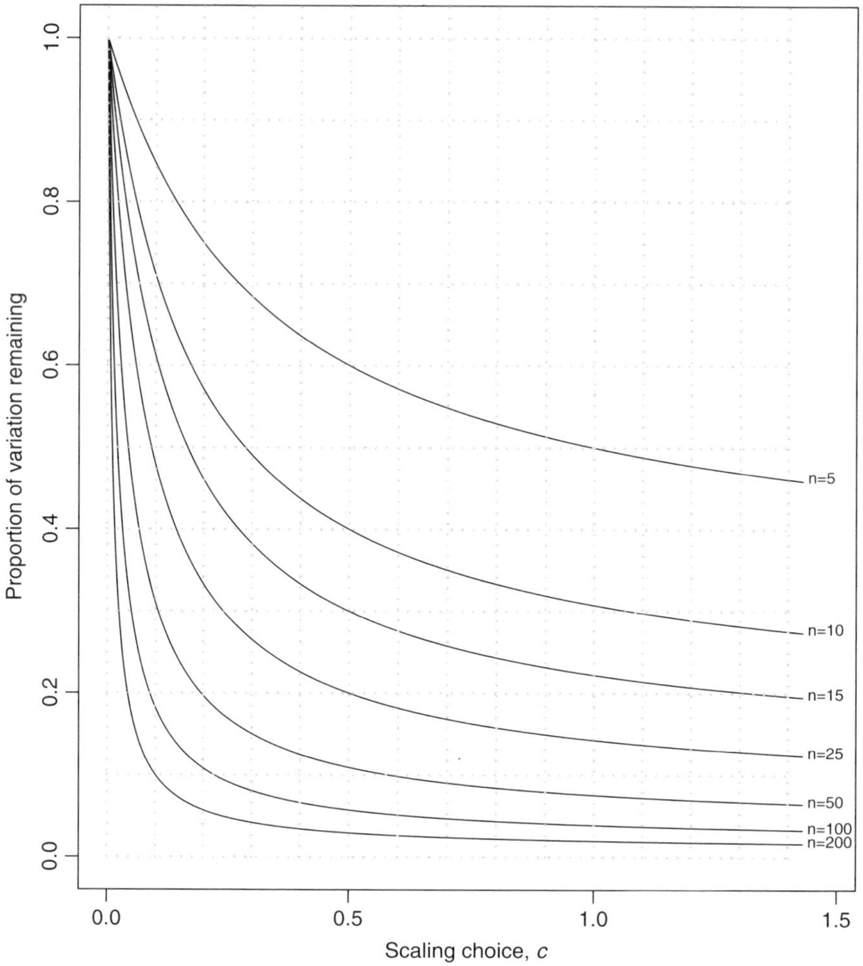

Figure 8.1 The proportion of prior variance remaining in $\mathcal{M}(V)$ after adjusting $\mathcal{M}(V)$ by s_n^2, for $\kappa = 2$ and a range of sample sizes, as a function of c. For $\kappa \neq 2$, replace n by $n' = (n-1)\kappa/2 + 1$.

expectation for $\mathcal{M}(V)$ via (8.21) with relative weighting according to the notional prior and actual sample sizes, i.e. using $\alpha = n/(m+n)$. These two methods turn out to be equivalent. The relationship between them is given by

$$m = \frac{\kappa n(c+1)}{(n-1)c} \approx \kappa + \frac{\kappa}{c}, \quad c \approx \frac{\kappa}{m-\kappa}, \tag{8.22}$$

where the approximation is reasonable for larger n. Finally, note that if we judge each Z_i to have the same kurtosis as a standard normal quantity, then $\text{Var}(Z_i^2) = 2$.

In this case, we would obtain

$$V_T = \frac{1}{n-1} V_{R(V)}, \tag{8.23}$$

by combining (8.11) and (8.17). Comparing (8.5) with (8.15), the implication is that, for roughly normal-shaped distributions for the scale parameter, knowledge of the population mean is worth roughly one observation in the adjusted variance.

8.4 Example: oral glucose tolerance test

For an example, we return to the oral glucose tolerance test. The most careful way to assess variances in this example would be through the representation that we provided in §2.4. However, to simplify the account we will make the specification directly. Belief specifications for this example were given in §6.6, and in particular we had a residual variance matrix given as (6.36), and data shown in Table 6.7. We will learn about variances for the initial glucose measurement, G_0. From (6.36), the residual variance for observation G_{0i} is

$$\mathrm{Var}(\mathcal{R}_i(G_0)) = 0.50 = V_R,$$

so that $\mathrm{E}(M(V)) = V_R = 0.5$. We will assume that a normal distribution is appropriate for the scale parameter Z_i, and so choose $\kappa = 2$ via (8.19).

To specify V_M we specify the value c, where $V_M = cV_R^2$. We examine Figure 8.1 to explore the relationship between c and sample size in the context of this example. For the purpose of variance learning, the smallest practicable sample size is $n = 4$: for this example, we feel that this sample size would not deliver a substantial reduction in variance – we judge that reduction in variance of around 20–25% might be achieved. Alternatively, we feel that a sample size of, say, $n = 100$ should resolve most of the reducible variation remaining in $\mathcal{M}(V)$. Thus, for this problem, it appears reasonable to us to choose a value of $c = 0.25$. From (8.22), this corresponds to considering our prior information about $\mathcal{M}(V)$ to be worth about $m = 10$ observations. The actual sample size for this example is $n = 15$, so our beliefs lead us to place slightly more emphasis on the sample information. We could, of course, carry out sensitivity analyses to test for sensitivity of adjusted expectations and adjusted variances to changes in c. For the chosen value $c = 0.25$, we obtain $V_M = 0.0625$ and $V_T = 0.0446$.

We may now carry out the adjustment of $\mathcal{M}(V)$ by s_n^2. This proceeds as a standard observed adjustment, as described in Chapters 3 and 4. The observed squared residuals $\mathcal{R}_i(G_0)$ are shown in Table 8.1, with the observed value of s_n^2 being

$$s_{15}^2 = \frac{1}{14}\sum_{i=1}^{15}[g_{0i} - \bar{g_0}]^2 = 0.3941.$$

One of the squared residuals is rather larger than the others: we are free to explore the implications using the diagnostic procedures described in earlier chapters. The

Table 8.1 Squared residuals, $[g_{0i} - \bar{g_0}]^2$, in ascending order.

0.0002	0.0075	0.0075	0.0128	0.0348
0.0455	0.0455	0.0822	0.2635	0.3442
0.4715	0.4715	0.5088	0.6188	2.6028

observed adjusted expectation for $\mathcal{M}(V)$ is

$$\mathrm{E}_{s_n^2}(\mathcal{M}(V)) = 0.4382, \tag{8.24}$$

with adjusted variance

$$\mathrm{Var}_{s_n^2}(\mathcal{M}(V)) = 0.0260,$$

representing a reduction of 0.0365, or about 58% of prior. The standardized change in expectation is -0.3253. Thus, the adjusted expectation is slightly smaller than the prior expectation, and the magnitude of change is unsurprising. The remaining variance of around 42% is as suggested in Figure 8.1 for the chosen value of $c = 0.25$. We conclude that the G_0 measurements are about as variable as we expected.

Choosing different values for c has little effect on the adjusted expectation for $\mathcal{M}(V)$ for this example, but does affect the adjusted variation. The choice $c = 0.1$ leads to adjusted expectation 0.4588 but with about 60% of prior variation remaining, whilst the choice $c = 0.5$ leads to adjusted expectation 0.4259 and about 30% of prior variation remaining.

We repeat the analysis for G_2, the 2-hour measurement. There is reason to suppose that we are rather more uncertain about the 2-hour measurements than for the fasting measurements: our beliefs related to the 2-hour value combine to some extent uncertainties for the baseline and for the oral glucose test effect, so that m in this case should be smaller. As such, we shall compare the results which we obtain using the same value of $c = 0.25$ as used for the fasting measurement, and the value $c = 0.5$ corresponding approximately to $m = 6$.

For the choice $c = 0.25$, the prior expectation for the population variance $\mathcal{M}(V)$ of the residuals for G_2 is

$$\mathrm{Var}(\mathcal{R}_i(G_2)) = 2.00 = V_R,$$

so that we choose $V_M = cV_R^2 = 1.0$. The observed squared residuals $\mathcal{R}_i(G_2)$ are shown in Table 8.2, with the observed value of s_n^2 being

$$s_{15}^2 = \frac{1}{14}\sum_{i=1}^{15}[g_{2i} - \bar{g_2}]^2 = 4.9512.$$

Clearly, these residuals are typically larger than those for G_0, and many are larger than their expected value of 2.00, including one squared residual nearly ten times larger than expected. (Particularly large values have been observed both for the G_0

Table 8.2 Squared residuals, $[g_{2i} - \bar{g_2}]^2$, in ascending order.

0.0022	0.0608	0.5675	0.8962	1.3148
1.8135	2.7115	2.7335	3.0742	3.8155
4.6082	9.2822	9.2822	9.3228	19.8322

and G_2 measurements. However, these do not correspond to the same individual.) The prior variance for $\mathcal{M}(V)$ for the choice $c = 0.25$ is $\text{Var}(\mathcal{M}(V)) = 1$. The adjusted expectation turns out to be

$$\text{E}_{s_n^2}(\mathcal{M}(V)) = 3.7216,$$

with adjusted variance

$$\text{Var}_{s_n^2}(\mathcal{M}(V)) = 0.4167,$$

representing a reduction of 0.5833, or about 58% of prior, as for G_0 because of our choice for c. The standardized change in expectation is 2.2540. Thus, the adjusted expectation is rather larger than the prior expectation, and we conclude that the G_2 measurements are rather more variable than foreseen.

If instead we downgrade the value of our prior information to $c = 0.5$, $m \approx 6$, this leads to a higher prior variance specification, $\text{Var}(\mathcal{M}(V)) = 2$. We then obtain instead an adjusted expectation and adjusted variance of

$$\text{E}_{s_n^2}(\mathcal{M}(V)) = 4.0659, \tag{8.25}$$

$$\text{Var}_{s_n^2}(\mathcal{M}(V)) = 0.6000,$$

representing a reduction of 70% of prior variance. The standardized change in expectation is about 1.75 standard deviations. In summary, we obtain a slightly higher variance estimate, we remain rather more uncertain about it, and the change in adjustment is about in line with what we expected. This appears to bear out our suspicion that the 2-hour measurements are more variable than the earlier measurements.

8.5 Adjusting the population residual variance in multiple linear regression: uncorrelated errors

Learning about the population variance from an exchangeable sample with unknown mean can be viewed as a special case of the more general problem of learning about the population variability of a quantity Y when the unknown mean is a linear function of a collection of explanatory variables $X_1, \ldots, X_p$. We have an in principle infinite collection of values $(y_i, x_{i1}, \ldots, x_{ip})$. We consider that there are unknown regression coefficients $\beta = (\beta_1, \ldots, \beta_p)^T$, such that we consider the derived quantities $\epsilon_1, \epsilon_2, \ldots$ to form an uncorrelated exchangeable sequence, where

$$\epsilon_i = y_i - \beta_1 x_{i1} - \ldots - \beta_p x_{ip}, \tag{8.26}$$

$$\mathrm{E}(\epsilon_i) = 0, \tag{8.27}$$

$$\mathrm{Var}(\epsilon_i) = V_R. \tag{8.28}$$

To learn about the population residual variance, we construct the representation for the corresponding quantity. Thus, suppose that we judge the sequence ϵ_k^2 to be second-order exchangeable with representation

$$\epsilon_k^2 = V_{\epsilon k} = \mathcal{M}(V_\epsilon) + \mathcal{R}_k(V_\epsilon), \tag{8.29}$$

where $\mathrm{E}(\mathcal{M}(V_\epsilon)) = V_R$, and the sequence $\mathcal{R}_1(V_\epsilon), \mathcal{R}_2(V_\epsilon), \ldots$ is uncorrelated with zero mean and constant variance $V_{R(V_\epsilon)}$ and each element $\mathcal{R}_k(V_\epsilon)$ is uncorrelated with $\mathcal{M}(V_\epsilon)$. Here, $\mathcal{M}(V_\epsilon)$ represents the population residual variance. We denote the variance of $\mathcal{M}(V_\epsilon)$ by

$$\mathrm{Var}(\mathcal{M}(V_\epsilon)) = V_{M_\epsilon}.$$

Suppose that we make the corresponding assessments to those of (8.8) and (8.9), which are:

$$\mathrm{Cov}(\mathcal{M}(V_\epsilon), \epsilon_k\epsilon_j) = \mathrm{Cov}(\epsilon_k\epsilon_j, \mathcal{R}_i(V_\epsilon)) = 0, \quad k \neq j \neq i; \tag{8.30}$$

and if $k > j$, $w > u$, then

$$\mathrm{Cov}(\epsilon_k\epsilon_j, \epsilon_w\epsilon_u) = 0, \quad \text{unless } k = w,\ j = u. \tag{8.31}$$

8.5.1 Sample information

Given $D_n = [(y_1, x_{11}, \ldots, x_{1p}), \ldots, (y_n, x_{n1}, \ldots, x_{np})]$, a sample of n individuals, we may adjust beliefs about both the regression coefficients and the population residual variance. In §8.2, we constructed a simple adjustment for the population variance based on the sample variance given an exchangeable sample. We may similarly adjust beliefs about the population residual variance based on the corresponding unbiased estimator for the population variance. We denote $y = (y_1, \ldots, y_n)^T$ and X as the $n \times p$ matrix whose (i, j)th value is x_{ij}, and we suppose that the matrix $X^T X$ is invertible. In the linear model $y = X\beta + \epsilon$, the least squares estimator for β is $\hat{\beta} = (X^T X)^{-1} X^T y$. We shall write

$$H = X(X^T X)^{-1} X^T,$$

where H is idempotent with $\mathbf{tr}\{H\} = p$. The vector of fitted values is then

$$\hat{y} = X\hat{\beta} = Hy,$$

and the residual vector is

$$\hat{r} = y - \hat{y}.$$

The usual unbiased estimator for σ^2 is the residual mean square,

$$\hat{\sigma}^2 = \frac{1}{n-p}\hat{r}^T\hat{r}.$$

We construct the representation for $\hat{\sigma}^2$ as follows:

$$\begin{aligned}\hat{\sigma}^2 &= \frac{1}{n-p}\hat{r}^T\hat{r} = \frac{1}{n-p}\epsilon^T(I-H)\epsilon \\ &= \frac{1}{n-p}\left[\sum_k(1-h_{kk})\epsilon_k^2 - 2\sum_{k<j}h_{kj}\epsilon_k\epsilon_j\right] \\ &= \mathcal{M}(V_\epsilon) + T_\epsilon,\end{aligned}$$

where, from (8.29), as $\sum_k(1-h_{kk}) = n-p$,

$$T_\epsilon = \frac{1}{n-p}\left[\sum_k(1-h_{kk})\mathcal{R}_k(V_\epsilon) - 2\sum_{k<j}h_{kj}\epsilon_k\epsilon_j\right].$$

We therefore have, from (8.29), (8.30), (8.31) that

$$\mathrm{E}(T_\epsilon) = 0, \tag{8.32}$$

$$\mathrm{Cov}(\mathcal{M}(V_\epsilon), T_\epsilon) = 0, \tag{8.33}$$

$$\begin{aligned}V_{T_\epsilon} = \mathrm{Var}(T_\epsilon) &= \frac{1}{(n-p)^2}\left[\sum_{k=1}^n(1-h_{kk})^2V_{R(V_\epsilon)} + 4\sum_{k<j}h_{kj}^2(V_{M_\epsilon}+V_R^2)\right] \\ &= \frac{1}{(n-p)^2}\left[V_{R(V_\epsilon)}\sum_{k=1}^n(1-h_{kk})^2\right. \\ &\quad \left. -2(V_{M_\epsilon}+V_R^2)\sum_{k=1}^n h_{kk}^2 + 2p(V_{M_\epsilon}+V_R^2)\right].\end{aligned} \tag{8.34}$$

It follows that

$$\mathrm{E}(\hat{\sigma}^2) = V_R, \quad \mathrm{Var}(\hat{\sigma}^2) = V_{M_\epsilon} + V_{T_\epsilon}, \quad \mathrm{Cov}(\hat{\sigma}^2, \mathcal{M}(V_\epsilon)) = V_{M_\epsilon}.$$

With these specifications, the adjusted mean and variance for $\mathcal{M}(V_\epsilon)$ given $\hat{\sigma}^2$ are

$$\mathrm{E}_{\hat{\sigma}^2}(\mathcal{M}(V_\epsilon)) = \frac{V_{M_\epsilon}\hat{\sigma}^2 + V_{T_\epsilon}V_R}{V_{M_\epsilon}+V_{T_\epsilon}}, \tag{8.35}$$

$$\mathrm{Var}_{\hat{\sigma}^2}(\mathcal{M}(V_\epsilon)) = \frac{V_{M_\epsilon}V_{T_\epsilon}}{V_{M_\epsilon}+V_{T_\epsilon}}. \tag{8.36}$$

These compare to (8.14) and (8.15), with s_n^2 replaced by $\hat{\sigma}^2$ and V_M and V_T replaced by analogous quantities.

8.5.2 Choice of prior values

We are free to judge appropriate values for V_{M_ϵ} and V_{T_ϵ}, and so to generate results from directly specified beliefs. Alternatively, if we are prepared to adopt some simplifying assumptions, the heuristics discussed in §8.3 lead to reasonable values for them: the representation $\epsilon_i = \sqrt{\mathcal{M}(V_\epsilon)}Z_i$ and the kurtosis choice $\text{Var}(Z_i^2) = 2$ lead to

$$V_{R(V_\epsilon)} = 2(V_{M_\epsilon} + V_R^2),$$

from which (8.34) simplifies to

$$V_{T_\epsilon} = \frac{1}{(n-p)}[2(V_{M_\epsilon} + V_R^2)] = \frac{1}{n-p}V_{R(V_\epsilon)}, \tag{8.37}$$

a result which corresponds with (8.23), but with fewer degrees of freedom. We may now apply the results of §8.3 directly, replacing the sample size n there by $\tilde{n} = n - p + 1$.

8.6 Example: Anscombe data sets

To apply some of these ideas in practice, we return to the first of the Anscombe data sets discussed in §5.14.1 and tabulated in Table 5.3. The scatter plot shown in Figure 5.1(a) gives us no reason to doubt the assumptions made above about the behaviour of the residuals. These data are artificial and so we shall, simply for illustration, suppose that our prior information is worth $m = 4$ observations, with Gaussian kurtosis choice for the scale parameter giving $\kappa = 2$ via (8.19). There are $p = 2$ coefficients and the actual sample size is $n = 11$. Thus, the choice $m = 4$ and sample size parameter

$$\tilde{n} = n - p + 1 = 10$$

correspond to a scaling choice $c = 1.25$ using (8.22). The nomogram (Figure 8.1) shows that $c = 1.25$, $\tilde{n} = 10$ corresponds to quite slow variance learning as a proportion of prior. This seems fair enough for this example: we do not know much at the start, and do not expect to learn very much more from a sample of this size. In (5.66) we specified

$$\text{Var}(\epsilon_i) = 1 = V_R = \text{E}(\mathcal{M}(V_\epsilon)).$$

The choices $m = 4$, $c = 1.25$ thus lead to the specifications

$$V_{M_\epsilon} = cV_R^2 = 1.25,$$

$$V_{T_\epsilon} = \frac{1}{n-p}[2(V_{M_\epsilon} + V_R^2)] = 0.5.$$

The residuals from the least squares fit are as follows:

$$\begin{array}{cccccc} 0.039, & -0.051, & -1.921, & 1.309, & -0.171, & -0.041, \\ 1.239, & -0.740, & 1.839, & -1.681, & 0.179 & \end{array}$$

giving a residual mean square of $\hat{\sigma}^2 = 1.529$. Our estimated residual variance is thus

$$\mathrm{E}_{\hat{\sigma}^2}(\mathcal{M}(V_\epsilon)) = \frac{V_{M_\epsilon}\hat{\sigma}^2 + V_{T_\epsilon}V_R}{V_{M_\epsilon} + V_{T_\epsilon}} = \frac{1.25 \times 1.529 + 0.5 \times 1}{1.25 + 0.5} = 1.378,$$

$$\mathrm{Var}_{\hat{\sigma}^2}(\mathcal{M}(V_\epsilon)) = \frac{V_{M_\epsilon}V_{T_\epsilon}}{V_{M_\epsilon} + V_{T_\epsilon}} = 0.357.$$

Our update for the residual variation, $\mathrm{Var}(\epsilon_i)$, is thus 1.378, rather larger than the prior specification of 0.5.

8.7 Adjusting the population residual variance in multiple linear regression: correlated errors

Suppose, as previously, that we wish to update our beliefs about the population residual variance for a linear model, where the residuals are as defined in (8.26), with $\mathrm{E}(\epsilon_i) = 0$, but are correlated in the form $\mathrm{Var}(\epsilon) = V_R G$, where G is a known $n \times n$ non-negative definite matrix. The correlatedness now makes it more difficult to exploit a representation such as (8.29). One possibility is as follows, but requires further assumptions.

In least squares multiple regression, correlated errors may be handled via generalized least squares (Draper and Smith 1998). Decompose non-negative definite G via its principal components into $G = Q\Lambda Q^T$, where Q is an $n \times r$ eigenvector matrix and Λ is the $r \times r$ diagonal matrix of corresponding positive eigenvalues, $0 < r \leq n$. From the linear model $Y = X\beta + \epsilon$, define

$$Y^* = \Lambda^{-\frac{1}{2}}Q^T Y, \quad X^* = \Lambda^{-\frac{1}{2}}Q^T X, \quad \epsilon^* = \Lambda^{-\frac{1}{2}}Q^T \epsilon. \tag{8.38}$$

We may then restate the linear model as $Y^* = X^*\beta + \epsilon^*$, and with uncorrelated error terms, $\mathrm{Var}(\epsilon^*) = V_R I_r$. This is in the form described in the previous section, with possibly reduced dimension depending on degeneracy in G, but with the following important difference. If we express beliefs about third- and fourth-order relationships amongst the original residuals ϵ_i, such as are required for the methodology given in §8.5, these beliefs are not preserved by the linear transformation from ϵ to ϵ^*. Indeed, if $r < n$ we cannot even back-transform to recover the ϵ quantities from the ϵ^* quantities. We can, therefore, proceed further only at the cost of making an extra assumption:

- we might assume that higher-order beliefs about the ϵ quantities are preserved, to a good approximation, under linear transformation. This is tantamount to assuming a multivariate normal distribution for the ϵ quantities (Stuart and Ord 1994);

- we might assume that our exchangeability representation applies directly to the transformed quantities ϵ^*.

If we are prepared to allow, as an approximation, the assumption that we may work with the transformed residuals, we may make a Bayes linear update for the residual variance V_R as follows.

We have transformed to an in principle second-order uncorrelated exchangeable sequence $\epsilon_1^*, \epsilon_2^*, \ldots$ which has expectation $\mathrm{E}(\epsilon_1^*) = 0$ and variance $\mathrm{Var}(\epsilon_1^*) = V_R$. As the population mean is known, we may apply the methodology of §8.1, representing

$$\epsilon_i^{*2} = \mathcal{M}(V) + \mathcal{R}_i(V).$$

Observations on the ϵ^* quantities are available as the residuals $Y^* - \hat{Y}^*$.

8.8 Example: regression with correlated responses

We illustrate with the correlated response example discussed in earlier chapters, with data shown in Table 5.6 and plotted in Figure 5.6. We organize the linear model as $Y = X\beta + \epsilon$, where, in partitioned form,

$$\begin{bmatrix} Y_1 \\ \vdots \\ Y_{12} \\ Z_1 \\ \vdots \\ Z_{12} \end{bmatrix} = \begin{bmatrix} 1 & x_1 & 0 & 0 \\ \vdots & \vdots & \vdots & \vdots \\ 1 & x_{12} & 0 & 0 \\ 0 & 0 & 1 & x_1 \\ \vdots & \vdots & \vdots & \vdots \\ 0 & 0 & 1 & x_{12} \end{bmatrix} \begin{bmatrix} a \\ b \\ c \\ d \end{bmatrix} + \begin{bmatrix} e_1 \\ \vdots \\ e_{12} \\ f_1 \\ \vdots \\ f_{12} \end{bmatrix}$$

and

$$\mathrm{Var}(\epsilon) = \mathrm{Var}\left(\begin{bmatrix} e_1 \\ \vdots \\ e_{12} \\ f_1 \\ \vdots \\ f_{12} \end{bmatrix}\right) = \begin{bmatrix} 6.25I_{12} & 2.5I_{12} \\ 2.5I_{12} & 4I_{12} \end{bmatrix} = 6.25 \begin{bmatrix} I_{12} & 0.4I_{12} \\ 0.4I_{12} & 0.64I_{12} \end{bmatrix},$$

where I_{12} is the 12×12 identity matrix, so that

$$V_R = 6.25$$

and

$$G = \begin{bmatrix} I_{12} & 0.4I_{12} \\ 0.4I_{12} & 0.64I_{12} \end{bmatrix}.$$

G is full rank, so that there are $r = n = 24$ observable dimensions. We now form the eigendecomposition of G and thereby form Y^* and X^* as described in (8.38).

The corresponding errors for this model, ϵ^*, then are uncorrelated and have variance $V_R = 6.25$. We now obtain the least squares fit of Y^* on X^* for the transformed model. The observed residuals for the transformed model are:

1.15	−1.44	−0.54	1.34	−0.31	0.56
0.64	−0.56	1.56	1.83	1.61	−0.02
4.78	−8.15	2.86	4.22	1.27	5.54
2.98	1.36	0.46	−1.58	−6.67	1.32

Applying the methodology of §8.1, the mean of these squared values, which is $\bar{X}_n^{(2)} = 9.2763$, is Bayes linear sufficient for the observed squared residuals for adjusting $\mathcal{M}(V)$.

We now need to specify the variance components V_M and $V_{R(V)}$: we shall adopt the suggestion of (8.17) with Gaussian kurtosis choice leading to $\text{Var}(Z_i^2) = 2 = \kappa$, and so specify $V_{R(V)} = 2(V_M + V_R^2)$. The proportion of variance remaining in $\mathcal{M}(V)$, relative to prior, is given by (8.20) but with n replaced by $r + 1$. For illustration, we shall consider our prior information about residual variation to be worth about $m = 10$ observations. We have $r = 24$ observations, corresponding to using $n = 25$ for the suggested method (8.20), and this leads to a scaling choice of about $c = 0.25$. The nomogram (Figure 8.1) shows that this choice corresponds to the belief that the sample information will resolve about 75% of our prior uncertainty in $\mathcal{M}(V)$, which appears order-of-magnitude appropriate.

We now have all the ingredients for the update. We have $c = 0.25$, so that $V_M = cV_R^2 = 9.7656$. We then can compute

$$V_{R(V)} = 2(V_M + V_R^2) = 97.6562.$$

Using (8.4) and (8.5), we find that the adjusted expectation and adjusted variance for the population residual variance are

$$\text{E}_n(\mathcal{M}(V)) = \frac{V_M \bar{X}_n^{(2)} + \frac{1}{n} V_{R(V)} V_R}{V_M + \frac{1}{n} V_{R(V)}} = 8.38,$$

$$\text{Var}_n(\mathcal{M}(V)) = \frac{\frac{1}{n} V_{R(V)} V_M}{\frac{1}{n} V_{R(V)} + V_M} = 2.87.$$

As such, our sample estimate of population residual variance is a little higher than we expected, and the Bayes linear adjusted expectation is, as a consequence, also a little higher. The smallness of the adjusted variance implies that we have tied this down fairly confidently.

It is worth emphasizing that the method of this section allows us to learn about a single scale parameter only, assuming known correlation structure. An alternative methodology, useful when the correlation is unknown, is presented in §8.12 and thereafter.

8.9 Example: analysing exchangeable regressions

We illustrate further with the exchangeable regressions example considered in §6.7. The error specifications for this example, as set out in §6.7.2, are rather complicated, and so it is useful to be able to carry out checks on whether we have these specifications about right. A simple way of doing so is to assume that the correlation structure arising is appropriate, and to use the methods of this section to check whether the scale parameter is about right. We may write the model (6.37) as $Y = X\beta + \epsilon$, where

$$Y = \begin{bmatrix} Y_{1,1} & \ldots & Y_{13,1} & Y_{1,2} & \ldots & Y_{13,2} & Y_{1,3} & \ldots & Y_{13,3} \end{bmatrix}^T,$$

$$\epsilon = \begin{bmatrix} \epsilon_1^T & \epsilon_2^T & \epsilon_3^T \end{bmatrix}^T = \begin{bmatrix} \epsilon_{1,1} & \ldots & \epsilon_{13,1} & \epsilon_{1,2} & \ldots & \epsilon_{13,2} & \epsilon_{1,3} & \ldots & \epsilon_{13,3} \end{bmatrix}^T,$$

$$\beta = \begin{bmatrix} a_1 & a_2 & a_3 & b_1 & b_2 & b_3 \end{bmatrix}^T,$$

and

$$X = \begin{bmatrix} \mathbf{1} & \mathbf{0} & \mathbf{0} & \mathbf{x} & \mathbf{0} & \mathbf{0} \\ \mathbf{0} & \mathbf{1} & \mathbf{0} & \mathbf{0} & \mathbf{x} & \mathbf{0} \\ \mathbf{0} & \mathbf{0} & \mathbf{1} & \mathbf{0} & \mathbf{0} & \mathbf{x} \end{bmatrix},$$

where $\mathbf{1}$ and $\mathbf{0}$ are 13×1 vectors of ones and zeros respectively, and $\mathbf{x}$ is the vector of integers up to 13. The variance matrix $\text{Var}(\epsilon)$, which we construct as in §6.7.2, is 39×39 block diagonal, with three identical 13×13 blocks $\text{Var}(\epsilon_1) = \text{Var}(\epsilon_2) = \text{Var}(\epsilon_3)$, as the error variance matrix is the same for each run r, and the error terms are uncorrelated across runs. For this illustration we shall ignore any advantages to be gained through exploiting such repeated structure. We shall represent the variance structure as a scalar multiple of the pure noise term $\text{Var}(E_{rt})$, which we specified to be 0.01 in §6.7.2. The prior variance matrix is thus

$$\text{Var}(\epsilon) = (0.01) \begin{bmatrix} G & 0 & 0 \\ 0 & G & 0 \\ 0 & 0 & G \end{bmatrix},$$

where G is a non-negative definite matrix which turns out to be

$$\begin{bmatrix}
6.0 & 3.8 & 2.9 & 2.3 & 1.9 & 1.6 & 1.4 & 1.3 & 1.2 & 1.1 & 1.1 & 1.0 & 1.0 \\
3.8 & 7.0 & 4.8 & 3.9 & 3.3 & 2.9 & 2.6 & 2.4 & 2.3 & 2.2 & 2.1 & 2.1 & 2.0 \\
2.9 & 4.8 & 8.0 & 5.8 & 4.9 & 4.3 & 3.9 & 3.6 & 3.4 & 3.3 & 3.2 & 3.1 & 3.1 \\
2.3 & 3.9 & 5.8 & 9.0 & 6.8 & 5.9 & 5.3 & 4.9 & 4.6 & 4.4 & 4.3 & 4.2 & 4.1 \\
1.9 & 3.3 & 4.9 & 6.8 & 10.0 & 7.8 & 6.9 & 6.3 & 5.9 & 5.6 & 5.4 & 5.3 & 5.2 \\
1.6 & 2.9 & 4.3 & 5.9 & 7.8 & 11.0 & 8.8 & 7.9 & 7.3 & 6.9 & 6.6 & 6.4 & 6.3 \\
1.4 & 2.6 & 3.9 & 5.3 & 6.9 & 8.8 & 12.0 & 9.8 & 8.9 & 8.3 & 7.9 & 7.6 & 7.4 \\
1.3 & 2.4 & 3.6 & 4.9 & 6.3 & 7.9 & 9.8 & 13.0 & 10.8 & 9.9 & 9.3 & 8.9 & 8.6 \\
1.2 & 2.3 & 3.4 & 4.6 & 5.9 & 7.3 & 8.9 & 10.8 & 14.0 & 11.8 & 10.9 & 10.3 & 9.9 \\
1.1 & 2.2 & 3.3 & 4.4 & 5.6 & 6.9 & 8.3 & 9.9 & 11.8 & 15.0 & 12.8 & 11.9 & 11.3 \\
1.1 & 2.1 & 3.2 & 4.3 & 5.4 & 6.6 & 7.9 & 9.3 & 10.9 & 12.8 & 16.0 & 13.8 & 12.9 \\
1.0 & 2.1 & 3.1 & 4.2 & 5.3 & 6.4 & 7.6 & 8.9 & 10.3 & 11.9 & 13.8 & 17.0 & 14.8 \\
1.0 & 2.0 & 3.1 & 4.1 & 5.2 & 6.3 & 7.4 & 8.6 & 9.9 & 11.3 & 12.9 & 14.8 & 18.0
\end{bmatrix}.$$

The matrix G is full rank, and so there are $r = n = 39$ observable dimensions.

We now apply the methodology of §8.1. We find that the mean of the 39 squared transformed residuals is $\bar{X}_n^{(2)} = 0.0147$. Our prior information concerning the variance specifications is quite thin: the specifications were made carefully, but were complicated, and so it is not certain that they are as precise as we would have liked. As such, we judge that we should let the sample values dominate the prior information: we will suppose that the prior information is worth about $m = 8$ observations. We are also uncertain as to the distribution we should assume for the scale effects Z_i. As such, we try a range of distributions and compare the results, using (8.4) and (8.5) to compute the adjusted expectation and adjusted variance for the population residual variance.

Uniform: $\text{Var}(Z_i^2) = 0.8$. This choice corresponds to choosing $c = 0.16$, and so to fast variance learning, relative to the sample size of $n = 39$. With these choices we find that

$$\text{E}(\mathcal{M}(V)) = 0.01, \qquad \text{Var}(\mathcal{M}(V)) = 0.0034^2,$$
$$\text{E}_n(\mathcal{M}(V)) = 0.01395, \qquad \text{Var}_n(\mathcal{M}(V)) = 0.0014^2.$$

Gaussian: $\text{Var}(Z_i^2) = 2$. This choice corresponds to choosing $c = 0.33$, and so to moderately fast variance learning, relative to the sample size of $n = 39$. With these choices we find that

$$\text{E}(\mathcal{M}(V)) = 0.01, \qquad \text{Var}(\mathcal{M}(V)) = 0.0058^2,$$
$$\text{E}_n(\mathcal{M}(V)) = 0.01392, \qquad \text{Var}_n(\mathcal{M}(V)) = 0.0024^2.$$

Scaled t_{10}: $\text{Var}(Z_i^2) = 3$ by (8.18). This corresponds to choosing $c = 0.59$, and so to slower variance learning, relative to the sample size of $n = 39$. With these choices we find that

$$\text{E}(\mathcal{M}(V)) = 0.01, \qquad \text{Var}(\mathcal{M}(V)) = 0.0117^2,$$
$$\text{E}_n(\mathcal{M}(V)) = 0.01392, \qquad \text{Var}_n(\mathcal{M}(V)) = 0.0032^2.$$

The calculations suggest that the error variance matrix should be scaled up by about 40%. There is very little difference between the adjusted expectations; this is largely because we have allowed the sample information to dominate the prior. There are important differences amongst the variances calculated. The variance calculated under the uniform suggestion implies that we have estimated the population residual variance quite precisely, and that we are fairly sure that we should be multiplying the matrix G by more than the prior value of 0.01. The variance calculated under the t_{10} suggestion implies that we remain very uncertain as to the appropriate value for this multiplier.

8.10 Adjusting a collection of population variances and covariances

Now suppose that $X = (X_1, X_2, \ldots)$ is a second-order exchangeable sequence of r-vectors with representation

$$X_k = \mathcal{M}(X) + \mathcal{R}_k(X),$$

where $\mathrm{E}(X_k) = \mu, \mathrm{Var}(X_k) = \Sigma, \mathrm{Cov}(X_k, X_j) = \Gamma$, and where the sequence $\mathcal{R}_k(X)$ is uncorrelated and has expectation vector $\mathrm{E}(\mathcal{R}_k(X)) = 0$ and variance matrix

$$\mathrm{Var}(\mathcal{R}_k(X)) = \Sigma - \Gamma,$$

with μ and non-negative definite matrices Σ, Γ dimensioned appropriately.

We may write the vector form of (8.3) as

$$\mathcal{R}_k(X)\mathcal{R}_k(X)^T = V_k = \mathcal{M}(V) + \mathcal{R}_k(V), \tag{8.39}$$

where $\mathcal{M}(V_{ii})$ is the population residual variance for variable X_i, and $\mathcal{M}(V_{ij})$ is the population residual covariance between X_i and X_j. All of the elements of $\mathcal{M}(V)$ are uncorrelated with all of the elements of each $\mathcal{R}_k(V)$. The elements of each $\mathcal{R}_k(V)$ have expectation zero, variance and covariance specifications for the elements of $\mathcal{R}_k(V)$ are the same for each k, and the elements of $\mathcal{R}_j(V)$ are uncorrelated with the elements of $\mathcal{R}_k(V)$ for $j \neq k$. Note that we have

$$\mathrm{E}(\mathcal{M}(V)) = \Sigma - \Gamma.$$

We may write the sample variance matrix

$$S_n^2 = \frac{1}{n-1}\sum_k (X_k - \bar{X}_n)(X_k - \bar{X}_n)^T \tag{8.40}$$

in corresponding form to (8.7) as the sum of two elementwise uncorrelated random matrices, namely

$$S_n^2 = \mathcal{M}(V) + T, \tag{8.41}$$

where

$$T = \frac{1}{n}\sum_k \mathcal{R}_k(V) - \frac{1}{n(n-1)}\sum_{k \neq j} \mathcal{R}_k(X)\mathcal{R}_j(X)^T,$$

so that

$$\mathrm{E}(T) = 0 \quad \text{and} \quad \mathrm{Cov}(\mathcal{M}(V), T) = 0.$$

We may use representation (8.41) to generate a full set of variances and covariances between all elements of S_n^2 and all elements of $\mathcal{M}(V)$. However, this requires a rather detailed level of prior specification, many of whose judgements may be difficult and unfamiliar, such as quantifying beliefs about the relation between the residual variance of X_i and the residual covariance of X_j and X_k. Further, there is no guarantee that the adjusted expectation for the overall variance matrix that results from this calculation will be non-negative definite. For critical problems, such careful consideration may be worthwhile. However, in many problems, it will be sufficient to follow a simple alternative approach for updating beliefs over the residual variance matrix which we now describe.

8.11 Direct adjustment for a population variance matrix

Let $B = (B_1, B_2, \ldots)$ be a collection of random $r \times r$ real symmetric matrices (for example, population and sample variance matrices). Let $C = (C_1, C_2, \ldots)$ be a basis for the linear space of constant $r \times r$ real symmetric matrices. Now form the vector space on $L = B \cup C$ and define the inner product (over equivalence classes) on L as

$$(A, B) = \mathrm{E}(\mathbf{tr}\{AB\}), \quad \forall A, B \in L,$$

corresponding to the metric

$$\|A - B\|^2 = \mathrm{E}(\|A - B\|_F^2) \tag{8.42}$$

where $\|\cdot\|_F$ denotes the Frobenius norm of the matrix, namely the sum of the squares of the elements.

Belief adjustment for scalar random quantities corresponds to orthogonal projection into subspaces of random quantities that we observe, using the scalar version of the above norm. Similarly, we may adjust the expectation of random matrices within the matrix inner product space by orthogonal projection into subspaces of L spanned by collections of matrices that we observe. While there are many different features of the matrix about which we may learn within this construction, we will here only describe the simplest projection, as this is sufficient to give a simple update for the residual variance matrix.

Thus, suppose that we wish to adjust $\mathcal{M}(V)$ by the space spanned by the sample variance matrix S_n^2 and the constant matrices. Term this expectation $\mathrm{E}_{S_n^2}(\mathcal{M}(V))$. From (8.41), we have

$$\mathrm{E}_{S_n^2}(\mathcal{M}(V)) = (1 - \alpha)\mathrm{E}(\mathcal{M}(V)) + \alpha S_n^2, \tag{8.43}$$

where

$$\alpha = \frac{\|\mathcal{M}(V) - \mathrm{E}(\mathcal{M}(V))\|^2}{\|\mathcal{M}(V) - \mathrm{E}(\mathcal{M}(V))\|^2 + \|T - \mathrm{E}(T)\|^2}. \tag{8.44}$$

We may specify α by assessing each scalar variance and covariance which is used to assess the norms in (8.44). Alternatively, we may specify α directly. A simple approach is to adapt the equivalent sample size heuristic that we used in §8.3 for constructing prior beliefs. In this heuristic, we consider our prior information about $\mathcal{M}(V)$ to be comparable to the information that we would have obtained by observing a notional previous sample variance matrix for X based on a sample of size m. In this case, the relative weightings on $\mathrm{E}(\mathcal{M}(V))$ and S_n^2 in (8.43) are in the ratio of m to n, so that we would use $\alpha = n/(m + n)$.

8.12 Example: regression with correlated responses

The method for variance learning discussed in §8.8 for this example could address only a scale parameter for the population residual matrix, whereas we prefer to learn separately about the three variance components:

$$\text{Var}(e_i) = \sigma_e^2, \quad \text{Var}(f_i) = \sigma_f^2, \quad \text{Cov}(e_i, f_i) = \sigma_{ef}.$$

The structure in this example allows us to use (8.43) straightforwardly, as follows. Suppose that we calculate separately the least squares fits $\hat{Y}$ for Y on X, and $\hat{Z}$ for Z on X. Write $H = X(X^TX)^{-1}X^T$, as in §8.5.1. These fits depend on $p = 2$ parameters in each case. We may write the residuals as

$$Y - \hat{Y} = (I - H)e \quad \text{and} \quad Z - \hat{Z} = (I - H)f,$$

where e, f are the collections $(e_1, \ldots, e_n)$ and $(f_1, \ldots, f_n)$ respectively, with $n = 12$. Now form the scaled sum-of-squared-residuals matrix as

$$S_n^2 = \frac{1}{n-p}\begin{bmatrix}(Y-\hat{Y})^T(Y-\hat{Y}) & (Y-\hat{Y})^T(Z-\hat{Z})\\ (Z-\hat{Z})^T(Y-\hat{Y}) & (Z-\hat{Z})^T(Z-\hat{Z})\end{bmatrix}, \tag{8.45}$$

with observed value

$$\begin{bmatrix}7.51 & -2.63\\ -2.63 & 4.76\end{bmatrix}. \tag{8.46}$$

As $I - H$ is idempotent and as $\mathbf{tr}\{H\} = p$, it is straightforward to show that

$$\text{E}(S_n^2) = \begin{bmatrix}\sigma_e^2 & \sigma_{ef}\\ \sigma_{ef} & \sigma_f^2\end{bmatrix} = \text{E}(\mathcal{M}(V)) = \begin{bmatrix}6.25 & 2.5\\ 2.5 & 4\end{bmatrix}, \tag{8.47}$$

so that S_n^2 is unbiased sample information on these variance components. We chose, in §8.12, to regard our prior information as worth $m = 10$ observations. Thus we weight our prior and sample variance matrices in the ratio 10 : 12, giving weight $\alpha = 10/22$. Our updated variance matrix is thus

$$\frac{10}{22}\begin{bmatrix}6.25 & 2.5\\ 2.5 & 4\end{bmatrix} + \frac{12}{22}\begin{bmatrix}7.51 & -2.63\\ -2.63 & 4.76\end{bmatrix} = \begin{bmatrix}6.94 & -0.30\\ -0.30 & 4.41\end{bmatrix}. \tag{8.48}$$

There is very little difference between the prior and updated variances. However, the sample information about the covariance between the error terms (e_i, f_i) seems completely at odds with the prior covariance, being moderately negative rather than moderately positive, as we felt a priori. There could be several reasons for such contradiction. First, with hindsight we might now judge that it was inappropriate to have chosen a positive correlation between the error quantities. In this case, we might deem it reasonable to give more weight to the sample information. We might even do this differentially, giving more weight to the sample correlation information than to the sample variance information. We describe how this can be achieved in general in the next section. Secondly, we might judge that the model we have chosen does not adequately represent the quantities, in which case we might expect to see model lack of fit confounded with pure error. Finally, of course, such contradiction can arise via chance fluctuation.

We will often replace the original population residual variance matrix by its updated version, usually so that we can perform two-stage Bayes linear analysis, which we describe in §8.15. In all these cases, it is natural to assess the sensitivity of the results to changes in residual variance matrix. As it happens, for this example there is very little difference between using the prior residual variance matrix (8.47) and the updated version (8.48).

8.13 Separating direct adjustment for population variances and for correlation structure

The method described in §8.11 is simple, and ensures a non-negative definite form for the adjusted expectation matrix. However, we may wish to input more aspects of our prior beliefs, and in particular we may feel more confident about our assessments of some of the population variances than about others. We now describe a simple modification to the above approach which may be appropriate when we are able to make more detailed specifications of our uncertainties for the individual residual variances than we may make for the joint residual covariance structure. We may carry out the adjustment in stages, as follows.

1. We adjust beliefs about the residual variance for each X_i individually. We assess the adjusted expectation for each $\mathcal{M}(V_{ii})$, based on the sample variance $s^2_{n(i)}$ for X_i using the corresponding adjusted expectation (8.14). This requires a full prior assessment for each of the variances required to specify $\mathrm{E}_{s^2_{n(i)}}(\mathcal{M}(V_{ii}))$ for each i. Suppose that we collect together these adjusted expectations into the diagonal matrix $\tilde{V}$, where

$$\tilde{V}_{ii} = \mathrm{E}_{s^2_{n(i)}}(\mathcal{M}(V_{ii})). \tag{8.49}$$

2. We specify our prior correlation matrix for $\mathcal{M}(V)$. Let this prior correlation matrix be $\mathrm{CR}(\mathcal{M}(V))$. We may assess $\mathrm{CR}(\mathcal{M}(V))$ directly. It may be difficult to specify initial correlations or covariances between the elements of $\mathcal{M}(V)$ which are required in order to specify $\mathrm{CR}(\mathcal{M}(V))$. In such cases, we may fall back on simple heuristics which suggest plausible order-of-magnitude values, while avoiding the choice of a multivariate correlation structure with unpleasant hidden consequences. A simple order-of-magnitude approximation for $\mathrm{CR}(\mathcal{M}(V))$ is to derive this matrix from the prior residual variance matrix $\mathrm{E}(\mathcal{M}(V))$. Denote by $\mathrm{CF}(V)$ the correlation matrix derived from the variance matrix V. We may thus determine, as a reasonable starting point, the prior correlation matrix for $\mathcal{M}(V)$ as

$$\mathrm{CR}(\mathcal{M}(V)) = \mathrm{CF}(\mathrm{E}(\mathcal{M}(V))) = \mathrm{CF}(\Sigma - \Gamma).$$

3. We derive the sample correlation matrix $\text{CF}(S_n^2)$ from S_n^2. We then form the updated version of the correlation matrix, using the corresponding form to (8.43), as

$$\text{CR}_n(\mathcal{M}(V)) = (1-\beta)\text{CR}(\mathcal{M}(V)) + \beta\text{CF}(S_n^2). \tag{8.50}$$

   As before, we may use an equivalent sample size argument to value the amount of information contained in the prior assessment for the correlation matrix as corresponding to an equivalent sample size m, which suggests an appropriate value for β to be $n/(m+n)$. Heuristics for this assessment are discussed in §8.13.1.

We reassemble the residual variance matrix, with variances given by stage 1, and correlations by stages 2 and 3, to give the following.

Definition 8.1 *The* ***semi-adjusted residual variance matrix*** *is the non-negative definite matrix*

$$\text{E}_{(n)}(\mathcal{M}(V)) = \tilde{V}^{\frac{1}{2}}\text{CR}_n(\mathcal{M}(V))\tilde{V}^{\frac{1}{2}}. \tag{8.51}$$

In forming this adjustment, we have taken various heuristic short-cuts and so use the term *semi-adjusted* rather than adjusted.

8.13.1 Assessing the equivalent sample size

We may choose the equivalent sample size m directly. For example, if, in a particular application, our prior knowledge about the correlation structure is quite vague, then we could choose a small equivalent sample size. As a sample size of four is about the smallest practicable actual sample size for learning about variances, we might choose $m = 4$ as indicating vague prior knowledge, leading to $\beta = n/(n+4)$.

Alternatively, the following arguments lead to a suggestion for m. Suppose that we are prepared to specify with some confidence the sign of a particular correlation, in addition to its magnitude. An approximate 95% classical confidence interval for ρ, the population correlation coefficient between two bivariate normally distributed quantities, can be obtained via Fisher's z transformation as

$$\frac{1}{2}\ln\left(\frac{1+r}{1-r}\right) \pm 2\sqrt{\frac{1}{m-3}},$$

where r is the sample correlation coefficient and m the sample size. For positive (negative) r, this interval has lower (upper) boundary zero for

$$m = 3 + 16\left[\ln\left(\frac{1+|r|}{1-|r|}\right)\right]^{-2}. \tag{8.52}$$

Consequently, if we are prepared to accept these simplifying assumptions and can specify with some confidence the sign of the correlation coefficient, we may take this value of m as providing a reasonable equivalent sample size. Note that, as we might expect, small (large) values of r lead to large (small) values of m.

This method works well when the dimension of $\mathcal{M}(V)$ is small, and, in particular, when $\Sigma - \Gamma$ is two-dimensional, there is a single prior correlation to consider, and the value of r to be used for (8.52) is known. When $\mathcal{M}(V)$ is multivariate, the choice of r is less obvious. Two possibilities for r are then:

(a) if the correlation terms in $\Sigma - \Gamma$ are roughly the same in magnitude, we might use their mean absolute value;

(b) if we are sure about the signs of all the correlations, we can choose as r the smallest correlation in $\text{CR}(\mathcal{M}(V))$, leading to a conservative choice for m.

8.14 Example: oral glucose tolerance test

We continue the example of §8.4, but now treating $B = [G_0, G_2]$ jointly rather than separately. The ingredients for the calculations are as follows. We have already found adjusted expectations for the individual residual variances for G_0 and G_2 as 0.4382 and 4.0659, respectively; see (8.24) and (8.25). In doing so, note that we judged our prior information to be rather stronger for learning about the residual variance for G_0. The original residual variance matrix for the collection B is $\text{E}(\mathcal{M}(V)) = \Sigma - \Gamma$, given in (6.36). The correlation form for this matrix is

$$\text{CF}(\text{E}(\mathcal{M}(V))) = \begin{bmatrix} 1 & 0.42 \\ 0.42 & 1 \end{bmatrix}. \tag{8.53}$$

The observed variance–covariance and correlation matrices are

$$S_n^2 = \begin{bmatrix} 0.3941 & 0.6372 \\ 0.6372 & 4.9512 \end{bmatrix}, \quad \text{CF}(S_n^2) = \begin{bmatrix} 1 & 0.4562 \\ 0.4562 & 1 \end{bmatrix},$$

based on a sample of size $n = 15$.

Next, we specify the prior correlation form as $\text{CF}(\text{E}(\mathcal{M}(V)))$. If we are reasonably confident that the correlation between G_0 and G_2 is non-negative and if we have specified a correlation $r = 0.42$ in (8.53), the argument given in §8.13.1 leads to deeming this information as though it originates from a sample of size

$$m = 3 + 16\left[\ln\left(\frac{1 + 0.42}{1 - 0.42}\right)\right]^{-2} \approx 23,$$

by (8.52). Thus, we revise our correlation using (8.50) as

$$\text{CR}_n(\mathcal{M}(V_{12})) = \frac{23}{23 + 15}0.42 + \frac{15}{23 + 15}0.4562 = 0.4343.$$

For this example, there is only a minor difference between the prior and revised correlations. Note that there is no necessary relationship between the notions of equivalent sample size for variance updating on the one hand (we used $m \approx 10$ for G_0 and $m \approx 6$ for G_2) and correlation updating (we used $m = 23$) on the other hand. This is made plain if we consider updating for a two-dimensional case with variables X and Y, and with X known to be very close to Y. We might have little prior information about the variances of X and Y, in which case the equivalent sample sizes for their updating should be quite small. On the other hand, we might judge that the correlation between X and Y is close to one, and this would be reflected in a very high prior equivalent sample size for updating the correlation.

We now rescale back to variance–covariance form by scaling according to the adjusted population variances for G_0 and G_2, to give

$$\mathrm{E}_{(n)}(\mathcal{M}(V)) = \begin{bmatrix} 0.4382 & 0.5797 \\ 0.5797 & 4.0659 \end{bmatrix}. \tag{8.54}$$

The semi-adjusted residual variance matrix (8.54) should be compared to the original specification given as (6.36):

$$\mathrm{E}(\mathcal{M}(V)) = \Sigma - \Gamma = \begin{bmatrix} 0.50 & 0.42 \\ 0.42 & 2.00 \end{bmatrix},$$

with correlation form shown in (8.53). So far we have commented on differences piecemeal, as there are only two variances and a correlation to think about. For larger-scale problems, we can explore differences between the original and semi-adjusted residual variance matrix either simply, through the matrix norm described above (8.42), or more generally by examining detailed structural differences using the methods which we develop in Chapter 9.

8.15 Two-stage Bayes linear analysis

The above account offers a brief introduction to the Bayes linear analysis of variance structures. This is an important but largely unexplored problem. In general, we have identified three features which often complicate such analyses. First, the quantities that form the basis of the exchangeability representations are often not directly observable. Secondly, the required uncertainty judgements are relatively unfamiliar. Thirdly, the constraints on the collection of assessments, such as non-negative definiteness, do not fit naturally into our framework. Thus, rather than carrying out a full linear analysis on the variance structure, we may often prefer to fit together certain, individually plausible, Bayes linear components which exploit our key prior judgements in an intuitively sensible way.

In particular, we may prefer simple approaches to variance estimation for problems where learning about the population variances is not our principle interest, but rather we are learning about the variances in order to improve our ability to learn about the population means. Thus, for many problems based on exchangeable observations, we may carry out the analysis in two stages.

In the first stage, we carry out the variance assessment as above, resulting, for example, in an assessment of the semi-adjusted residual variance matrix

$\mathrm{E}_{(n)}(\mathcal{M}(V))$. In the second stage, we carry out the Bayes linear analysis for the mean vector. This is exactly as we have described, with the sole difference that in the representation theorem for exchangeable random vectors (Theorem 6.3) we replace the prior residual variance matrix $\mathrm{Var}(\mathcal{R}_j(X)) = \Sigma - \Gamma$ by the semi-adjusted version $\mathrm{Var}(\mathcal{R}_j(X)) = \mathrm{E}_{(n)}(\mathcal{M}(V))$.

Provided that we make the judgement that our beliefs about the population mean vector are independent of our beliefs about population residual variation, then we may carry out the mean analysis exactly as previously described, with the new residual variance matrix, as heuristically this matrix now expresses our judgements about residual variation. We call this procedure a **two-stage Bayes linear analysis**. The Bayes linear assessments for the mean are termed **variance-modified Bayes linear assessments**.

The simplest such assessment derives from the analysis of a scalar exchangeable sample $X_1, X_2, \ldots,$ as discussed in §6.11. The Bayes linear adjustment of the population mean $\mathcal{M}(X)$ by the sample mean $\bar{X}_n$ is given, as in (6.55), by

$$\mathrm{E}_n(\mathcal{M}(X)) = \frac{\gamma \bar{X}_n + \frac{1}{n}\eta\mu}{\gamma + \frac{1}{n}\eta}.$$

If we carry out a two-stage analysis, we first reassess the value of η, replacing the prior value by the adjusted value given s_n^2, as given by (8.14), namely

$$\eta^* = \frac{V_M s_n^2 + V_T V_R}{V_M + V_T},$$

so that the variance modified Bayes linear adjustment is

$$\mathrm{E}_{n*}(\mathcal{M}(X)) = \frac{\gamma \bar{X}_n + \frac{1}{n}\eta^*\mu}{\gamma + \frac{1}{n}\eta^*}, \tag{8.55}$$

with corresponding modified variance

$$\mathrm{Var}_{n*}(\mathcal{M}(X)) = \frac{\frac{1}{n}\eta^*\gamma}{\frac{1}{n}\eta^* + \gamma}. \tag{8.56}$$

Provided that we consider that η^* is a good representation of our current residual uncertainty, and that this variance analysis did not carry substantial additional information which would now cause us to modify the values for μ or γ, then the variance modified forms (8.55), (8.56) will give an improved form for our adjusted mean and variance for $\mathcal{M}(X)$. Certain conventional Bayesian prior distributions, such as the normal gamma prior for normal sampling, would not meet these requirements, and each problem must be considered carefully to judge whether the necessary requirements for our analysis will be met.

8.16 Example: oral glucose tolerance test

We now carry out a two-stage Bayes linear analysis, continuing the example of this chapter. We repeat the analysis of §6.16, and in particular the example of §6.16.4. To recapitulate, there we were concerned with making a predictive adjustment of fasting and 2-hour measurements for a new individual, given measurements on $n = 15$ other individuals. A summary of the adjustment was shown in Tables 6.8 and 6.9. For the two-stage Bayes linear analysis, we replace the residual variance matrix $\Sigma - \Gamma$ (6.36) by the semi-adjusted residual variance matrix (8.54), but retain the prior variance matrix for the mean components, Γ (6.34). The prior variance matrix Σ given by (6.35) for a single vector observation needs to be replaced by

$$\Sigma^* = \Gamma + \mathrm{E}_{(n)}(\mathcal{M}(V)) = \begin{bmatrix} 1.0582 & 0.8797 \\ 0.8797 & 4.4959 \end{bmatrix}.$$

As we noted above, the main difference is the increase in variation for an individual's 2-hour measurement. The analysis now proceeds to the second stage, a standard adjustment of beliefs as in §6.16.4. A summary of the adjustment is shown in Table 8.3. Comparing this to Table 6.9, the main differences are that the proportions of variation explained for another individual's measurements G_2 and G_h are much smaller than for the unmodified assessments, to reflect that the residual variance assessment is higher than before. The variance-modified adjusted expectations are little changed. Overall, the two-stage analysis better takes into account the variability in the residual variation, something we probably understated initially.

8.17 Example: analysing exchangeable regressions

In §8.9, we found that the overall error variance matrix for this problem is rather understated. Consequently, we might reanalyse the data by first updating the error variance matrix, and then duplicating the analyses we have already shown. We will discuss one such reanalysis. First, we scale up the overall error matrix $\mathrm{Var}(\epsilon_{rt})$ by a factor of 1.39, corresponding to our updated value $\mathrm{E}_n(\mathcal{M}(V))$ from §8.9. We

Table 8.3 Summary of the adjustment by 15 observations, following a first stage to obtain a semi-adjusted residual variance matrix. Shown are the variance-modified assessments: prior and adjusted expectations with standardized change in adjustment, relative to variance resolved; and prior and adjusted variances with resolutions.

	Expectation			Variation		
	Prior	Adjusted	Change	Prior	Adjusted	Resolution
G_0	4.16	4.6528	0.64	1.0582	0.4651	0.5605
G_2	6.25	6.6126	0.68	4.4959	4.2416	0.0626
G_h	2.09	1.9598	−0.23	3.7947	3.4689	0.0859

Table 8.4 Adjusted expectations, standardized changes, and variances for the mean components, with increased prior variation.

Component	Adjusted expectations		Prior	Variance resolutions	
	Original	Two-stage	Variance	Original	Two-stage
$\mathcal{M}(Y_1)$	1.6031	1.5933	0.0396	0.6528	0.6016
$\mathcal{M}(Y_2)$	1.7013	1.6924	0.0444	0.7009	0.6550
$\mathcal{M}(Y_3)$	1.7994	1.7915	0.0524	0.7418	0.6992
$\mathcal{M}(Y_4)$	1.8976	1.8906	0.0636	0.7719	0.7311
$\mathcal{M}(Y_5)$	1.9958	1.9898	0.0780	0.7921	0.7520
$\mathcal{M}(Y_6)$	2.0939	2.0889	0.0956	0.8050	0.7649
$\mathcal{M}(Y_7)$	2.1921	2.1880	0.1164	0.8127	0.7722
$\mathcal{M}(Y_8)$	2.2902	2.2872	0.1404	0.8171	0.7759
$\mathcal{M}(Y_9)$	2.3884	2.3863	0.1676	0.8193	0.7775
$\mathcal{M}(Y_{10})$	2.4866	2.4854	0.1980	0.8201	0.7776
$\mathcal{M}(Y_{11})$	2.5847	2.5846	0.2316	0.8201	0.7769
$\mathcal{M}(Y_{12})$	2.6829	2.6837	0.2684	0.8195	0.7757
$\mathcal{M}(Y_{13})$	2.7811	2.7828	0.3084	0.8185	0.7742

will explore the effects of the variance revision for adjusting the mean component quantities. The original adjustment is summarized in Table 6.2. The revised, two-stage, adjustment is summarized in Table 8.4. There are no major differences. The adjusted expectations are slightly smaller (larger) for components at the beginning (end) of the experiment. The prior variances for the mean components are unchanged, but the corresponding variance resolutions are slightly smaller, as we would expect from having made the data noisier.

Table 8.5 Variances for predicting future observables $Y_{1,F}, \ldots, Y_{13,F}$, with increased prior variation.

Quantity	Original			Two-stage		
	Prior	Adjusted	Resolution	Prior	Adjusted	Resolution
$Y_{1,F}$	0.1197	0.0939	0.2160	0.1434	0.1196	0.1661
$Y_{2,F}$	0.1348	0.1037	0.2309	0.1625	0.1334	0.1790
$Y_{3,F}$	0.1533	0.1144	0.2535	0.1849	0.1483	0.1982
$Y_{4,F}$	0.1752	0.1261	0.2802	0.2108	0.1642	0.2206
$Y_{5,F}$	0.2005	0.1387	0.3082	0.2400	0.1813	0.2444
$Y_{6,F}$	0.2292	0.1522	0.3358	0.2727	0.1995	0.2682
$Y_{7,F}$	0.2613	0.1667	0.3620	0.3087	0.2188	0.2912
$Y_{8,F}$	0.2968	0.1821	0.3865	0.3482	0.2392	0.3129
$Y_{9,F}$	0.3357	0.1984	0.4090	0.3910	0.2607	0.3333
$Y_{10,F}$	0.3780	0.2156	0.4296	0.4373	0.2833	0.3521
$Y_{11,F}$	0.4237	0.2338	0.4483	0.4869	0.3070	0.3695
$Y_{12,F}$	0.4728	0.2529	0.4652	0.5400	0.3318	0.3856
$Y_{13,F}$	0.5253	0.2729	0.4806	0.5964	0.3577	0.4003

We would expect to see more differences between the original and two-stage analyses for the predictive components, which we analysed in §6.17. The differences are summarized in Table 8.5. The prior variances are naturally rather larger under the model with scaled-up error variances. Furthermore, the data are noisier, and so the variance resolutions are not as high as for the original analysis. The consequence is higher adjusted variances for these components, and especially for those at the end of the series.

8.18 Further reading

Basic ideas concerning learning about a population variance and using this information to modify the adjustment of a population mean are given in Goldstein (1979). The two-stage adjustment procedure raises interesting coherence questions which are explored in Goldstein (1983a). Variance learning for the univariate, locally linear dynamic linear model is developed and applied in Wilkinson (1997). Issues arising in adjusting beliefs about variance matrices are covered in Wilkinson (1995) and Wilkinson and Goldstein (1996). The general form for the adjustment of matrix-like objects is treated in Goldstein and Wilkinson (2001).

9

Belief comparison

In this chapter we discuss the comparison of collections of belief specifications. It is rare, in complex situations, for any proposed belief specification to correspond exactly to the uncertainties in the system. The requirements of our specification will therefore depend on the context for the analysis. For example, we might construct different possible belief specifications from consideration of competing theoretical judgements about a physical system, so that accepting one set of judgements could lead to rejecting the other collections, though we might reserve judgement or end up rejecting each specification. Alternatively, we might have access to different specifications, each corresponding to the judgement of a different expert, where each expert might perform well for certain aspects of the specification reflecting their area of expertise, while there might be aspects for which each expert would perform poorly. As a further alternative, we often seek to simplify our specification in order to simplify the resulting uncertainty analysis. In such cases, we might be fairly confident that the more complex specification would be more accurate than simpler forms, but our concern would be that our approximations did not impact too much on our subsequent analysis of the system.

In each case, we are rarely concerned with a single summary measure for the comparison. Instead, it is more informative to decompose the full model comparison into a sequence of simple comparisons which express the main differences between the specifications. Such methods allow us to judge how successful we are likely to be in using historical data to distinguish between the competing specifications, by identifying whether there are areas of substantial disagreement between the specifications. We may also identify further observations that we expect to be useful in distinguishing between the specifications. The analysis will also reveal whether it is important to be able to choose between the specifications in terms of our ability to make reliable predictions for the future behaviour of the system. When we make such an informed comparison, we may find that certain aspects of the data support one specification, other aspects of the data support another

Bayes Linear Statistics: Theory and Methods M. Goldstein and D.A. Wooff

specification, and further aspects of the data may appear to contradict all suggested specifications.

We shall describe a simple and natural geometric representation for the comparison of uncertainty specifications which is well suited to informative graphical display. The approach is general, but is especially well suited to comparing alternative specifications for exchangeable multivariate samples, where the displays may be interpreted as multi-variable residual plots on carefully chosen axes of variation. Our basic tool for developing such informative comparisons is the belief transform for the comparison. Much of the general development may be illustrated by comparing variance matrices. Thus, we first present the various ideas in this context and discuss the geometric considerations underlying this comparison. We then extend the methodology to cover comparisons between competing expectation and variance specifications.

9.1 Comparing variance specifications

Suppose that X is a random p-dimensional vector with prior mean zero. We want to specify the variance matrix for X, and we have two alternative possibilities, the matrices U_1, U_2 say, each of full rank, for this choice. For example, U_1 might be a full, careful specification of the variance matrix, while U_2 might be some pragmatic simplification, which will greatly simplify certain analyses. Alternatively, U_1 and U_2 might correspond to competing physical hypotheses for the distribution of X.

In either case, there are various comparisons that we may wish to make between the effects of the different specifications on our beliefs about X. For each element $Y \in \langle X \rangle$, we define

$$V_{\frac{2}{1}}(Y) = \frac{\mathrm{Var}_2(Y)}{\mathrm{Var}_1(Y)}, \tag{9.1}$$

where $\mathrm{Var}_1(Y)$, $\mathrm{Var}_2(Y)$ are the variances that we assign for Y using U_1, U_2, respectively. A simple comparison that we may make is to evaluate the maximal value of $V_{\frac{2}{1}}$ over $\langle X \rangle$, which we denote by

$$\mathrm{DV}_{\frac{2}{1}}(X) = \max_{Y \in \langle X \rangle} V_{\frac{2}{1}}(Y). \tag{9.2}$$

Similarly, $\mathrm{DV}_{\frac{1}{2}}(X)$ is the maximal value of $V_{\frac{1}{2}}$, and so the minimal value of $1/V_{\frac{2}{1}}$ over $\langle X \rangle$. Thus, if $\mathrm{DV}_{\frac{2}{1}}(X)$ and $\mathrm{DV}_{\frac{1}{2}}(X)$ are both near 1, then there is little predictive difference for X between U_1 and U_2. If either $\mathrm{DV}_{\frac{2}{1}}(X)$ or $\mathrm{DV}_{\frac{1}{2}}(X)$ is large, then we may identify the corresponding elements of $\langle X \rangle$, which have been assigned very different variances by the two specifications. The forms of these quantities may give insights into the nature of the differences between the two variance specifications, while the observed values of these quantities may be informative in choosing between the two specifications.

To identify these quantities, and develop more detailed comparisons, we introduce the **belief transform matrix** which we define as

$$W_{\frac{2}{1}} = U_1^{-1} U_2. \tag{9.3}$$

This matrix has the following property. For any constant p-vectors, a, b, we have

$$\mathrm{Cov}_2(a^T X, b^T X) = \mathrm{Cov}_1(a^T X, (W_{\frac{2}{1}} b)^T X) \tag{9.4}$$

as

$$\mathrm{Cov}_2(a^T X, b^T X) = a^T U_2 b = a^T U_1 U_1^{-1} U_2 b = a^T U_1 W_{\frac{2}{1}} b$$
$$= \mathrm{Cov}_1(a^T X, (W_{\frac{2}{1}} b)^T X).$$

We may therefore analyse the differences between the two variance specifications by considering the eigenstructure of $W_{\frac{2}{1}}$. Note that $W_{\frac{2}{1}}$ has a full set of orthogonal eigenvectors, $b_1, \ldots, b_p$, corresponding to eigenvalues

$$\lambda_1 \geq \lambda_2 \geq \ldots \geq \lambda_p > 0.$$

This follows as

$$W_{\frac{2}{1}} b = \lambda b,$$

if and only if

$$U_1^{-1/2} U_2 U_1^{-1/2} c = \lambda c,$$

where $c = U_1^{1/2} b$. We construct the quantities $Z_i = b_i^T X$, $i = 1, \ldots, p$, norming each Z_i so that $\mathrm{Var}_1(Z_i) = 1$. We call $Z = (Z_1, \ \ldots, \ Z_p)$ the **canonical quantities** for the comparison between U_1 and U_2, with **canonical values** $(\lambda_1, \ \ldots, \ \lambda_p)$. As, for each i, $U_2 b_i = \lambda_i U_1 b_i$, we have $b_i^T U_1 b_j = b_i^T U_2 b_j = 0$ for $\lambda_i \neq \lambda_j$, so that we may construct the canonical quantities to be mutually uncorrelated under each of the variance specifications U_1 and U_2.

For each r and $Y \in \langle X \rangle$, we have from (9.4) that

$$\mathrm{Cov}_2(Y, Z_r) = \lambda_r \mathrm{Cov}_1(Y, Z_r). \tag{9.5}$$

Conversely, any mutually uncorrelated collection of quantities, Z, satisfying (9.5) must be the canonical quantities for the comparison. In particular, for each r, $\mathrm{Var}_2(Z_r) = \lambda_r$. We may write each $Y \in \langle X \rangle$ as

$$Y = \sum_{i=1}^{p} \alpha_i(Y) Z_i,$$

where $\alpha_i(Y) = \mathrm{Cov}_1(Y, Z_i)$. Therefore, for any pair of elements $\tilde{Y}, Y \in \langle X \rangle$, we have from (9.5) that

$$\mathrm{Cov}_1(\tilde{Y}, Y) = \sum_{i=1}^{p} \alpha_i(\tilde{Y}) \alpha_i(Y), \tag{9.6}$$

$$\mathrm{Cov}_2(\tilde{Y}, Y) = \sum_{i=1}^{p} \lambda_i \alpha_i(\tilde{Y}) \alpha_i(Y). \tag{9.7}$$

For any $Y \in \langle X \rangle$, we have

$$V_{\frac{2}{1}}(Y) = \frac{\sum_{i=1}^{p} \lambda_i \alpha_i^2(Y)}{\sum_{i=1}^{p} \alpha_i^2(Y)}. \quad (9.8)$$

Therefore, $V_{\frac{2}{1}}(Y)$ will be large (small) if and only if Y has large components corresponding to eigenvectors with large (small) eigenvalues. In particular,

$$\mathrm{DV}_{\frac{2}{1}}(X) = \lambda_1, \quad \mathrm{DV}_{\frac{1}{2}}(X) = 1/\lambda_p,$$

corresponding to the elements Z_1, Z_p. More generally,

$$V_{\frac{2}{1}}(Z_j) = \lambda_j = \max V_{\frac{2}{1}}(Y) : \mathrm{Cov}_i(Y, Z_k) = 0, \quad i = 1, 2,\ k < j.$$

We may make the following canonical decomposition of $\langle X \rangle$. We collect the eigenvectors of $W_{\frac{2}{1}}$ into three groups. We denote by $\langle X \rangle_+$ the linear space spanned by all the eigenvectors of $W_{\frac{2}{1}}$ which correspond to eigenvalues greater than 1. Each element $Y_+ \in \langle X \rangle_+$ has $V_{\frac{2}{1}}(Y_+) > 1$, from (9.8). Similarly, we construct $\langle X \rangle_*$ and $\langle X \rangle_-$ as the corresponding linear spaces for eigenvectors with eigenvalues equal to 1 and less than 1, respectively. Then, any $Y_* \in \langle X \rangle_*$ has $V_{\frac{2}{1}}(Y_*) = 1$, and any $Y_- \in \langle X \rangle_-$ has $V_{\frac{2}{1}}(Y_-) < 1$. $\langle X \rangle$ is the orthogonal sum of the three spaces, so that each $Y \in \langle X \rangle$ may be uniquely written as the sum of three uncorrelated quantities

$$Y = Y_+ + Y_* + Y_-,$$

where $Y_+ \in \langle X \rangle_+$, $Y_* \in \langle X \rangle_*$, and $Y_- \in \langle X \rangle_-$. We term $\langle X \rangle_+$, $\langle X \rangle_*$, and $\langle X \rangle_-$ the **expansion**, **unit**, and **contraction** spaces for the comparison of U_1 and U_2. It is often natural to compare the two variance specifications by separately analysing the three subspaces. In particular, as the variance specification over the unit space is the same for U_1, U_2, observations on quantities in this space cannot be used to distinguish between the two specifications. However, such observations may be particularly useful for assessing whether the two specifications are both inappropriate.

9.1.1 Rank-degenerate case

Consider now the more general case where either or both of the variance matrices U_1, U_2 contain a null space, possibly shared. In this case it is possible to obtain a full set of eigenvectors $Z_1, Z_2, \ldots, Z_p$, but for which variances under either or both of the specifications may be zero. For the general, possibly rank-deficient case, we need to solve the generalized eigenvalue problem

$$U_1 z = \lambda U_2 z,$$

as described in §11.11, where the generalized eigenvectors z provide the canonical quantities. This corresponds to solving for the belief transform $W_{\frac{2}{1}}$, where

$U_1 W_{\frac{2}{1}} = U_2$. Once we have obtained the generalized eigenvalues and eigenvectors, we normalize and arrange them as follows. For convenience, we shall label the two alternative specifications as $\mathbf{H}_1$ and $\mathbf{H}_2$.

- Let Z^{++} be the set of orthogonal eigenvectors such that every $Z_i \in Z^{++}$ has positive variance under both specifications. Conventionally, we normalize each such Z_i to have variance 1 under $\mathbf{H}_1$, and variance λ_i under $\mathbf{H}_2$.

- Let Z^{+0} be the set of orthogonal eigenvectors such that every $Z_i \in Z^{+0}$ has positive variance under $\mathbf{H}_1$ and zero variance under $\mathbf{H}_2$. Conventionally, we normalize each such Z_i to have variance 1 under $\mathbf{H}_1$.

- Let Z^{0+} be the set of orthogonal eigenvectors such that every $Z_i \in Z^{0+}$ has positive variance under $\mathbf{H}_2$ and zero variance under $\mathbf{H}_1$. Conventionally, we normalize each such Z_i to have variance 1 under $\mathbf{H}_2$.

- Let Z^{00} be the set of orthogonal eigenvectors such that every $Z_i \in Z^{00}$ has zero variance under both specifications.

The results (9.5)–(9.8) must now be amended as follows. For $Y \in \langle X \rangle$, Y has different representations under the two specifications. However, for every $Z_r \in Z^{++}$, we have

$$\mathrm{Cov}_2(Y, Z_r) = \lambda_r \mathrm{Cov}_1(Y, Z_r). \tag{9.9}$$

Excluding the constant term, under $\mathbf{H}_1$ we may write each $Y \in \langle X \rangle$ as

$$Y^{(1)} = \sum_{Z_i \in Z^{++}} \alpha_{+i} Z_i + \sum_{Z_i \in Z^{+0}} \alpha_{0i} Z_i, \tag{9.10}$$

where

$$\begin{aligned} \alpha_{+i} &= \mathrm{Cov}_1(Y, Z_i), \qquad Z_i \in Z^{++}, \\ \alpha_{0i} &= \mathrm{Cov}_1(Y, Z_i), \qquad Z_i \in Z^{+0}. \end{aligned}$$

Under $\mathbf{H}_2$ we may write each $Y \in \langle X \rangle$ as

$$Y^{(2)} = \sum_{Z_i \in Z^{++}} \beta_{+i} Z_i + \sum_{Z_i \in Z^{0+}} \beta_{0i} Z_i, \tag{9.11}$$

where

$$\begin{aligned} \beta_{+i} &= \frac{1}{\lambda_i} \mathrm{Cov}_2(Y, Z_i) = \alpha_{+i}, \qquad Z_i \in Z^{++}, \\ \beta_{0i} &= \mathrm{Cov}_2(Y, Z_i), \qquad Z_i \in Z^{0+}, \end{aligned}$$

so that the first components in (9.10) and (9.11) are common. It follows that, for any $Y \in \langle X \rangle$, we have

$$V_{\frac{2}{1}}(Y) = \frac{\sum \lambda_i \alpha_{+i}^2 + \sum \beta_{0i}^2}{\sum \alpha_{+i}^2 + \sum \alpha_{0i}^2}. \tag{9.12}$$

The summary statistics $\mathrm{DV}_{\frac{2}{1}}(\cdot)$ and $\mathrm{DV}_{\frac{1}{2}}(\cdot)$ are not finite in the degenerate rank case, so that we need to make the comparison separately over Z^{++}, Z^{+0}, Z^{0+}, and Z^{00}.

9.1.2 Comparison of orthogonal subspaces

Many variance comparisons preserve certain qualitative features of the variance structure. In particular, suppose that we have two vectors X_A, X_B of lengths p, q, respectively, for which

$$\mathrm{Cov}_1(X_A, X_B) = \mathrm{Cov}_2(X_A, X_B) = 0. \tag{9.13}$$

Let $U_A = (U_1, \ldots, U_p)$ be the canonical quantities for the comparison between specifications 1 and 2 for X_A, with canonical values $\lambda_{A1}, \ldots, \lambda_{Ap}$, and $W_B = (W_1, \ldots, W_q)$ and $\lambda_{B1}, \ldots, \lambda_{Bq}$ be the corresponding quantities for X_B. Let $X = (X_A, X_B)^T$. Any element $Y \in \langle X \rangle$ is of the form $a^T Y_A + b^T Y_B$, for vectors a, b and where $Y_A \in \langle X_A \rangle$, $Y_B \in \langle X_B \rangle$, and therefore, for all $Y \in \langle X \rangle$, we have

$$\begin{aligned} \lambda_{Ai}\mathrm{Cov}_1(Y, U_i) &= \mathrm{Cov}_2(Y, U_i), \quad \forall i, \\ \lambda_{Bj}\mathrm{Cov}_1(Y, W_j) &= \mathrm{Cov}_2(Y, W_j), \quad \forall j. \end{aligned} \tag{9.14}$$

Therefore, from (9.14), the collection (U_A, W_B) is a sequence of $p+q$ elements of $\langle X \rangle$, mutually uncorrelated under both belief specifications, satisfying the relation (9.5) for all $Y \in \langle X \rangle$, and so must be the canonical collection for the comparison, with canonical values the collection $(\lambda_{Ai}, \lambda_{Bj})$. Thus, for any two vectors satisfying (9.13), the canonical quantities and values for the combined vector are the corresponding quantities and values for the individual sub-vectors.

9.2 Example: variance comparison

To illustrate the canonical structure for the comparison of variance matrices, and its interpretation, we construct the following quantities. We consider two alternative specifications labelled $\mathbf{H}_1$ and $\mathbf{H}_2$. Suppose that there are quantities $X = X_1, X_2, \ldots, X_7$ with alternative variance matrices $\mathrm{Var}_{\mathbf{H}_1}(X) = U_1$ and $\mathrm{Var}_{\mathbf{H}_2}(X) = U_2$, where

$$U_1 = \begin{bmatrix} 2 & 1 & 5 & 1 & 0 & 0 & 0 \\ 1 & 4 & 6 & -3 & 0 & 0 & 0 \\ 5 & 6 & 16 & -1 & 0 & 0 & 0 \\ 1 & -3 & -1 & 4 & 0 & 0 & 0 \\ 0 & 0 & 0 & 0 & 6 & -4 & 7 \\ 0 & 0 & 0 & 0 & -4 & 6 & -5 \\ 0 & 0 & 0 & 0 & 7 & -5 & 9 \end{bmatrix}, \tag{9.15}$$

$$U_2 = \begin{bmatrix} 2 & 1 & 3 & 1 & 0 & 0 & 0 \\ 1 & 2 & 3 & -1 & 0 & 0 & 0 \\ 3 & 3 & 6 & 0 & 0 & 0 & 0 \\ 1 & -1 & 0 & 2 & 0 & 0 & 0 \\ 0 & 0 & 0 & 0 & 6 & 0 & 7 \\ 0 & 0 & 0 & 0 & 0 & 6 & -1 \\ 0 & 0 & 0 & 0 & 7 & -1 & 9 \end{bmatrix}. \tag{9.16}$$

We assume, without loss of generality, that each X_i has expectation zero under both specifications.

9.2.1 Canonical structure for the comparison

We now examine the canonical structure for comparing the two specifications. The comparison is summarized in Table 9.1, which lists the canonical quantities Z_i, and Table 9.2, which shows the variance $\mathrm{Var}_{\mathbf{H}_1}(Z_i)$ of each canonical quantity under $\mathbf{H}_1$, and the variance $\mathrm{Var}_{\mathbf{H}_2}(Z_i)$ of each canonical quantity under $\mathbf{H}_2$, namely the canonical variate λ_i. The signs of the canonical quantities are arbitrary; for this example we have chosen signs to accord with those obtained when we extend the example in §9.10.

The canonical quantities, Z_i, shown are orthogonal for both belief specifications, and are normed to have variance unity under $\mathbf{H}_1$, if possible, and otherwise to have variance unity in $\mathbf{H}_2$, if possible. The canonical quantity, Z_7, given for the common null space is normed so that its squared coefficients sum to unity. We could instead have switched the normalization so that the canonical quantities have variance unity under $\mathbf{H}_2$, in which case the corresponding canonical quantities are rescaled versions of the Z_i shown, and the corresponding eigenvalues are inverted. The canonical quantities are located to have expectation zero:

$$\mathrm{E}_{\mathbf{H}_1}(Z_i) = 0, \quad \forall i,$$

Table 9.1 Canonical quantities for the comparison of variance specifications.

$$\begin{aligned} Z_1 &= -\tfrac{1}{\sqrt{2}}(X_1 + X_2 - X_3) \\ Z_2 &= -\tfrac{1}{\sqrt{14}}(X_2 - X_4) \\ Z_3 &= \tfrac{1}{4}(3X_5 - X_6 - 2X_7) \\ Z_4 &= X_5 - X_7 \\ Z_5 &= \tfrac{1}{2}(X_5 + X_6) \\ Z_6 &= -\tfrac{1}{\sqrt{18}}(5X_1 + 4X_2 - 3X_3 + X_4) \\ Z_7 &= -\tfrac{1}{\sqrt{3}}(X_1 - X_2 - X_4) \end{aligned}$$

Table 9.2 Variances and covariances for the canonical quantities for the comparison of variance specifications. Whenever $\text{Var}_{\mathbf{H}_1}(Z_i) = 1$, $\text{Var}_{\mathbf{H}_2}(Z_i) = \lambda_i$.

			$Y = \sum X_i$	
i	$\text{Var}_{\mathbf{H}_2}(Z_i)$	$\text{Var}_{\mathbf{H}_1}(Z_i)$	$\text{Cov}_{\mathbf{H}_1}(Z_i, Y)$	$\text{Cov}_{\mathbf{H}_2}(Z_i, Y)$
1	0	1	$\sqrt{\frac{81}{2}}$	0
2	$\frac{3}{7}$	1	$-\sqrt{\frac{7}{2}}$	$-\sqrt{\frac{9}{14}}$
3	$\frac{1}{2}$	1	2	1
4	1	1	−2	−2
5	3	1	3	9
6	1	0	0	$-\sqrt{\frac{49}{2}}$
7	0	0	0	0

although this is irrelevant as far as the variance comparison is concerned. The main features of the comparison are as follows.

- The canonical structures for the two sets of quantities are orthogonal, and could have been obtained via separate comparisons of their respective variance matrices, as described in §9.1.2. The two orthogonal sets are $X_A = \{X_1, X_2, X_3, X_4\}$ and $X_B = \{X_5, X_6, X_7\}$, as is obvious from the direct specification (9.15). Equivalently, from Table 9.1, the collections of canonical quantities

$$Z_A = \{Z_1, Z_2, Z_6, Z_7\} \quad \text{and} \quad Z_B = \{Z_3, Z_4, Z_5\}$$

  contain respectively only elements from X_A and X_B.

- The quantity $Z_1 \propto X_1 + X_2 - X_3$ has been assigned a variance of unity under $\mathbf{H}_1$ and zero under $\mathbf{H}_2$, indicating a qualitative difference in specification.

- The quantities $Z_2 \propto X_2 - X_4$ and $Z_3 \propto 3X_5 + X_6 - 2X_7$ have been assigned a smaller variance under $\mathbf{H}_2$ than under $\mathbf{H}_1$.

- The quantity $Z_4 \propto X_5 - X_7$ has been assigned a variance of unity under both $\mathbf{H}_1$ and $\mathbf{H}_2$, so that the specifications match in this respect.

- The quantity $Z_5 \propto X_5 + X_6$ has been assigned a larger variance under $\mathbf{H}_2$ than under $\mathbf{H}_1$.

- The quantity $Z_6 \propto 5X_1 + 4X_2 - 3X_3 + X_4$ has been assigned a variance of unity under $\mathbf{H}_2$ and zero under $\mathbf{H}_1$, indicating a qualitative difference in specification.

- The quantity $Z_7 \propto X_1 - X_2 - X_4$ has been assigned a variance of zero under both $\mathbf{H}_1$ and $\mathbf{H}_2$, so that the specifications match in this respect.

For this example, we have eigenvalues equal to zero in one space and non-zero in the alternative space, so that $\mathrm{DV}_{\frac{2}{1}}(X)$ and $\mathrm{DV}_{\frac{1}{2}}(X)$ are infinitely large. For the non-degenerate part of the specification, the expansion space $\langle X\rangle_+$ is spanned by Z_5; the unit space $\langle X\rangle_*$ is spanned by Z_4; and the contraction space $\langle X\rangle_-$ is spanned by $\{Z_2, Z_3\}$. With respect to the separation for degeneracy, in the notation of §9.1.1 we have

$$\mathrm{Z}^{++} = \{Z_2, Z_3, Z_4, Z_5\}, \quad \mathrm{Z}^{+0} = \{Z_1\}, \quad \mathrm{Z}^{0+} = \{Z_6\}, \quad \mathrm{Z}^{00} = \{Z_7\}.$$

9.2.2 Consistency checks

We will consider more specifically the implication of observed data later in this chapter. However, a canonical quantity which has a variance of zero under one specification and a positive variance under another specification corresponds to a qualitative difference between specifications which can be directly checked when there are observed data available. This follows as data must be consistent with beliefs specified about them (§4.2, §12.3), so that a variance of zero for a canonical quantity implies that its observed value cannot differ from its expectation. For example, an observation of Z_6 not equal to zero would immediately contradict specification $\mathbf{H}_1$, but would be consistent with specification $\mathbf{H}_2$, whereas an observation of Z_1 not equal to its expectation under $\mathbf{H}_2$ would contradict specification $\mathbf{H}_2$, but would be consistent with specification $\mathbf{H}_1$. An observation of Z_7 not equal to its expectation under both $\mathbf{H}_1$ and $\mathbf{H}_2$ would contradict both specifications.

There may be occasions where some eigenvalues λ_i are calculated to be tiny or huge, perhaps as a result of rounding error. Because of machine computation limitations, it can be impossible to determine whether a small calculated eigenvalue represents a small but genuine positive variance, or a zero variance plus rounding error. In such cases the information that the canonical structure can provide, in relation to the distinction between qualitative and quantitative structure, is not so sharp. For some of these occasions it may be possible to re-examine the inputs to the prior specification process, in order to ascertain whether there is genuine intended structural degeneracy.

9.2.3 Comparisons for further constructed quantities

It is simple to compare specifications for any quantity $Y \in \langle X\rangle$, via (9.8) or, in the degenerate rank case, via (9.12). For example, suppose that our main interest is in the sum $Y = X_1 + \ldots + X_7$. The latter columns of Table 9.2 show the covariances of Y with each canonical quantity Z_i under both specifications. Observe that for the quantities $Z_i \in \mathrm{Z}^{++}$ with positive variances under both specifications we can verify (9.9), that

$$\mathrm{Cov}_{\mathbf{H}_1}(Y, Z_i) = \lambda_i \mathrm{Cov}_{\mathbf{H}_2}(Y, Z_i).$$

We may also calculate, via (9.12), that

$$V_{\frac{2}{1}}(Y) = \frac{[\frac{3}{7} \times \frac{7}{2} + \frac{1}{2} \times 4 + 1 \times 4 + 3 \times 9] + \frac{49}{2}}{[\frac{7}{2} + 4 + 4 + 9] + \frac{81}{2}} = \frac{34.5 + 24.5}{20.5 + 40.5}$$
$$= 0.9672.$$

Thus, overall the variance for Y is similar under the two specifications, but notice that the variance in the shared space Z^{++} is rather higher (34.5 compared to 20.5) under $\mathbf{H}_2$; and that this is counterbalanced by differences in variance (24.5, 40.5) for qualitatively different aspects of the specifications.

9.2.4 Construction of specifications

The specifications (9.15) were deliberately chosen to exhibit a number of features, and were constructed as follows. For both specifications we have $X_A = \{X_1, X_2, X_3, X_4\}$ uncorrelated with $X_B = \{X_5, X_6, X_7\}$. For X_B, we constructed three uncorrelated quantities F_1, F_2, F_3 with respective variances $1, 1, 1$ under $\mathbf{H}_1$ and $3, 1, 0.5$ under $\mathbf{H}_2$, and then constructed

$$X_5 = F_1 + F_2 + 2F_3, \quad X_6 = F_1 - F_2 - 2F_3, \quad X_7 = F_1 + 2F_2 + 2F_3.$$

For the set X_B, the specifications differ quantitatively (the underlying variances differ) but not qualitatively (the canonical structure is the same). For the set X_A we began with quantities X_1 and X_2 with specifications as shown in (9.15), and then assigned

$$\mathbf{H}_1: \quad X_3 = 2X_1 + X_2, \quad X_4 = X_1 - X_2,$$
$$\mathbf{H}_2: \quad X_3 = X_1 + X_2, \quad X_4 = X_1 - X_2,$$

so that each of the specifications for the set X_A is rank-deficient. Note that the alternative specifications for the set X_A differ quantitatively (variances differ) and qualitatively (the eigenstructures of the variance matrices U_1 and U_2 differ). The constructions are, as might be expected, clearly exhibited in the canonical structure summarized in Table 9.1.

9.3 Comparing many variance specifications

Suppose now that we want to compare k possible variance specifications, $U_1, \ldots, U_k$, each of full rank. For any pair U_i, U_j we construct the belief transform matrix $W_{\frac{j}{i}}$. The basic property of such collections of belief transform matrices is that they are multiplicative, namely, for any i, j, k,

$$W_{\frac{k}{i}} = W_{\frac{j}{i}} W_{\frac{k}{j}}, \tag{9.17}$$

as

$$W_{\frac{k}{i}} = U_i^{-1}U_k = U_i^{-1}U_jU_j^{-1}U_k = W_{\frac{j}{i}}W_{\frac{k}{j}}.$$

In particular, if Z is an eigenvector of both $W_{\frac{j}{i}}$ and $W_{\frac{k}{j}}$, with eigenvalues $\lambda_{\frac{j}{i}}, \lambda_{\frac{k}{j}}$ respectively, then Z is also an eigenvector of $W_{\frac{k}{i}}$, with eigenvalue $\lambda_{\frac{k}{i}} = \lambda_{\frac{j}{i}}\lambda_{\frac{k}{j}}$.

The relationship is the same in the rank-deficient case: if $U_iW_{\frac{j}{i}} = U_j$, then $U_iW_{\frac{j}{i}}W_{\frac{k}{j}} = U_jW_{\frac{k}{j}} = U_k$, so that $W_{\frac{j}{i}}W_{\frac{k}{j}} = W_{\frac{k}{i}}$.

In general, to compare the collection of variance matrices we need to compare all pairs. For example, if we decide that there are substantial differences between U_1 and U_2, and that there are substantial differences between U_2 and U_3, it does not follow, without further analysis, that there are substantial differences between U_1 and U_3. However, there are certain special cases where we can draw such conclusions, as follows.

We say that the sequence of variance matrices $U_1, \ldots, U_k$ is a **nested sequence** if we can choose a basis $Z_1, \ldots, Z_p$ for $\langle X \rangle$ with the property that each Z_j is an eigenvector of each $W_{\frac{i+1}{i}}$ with eigenvalue $\lambda_{\frac{i+1}{i}j}$ and either $\lambda_{\frac{i+1}{i}j} \geq 1, i = 1, \ldots, p-1$, or $\lambda_{\frac{i+1}{i}j} \leq 1, i = 1, \ldots, p-1$. In such cases, we can form three disjoint linear subspaces: $\langle X \rangle_{+(1\ldots k)}$ with basis vectors all those Z_j for which $\lambda_{\frac{i+1}{i}j} > 1$, for some i, $\langle X \rangle_{*(1\ldots k)}$ whose basis is those Z_j for which $\lambda_{\frac{i+1}{i}j} = 1$, for all i, and $\langle X \rangle_{-(1\ldots k)}$ whose basis is those Z_j for which $\lambda_{\frac{i+1}{i}j} < 1$, for some i.

We term $\langle X \rangle_{+(i,j)}, \langle X \rangle_{*(i,j)}, \langle X \rangle_{-(i,j)}$ the expansion, unit and contraction spaces for the comparison of U_i and U_j. It follows from (9.17) that if the sequence is nested, then,

(i) for each $i < j$, $\langle X \rangle_{+(i,j)} \subseteq \langle X \rangle_{+(1\ldots k)}$, $\langle X \rangle_{-(i,j)} \subseteq \langle X \rangle_{-(1\ldots k)}$;

(ii) for any $Y \in \langle X \rangle_{+(1\ldots k)}, \langle X \rangle_{-(1\ldots k)}, \langle X \rangle_{*(1\ldots k)}$, respectively, it is the case that the sequence $\text{Var}_1(Y), \ldots, \text{Var}_k(Y)$ is monotone non-decreasing, monotone non-increasing, and constant, respectively.

Therefore, a nested sequence of variance matrices may be compared by the sequence of ordered pairwise comparisons.

A special type of nested comparison is the orthogonal comparison. We say that the sequence of variance matrices $U_1, \ldots, U_k$ is **orthogonal** if the sequence is nested with the further property that each collection

$$\langle X \rangle_{+(i,i+1)}, \quad i = 1, \ldots, k-1,$$

and each collection

$$\langle X \rangle_{-(i,i+1)}, \quad i = 1, \ldots, k-1,$$

is uncorrelated with each other collection of either type. If $U_1, \ldots, U_k$ are orthogonal, then the comparison between U_1 and U_k can be further decomposed into the $k-1$ separate pairwise comparisons, over the collections of mutually orthogonal quantities.

9.4 Example: comparing some simple nested hypotheses

To illustrate the above ideas, we develop the following comparisons. We have a vector of observations $Y = (Y_1, \ldots, Y_n)$. We have three possible belief specifications for Y. In each case, $\mathrm{E}(Y) = 0$. The three variance specifications are as follows.

H$_1$: The sequence is uncorrelated, with constant variance v_1.

H$_2$: The sequence is exchangeable, with constant variance v_2, and positive covariance c_2 between each pair of elements of Y.

H$_3$: Suppose that each observation Y_i was made at time point t_i. The origin for time is chosen so that $\sum t_i = 0$. We consider that there might be a trend in the observations. Therefore, beliefs about Y are given by the regression model

$$Y_i = m + bt_i + e_i,$$

where $m, b, e_1, \ldots, e_n$ are uncorrelated random quantities with, for all i,

$$\mathrm{E}(m) = \mathrm{E}(b) = \mathrm{E}(e_i) = 0, \quad \mathrm{Var}(m) = v_m, \quad \mathrm{Var}(b) = v_b, \quad \mathrm{Var}(e_i) = v_e.$$

We first compare **H**$_1$ and **H**$_2$. Under **H**$_1$ and **H**$_2$ the variance matrices for Y are V_1 and V_2, given by

$$V_1 = v_1\mathbf{I}_n, \qquad V_2 = w_2\mathbf{I}_n + c_2\mathbf{J}_n,$$

where $w_2 = v_2 - c_2$ and $\mathbf{J}_n$ is the matrix each of whose entries is 1. The corresponding belief transform matrix is

$$T_{\frac{2}{1}} = V_1^{-1}V_2 = \frac{1}{v_1}(w_2\mathbf{I}_n + c_2\mathbf{J}_n).$$

The eigenvalues of $T_{\frac{2}{1}}$ are

$$\lambda_1 = \frac{w_2 + nc_2}{v_1}, \qquad \lambda_2 = \lambda_3 = \ldots = \lambda_n = \frac{v_2 - c_2}{v_1}.$$

The normalized eigenvector corresponding to λ_1 is

$$Z_1 = \sqrt{\frac{n}{v_1}}\bar{Y}_n,$$

where $\bar{Y}_n = (Y_1 + \ldots + Y_n)/n$. The remaining eigenvectors are any $(n - 1)$ mutually uncorrelated linear combinations $\sum_{i=1}^n a_i Y_i$ with $\sum_{i=1}^n a_i = 0$. For example, we might choose the cumulative residuals

$$R_j = Y_j - \frac{Y_1 + \ldots + Y_{j-1}}{j - 1},$$

normalized to variance 1 under $\mathbf{H}_1$; these are the contrasts deriving from the Helmert matrix: see Definition 11.58.

In particular, if $w_2 = v_1$, then there is a single informative direction, Z_1, for distinguishing between $\mathbf{H}_1$ and $\mathbf{H}_2$. Note that w_2 is the residual variance for each Y_i, namely the common value of $\text{Var}(\mathcal{R}_i(Y))$ in the exchangeability representation $Y_i = \mathcal{M}(Y) + \mathcal{R}_i(Y)$. Thus the case where $w_2 = v_1$ corresponds to the standard location shift problem, where under $\mathbf{H}_2$ an unknown fixed constant, with prior mean zero, has been added to each Y_i.

We now compare $\mathbf{H}_2$ and $\mathbf{H}_3$. Let us suppose that $v_1 = w_2 = v_e$, for simplicity. Under $\mathbf{H}_3$, the variance matrix for Y is

$$V_3 = v_e\mathbf{I}_n + v_m\mathbf{J}_n + w_b tt^T,$$

where $w_b = v_b + \text{Var}(b)$. Therefore, the matrix form for the transform $T_{\frac{3}{2}}$ is

$$T_{\frac{3}{2}} = V_2^{-1}V_3 = \frac{1}{v_e}(\mathbf{I}_n - k\mathbf{J}_n)(v_e\mathbf{I}_n + v_m\mathbf{J}_n + w_b tt^T),$$

where $k = u_m/(nu_m + v_e)$. Therefore, as $\mathbf{J}_n tt^T = 0$,

$$T_{\frac{3}{2}} = \mathbf{I}_n + d\mathbf{J}_n + rtt^T,$$

where $d = (v_m - u_m)/(nu_m + v_e)$ and $r = w_b/v_e$. There are two eigenvalues not equal to one, namely

$$\lambda_t = 1 + t_{(2)}r, \quad \lambda_d = 1 + nd,$$

corresponding to eigenvectors $\bar{Y}_t = (t_1Y_1 + \ldots + t_nY_n)/n$, and $\bar{Y}_n$, respectively.

Finally, we compare $\mathbf{H}_1$ with $\mathbf{H}_3$, which we shall form from the comparison between $\mathbf{H}_1$ and $\mathbf{H}_2$ and the comparison between $\mathbf{H}_2$ and $\mathbf{H}_3$, using the relation

$$T_{\frac{3}{1}} = T_{\frac{2}{1}}T_{\frac{3}{2}}.$$

As $T_{\frac{2}{1}}$ and $T_{\frac{3}{2}}$ have the same eigenvectors, $T_{\frac{3}{1}}$ also has the same eigenvectors. Therefore, the eigenvectors of $T_{\frac{3}{1}}$ with non-unit eigenvalues are $\bar{Y}_t, \bar{Y}_n$ with eigenvalues $1 + t_{(2)}r$ and $(1 + n(v_m/v_e))(1 + nd)$, respectively. Note that the comparisons are nested, so that differences between $\mathbf{H}_1$ and $\mathbf{H}_2$ or between $\mathbf{H}_2$ and $\mathbf{H}_3$ imply differences between $\mathbf{H}_1$ and $\mathbf{H}_3$. If, further, the uncertainty for the level is the same under both $\mathbf{H}_2$ and $\mathbf{H}_3$, i.e. if $v_m = u_m$, then $\lambda_d = 1$ and the comparisons are orthogonal. In this case, the comparison between $\mathbf{H}_1$ and $\mathbf{H}_3$ can be decomposed into the comparison based solely on $\bar{Y}_n$, which distinguishes between $\mathbf{H}_1$ and $\mathbf{H}_2$, and the comparison based solely on $\bar{Y}_t$, which distinguishes between $\mathbf{H}_2$ and $\mathbf{H}_3$.

In this comparison, $\bar{Y}_t, \bar{Y}_n$ form the expansion space, there is no contraction space and the unit space is spanned by the residuals

$$R_i = Y_i - \bar{Y}_n - t_i/t_{(2)}\bar{Y}_t,$$

which may be plotted for diagnostic purposes. The eigenvectors do not change if we allow the error variances in each hypothesis to differ. For example, if we reduce the variance of each e_i in $\mathbf{H}_3$ by some fixed amount, then the comparison is still nested but all of the eigenvectors within the residual space now have a common eigenvalue less than one, and so comprise the contraction space for the comparison.

Note that the qualitative features of the analysis do not depend on the precise numerical quantification for the various constants in the three belief specifications. Thus, the belief transform directs us qualitatively, through the form of the eigenvectors, to the general types of features which distinguish between the various hypotheses and then, through the magnitudes of the eigenvalues, suggests quantitative guidelines for assessing these differences.

9.5 General belief transforms

The comparison of variance matrices is a special case of a general geometric form for comparing inner products which also includes the adjusted belief transforms that we have discussed in earlier chapters. We now describe this general form.

9.5.1 General belief transforms

Suppose that we have a closed inner product space, $[B]$, of linear combinations of random quantities with inner product $(\cdot, \cdot)$. For example, the inner product might be covariance, having identified with zero all elements of $[B]$ with zero variance. Using this inner product, we define the norm $\|X\| = \sqrt{(X, X)}$, for example standard deviation. Suppose that we want to compare the inner product with a symmetric positive semi-definite sesquilinear (SPSDS) functional $\{\cdot, \cdot\}$ on $[B]$. An SPSDS functional $\{\cdot, \cdot\}$ satisfies all of the properties of an inner product over $[B]$, with the exception that we only require that $\{Y, Y\} \geq 0$ for each $Y \in [B]$; for example, $\{Y, Z\}$ might be the covariance between Y and Z for some variance specification of less than full rank.

We say that $\{\cdot, \cdot\}$ is bounded if the infimum of the values k for which

$$k\|Y\|\|Z\| \geq |\{Y, Z\}|, \quad \forall Y, Z \in [B]$$

is finite. This infimum is the norm of $\{\cdot, \cdot\}$. A necessary and sufficient condition for $\{\cdot, \cdot\}$ to be a bounded SPSDS functional on $[B]$ is that $\{\cdot, \cdot\}$ is of the form

$$\{X, Y\} = (X, SY), \tag{9.18}$$

where S is a bounded self-adjoint operator over $[B]$. The norm of S in (9.18) is equal to the norm of $\{\cdot, \cdot\}$. Therefore each choice $\{\cdot, \cdot\}$ may be uniquely identified with an operator S which we term the **(belief) transform for** $(\cdot, \cdot)$ **associated with** $\{\cdot, \cdot\}$.

We showed, in §9.1, how to construct the matrix representation for such a comparison, when comparing two variance specifications for a finite vector via the belief transform matrix, and property (9.4) is the finite-dimensional representation of the general relation (9.18) in this case. Provided that $[B]$ is finite-dimensional, we may always construct S by a similar approach to (9.3), as follows. We select a minimal basis $H = (H_1, \ldots, H_r)$ for $[B]$. We construct the matrix representation V for $(\cdot, \cdot)$ with respect to H, so that $V = (v_{ij})$, the $r \times r$ matrix whose (i, j)th entry is $v_{ij} = (H_i, H_j)$. We similarly construct the matrix representation U for $\{\cdot, \cdot\}$ with respect to H, namely $U = (u_{ij})$, where $u_{ij} = \{H_i, H_j\}$. Let

$$W = V^{-1}U.$$

W is the matrix representation of S with respect to the basis H. To see this, write any $X, Y \in \langle B \rangle$ in the coordinate system of H, i.e. if

$$X = \sum_i x_i H_i \quad \text{and} \quad Y = \sum y_i H_i,$$

then represent X, Y as

$$X = (x_1, \ldots, x_r) \quad \text{and} \quad Y = (y_1, \ldots, y_r).$$

We then have, as required,

$$\{X, Y\} = X^T U Y = X^T V V^{-1} U Y = X^T V W Y = (X, SY).$$

9.5.2 Properties of general belief transforms

If $[B]$ has finite dimension r, then we can choose as basis for $[B]$ the r orthogonal eigenvectors, $Z_1, \ldots, Z_r$, of S, each with norm 1. For any $X \in [B]$, we have $X = \sum_{i=1}^r (X, Z_i) Z_i$, so that

$$\{X, Y\} = \sum_{i=1}^r (X, Z_i)(Y, Z_i)\{Z_i, Z_i\} = \sum_{i=1}^r \lambda_i (X, Z_i)(Y, Z_i).$$

We may construct the ratio of quadratic forms as

$$V(X, Y) = \frac{\{X, Y\}}{(X, Y)} = \frac{\sum_{i=1}^r \lambda_i (X, Z_i)(Y, Z_i)}{\sum_{i=1}^r (X, Z_i)(Y, Z_i)}. \tag{9.19}$$

In particular, $V(X) = V(X, X)$ will be large (small) if and only if X has large components corresponding to eigenvectors with large (small) eigenvalues. In particular, the largest value of $V(X)$ over all elements of $[B]$ which are orthogonal to $Z_1, \ldots, Z_s$ is λ_{s+1} corresponding to Z_{s+1}, and the smallest value of $V(X)$ over all elements of $[B]$ orthogonal to $Z_s, \ldots, Z_r$ is λ_{s-1} corresponding to Z_{s-1}. The norm of S is λ_1.

If S has a full set of eigenvectors, then we may make the canonical decomposition of $[B]$ into the expansion, unit, and contraction subspaces $[B]_+$, $[B]_*$, and $[B]_-$, corresponding to the subsets of eigenvectors of S with eigenvalues greater than, equal to, or less than one, respectively. The value of $V(X)$ is greater than 1, equal to 1, and less than 1 for X in $[B]_+$, $[B]_*$, and $[B]_-$, respectively.

Now suppose that we have two SPSDS functionals $\{\cdot,\cdot\}_1, \{\cdot,\cdot\}_2$, that we want to compare with the inner product $(\cdot,\cdot)$. We may define the two transforms S_1, S_2 to compare $\{\cdot,\cdot\}_1, \{\cdot,\cdot\}_2$ with $(\cdot,\cdot)$. We may further define the transform $S_{\frac{2}{1}}$ for comparing $\{\cdot,\cdot\}_1$ with $\{\cdot,\cdot\}_2$, provided that $\{\cdot,\cdot\}_2$ is bounded with respect to $\{\cdot,\cdot\}_1$ and in particular that the zero elements of $\{\cdot,\cdot\}_1$ are contained in the zero elements of $\{\cdot,\cdot\}_2$. The belief transforms are multiplicative, namely

$$S_2 = S_1 S_{\frac{2}{1}}. \tag{9.20}$$

This follows as, for all $X, Y \in [B]$,

$$(X, S_2(Y)) = \{X, Y\}_2 = \{X, S_{\frac{2}{1}}(Y)\}_1 = (X, S_1(S_{\frac{2}{1}}(Y))).$$

Thus, an eigenvector of S_1 and $S_{\frac{2}{1}}$ with eigenvalues λ and μ respectively must also be an eigenvector of S_2, with eigenvalue $\lambda\mu$. As such, the transform $S_{\frac{2}{1}}$ allows us to relate the SPSDS functional $\{\cdot,\cdot\}_2$ to the original inner product $(\cdot,\cdot)$ through the intermediate functional $\{\cdot,\cdot\}_1$. This is important when we wish to create the transform S_2 incrementally or when we wish to separate the comparison between $\{\cdot,\cdot\}_2$ and $(\cdot,\cdot)$ into component parts.

Suppose that we wish to compare a sequence of SPSDS functionals

$$\{\cdot,\cdot\}_1, \ldots, \{\cdot,\cdot\}_k.$$

We term $[B]_{+(i,j)}$, $[B]_{*(i,j)}$, and $[B]_{-(i,j)}$ the expansion, unit, and contraction spaces for the comparison of $\{\cdot,\cdot\}_i$ and $\{\cdot,\cdot\}_j$.

As for the comparison of variance matrices, we say that the sequence of SPSDS functionals is a **nested sequence** if we can choose a basis $Z_1, \ldots, Z_p$ for $[B]$ with the property that each Z_j is an eigenvector of each $S_{\frac{i+1}{i}}$ with eigenvalue λ_{ij} and either

$$\lambda_{ij} \geq 1, \quad i = 1, \ldots, p-1, \quad \text{or} \quad \lambda_{ij} \leq 1, \quad i = 1, \ldots, p-1.$$

In such cases, we can form three disjoint linear subspaces: $\langle B\rangle_{+(1\ldots k)}$ with basis vectors all those Z_j for which $\lambda_{ij} > 1$, for some i, $\langle B\rangle_{*(1\ldots k)}$ whose basis is those Z_j for which $\lambda_{ij} = 1$, for all i, and $\langle B\rangle_{-(1\ldots k)}$ whose basis is those Z_j for which $\lambda_{ij} < 1$, for some i.

Property 9.1 (Properties of nested sequences) *If the sequence is nested, then, from* (9.20),

9.1.1: *for each* $i < j$,

$$[B]_{+(i,j)} \subseteq [B]_{+(1\ldots k)},\ [B]_{-(i,j)} \subseteq [B]_{-(1\ldots k)};$$

9.1.2: *for any* $Y \in [B]_{+(1\ldots k)}, \langle B\rangle_{-(1\ldots k)}, \langle B\rangle_{*(1\ldots k)}$, *respectively, the sequence*

$$\{Y, Y\}_1, \ldots, \{Y, Y\}_k$$

is monotone non-decreasing, monotone non-increasing, and constant, respectively.

Therefore, as for the comparison of variance matrices, a nested sequence of SPSDS functionals may be compared by the sequence of ordered pairwise comparisons.

We say that the sequence of SPSDS functionals is **orthogonal** if the sequence is nested with the further property that each collection

$$\langle B\rangle_{+(i,i+1)}, \quad i = 1, \ldots, k-1,$$

and each collection

$$\langle B\rangle_{-(i,i+1)}, \quad i = 1, \ldots, k-1,$$

is uncorrelated with each other collection of either type. If the functionals are orthogonal, then the comparison between functional 1 and k can be further decomposed into the $k-1$ separate pairwise comparisons, over the collections of mutually orthogonal quantities.

9.5.3 Adjusted belief transforms as general belief transforms

In previous chapters, we introduced the adjusted belief transform as a way to summarize the effect of a belief adjustment. If we are adjusting beliefs about a collection B by observation of a collection D, then the adjusted belief transform $\mathbb{S}_{B:D}$ is defined by the relation that, for any $X, Y \in [B]$,

$$\text{Cov}_D(X, Y) = \text{Cov}(X, \mathbb{S}_{B:D}(Y)).$$

By comparison with (9.18), $\mathbb{S}_{B:D}$ is the belief transform for the inner product $(X, Y) = \text{Cov}(X, Y)$, associated with the SPSDS form $(X, Y)_D = \text{Cov}_D(X, Y)$. Therefore, adjusted belief transforms are a special case of the general transforms described above. Further, we see that we may define such an adjustment transform however we choose to assess the adjusted covariance function.

Adjusted belief transforms inherit all of the properties of general belief transforms. In particular, suppose that we adjust B by the two collections D_1 and D_2. We have, from (5.15), the additive representation

$$\mathbb{T}_{B:(D_1 \cup D_2)} = \mathbb{T}_{B:D_1} + \mathbb{T}_{B:[D_2/D_1]}, \tag{9.21}$$

where $\mathbb{T}_{B:D} = \mathbb{I} - \mathbb{S}_{B:D}$ is the resolution transform. We have also, from (5.27), the multiplicative form corresponding to the general form (9.20),

$$\mathbb{S}_{B:(D_1 \cup D_2)} = \mathbb{S}_{B:D_1} \mathbb{S}_{B:D_2(D_1)}, \tag{9.22}$$

where $\mathbb{S}_{B:D_2(D_1)}$ is the relative adjusted belief transform for $[B/D_1]$ given D_2, satisfying

$$\text{Cov}_{D_1 \cup D_2}(X, Y) = \text{Cov}_{D_1}(X, \mathbb{S}_{B:D_2(D_1)}(Y)). \tag{9.23}$$

Note that combining (9.21) and (9.22) gives the relation

$$\mathbb{S}_{B:D_1} \mathbb{T}_{B:D_2(D_1)} = \mathbb{T}_{B:[D_2/D_1]} \tag{9.24}$$

where $\mathbb{T}_{B:D_2(D_1)} = \mathbb{I} - \mathbb{S}_{B:D_2(D_1)}$. Therefore, relations (9.21), (9.22), (9.24) allow us to move between the additive and multiplicative representations as required. In particular, we may recreate the adjusted belief transform for the combined effect of adjustment by $D_1 \cup D_2$ from the marginal effects of D_1 and the adjusted effect of D_2, which allows efficient local computation of such transforms, as we shall describe in the following chapter.

9.5.4 Example: adjustment of exchangeable structures

As a simple example of the representation of multiplicative forms, suppose that we have an exchangeable sample of p-dimensional vectors $Y_1, \ldots, Y_n$ from the representation $Y_i = \mathcal{M}(Y) + \mathcal{R}_i(Y)$. We adjust $[\mathcal{M}(Y)]$ by $\bar{Y}_n$, the sample average of the n vectors. We then take a further sample, $Y_{n+1}, \ldots, Y_{n+m}$. Let $\mathbb{S}_r$ be the adjustment transform based on a sample of size r, and let $\mathbb{S}_{s[r]}$ be the adjustment transform based on a sample of size s, given a prior sample of size r, as defined by (9.23). We have, from (9.22), for any m, n, that

$$\mathbb{S}_{n+m} = \mathbb{S}_n \mathbb{S}_{m[n]}. \tag{9.25}$$

As from Theorem 6.5, $\mathbb{S}_n$, $\mathbb{S}_{n+m}$ have the same eigenvectors, it follows from (9.25) that $\mathbb{S}_{m[n]}$ must have the same set of eigenvectors as each $\mathbb{S}_n$. Equivalently the eigenvectors of $\mathbb{T}_{m[n]} = \mathbb{I} - \mathbb{S}_{m[n]}$ are the same as the eigenvectors of $\mathbb{T}_n$, for each m, n. If, for $\mathbb{T}_n$, $\mathbb{T}_{m[n]}$, the eigenvalue corresponding to eigenvector W is $\lambda_{(n)}, \lambda_{m[n]}$, respectively, then from (9.25), we have

$$(1 - \lambda_{(m+n)}) = (1 - \lambda_{(n)})(1 - \lambda_{(m[n])}). \tag{9.26}$$

As, from (6.57),

$$\lambda_{(n)} = \frac{n\lambda_{(1)}}{(n-1)\lambda_{(1)} + 1}, \tag{9.27}$$

we obtain, from (9.26), that, for each n, the eigenvalue of the resolution transform $\mathbb{T}_{1[n]}$ corresponding to W is

$$\lambda_{1[n]} = \frac{\lambda_{(1)}}{n\lambda_{(1)} + 1}, \tag{9.28}$$

and the corresponding eigenvalue of $\mathbb{T}_{m[n]}$ is

$$\lambda_{m[n]} = \frac{m\lambda_{1[n]}}{(m-1)\lambda_{1[n]} + 1}. \tag{9.29}$$

Comparing (9.27) with (9.29), we see that the only difference between $\mathbb{T}_m$ and $\mathbb{T}_{m[n]}$ is that, for the latter, each eigenvalue $\lambda_{(1)}$ is replaced by the corresponding value $\lambda_{1[n]}$. As $\lambda_{1[n]}$ is a decreasing function of n, the proportion of the remaining variance which we may remove with a further sample of m, given a previous sample of n, is a decreasing function of n for each eigenvector. An interesting feature of this function is that the change in the resolved precision, i.e. the difference in the reciprocal of the eigenvalue, is the same in all components, namely, for each λ,

$$n = \frac{1}{\lambda_{1[n]}} - \frac{1}{\lambda}.$$

9.5.5 Example: analysing exchangeable regressions

In §6.14.3 we found the canonical resolutions for this example: $\lambda_{1(1)} = 0.6032$, $\lambda_{2(1)} = 0.2976$. Our actual sample size for this problem is $n = 3$, corresponding to canonical resolutions $\lambda_{1(3)} = 0.8202$ (6.83) and $\lambda_{2(3)} = 0.5597$ (6.84). These imply that we resolve at least 56% of the variation for every linear combination of the mean components. Suppose that for planning purposes we wish to find a further sample of size m to improve this percentage to 90% of original variation. We can simply follow the method based on simple exchangeability, giving (6.89). This requires an overall sample size of 22 to achieve 90% reduction in variation across the board, suggesting that we need to take $m = 22 - 3 = 19$.

Alternatively, we can use the methods of this section. Our interest is in the minimal canonical resolution, so we drop the subscript denoting that this is the second of the two canonical resolutions for the two-dimensional space of mean components of interest. We have $\lambda_{(1)} = 0.2976$ and so, using (9.28),

$$\lambda_{1[3]} = \frac{\lambda_{(1)}}{3\lambda_{(1)} + 1} = 0.1572.$$

This is thus the minimal canonical resolution for $\mathbb{T}_{1[3]}$, the resolution transform for a further adjustment by a sample of size $m = 1$ given an existing adjustment by a sample of size $n = 3$. Relatively, we wish to resolve at least 90% of original variation. Given our present resolution of 55.97% given $n = 3$, we thus need to resolve a further

$$\frac{0.9 - 0.5597}{1 - 0.5597} = 77.29\%$$

of remaining variation. Corollary 6.6 suggests that we need to take

$$m > \frac{0.7729}{1 - 0.7729} \frac{1 - 0.1572}{0.1572} = 18.24, \tag{9.30}$$

i.e. a sample of size $m = 19$. This, of course, the answer we obtained above. Note, in particular, the similarity with (6.89). This example emphasizes two features. First, once we have obtained the relative transform $\mathbb{T}_{1[\cdot]}$ and its canonical structure, we have available the full array of methods offered by exploiting exchangeability, as result (9.27) is functionally identical to result (9.29). Secondly, this relative transform is itself deduced straightforwardly from the original, so that we may move easily between initial and relative adjustments.

9.6 Comparing expectations and variances

We now extend the comparison of beliefs to quantities with different expectations and variances under each belief specification. Suppose that X is a random p-vector with expectation $\mathrm{E}_{\mathbf{H}_1}(X)$ and variance $\mathrm{Var}_{\mathbf{H}_1}(X)$ under specification $\mathbf{H}_1$, and expectation $\mathrm{E}_{\mathbf{H}_2}(X)$ and variance $\mathrm{Var}_{\mathbf{H}_2}(X)$ under specification $\mathbf{H}_2$. Initially we suppose that the variance matrices $\mathrm{Var}_{\mathbf{H}_1}(X)$ and $\mathrm{Var}_{\mathbf{H}_2}(X)$ are full rank. A natural comparison of the two expectation specifications is as follows.

Definition 9.2 *The* ***bearings for the belief comparison*** *are the elements* $\mathrm{G}_{\frac{2}{1}}, \mathrm{G}_{\frac{1}{2}} \in \langle X \rangle$ *with the properties*

$$\frac{[\mathrm{E}_{\mathbf{H}_2}(\mathrm{G}_{\frac{2}{1}}) - \mathrm{E}_{\mathbf{H}_1}(\mathrm{G}_{\frac{2}{1}})]^2}{\mathrm{Var}_{\mathbf{H}_1}(\mathrm{G}_{\frac{2}{1}})} = \max_{Y \in \langle X \rangle} \frac{[\mathrm{E}_{\mathbf{H}_2}(Y) - \mathrm{E}_{\mathbf{H}_1}(Y)]^2}{\mathrm{Var}_{\mathbf{H}_1}(Y)} = \mathrm{DE}_{\frac{2}{1}}(X), \tag{9.31}$$

$$\frac{[\mathrm{E}_{\mathbf{H}_1}(\mathrm{G}_{\frac{1}{2}}) - \mathrm{E}_{\mathbf{H}_2}(\mathrm{G}_{\frac{1}{2}})]^2}{\mathrm{Var}_{\mathbf{H}_2}(\mathrm{G}_{\frac{1}{2}})} = \max_{Y \in \langle X \rangle} \frac{[\mathrm{E}_{\mathbf{H}_1}(Y) - \mathrm{E}_{\mathbf{H}_2}(Y)]^2}{\mathrm{Var}_{\mathbf{H}_2}(Y)} = \mathrm{DE}_{\frac{1}{2}}(X). \tag{9.32}$$

Informally, if $\mathrm{DE}_{\frac{2}{1}}(X)$, $\mathrm{DE}_{\frac{1}{2}}(X)$ are both very large, then we would expect to be able to distinguish between the two belief specifications by observation of the vector X, and $\mathrm{G}_{\frac{2}{1}}, \mathrm{G}_{\frac{1}{2}}$ identify the aspects of X which are most informative for the comparison.

We construct the quantities $\mathrm{G}_{\frac{2}{1}}, \mathrm{G}_{\frac{1}{2}}$ which achieve these maximal changes as linear combinations of the canonical quantities for the comparison. Let $Z_1, \ldots, Z_p$ be the canonical quantities for the comparison between the variance matrices under $\mathbf{H}_1$ and $\mathbf{H}_2$, normed to variance one under $\mathbf{H}_1$, with corresponding canonical values $\lambda_1, \ldots, \lambda_p$. The signs of the canonical quantities are arbitrary. For future convenience we choose the sign of each Z_i so that $\mathrm{E}_{\mathbf{H}_2}(Z_i) \geq \mathrm{E}_{\mathbf{H}_1}(Z_i)$. We shall centre each Z_i so that $\mathrm{E}_{\mathbf{H}_1}(Z_i) = 0$. We address the algebraic implementation in §12.13.1. Similarly, let $\tilde{Z}_1, \ldots, \tilde{Z}_p$ be the canonical quantities for the reverse comparison between $\mathbf{H}_2$ and $\mathbf{H}_1$, normed to variance one under $\mathbf{H}_2$, centred so that each $\mathrm{E}_{\mathbf{H}_2}(\tilde{Z}_i) = 0$, with corresponding canonical values $\tilde{\lambda}_1, \ldots, \tilde{\lambda}_p$. The sign of each $\tilde{Z}_i$ is chosen so that $\mathrm{E}_{\mathbf{H}_1}(\tilde{Z}_i) \geq \mathrm{E}_{\mathbf{H}_2}(\tilde{Z}_i)$. The relationship between the canonical quantities for the two comparisons is that

$$\tilde{\lambda}_i = \frac{1}{\lambda_i},$$

$$\tilde{Z}_i = \frac{-1}{\sqrt{\lambda_i}}(Z_i - \mathrm{E}_{\mathbf{H}_2}(Z_i)).$$

Note that we have

$$\mathrm{E}_{\mathbf{H}_1}(\tilde{Z}_i) = \frac{1}{\sqrt{\lambda_i}} \mathrm{E}_{\mathbf{H}_2}(Z_i).$$

We construct $\mathrm{G}_{\frac{2}{1}}$ and $\mathrm{G}_{\frac{1}{2}}$ as follows:

$$\mathrm{G}_{\frac{2}{1}} = \sum_{i=1}^{p} [\mathrm{E}_{\mathbf{H}_2}(Z_i) - \mathrm{E}_{\mathbf{H}_1}(Z_i)] Z_i = \sum_{i=1}^{p} \mathrm{E}_{\mathbf{H}_2}(Z_i) Z_i \tag{9.33}$$

and

$$\mathrm{G}_{\frac{1}{2}} = \sum_{i=1}^{p} [\mathrm{E}_{\mathbf{H}_1}(\tilde{Z}_i) - \mathrm{E}_{\mathbf{H}_2}(\tilde{Z}_i)] \tilde{Z}_i = \sum_{i=1}^{p} \mathrm{E}_{\mathbf{H}_1}(\tilde{Z}_i) \tilde{Z}_i \tag{9.34}$$

$$= -\sum_{i=1}^{p} \frac{1}{\lambda_i} \mathrm{E}_{\mathbf{H}_2}(Z_i) Z_i + \sum_{i=1}^{p} \frac{1}{\lambda_i} \mathrm{E}_{\mathbf{H}_2}(Z_i)^2. \tag{9.35}$$

To show that $\mathrm{G}_{\frac{2}{1}}$ and $\mathrm{G}_{\frac{1}{2}}$, constructed as above, do indeed maximize the normed expectation differences, we first derive the following basic property of our construction. We can write any $Y \in \langle X \rangle$ as $Y = \sum_{i=1}^{p} a_i Z_i$, so that we have

$$\begin{aligned} \mathrm{Cov}_{\mathbf{H}_1}(\mathrm{G}_{\frac{2}{1}}, Y) &= \sum_{i=1}^{p} a_i \mathrm{Cov}_{\mathbf{H}_1}([\mathrm{E}_{\mathbf{H}_2}(Z_i) - \mathrm{E}_{\mathbf{H}_1}(Z_i)] Z_i, Z_i) \\ &= \sum_{i=1}^{p} a_i [\mathrm{E}_{\mathbf{H}_2}(Z_i) - \mathrm{E}_{\mathbf{H}_1}(Z_i)] = \mathrm{E}_{\mathbf{H}_2}(Y) - \mathrm{E}_{\mathbf{H}_1}(Y). \end{aligned}$$

We may make a similar calculation for $\mathrm{G}_{\frac{1}{2}}$, so that we have the twin properties that, for all $Y \in \langle X \rangle$,

$$\mathrm{E}_{\mathbf{H}_2}(Y) - \mathrm{E}_{\mathbf{H}_1}(Y) = \mathrm{Cov}_{\mathbf{H}_1}(Y, \mathrm{G}_{\frac{2}{1}}) \tag{9.36}$$

$$= -\mathrm{Cov}_{\mathbf{H}_2}(Y, \mathrm{G}_{\frac{1}{2}}). \tag{9.37}$$

We may therefore deduce that $\mathrm{G}_{\frac{2}{1}}$, as constructed by (9.33), is indeed the element of $\langle X \rangle$ with maximal normed difference in expectation, under $\mathbf{H}_1$ as defined by (9.31), as we have

$$\begin{aligned} \mathrm{DE}_{\frac{2}{1}}(X) &= \max_{Y \in \langle X \rangle} \frac{[\mathrm{E}_{\mathbf{H}_2}(Y) - \mathrm{E}_{\mathbf{H}_1}(Y)]^2}{\mathrm{Var}_{\mathbf{H}_1}(Y)} \\ &= \max_{Y \in \langle X \rangle} \frac{\mathrm{Cov}_{\mathbf{H}_1}(Y, \mathrm{G}_{\frac{2}{1}})^2}{\mathrm{Var}_{\mathbf{H}_1}(Y)} = \mathrm{Var}_{\mathbf{H}_1}(\mathrm{G}_{\frac{2}{1}}). \end{aligned}$$

Note, further, that

$$\begin{aligned}\mathrm{Var}_{\mathbf{H}_1}(\mathrm{G}_{\frac{2}{1}}) &= \sum_{i=1}^{p}[\mathrm{E}_{\mathbf{H}_2}(Z_i) - \mathrm{E}_{\mathbf{H}_1}(Z_i)]^2 \\ &= \mathrm{E}_{\mathbf{H}_2}(\mathrm{G}_{\frac{2}{1}}) - \mathrm{E}_{\mathbf{H}_1}(\mathrm{G}_{\frac{2}{1}}) \\ &= -\mathrm{Cov}_{\mathbf{H}_2}(\mathrm{G}_{\frac{2}{1}}, \mathrm{G}_{\frac{1}{2}}). \end{aligned} \tag{9.38}$$

Similarly, $\mathrm{G}_{\frac{1}{2}}$ is the element of $\langle X\rangle$ with maximal normed difference in expectation, under $\mathbf{H}_2$ and

$$\begin{aligned}\mathrm{DE}_{\frac{1}{2}}(X) = \mathrm{Var}_{\mathbf{H}_2}(\mathrm{G}_{\frac{1}{2}}) &= \sum_{i=1}^{p}\frac{1}{\lambda_i}[\mathrm{E}_{\mathbf{H}_2}(Z_i) - \mathrm{E}_{\mathbf{H}_1}(Z_i)]^2 \\ &= \mathrm{E}_{\mathbf{H}_1}(\mathrm{G}_{\frac{1}{2}}) - \mathrm{E}_{\mathbf{H}_2}(\mathrm{G}_{\frac{1}{2}}) && (9.39) \\ &= -\mathrm{Cov}_{\mathbf{H}_1}(\mathrm{G}_{\frac{2}{1}}, \mathrm{G}_{\frac{1}{2}}). && (9.40)\end{aligned}$$

We call $\mathrm{G}_{\frac{2}{1}}, \mathrm{G}_{\frac{1}{2}}$ the bearings for the belief comparison by analogy with the development described in §4.6, as, from (9.36), (9.37), for any Y in $\langle X\rangle$ which is uncorrelated with $\mathrm{G}_{\frac{2}{1}}$, under specification $\mathbf{H}_1$, there is no change in the expectation of Y in moving from specification $\mathbf{H}_1$ to specification $\mathbf{H}_2$, and similarly for $\mathrm{G}_{\frac{1}{2}}$.

Note the formal similarity of (9.33), (9.34) to (4.56), the construction of the bearing for a belief adjustment. This corresponds to the interpretation of the bearing for a belief adjustment as the bearing for the belief comparison between the prior and the adjusted version of beliefs over X.

9.7 Geometric interpretation

The bearings for the belief comparison arise naturally within the Hilbert space formalism as follows. Variance and covariance for full rank variance specifications are represented through the covariance inner products given by

$$(X_j, X_k)_{\mathbf{H}_i} = \mathrm{Cov}_{\mathbf{H}_i}(X_j, X_k), \qquad X_j, X_k \in \langle X\rangle,\ i = 1, 2. \tag{9.41}$$

We write $\mathcal{I}(X, \mathbf{H}_1)$ and $\mathcal{I}(X, \mathbf{H}_2)$ to denote these inner product spaces. Define $f_{\frac{2}{1}}$ by

$$f_{\frac{2}{1}}(Y) = \mathrm{E}_{\mathbf{H}_2}(Y) - \mathrm{E}_{\mathbf{H}_1}(Y), \tag{9.42}$$

so that $f_{\frac{2}{1}}$ is a linear functional on $\mathcal{I}(X, \mathbf{H}_1)$. Provided that $f_{\frac{2}{1}}$ is bounded, from the Riesz representation for linear functionals (see §4.10) we can construct a unique vector $\mathrm{G}_{\frac{2}{1}}$ in $\mathcal{I}(X, \mathbf{H}_1)$ satisfying

$$f_{\frac{2}{1}}(Y) = (Y, \mathrm{G}_{\frac{2}{1}})_{\mathbf{H}_1}, \quad \forall Y \in \langle X\rangle. \tag{9.43}$$

In particular, for the Bayes linear problems that we are mainly concerned with in the present work, $\langle X\rangle$ has a finite number of elements, so that the functional $f_{\frac{2}{1}}$ must be bounded, and $\mathrm{G}_{\frac{2}{1}}$ satisfying (9.43) must exist. Then, as $\mathrm{G}_{\frac{2}{1}} \in \mathcal{I}(X, \mathbf{H}_1)$, we must be able to write $\mathrm{G}_{\frac{2}{1}} = \sum_{i=1}^{p} a_i Z_i$, for scalars $a_1, \ldots, a_p$. We have, from this representation, that

$$\mathrm{Cov}_{\mathbf{H}_1}(\mathrm{G}_{\frac{2}{1}}, Z_j) = \sum_{i=1}^{p} a_i \mathrm{Cov}_{\mathbf{H}_1}(Z_i, Z_j) = a_j,$$

so that

$$a_j = \mathrm{Cov}_{\mathbf{H}_1}(\mathrm{G}_{\frac{2}{1}}, Z_j) = f_{\frac{2}{1}}(Z_j)$$

by (9.36). This gives the representation

$$\mathrm{G}_{\frac{2}{1}} = \sum_{i=1}^{p} f_{\frac{2}{1}}(Z_i) Z_i. \tag{9.44}$$

9.8 Residual forms for mean and variance comparisons

It is often helpful to transform X so as to separate out, as far as is possible, the differences in the specifications which arise from differences in the mean specification from those corresponding to differences in the variance specification. A simple construction for this purpose is as follows.

Let $f = (f_1, \ldots, f_p)$, where $f_i = f_{\frac{2}{1}}(Z_i) = \mathrm{E}_{\mathbf{H}_2}(Z_i) - \mathrm{E}_{\mathbf{H}_1}(Z_i)$. Construct $W = (W_1, \ldots, W_{p-1})$ as any vector of $(p-1)$ linear combinations $\sum_i c_{ji} Z_i$, where $c_1, \ldots, c_{p-1}$ are a set of $p-1$ mutually orthogonal vectors, $c_j = (c_{j1}, \ldots, c_{jp})$, each normed so that $c_j^T c_j = 1$ and chosen so that $c_j^T f = 0$ for each j. W is uncorrelated with $\mathrm{G}_{\frac{2}{1}}$ under $\mathbf{H}_1$, but not necessarily under $\mathbf{H}_2$. We have

$$\mathrm{E}_{\mathbf{H}_1}(W) = \mathrm{E}_{\mathbf{H}_2}(W),$$

as, for each i,

$$f_{\frac{2}{1}}(W_i) = \mathrm{E}_{\mathbf{H}_2}(W_i) - \mathrm{E}_{\mathbf{H}_1}(W_i) = c_i^T \mathrm{E}_{\mathbf{H}_2}(Z) - c_i^T \mathrm{E}_{\mathbf{H}_1}(Z) = c_i^T f = 0.$$

Therefore, instead of assessing the canonical quantities for the comparison between the two specifications over X, we may prefer to transform X to $(\mathrm{G}_{\frac{2}{1}}, W)$, so that all the differences in the mean specification are expressed in the first component, $\mathrm{G}_{\frac{2}{1}}$. We may then derive the canonical quantities for the comparison between $\mathbf{H}_1$ and $\mathbf{H}_2$ over W, and make a separate comparison for the scalar $\mathrm{G}_{\frac{2}{1}}$ comparing the mean and variance under each specification. Equivalently, if our preference is to make a comparison relative to $\mathbf{H}_2$, we can transform X using instead $\mathrm{G}_{\frac{1}{2}}$.

The above construction is a special case of a general approach for separating out mean and variance comparisons. We say that X is expressed in **(mean, residual)** or (M, R) form, $X^+ = (M, R_1, \ldots, R_{p-1})^T$, for the comparison of the two specifications, if X^+ is an orthogonal transformation of X for which $\mathrm{E}_{\mathbf{H}_1}(R_i) = \mathrm{E}_{\mathbf{H}_2}(R_i)$, $i = 1, 2, \ldots, p-1$. Thus, all the differences in the mean specification are expressed in the first component, M, of X^+. We call R a **residual form** for the comparison between the two specifications. For any such representation, we may derive the canonical quantities for the comparison over R, and separately compare the mean and variance specifications over M.

We may generate a wide class of such representations as follows. Choose a variance matrix, V, for X, Choose any collection of p quantities $J_i = j_i^T X$, which are mutually uncorrelated under V and scaled so that each has variance one under V. Then, we may proceed as in the construction of $(\mathrm{G}_{\frac{2}{1}}, W)$ and define

$$M_V = \sum_i f_{\frac{2}{1}}(J_i) J_i. \tag{9.45}$$

We term M_V the **mean direction with respect to** V, as, for any $Y \in \langle X \rangle$, we have

$$\mathrm{Cov}_V(M_V, Y) = \mathrm{E}_{\mathbf{H}_2}(Y) - \mathrm{E}_{\mathbf{H}_1}(Y) \tag{9.46}$$

so that

$$\mathrm{Cov}_V(Y, M_V) = 0 \Leftrightarrow \mathrm{E}_{\mathbf{H}_1}(Y) = \mathrm{E}_{\mathbf{H}_2}(Y). \tag{9.47}$$

Therefore, M_V is the element J which maximizes

$$D_V(J) = \frac{[\mathrm{E}_{\mathbf{H}_2}(H) - \mathrm{E}_{\mathbf{H}_1}(J)]^2}{\mathrm{Var}_V(J)} \tag{9.48}$$

over all $J \in \langle X \rangle$ with $\mathrm{Var}_V(J) > 0$, as, for each such J,

$$D_V(J) = \frac{[\mathrm{Cov}_V(M_V, J)]^2}{\mathrm{Var}_V(J)}. \tag{9.49}$$

Thus, suppose that we select any variance matrix V, construct the quantity M_V and construct another collection of $p-1$ random quantities $R_{V_1}, \ldots, R_{V_{p-1}}$ uncorrelated with M_V under V. Then, from (9.47), the vector

$$(M_V, R_{V_1}, \ldots, R_{V_{p-1}})$$

must be in (M, R) form for comparing the two mean and variance specifications. Informally, this representation may be viewed as the particular choice of (M, R) forms which gives the sharpest mean discrepancy evaluation according to variance specification V.

A particular class of choices of interest is the collection

$$V_\alpha = \alpha \mathrm{Var}_{\mathbf{H}_1}(X) + (1-\alpha)\mathrm{Var}_{\mathbf{H}_2}(X), \tag{9.50}$$

which allows us to norm according to a weighted combination of the two variance specifications. The vector $M_\alpha = M_{V_\alpha}$ may be constructed as

$$M_\alpha = \sum_i \frac{f_i}{\alpha + (1-\alpha)\lambda_i} Z_i, \tag{9.51}$$

with

$$\mathrm{E}_1(M_\alpha) = 0, \qquad \mathrm{Var}_1(M_\alpha) = \sum \left[\frac{f_i}{\alpha + (1-\alpha)\lambda_i}\right]^2,$$

$$\mathrm{E}_2(M_\alpha) = \sum \frac{f_i^2}{\alpha + (1-\alpha)\lambda_i}, \qquad \mathrm{Var}_2(M_\alpha) = \sum \lambda_i \left[\frac{f_i}{\alpha + (1-\alpha)\lambda_i}\right]^2.$$

Observe that the construction $(\mathrm{G}_{\frac{2}{1}}, W)$ corresponds to the choice $M_\alpha = M_1$, corresponding to $V = \mathrm{Var}_{\mathbf{H}_1}(X)$. Similarly, setting $\alpha = 0$ corresponds to the choice $V = \mathrm{Var}_{\mathbf{H}_2}(X)$. Varying α from one to zero therefore provides comparisons which are sharp for $\mathbf{H}_1$ and $\mathbf{H}_2$ separately, or for both $\mathbf{H}_1, \mathbf{H}_2$ jointly, and $\alpha = 0.5$ is a natural choice for such an intermediate display. Sometimes, the choices for M_V will be roughly similar for each α. However, if f_i decreases with i then it may be useful to compare different choices of α. As a somewhat extreme example, if $p = 3$ and

$$\lambda_1 = 100^2, \qquad \lambda_2 = 1, \qquad \lambda_3 = 0.01^2,$$

$$f_1 = 20, \qquad f_2 = 5, \qquad f_3 = 0.2,$$

then, to a good approximation, we would have

$$M_1 \approx X_1^*, \qquad M_{0.5} \approx X_2^*, \qquad M_0 \approx X_3^*,$$

reflecting the different types of information obtainable from each mean representation. Figure 9.1 plots the expectation comparison as a function of α: envelopes corresponding to $\mathrm{E}(M_\alpha) \pm 2\sqrt{\mathrm{Var}(M_\alpha)}$ under each specification are plotted, together with the observed value of M_α where, for illustration, we suppose that the canonical quantities are observed to be $x_1^* = 5$, $x_2^* = 0$, and $x_3^* = -5$. We see that, depending on our choice of standardization, the data plot suggests consistency with neither, one, or both specifications. For $\alpha = 1$, the data are consistent with $\mathbf{H}_2$ but not $\mathbf{H}_1$. For $\alpha = 0$, the data are consistent with neither $\mathbf{H}_2$ nor $\mathbf{H}_1$. For some intermediate values of α, the data tend to be more consistent with $\mathbf{H}_1$.

9.8.1 Rank-degenerate case

When the variance specifications exhibit some form of rank deficiency, it is necessary to separate the canonical quantities as described in §9.1.1 into the collections Z^{++}, Z^{+0}, Z^{0+}, and Z^{00}. For the collection Z^{++} we form $\mathrm{G}_{\frac{2}{1}}, \mathrm{G}_{\frac{1}{2}}$ and $\mathrm{DE}_{\frac{2}{1}}(Z^{++}), \mathrm{DE}_{\frac{1}{2}}(Z^{++})$ as above, limiting to the canonical quantities $Z_i \in Z^{++}$. These quantities then compare expectations under the two specifications $\mathbf{H}_1$, $\mathbf{H}_2$ constrained to the linear combinations with positive variance under both. For all remaining canonical quantities, these correspond to qualitative (and possibly also quantitative) differences in specification, and we may directly comment on the differences in expectation,

$$f_{\frac{2}{1}}(Z_i), \quad Z_i \in \{Z^{+0}, Z^{0+}, Z^{00}\}.$$

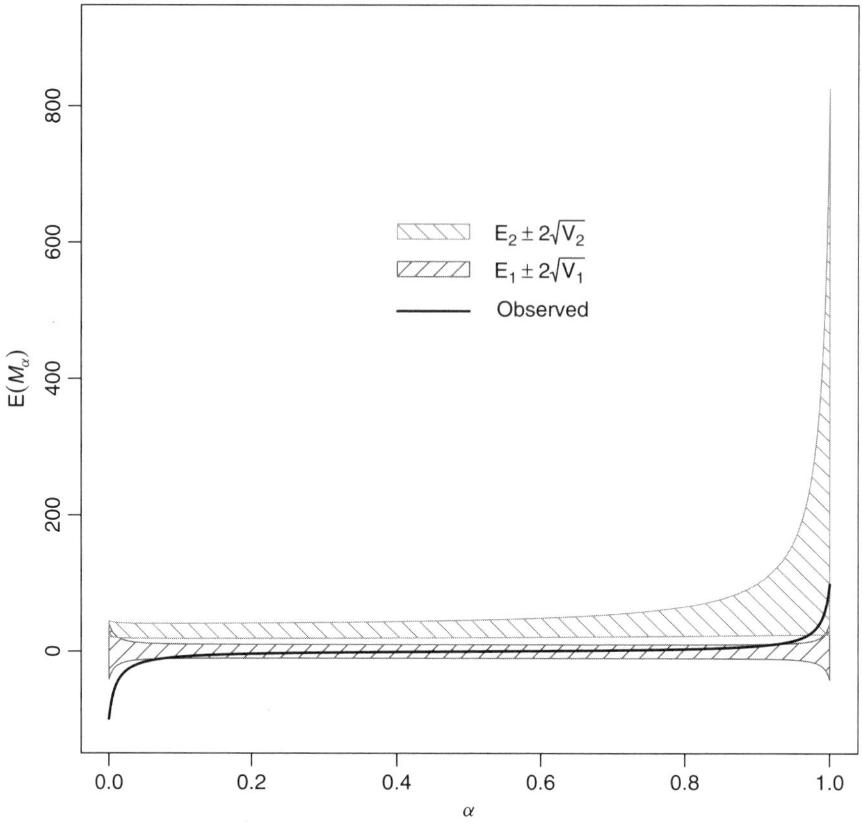

Figure 9.1 Expectation comparison: plotted versus α are two-standard-deviation boundaries for M_α under $\mathbf{H}_1$ and $\mathbf{H}_2$, and the observed value of M_α.

9.9 The observed comparison

Suppose that data become available, so that we observe $X = x$, and thereby observe each canonical quantity Z_i to be z_i. If one or more of the belief specifications have rank-degenerate variance matrices, then we can use the observed values of the eigenvectors corresponding to zero eigenvalues as consistency checks to help discriminate between the specifications. In particular, any canonical quantity which has variance zero under a specification should have an observed value equal to its expectation under that specification; otherwise the specification is contradicted.

For canonical quantities with positive variance under both specifications, define standardized canonical residuals under each specification as follows. For each such canonical quantity Z_i, let

$$R_{i1} = Z_i, \qquad \text{where } \mathrm{E}_{\mathbf{H}_1}(R_{i1}) = 0 \quad \text{and} \quad \mathrm{Var}_{\mathbf{H}_1}(R_{i1}) = 1, \tag{9.52}$$

$$R_{i2} = \frac{Z_i - \mathrm{E}_{\mathbf{H}_2}(Z_i)}{\sqrt{\lambda_i}}, \qquad \text{where } \mathrm{E}_{\mathbf{H}_2}(R_{i2}) = 0 \quad \text{and} \quad \mathrm{Var}_{\mathbf{H}_2}(R_{i2}) = 1. \tag{9.53}$$

The observed values

$$r_{i1} = z_i, \tag{9.54}$$

$$r_{i2} = \frac{z_i - \mathrm{E}_{\mathbf{H}_2}(Z_i)}{\sqrt{\lambda_i}} \tag{9.55}$$

of the standardized residuals provide evidence for the comparison of the two specifications, in that small values of r_{i1} are consistent with $\mathbf{H}_1$, and small values of r_{i2} are consistent with $\mathbf{H}_2$.

This comparison is more straightforward if it is based on the residual form as constructed in §9.8, as the canonical comparison of the residual form for W only involves comparison of quantities $r_{i1} = z_i$, against $r_{i2} = z_i/\sqrt{\lambda_i}$.

For variance comparison, we may also evaluate the observed values of the squared standardized canonical residuals, r_{i1}^2 and r_{i2}^2. Values of r_{i1}^2 close to unity are consistent with specification $\mathbf{H}_1$, whereas values of r_{i2}^2 close to unity are consistent with specification $\mathbf{H}_2$. For both expectation and variance comparisons, it might be that both specifications are consistent with the data, or that neither is, or that certain aspects of the data are consistent with one specification and other aspects consistent with the other specification.

9.9.1 Combined directions

For very large systems, it may be convenient to compare specifications over subspaces, combining directions with similar eigenvalues. In particular, if several eigenvalues for the comparison are the same, say

$$\lambda_{i+1} = \lambda_{i+2} = \ldots = \lambda_{i+m} = \lambda,$$

then the corresponding canonical quantities are not uniquely defined. Instead, a subspace of dimension m, $\langle \tilde{X} \rangle$, is identified, and any element of this subspace is also a canonical quantity with eigenvalue λ. In this case, there is no unique representation in the above form. Instead, a natural reduction of the data is given by combining the corresponding directions. We now discuss how we might combine the information from directions with similar eigenvalues.

Suppose that we wish to combine over the directions

$$\tilde{Z} = \{Z_{(1)}, \ldots, Z_{(m)}\}.$$

In principle, this might be any subset of directions, but we will restrict attention to those with positive variance under both specifications; there is no simple useful summary when combining directions which exhibit different kinds of degeneracy. To make the comparison, we may define summaries based on the mean variance under $\mathbf{H}_2$ and the means of the squared standardized residuals. We define

$$\lambda_{\tilde{Z}} = \frac{1}{m}\sum_{i=1}^{m} \lambda_{(i)}, \tag{9.56}$$

$$R^2_{\tilde{z}1} = \frac{1}{m}\sum_{i=1}^{m} R^2_{(i)1}, \tag{9.57}$$

$$R^2_{\tilde{z}2} = \frac{1}{m}\sum_{i=1}^{m} R^2_{(i)2}. \tag{9.58}$$

Summary residuals (9.57) and (9.58) have expectation unity under $\mathbf{H}_1$ and $\mathbf{H}_2$ respectively, and so we may interpret their observed values as above.

Alternatively, consider any scaled unit linear combination of the elements of the form

$$\tilde{Z}_u = \frac{1}{\sqrt{m}}\sum_{i=1}^{m} u_i Z_{(i)},$$

where each $|u_i| = 1$. We have

$$\text{Var}_1(\tilde{Z}_u) = 1, \quad \text{Var}_2(\tilde{Z}_u) = \lambda_{\tilde{Z}},$$

so that we can transform $\tilde{Z}$ into a new vector all of whose components inherit the eigenvalue $\lambda_{\tilde{Z}}$. When we observe the value of $\tilde{Z}$, then we may evaluate several elements of form $u^T_{(i)}\tilde{Z}$, selected so that each $u^T_{(i)}u_{(j)} = 0$, $i \neq j$.

9.10 Example: mean and variance comparison

Let us return to the example of §9.2. Suppose that there are alternative expectation specifications $\text{E}_{\mathbf{H}_1}(X_5) = -1$ and $\text{E}_{\mathbf{H}_2}(X_5) = 1$, with all other quantities having the same, unchanged, expectation of zero. In relation to the underlying quantities of §9.2.4, the alternative specifications derive from assigning expectations $-0.5, 1, -0.75$ under $\mathbf{H}_1$ and $0.5, -1, 0.75$ under $\mathbf{H}_2$ for F_1, F_2, F_3, respectively.

The canonical directions remain essentially as in Table 9.1, but with one difference. We conventionally locate each direction to have expectation zero under $\mathbf{H}_1$, so that offsets have been introduced for Z_3, Z_4, Z_5. Note that we now insist on choosing the signs of the canonical directions so that $\text{E}_{\mathbf{H}_2}(Z_i) \geq \text{E}_{\mathbf{H}_1}(Z_i) = 0$; this motivated the choice of sign made in displaying Table 9.1. The modified directions are shown in Table 9.3. To compare the specifications we concentrate on the expectations under $\mathbf{H}_2$, relative to the alternative variance specifications. These are summarized in Table 9.4, together with the canonical variances. We can compare the canonical differences in expectation directly, as follows.

- There are four directions (Z_1, Z_2, Z_6, Z_7) where the expectations match. Any $Y \in \langle X \rangle$ which can be represented as a linear combination of these quantities will similarly have matching expectations under the two specifications.

- Expectations match for the qualitatively different structure, as summarized by the quantities Z_1, Z_6, and for the direction Z_7 with zero variance under both specifications. If an observed value of Z_1 or Z_6 becomes available, we may immediately be able to distinguish between $\mathbf{H}_1$ and $\mathbf{H}_2$ as an observation for a linear combination with variance zero cannot differ from its expectation.

Table 9.3 Canonical quantities for the comparison of variance specifications.

$$
\begin{aligned}
Z_1 &= -\frac{1}{\sqrt{2}}(X_1 + X_2 - X_3) \\
Z_2 &= -\frac{1}{\sqrt{14}}(X_2 - X_4) \\
Z_3 &= \frac{1}{4}(3X_5 - X_6 - 2X_7 + 3) \\
Z_4 &= X_5 - X_7 + 1 \\
Z_5 &= \frac{1}{2}(X_5 + X_6 + 1) \\
Z_6 &= -\frac{1}{\sqrt{18}}(5X_1 + 4X_2 - 3X_3 + X_4) \\
Z_7 &= -\frac{1}{\sqrt{3}}(X_1 - X_2 - X_4)
\end{aligned}
$$

Table 9.4 Canonical comparison of expectation specifications: variances, expectations, and residuals for each canonical direction.

i	$\text{Var}_{\mathbf{H}_2}(Z_i)$	$\text{Var}_{\mathbf{H}_1}(Z_i)$	$\text{E}_{\mathbf{H}_2}(Z_i)$	$\text{E}_{\mathbf{H}_1}(Z_i)$	$\lvert r_{i1}\rvert$	$\lvert r_{i2}\rvert$
1	0	1	0	0	0	0
2	$\frac{3}{7}$	1	0	0	0.2673	0.4082
3	$\frac{1}{2}$	1	$\frac{3}{2}$	0	4.7500	4.5962
4	1	1	2	0	0	2.0000
5	3	1	1	0	4.5000	2.0207
6	1	0	0	0	2.1213	2.1213
7	0	0	0	0	0	0

- There remain three quantities Z_3, Z_4, Z_5, which have different expectations under $\mathbf{H}_2$, of which the largest difference in expectation is in direction Z_4, which turns out to have the same variance under both specifications.

Over the space $\langle Z^{++}\rangle$, i.e. Z_2, Z_3, Z_4, Z_5, we construct our summary quantities as follows. The bearing for the comparison, norming under $\mathbf{H}_1$ (9.33), is

$$\begin{aligned} \mathrm{G}_{\frac{2}{1}} &= \sum_{Z_i \in Z^{++}} \mathrm{E}_{\mathbf{H}_2}(Z_i) Z_i = \sum_{i=3}^{5} \mathrm{E}_{\mathbf{H}_2}(Z_i) Z_i \\ &= 0 + \frac{3}{2} \times \frac{1}{4}(3X_5 - X_6 - 2X_7 + 3) + \ldots \\ &= 3.625X_5 + 0.125X_6 - 2.75X_7 + 3.625, \end{aligned}$$

corresponding to maximum squared change in expectation

$$\begin{aligned} \mathrm{DE}_{\frac{2}{1}}(Z^{++}) &= \sum_{Z_i \in Z^{++}} \mathrm{E}_{\mathbf{H}_2}(Z_i)^2 \\ &= 0 + 1.5^2 + 2^2 + 1^2 = 7.25 \\ &= \mathrm{E}_{\mathbf{H}_2}(\mathrm{G}_{\frac{2}{1}}). \end{aligned}$$

The summaries, norming relative to variance under $\mathbf{H}_2$, are

$$\begin{aligned} \mathrm{G}_{\frac{1}{2}} &= -\sum_{Z_i \in Z^{++}} \frac{1}{\lambda_i} \mathrm{E}_{\mathbf{H}_2}(Z_i) Z_i + \sum_{Z_i \in Z^{++}} \frac{1}{\lambda_i} \mathrm{E}_{\mathbf{H}_2}(Z_i)^2 \\ &= -4.4167X_5 + 0.5833X_6 + 3.5X_7 + 4.4167. \end{aligned}$$

$$\begin{aligned} \mathrm{DE}_{\frac{1}{2}}(Z^{++}) &= \sum_{Z_i \in Z^{++}} \frac{1}{\lambda_i} \mathrm{E}_{\mathbf{H}_2}(Z_i)^2 = 8.8333, \\ &= \mathrm{E}_{\mathbf{H}_1}(\mathrm{G}_{\frac{1}{2}}). \end{aligned}$$

Thus, the maximal squared difference in expectation, relative to variation in $\mathbf{H}_1$ is 7.25 (2.69 standard deviations), or about 8.83 (2.97 standard deviations) relative to variation in $\mathbf{H}_2$.

It might come as a surprise that the difference in expectation is of such magnitude. The only difference in expectation over the quantities $X_1, \ldots, X_7$ is for $\mathrm{E}(X_5) = \pm 1$, where $\mathrm{Var}(X_5) = 6$ under both specifications, so that the maximum squared change in expectation is only $\frac{4}{6}$ standard deviation for X_5 alone. This can be explained as follows. X_5 is correlated with the other quantities, and so changes in expectation for X_5 do have hidden implications for the remaining quantities, whether or not their expectations have changed. In other words, it is not sufficient simply to assess those quantities which have changes in expectation; we need to include any other quantities correlated with them.

Indeed, we can show that $G_{\frac{2}{1}}$ is proportional to the adjusted version of X_5 given X_6, X_7, namely $\mathbb{A}_{(X_6,X_7)}(X_5)$, where the adjustment is with respect to variation under $\mathbf{H}_1$. This follows because the addition of linear combinations of X_6, X_7 to X_5 cannot change the difference in expectation between $\mathbf{H}_1$ and $\mathbf{H}_2$, so that we want to construct the combination with smallest variance: this combination is $\mathbb{A}_{(X_6,X_7)}(X_5)$. The summary $\mathrm{DE}_{\frac{2}{1}}(Z^{++})$ is then the ratio of the squared differences between the two expectations, normed by the adjusted variance $\mathrm{Var}_{X_6\cup X_7}(X_5)$ calculated under $\mathbf{H}_1$. Given this perspective, we can see clearly that it is a sharper comparison to examine the adjusted version of X_5, rather than the raw version of X_5, in distinguishing between $\mathbf{H}_1$ and $\mathbf{H}_2$. Our above construction extends this idea to a general vector of different expectations. Similar arguments hold whenever we can partition the collection into $X = (X_A, X_B)$ where X_B has the same expectation under both specifications, as then the minimum relevant variance matrix is $\mathrm{Var}_{X_B}(X_A)$ with respect to the adjusted version $\mathbb{A}_{X_B}(X_A)$.

For a further quantity constructed from $\langle Z^{++}\rangle$, we may deduce its change in expectation directly from the bearing $G_{\frac{2}{1}}$ using (9.36). For example, for $Y = X_5 + X_6 + X_7$ it is straightforward to show that

$$\mathrm{Cov}_{\mathbf{H}_1}(Y, G_{\frac{2}{1}}) = 2 = \mathrm{E}_{\mathbf{H}_2}(Y) - \mathrm{E}_{\mathbf{H}_1}(Y).$$

9.10.1 The observed comparison

Suppose that the seven quantities are observed to be

$$x = \begin{bmatrix} 3 & 1 & 4 & 2 & 13 & -5 & 14 \end{bmatrix}^T.$$

The absolute values of the residuals under each specification for each canonical quantity are shown in the final columns of Table 9.4. Two of the residuals under $\mathbf{H}_1$ and one of the residuals under $\mathbf{H}_2$ are particularly large. To assess the data in relation to the direction for which expectations differ, the observed value of $G_{\frac{2}{1}}$ is

$$3.625X_5 + 0.125X_6 - 2.75X_7 + 3.625 = 11.625,$$

with standardized values

$$\mathbf{H}_1 : \frac{11.625 - \mathrm{E}_{\mathbf{H}_1}(G_{\frac{2}{1}})}{\sqrt{\mathrm{Var}_{\mathbf{H}_1}(G_{\frac{2}{1}})}} = \frac{11.625}{\sqrt{7.25}} = 4.31,$$

$$\mathbf{H}_2 : \frac{11.625 - \mathrm{E}_{\mathbf{H}_2}(G_{\frac{2}{1}})}{\sqrt{\mathrm{Var}_{\mathbf{H}_2}(G_{\frac{2}{1}})}} = \frac{11.625 - 7.25}{\sqrt{8.125}} = 1.53,$$

where we have needed to calculate

$$\mathrm{Var}_{\mathbf{H}_2}(G_{\frac{2}{1}}) = \sum_{Z_i \in Z^{++}} \lambda_i \mathrm{E}_{\mathbf{H}_2}(Z_i)^2 = 0 + 0.5 \times 1.5^2 + 1 \times 2^2 + 3 \times 1^2 = 8.125.$$

This observation is thus consistent with $\mathbf{H}_2$ (the residual under $\mathbf{H}_2$ is small and its square is not too far from one) but appears abnormal under $\mathbf{H}_1$ (the residual is far from zero and its square is much larger than one).

With regard to changing the normalization from $\mathbf{H}_1$ to $\mathbf{H}_2$ (9.50), Figure 9.2 plots the expectation comparison as a function of α. There is some overlap between the two specifications. However, we see that, regardless of our choice of standardization, the data plot is far more consistent with specification $\mathbf{H}_2$ than $\mathbf{H}_1$. Indeed, the data are consistent with $\mathbf{H}_2$, albeit a little less so for α near zero. The choice $\alpha = 1$ corresponds to the normalization used for the comparison made in the previous section.

9.11 Graphical comparison of specifications

Much of the Bayes linear methodology that we have described concerns summary measures which give interpretative and diagnostic insights into the specification

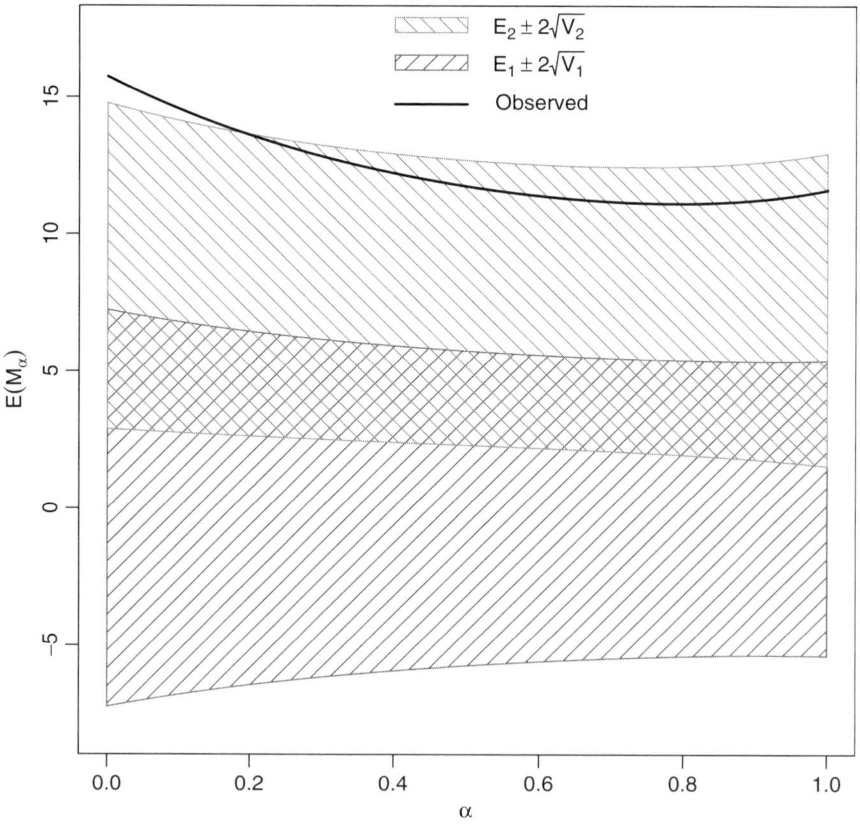

Figure 9.2 Expectation comparison plot.

and analysis of beliefs. Many of these measures are particularly suited to graphical display. In this section, we introduce the notion of such representations by suggesting a visual approach to the comparison of competing prior specifications through the graphical display of the eigenstructure of the corresponding belief transform. There are many different ways to develop such graphical representations. We shall restrict attention here to displays which correspond directly to the various geometric features that we have described.

In particular, we are concerned with constructing simple graphical displays for potentially complicated specifications. Simple visual displays are particularly important when we are comparing two competing specifications over a complex interconnected system. Any individual segment of the problem may be scrutinized by a variety of stringent comparisons. However, if our primary interest is in getting some qualitative feeling for how well the two specifications are doing over all of the various aspects of the system, then we need pictures which are designed to give this kind of overall visual summary. We now describe one such display.

9.11.1 Belief comparison diagram

We now describe a graphical method for comparing, in detail, features of the two specifications. We suppose that the specifications have p canonical quantities, some of which might correspond to rank degeneracy under one or both specifications. For each specification we consider a semicircle divided into p sectors of equal size. Each sector corresponds to one canonical quantity. Under specification $\mathbf{H}_2$, we take the sector corresponding to Z_1 to be the sector starting at 0°, proceeding anticlockwise with the remaining canonical directions, $Z_2, Z_3, \ldots, Z_p$, with the sector for Z_p ending at 180°. Under specification $\mathbf{H}_1$, we take the sector corresponding to Z_1 to be the sector starting at 180°, proceeding anticlockwise with the remaining canonical directions, $Z_2, Z_3, \ldots, Z_p$, with the sector for Z_p ending at 360°. We place the semicircles within a circle so that diagonally opposite sectors correspond to the same canonical quantity. Note that the upper semicircle corresponds to $\mathbf{H}_2$ and the lower semicircle corresponds to $\mathbf{H}_1$, reflecting our convention that our basic standardization is with variances under $\mathbf{H}_1$ in the denominator.

Now consider the two sectors corresponding to Z_i. In each of the two sectors we draw an inner arc, where the radius ρ_i of the inner arc is chosen according to the value of $\mathrm{Var}_{\mathbf{H}_1}(Z_i)$ and $\mathrm{Var}_{\mathbf{H}_2}(Z_i) = \lambda_i$. There are six cases to consider, as shown in Table 9.5. We also shade the outer sector with dark or light shading as shown in Table 9.5. The inner arc radius summarizes the difference between the two variance specifications for a canonical quantity, with small radii indicating large differences. As for the shading,

- dark shading for a lower/upper pair of sectors indicates that the canonical quantity has a higher variance under $\mathbf{H}_1$;
- light shading for a lower/upper pair of sectors indicates that the canonical quantity has higher variance under $\mathbf{H}_2$;

Table 9.5 Inner arc radius and shading choices for the pair of sectors corresponding to Z_i for the belief comparison diagram.

			Radius	Outer sector shading	
Case	$\mathrm{Var}_{\mathbf{H}_1}(Z_i)$	$\mathrm{Var}_{\mathbf{H}_2}(Z_i)$	ρ_i	Upper	Lower
1	1	0	0	none	dark
2	0	1	0	light	none
3	0	0	0	none	none
4	1	$\lambda_i > 1$	$1/\sqrt{\lambda_i}$	light	light
5	1	$\lambda_i = 1$	1	none	none
6	1	$\lambda_i < 1$	$\sqrt{\lambda_i}$	dark	dark

- no shading for a lower/upper pair of sectors indicates that the specifications match under $\mathbf{H}_1$ and $\mathbf{H}_2$, including the case where the canonical quantity has zero variance under both specifications;

- no shading for the upper sector and full dark shading for the lower sector indicates the case where a canonical quantity has variance zero under $\mathbf{H}_2$ but not under $\mathbf{H}_1$;

- no shading for the lower sector and full light shading for the lower sector indicates the case where a canonical quantity has variance zero under $\mathbf{H}_1$ but not under $\mathbf{H}_2$.

Generally, a large amount of dark shading within the circle indicates that specification $\mathbf{H}_1$ gives higher variance than specification $\mathbf{H}_2$, for many linear combinations of $\langle X \rangle$, whilst a large amount of light shading indicates that specification $\mathbf{H}_2$ gives higher variance than specification $\mathbf{H}_1$. We shall call diagrams constructed in this way **canonical wheels**.

9.11.1.1 Example

Consider Figure 9.3(a). This shows the belief comparison for the example discussed in §9.2, with canonical quantities summarized in Table 9.1. There are seven canonical quantities and so seven sectors in each semicircle. There are two sector pairs with no shading corresponding to Z_4 and Z_7 which have variances matching under each specification. There are two sector pairs with a degree of dark shading for the canonical quantities Z_2, Z_3 which have a higher variance under $\mathbf{H}_1$. There is one sector pair with a degree of light shading for the canonical quantity Z_5 which has a higher variance under $\mathbf{H}_2$. There are two further sector pairs for Z_1 and Z_6. Z_1 has positive variance under $\mathbf{H}_1$ and zero variance under $\mathbf{H}_2$, whilst the reverse is true for Z_6. Consequently, these have no shading in one sector and full shading in the paired sector. Overall, there appears slightly more dark than light shading, indicating slightly higher variances under $\mathbf{H}_1$.

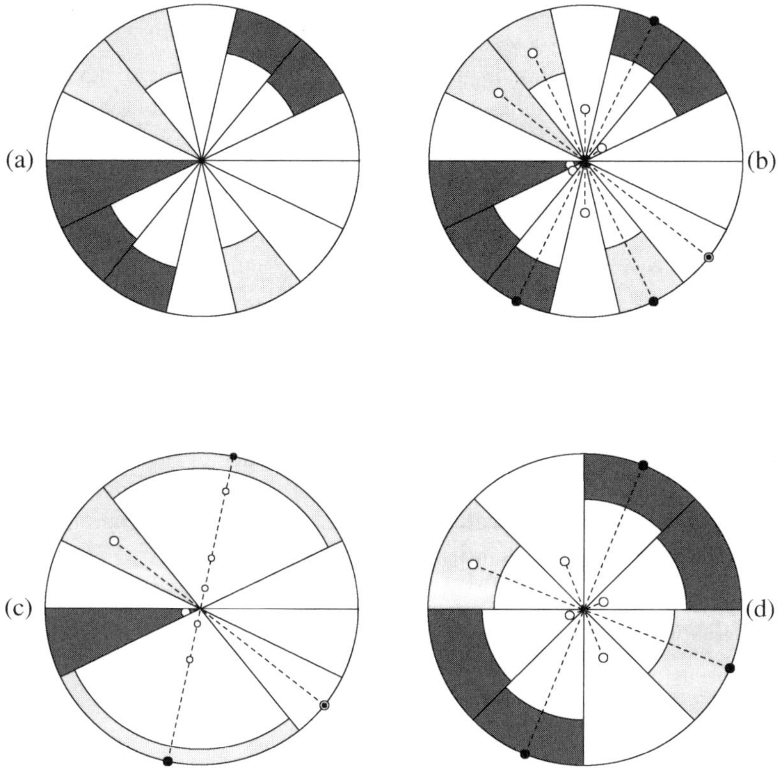

Figure 9.3 Belief comparison diagrams: (a) without data; (b) with a data set which is consistent with $\mathbf{H}_1$ but not $\mathbf{H}_2$; (c) as (b), but with some sectors combined; (d) as (b), but with some sectors omitted. The upper semicircle corresponds to $\mathbf{H}_2$, the lower to $\mathbf{H}_1$.

9.11.2 The observed comparison

Observation of each canonical quantity Z_i leads to a pair of residuals, one for each specification. We superimpose these residuals on the belief comparison diagram as follows. We plot the residual for Z_i corresponding to $\mathbf{H}_1$ in the lower sector corresponding to Z_i, and we plot the residual for Z_i corresponding to $\mathbf{H}_2$ in the upper sector corresponding to Z_i. Each observation is plotted as a roundel shaded white, and the pair of roundels corresponding to the same canonical quantity are connected by a dotted line to aid interpretation

With respect to the data comparison, we regard the radius c of the circle as representing $c = 3$ standard deviations. We can change this scaling if appropriate. Thus, we plot the observed value of $|r_{1i}|$ in the lower sector for Z_i at a distance of $|r_{1i}|$ standard deviations from the centre. Under $\mathbf{H}_1$, if $c = 3$ we expect to see each plotted observation about one-third of the distance from the centre. Similarly, we plot $|r_{2i}|$ in the upper sector for Z_i.

Occasionally these residuals may be more than c standard deviations distant from the centre. In this case we plot roundels shaded black on the outer boundary of the sector. More rarely we may come across observed values of canonical quantities which have a variance of zero under a given specification. In this case, the observation must equal its expectation under this specification. When this is true, we do not plot the residual. However, when the observation is not equal to its expectation, the specification is contradicted. We indicate such cases by plotting a white roundel with a black inner on the outer boundary of the sector.

For an informal interpretation, for any choice c, a particular specification is consistent with the data if, in most sectors of the semicircle for that specification, the corresponding roundel appears at a distance roughly $1/c$ of the way between the centre of the circle and the boundary. The appearance of many roundels on the circumference suggests that the specification might have underestimated variability or misspecified the mean. Points clustered in the centre of the diagram may suggest that the specification has inflated the variability.

9.11.2.1 Example

We continue using the example of §9.2 with matching expectations

$$\mathrm{E}_{\mathbf{H}_1}(X) = \mathrm{E}_{\mathbf{H}_2}(X) = 0,$$

and with X observed as in §9.10.1. The variances for the canonical directions are shown in Table 9.6, together with their standardized observations. Note that these differ slightly from those shown in Table 9.4, which were calculated for different expectation specifications.

Table 9.6 Canonical comparison of variance specifications: variances and residuals for each canonical direction. Expectations are zero under each specification.

i	$\mathrm{Var}_{\mathbf{H}_2}(Z_i)$	$\mathrm{Var}_{\mathbf{H}_1}(Z_i)$	$\lvert r_{i1}\rvert$	$\lvert r_{i2}\rvert$
1	0	1	0	0
2	$\frac{3}{7}$	1	0.2673	0.4082
3	$\frac{1}{2}$	1	4.0000	5.6569
4	1	1	1.0000	1.0000
5	3	1	4.0000	2.3094
6	1	0	2.1213	2.1213
7	0	0	0	0

The residuals are plotted in Figure 9.3(b), which has the same variance comparison as in Figure 9.3(a). The plot is interpreted as follows. The direction Z_6 has the residual $|r_{61}| = 2.1213$ under $\mathbf{H}_1$. However, this contradicts the variance specification as $\text{Var}_{\mathbf{H}_1}(Z_6) = 0$ and this in turn implies a residual of zero as an observation of Z_6 must match its expectation under $\mathbf{H}_1$. The inconsistency is marked by plotting a white roundel with a black inner on the outer boundary of the sector corresponding to Z_6 under $\mathbf{H}_1$. The observation for Z_6 does, however, appear consistent with $\mathbf{H}_2$. Direction Z_3 has data which are abnormal, beyond three standard deviations, under both specifications: these are indicated by black roundels on the outer boundary, where the outer boundary represents $c = 3$ standard deviations under the appropriate specification. The data appear abnormal for direction Z_5 under $\mathbf{H}_1$ but not $\mathbf{H}_2$. The data for direction Z_4 is as expected under both specifications: white roundels are plotted at one standard deviation from the centre. The residuals for direction Z_2 are rather smaller than expected, possibly indicating too large a variance in this direction under both specifications.

9.11.3 Combining information

When $\langle X \rangle$ has very high dimension, then there will be many sectors, and the display may become overly crowded. We may then use the methods described in §9.9.1 for producing summaries for combined directions. In particular, this may be appropriate when several eigenvalues are roughly similar. The size of the sector shown for several directions combined depends on whether our primary interest is in the detail of the other directions. If so, we represent any combined sector as having the same area as sectors for single directions. Otherwise, we draw the sector as having area proportional to the number of directions it represents.

9.11.3.1 Example

Figure 9.3(c) reproduces Figure 9.3(b), but with the sectors corresponding to Z_2, Z_3, Z_4, Z_5 combined; these are the directions with positive variance under both specifications. For the combination, two of the residuals under $\mathbf{H}_1$ (lower semicircle) exceed three standard deviations. For such cases, we plot a black roundel with area proportional to the number of abnormal residuals. On average, the four combined directions have slightly higher variance under $\mathbf{H}_2$ and the data better support $\mathbf{H}_2$.

We may also choose not to show some sectors in order to focus on certain aspects of the comparison. For example, Figure 9.3(d) reproduces Figure 9.3(b), except that the sectors corresponding to the rank-degenerate directions Z_1, Z_6, Z_7 have been intentionally omitted.

9.11.4 Residual belief comparison diagrams

When the expectation vector and the variance matrix each differ under the two specifications, then it may be preferable to choose a display which distinguishes, as much as is possible, those comparisons arising from differences in the mean specification from those arising from differences in the variance specification. In §9.8, we showed how X could be transformed into a mean quantity M and a residual vector R with the same expectation under each specification. Therefore, we may prefer to construct the belief comparison diagram from the residual form R, as each pair of opposing sectors in such a residual belief comparison diagram corresponds to a quantity with a common expectation in each specification. All differences in such a picture relate to variance comparisons, and so the picture is likely to be more straightforward to interpret. If we choose such a representation, then we must separately display the value of M. A natural display is a line plot on which we mark the locations of $\mathrm{E}_{\mathbf{H}_1}(M)$, $\mathrm{E}_{\mathbf{H}_2}(M)$ surrounded by, for example, three-standard-deviation bands under $\mathrm{Var}_{\mathbf{H}_1}(M)$, $\mathrm{Var}_{\mathbf{H}_2}(M)$. If we observe $X = x$, then we mark the observed value $M = m$ as a roundel plotted between the two lines given by these bands.

9.11.4.1 Example

To illustrate, we return to the example of §9.10.1, for which the variance specifications for the quantities in $\langle Z^{++} \rangle$ are as above, but with alternative expectation specifications for $\mathbf{H}_1$ and $\mathbf{H}_2$. For our mean component (9.51) we standardize initially according to $\mathrm{Var}_{\mathbf{H}_1}(\cdot)$ and so choose $\alpha = 1$. This choice corresponds to taking $M = M_1 = \mathrm{G}_{\frac{2}{1}}$. Summary statistics for $\mathrm{G}_{\frac{2}{1}}$ were shown in §9.10.1: under $\mathbf{H}_1$, $\mathrm{G}_{\frac{2}{1}}$ has mean zero and standard deviation 2.69, under $\mathbf{H}_2$ $\mathrm{G}_{\frac{2}{1}}$ has mean 7.25 and standard deviation 2.85, and the observed value is 11.625. These features are plotted in Figure 9.4(a), showing a broadly similar variance specification (the intervals are about the same width) but quite different expectation specifications, and with an observed value which appears compatible with $\mathbf{H}_2$ but not $\mathbf{H}_1$.

A residual wheel is shown in Figure 9.4(b). As $\langle Z^{++} \rangle$ is four-dimensional and as the mean component $\mathrm{G}_{\frac{2}{1}}$ has been extracted, the residual space is the

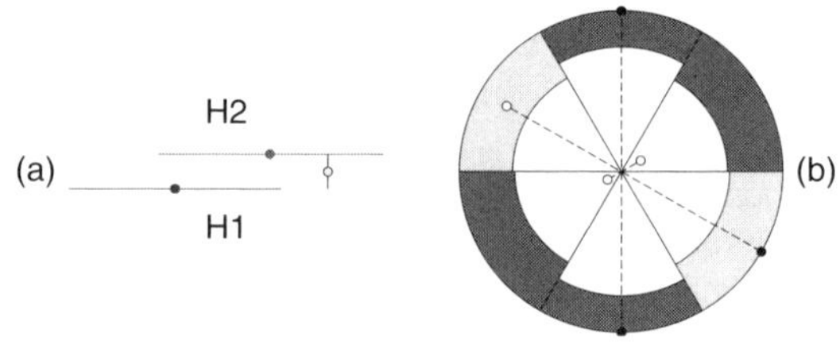

Figure 9.4 The mean component and the residual wheel, norming under $\mathrm{Var}_{\mathbf{H}_1}(\cdot)$. (a) Expectations and six-standard-deviation intervals for the mean component, with the observed mean component. (b) The residual wheel: each direction has expectation zero under $\mathbf{H}_1$ and $\mathbf{H}_2$. The upper semicircle corresponds to $\mathbf{H}_2$, the lower to $\mathbf{H}_1$.

three-dimensional orthogonal complement of $G_{\frac{2}{1}}$ in $\langle Z^{++}\rangle$. This gives rise to three residual directions, R_1, R_2, R_3, for the comparison, each of which has the same expectation, zero, under $\mathbf{H}_1$ and $\mathbf{H}_2$. Variances and standardized residuals for these directions are shown in Table 9.7. Notice that residual direction R_1 happens to coincide with canonical quantity Z_2 in Table 9.3 because of the way in which variances and expectations happen to have been made: the relation between R_1 and Z_2 corresponds to the orthogonality between (X_2, X_4) and (X_5, X_6, X_7). As this orthogonality is preserved under $\mathbf{H}_1$ and $\mathbf{H}_2$, we get this correspondence for each value of α. . The data do not much distinguish between the two specifications. There is one residual direction, R_2, having higher variance under $\mathbf{H}_1$, for which the data appear abnormal under both specifications, particularly $\mathbf{H}_2$; another residual direction, R_1, where the data appear consistent with the specifications; and finally a residual direction, R_3, where the data appear more consistent with $\mathbf{H}_2$.

If instead we standardize by norming according to variation in $\mathbf{H}_2$, and so choose $\alpha = 0$, we obtain the mean component and residual wheel shown in Figure 9.5. The main features are similar to those shown in Figure 9.4. The observed value for the bearing for this comparison, $G_{\frac{1}{2}}$, is consistent with its mean and standard deviation under $\mathbf{H}_2$, but not $\mathbf{H}_1$. For the residual wheel, the canonical directions and the amounts of shading are identical to those for the standardization under $\mathbf{H}_1$, but are here displayed with left–right and top–bottom reversal and with light and dark shading switched, as befits the norming under $\mathbf{H}_2$ rather than $\mathbf{H}_1$. Contrast the upper central sector corresponding to residual R_2 under $\mathbf{H}_1$ in Figure 9.5 with the lower central sector corresponding to residual R_2 under $\mathbf{H}_1$ in Figure 9.4. Under the standardization $\mathbf{H}_1$, this observed residual is more than three standard deviations from expectation, whilst under standardization $\mathbf{H}_2$, the observed value has been plotted within the three-standard-deviation boundary. Inspection of the observed residuals shows, however, that the differences are minor.

9.12 Example: exchangeable regressions

For an example of the comparison of full rank specifications we return to the exchangeable regressions example of §6.7. Specifications for this example were

Table 9.7 Comparison of residual directions. Expectations are zero under each specification. Variances are normed to be equal to one under $\mathbf{H}_1$ and residuals are standardized observed values.

i	$\text{Var}_{\mathbf{H}_2}(R_i)$	$\text{Var}_{\mathbf{H}_1}(R_i)$	$\lvert r_{i1}\rvert$	$\lvert r_{i2}\rvert$
1	0.4286	1	0.2673	0.4082
2	0.6000	1	3.2967	4.2560
3	2.2308	1	3.6475	2.4421

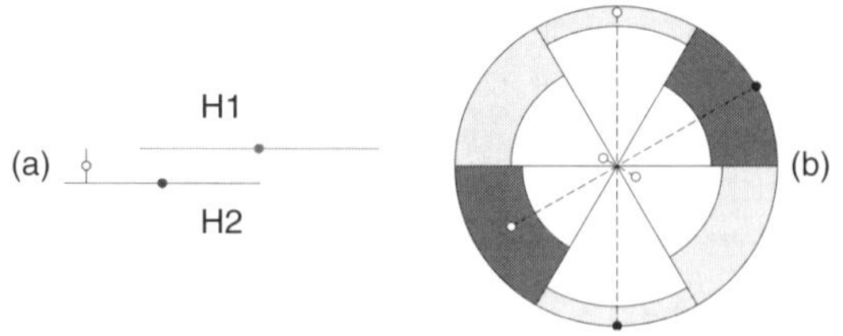

Figure 9.5 The mean component and the residual wheel, norming under $\text{Var}_{\mathbf{H}_2}(\cdot)$. (a) Expectations and six-standard-deviation intervals for the mean component, with the observed mean component. (b) The residual wheel: each direction has expectation zero under $\mathbf{H}_1$ and $\mathbf{H}_2$. The upper semicircle corresponds to $\mathbf{H}_1$, the lower to $\mathbf{H}_2$.

shown in §6.7.2 (error specifications) and §6.7.3 (regression coefficient specifications). We shall label these specifications $\mathbf{H}_1$. As an alternative specification, which we label $\mathbf{H}_2$, we consider the adequacy of a model with intercept only and with much simpler error structure:

$$Y_{rt} = a_r + E_{rt}, \tag{9.59}$$

where we compensate by specifying a variance for E_{rt} which is four times larger than under $\mathbf{H}_1$. This approximately maintains the magnitude of correlations between Y_{rt} and $Y_{r,t+1}$, but has rather smaller variance for each Y_{rt}.

For a given run r, we have for each specification a 13×13 variance matrix for the vector of observables, Y_r, together with an expectation vector under each specification. We now compare these specifications.

9.12.1 Basic canonical analysis

The key features are shown in Table 9.8. There are 13 canonical quantities $Z_1, \ldots, Z_{13}$. Comparing the variances, we see that most are similar in that the eigenvalues λ_i are not far from unity. The exception is for the first canonical quantity Z_1, which has a much smaller variance under $\mathbf{H}_2$ than $\mathbf{H}_1$. As a linear combination of the elements of the observation vector, Y, the quantity Z_1 is approximately

$$\begin{aligned} Z_1 = &+ 0.46Y_1 + 0.41Y_2 + 0.34Y_3 + 0.26Y_4 + 0.17Y_5 + 0.07Y_6 - 0.03Y_7 \\ &- 0.13Y_8 - 0.22Y_9 - 0.31Y_{10} - 0.39Y_{11} - 0.45Y_{12} - 0.50Y_{13} + 2.25. \end{aligned}$$

The pattern provided by the coefficients suggests a time feature. Comparing the two sets of expectations, we see that the expectations match fairly closely except in the direction Z_1. The maximal difference in expectation, relative to variation under $\mathbf{H}_1$,

$$\text{DE}_{\frac{2}{1}}(Y_r) = \sum \text{E}_{\mathbf{H}_2}(Z_i)^2 = 1.7896^2 + \ldots + 0.0030^2 = 3.4262,$$

Table 9.8 Variance and expectation summaries for the comparison of alternative specifications for the exchangeable regressions example. Signs of canonical quantities are chosen so that $\mathrm{E}_{\mathbf{H}_2}(Z_i) > 0$.

i	$\mathrm{Var}_{\mathbf{H}_2}(Z_i) = \lambda_i$	$\mathrm{Var}_{\mathbf{H}_1}(Z_i)$	$\max(\lambda_i, \frac{1}{\lambda_i})$	$\mathrm{E}_{\mathbf{H}_2}(Z_i)$	$\mathrm{E}_{\mathbf{H}_1}(Z_i)$	$\mathrm{G}_{\frac{2}{1}}$
1	0.0603	1	16.58	1.7896	0	1.04
2	0.3098	1	3.23	0.0490	0	0.29
3	0.4825	1	2.07	0.4188	0	0.12
4	0.6661	1	1.50	0.1092	0	0.06
5	0.8398	1	1.19	0.1317	0	0.02
6	1.0742	1	1.07	0.0892	0	-0.01
7	1.3061	1	1.31	0.0416	0	-0.05
8	1.5057	1	1.51	0.0667	0	-0.09
9	1.6778	1	1.68	0.0203	0	-0.13
10	1.8126	1	1.81	0.0396	0	-0.18
11	1.9167	1	1.92	0.0100	0	-0.27
12	1.9884	1	1.99	0.0183	0	-0.53
13	2.0311	1	2.03	0.0030	0	-1.79

is almost entirely in the direction Z_1 and it is obvious that the bearing for the comparison, $\mathrm{G}_{\frac{2}{1}} = \sum \mathrm{E}_{\mathbf{H}_2}(Z_i)Z_i$, will be highly correlated with Z_1. The bearing is summarized in column 7 of Table 9.8, except for its constant term which is 5.53, so that the bearing is

$$\mathrm{G}_{\frac{2}{1}} = 1.04Y_{r1} + 0.29Y_{r2} + \ldots - 1.79Y_{r13} + 5.53.$$

In short, most of the differences between $\mathbf{H}_1$ and $\mathbf{H}_2$ are captured by Z_1.

9.12.2 Mean and residual comparisons

For this example, we have different expectation specifications as well as different variance specifications and so we tease out differences by applying the methods of §9.8. In parallel, we examine the consistency of the data under the specifications.

Figure 9.6 shows the residual wheel and the expectation comparison for each run separately and for the overall mean of the three runs. We choose to take $\alpha = 1$, as we wish to compare relative to specification $\mathbf{H}_1$. The line plot displays the locations of $\mathrm{E}_1(M_1)$, $\mathrm{E}_2(M_1)$, with six-standard-deviation bands based on $\mathrm{Var}_1(M_1)$, $\mathrm{Var}_2(M_1)$, and marking between them the observed value m_1. We see that M_1 has a much higher variance under $\mathbf{H}_1$. For each run, the observed value is closer to its expectation under $\mathbf{H}_1$. It is a consequence of the normalization that $\mathrm{E}_2(M) > \mathrm{E}_1(M)$. In summary, the line plots suggest substantially more support for $\mathbf{H}_1$ than for $\mathbf{H}_2$. The residual wheels show the residual directions $R_1, \ldots, R_{12}$ for each run, these being identical and with identical variance summaries for each run. There are typically higher variances for the residual components under $\mathbf{H}_2$,

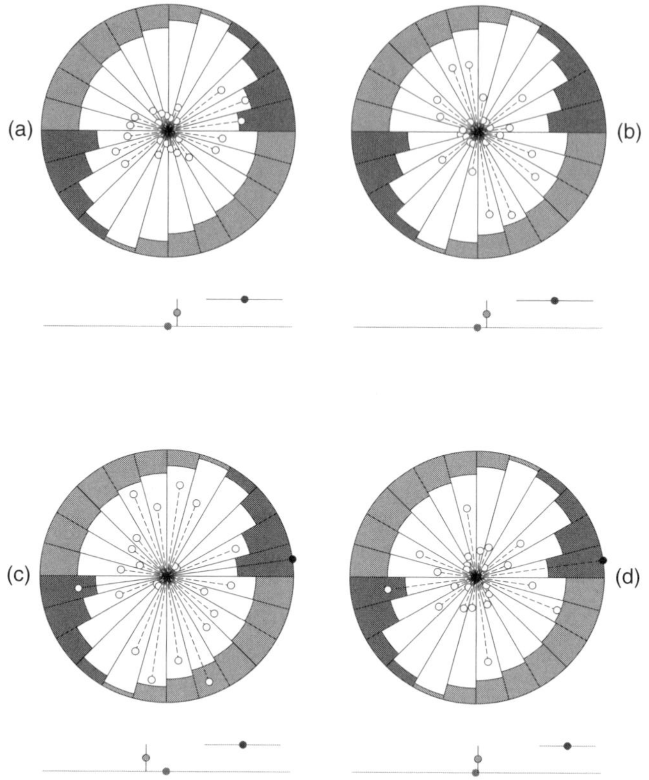

Figure 9.6 The residual wheel and, below each wheel, the corresponding expectation comparison. The upper bar and semicircle correspond to $\mathbf{H}_2$, the lower to $\mathbf{H}_1$. (a) For process run 1. (b) For process run 2. (c) For process run 3. (d) Comparison of the sample mean.

there being more light shading: the higher variances under $\mathbf{H}_1$ for the Y_{rt} are represented in the mean component rather than by the residuals. The data appear abnormal under specification $\mathbf{H}_2$ for direction R_1 for run 3.

The residual wheel and line plot for the mean $\bar{Y} = (Y_1 + Y_2 + Y_3)/3$ of the three runs (Figure 9.6(d)) has a similar interpretation, with $\mathbf{H}_1$ rather better supported than $\mathbf{H}_2$. Note that the variance matrices for $\bar{Y}$ take into account covariances such as $\text{Cov}_1(Y_1, Y_2)$, and so differ from the variance matrices for single runs. Thus, the canonical directions for the comparison for $\bar{Y}$ differ from those for the single runs, though not markedly so in this example.

Any lack of fit exhibited by the plot can be investigated further as desired. The actual residuals are shown in Table 9.9. For each run, the standardized residuals for the mean direction summarizing differences in expectation, $M_1 = \mathrm{G}_{\frac{2}{1}}$, are large under $\mathbf{H}_2$ but appear compatible with $\mathbf{H}_1$. Residuals here are calculated as described

Table 9.9 Standardized residuals under each specification for the mean direction $M_1 = \mathrm{G}_{\frac{2}{1}}$ and the residual directions R_i, for each of the process runs Y_r and the overall mean $\bar{Y}$.

	Y_1		Y_2		Y_3		$\bar{Y}$	
	$\lvert r_1\rvert$	$\lvert r_2\rvert$	$\lvert r_1\rvert$	$\lvert r_2\rvert$	$\lvert r_1\rvert$	$\lvert r_2\rvert$	$\lvert r_1\rvert$	$\lvert r_2\rvert$
M_1	0.22	5.31	0.22	5.31	0.48	7.60	0.05	9.63
R_1	0.96	1.73	0.41	0.74	2.09	3.76	2.10	3.74
R_2	1.32	1.94	0.07	0.10	1.18	1.74	0.55	0.79
R_3	1.28	1.57	1.06	1.31	0.18	0.23	1.12	1.30
R_4	0.10	0.11	0.20	0.21	0.10	0.11	0.05	0.06
R_5	0.61	0.59	0.11	0.10	1.93	1.86	0.79	0.76
R_6	0.30	0.26	0.94	0.82	2.48	2.17	0.72	0.63
R_7	0.43	0.35	1.97	1.60	2.02	1.65	2.02	1.65
R_8	0.56	0.43	2.12	1.64	2.69	2.08	0.68	0.53
R_9	0.79	0.58	0.16	0.12	1.53	1.14	0.49	0.36
R_{10}	0.31	0.22	1.70	1.23	1.42	1.03	0.00	0.00
R_{11}	1.19	0.85	1.36	0.96	0.95	0.68	2.04	1.44
R_{12}	1.28	0.90	0.51	0.36	1.53	1.07	1.32	0.93

in §9.9 for the corresponding linear combination of original quantities. For example, the standardized residuals for the mean direction M_1 are

$$\mathbf{H}_1\colon r_1 = \frac{\sum f_i z_i}{\sqrt{\sum f_i^2}},$$

$$\mathbf{H}_2\colon r_2 = \frac{\sum f_i z_i - \sum f_i^2}{\sqrt{\sum \lambda_i f_i^2}},$$

where $z_1, \ldots, z_{13}$, are the observed values of the canonical directions for the comparison. The abnormality noted for the process mean $\bar{Y}$ for residual direction R_1 under $\mathbf{H}_2$ is revealed as a standardized residual of 3.74, which is perhaps not so abnormal. Examining the correlations between the residual direction R_1^* and the original quantities, $(\bar{Y}_1, \ldots, \bar{Y}_{13})$, these show a pattern of rising correlations from about zero at time $t = 1$ to about 0.34 at $t = 7$, and falling correlations to about zero at $t = 13$. This suggests that the model $\mathbf{H}_2$ is failing to capture systematic features of the data. As $\mathbf{H}_2$ is a simplification of $\mathbf{H}_1$, lacking some time features, the conclusion is that the simplification is inappropriate.

For the mean of the three processes, $\bar{Y}$, we also inspect the plot shown in Figure 9.7. This plots, as a function of α, envelopes corresponding to

$$\mathrm{E}(M_\alpha) \pm 2\sqrt{\mathrm{Var}(M_\alpha)}$$

under each specification, together with the observed value of M_α. For this example, the observed value lies within the $\mathbf{H}_1$ envelope for all α, and outside the $\mathbf{H}_2$

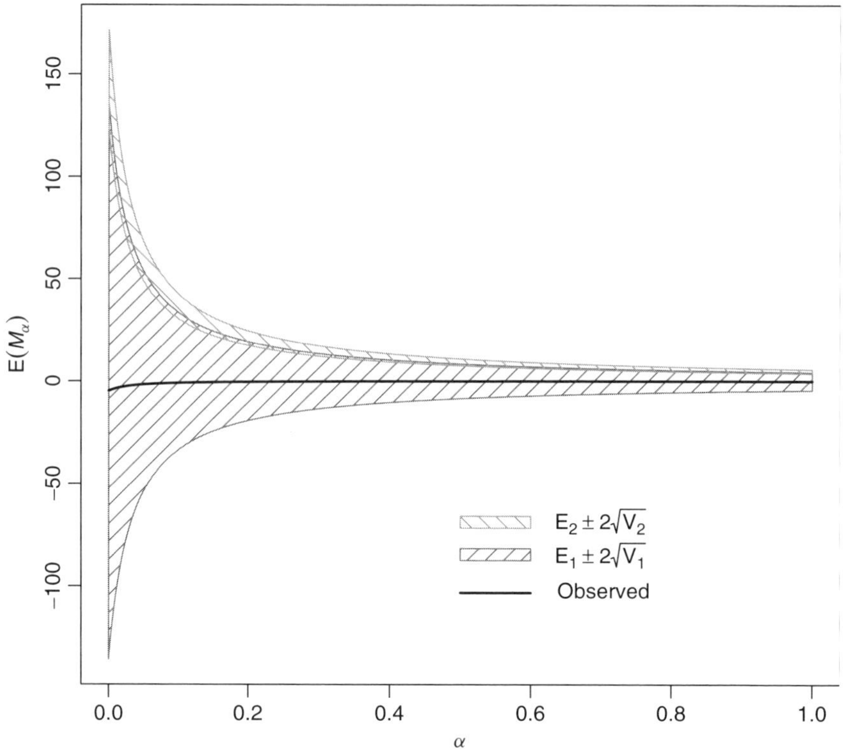

Figure 9.7 Expectation comparison plot for the aluminium extraction example. Plotted versus α are two-standard-deviation boundaries for M_α under each specification, together with the observed value of M_α.

envelope. We conclude that the data are consistent with $\mathbf{H}_1$ but not $\mathbf{H}_2$, irrespective of choice of standardization.

With respect to differences between runs under $\mathbf{H}_1$, the residuals appear to cluster near the centre for run 1. For run 2, there are a small number of residuals with larger values, but none are particularly unusual. For run 3 there are more large residuals, though none is larger than three standard deviations. We might suspect, if the runs were taken in time order, that there is some aspect of the run order (for example, changes in the physical environment) which is not captured by either specification. To investigate this more fully, we need a more sophisticated treatment that is capable of accounting for the underlying similarities between runs; this is the subject of the next section.

9.13 Comparisons for exchangeable structures

Suppose that we have a collection of second-order exchangeable (§6.3) vectors $X_1, X_2, \ldots,$ where each X_i is a vector of p random quantities. We specify the expectation, variance, and covariance for all $j \neq k$ as

$$\mathrm{E}(X_j) = \mu, \quad \mathrm{Var}(X_j) = \Sigma, \quad \mathrm{Cov}(X_j, X_k) = \Gamma,$$

as in (6.11), (6.12). We now suppose that we want to compare two specifications:

$$\mathbf{H}_1: \quad \mathrm{E}(X_j) = \mu_1, \quad \mathrm{Var}(X_j) = \Sigma_1, \quad \mathrm{Cov}(X_j, X_k) = \Gamma_1, \tag{9.60}$$

$$\mathbf{H}_2: \quad \mathrm{E}(X_j) = \mu_2, \quad \mathrm{Var}(X_j) = \Sigma_2, \quad \mathrm{Cov}(X_j, X_k) = \Gamma_2. \tag{9.61}$$

Suppose that we observe $X_1, \ldots, X_n$, so that we have n observations of p-dimensional vectors. The comparison for such exchangeable structures can be decomposed into separate comparisons over the mean vector and residual vectors as follows. Let

$$\bar{X} = \frac{1}{n}\sum_{j=1}^{n} X_j.$$

The two specifications give

$$\mathrm{E}_i(\bar{X}) = \mu_i, \quad \mathrm{Var}_i(\bar{X}) = \Gamma_i + \frac{1}{n}\Phi_i,$$

where $\Phi_i = \Sigma_i - \Gamma_i$. Let $\mathcal{R}$ be the residual space for the collection X, namely all linear combinations $\sum_{j=1}^{n} a_j X_j$ with weights a_j constrained by $\sum_j a_j = 0$. Note that, for each element $U \in \mathcal{R}$, we have $\mathrm{E}_1(U) = \mathrm{E}_2(U) = 0$. Thus, all of the mean comparison over X is expressed in the comparison for $\bar{X}$. Next, we choose any orthonormal basis for $\mathcal{R}$, namely any collection of $n - 1$ linear combinations

$$\mathcal{R}_i = \sum_j a_{ij} X_j, \quad i = 1, \ldots, n-1,$$

where, for each i, (i) $\sum_j a_{ij} = 0$, (ii) $\sum_j a_{ij}^2 = 1$, (iii) $\sum_j a_{ij}a_{kj} = 0, i \neq k$. It is straightforward to check that

$$\mathrm{Cov}_1(\bar{X}, \mathcal{R}_i) = \mathrm{Cov}_2(\bar{X}, \mathcal{R}_i) = 0, \qquad \forall i,$$

$$\mathrm{Cov}_1(\mathcal{R}_k, \mathcal{R}_i) = \mathrm{Cov}_2(\mathcal{R}_k, \mathcal{R}_i) = 0, \qquad \forall i \neq k,$$

so that, as in §9.1.2, the belief comparison is mutually orthogonal over the individual sub-vectors $\bar{X}, \mathcal{R}_1, \ldots, \mathcal{R}_{n-1}$. Therefore, the belief comparison can be carried out separately over $\bar{X}$, which is the only vector for which expectations differ, and the sequence $\mathcal{R}_1, \ldots, \mathcal{R}_{n-1}$.

For $\bar{X}$, we carry out the belief comparison described in §9.8, separating out differences between expectations and variances.

For each residual vector, $\mathcal{R}_i$, the canonical belief comparison is identical because

$$\mathrm{Var}_1(\mathcal{R}_j) = \Phi_1, \quad \mathrm{Var}_2(\mathcal{R}_j) = \Phi_2, \qquad j = 1, \ldots, n-1, \tag{9.62}$$

so that for each $\mathcal{R}_j$, the appropriate comparison is Φ_1 to Φ_2. Suppose we make this comparison, as described in §9.1, and so obtain the canonical quantities $\mathcal{R}_1^*, \ldots, \mathcal{R}_p^*$ corresponding to canonical values $\lambda_1, \ldots, \lambda_p$. Note that there are p such canonical quantities as the variance matrices Φ_i are p-dimensional. This canonical structure is duplicated for each residual vector $\mathcal{R}_j$, and so if we are making a graphical comparison, our graphs consist of the following elements:

(a) a line plot, to summarize differences in expectation and variance for the mean (M) part of the average, $\bar{X}$, of the exchangeable quantities;

(b) a residual wheel, to summarize differences in variation for the residual (R) part of $\bar{X}$;

(c) a single canonical wheel (combining the $n-1$ identical such wheels) to summarize differences in variance for the residuals of the exchangeable quantities, $\mathcal{R}_1, \ldots, \mathcal{R}_{n-1}$.

9.13.1 The observed comparison

The graphical comparison for $\bar{X}$ is as described earlier, augmented by observations. For the residual collection, a natural graphical display is to construct the canonical wheel for the comparison of Φ_1, Φ_2, and, for each observation j, to plot, in the arc corresponding to canonical quantity $\mathcal{R}_i^*$, its standardized value, on a line bisecting the arc and extending from the centre of the wheel. Alternatively, and especially when n is large, the collection of observed standardized values might be superimposed in the form of a box plot.

As a further refinement, we might select $n-1$ different symbols, or colours, $s_1, \ldots, s_{n-1}$, and plot each observed standardized value using symbol s_i, to allow us to identify, for example, whether large values in the different sectors correspond to the same observation $\mathcal{R}_j$. This may identify which linear contrasts showed most lack of fit to the belief specification.

9.13.1.1 Identifying temporal and other features

While the above construction applies equally to any choice of basis for $\mathcal{R}$, there will usually be natural choices for the basis which are sensitive to particular ways in which exchangeability might break down. For example, if the observations were taken in time order, and we thought that there might be a change in the process at some point, then we might choose the Helmert basis of order n (see Definition 11.58). The first column corresponds to the mean component $\bar{X}$. The remaining $n-1$ columns are used to construct the $\mathcal{R}_j$, leading to

$$\mathcal{R}_1 \propto Y_2 - Y_1, \tag{9.63}$$

$$\mathcal{R}_2 \propto Y_3 - \frac{Y_1 + Y_2}{2}, \tag{9.64}$$

$$\mathcal{R}_3 \propto Y_4 - \frac{Y_1 + Y_2 + Y_3}{3}, \tag{9.65}$$

and so forth. Alternatively, we might choose the orthonormal basis as the orthogonal polynomials in the index, so that the first residual combination is $\sum_i a_{1i} Y_i$, with $a_{1i} = ai + b$, where a, b are chosen so that $\sum a_{1i} = 0$, the second component is $\sum_i a_{2i} Y_i$, with $a_{2i} = ci^2 + di + e$, where c, d, e are chosen for orthogonality with the first component, and so forth. In comparison with the basis designed to identify individual break points in the series, the orthogonal polynomial basis explores general types of overall change across the sequence. There are many ways in which we could order the data before applying such transformations; for example, we might choose the order on the basis of a covariate to see whether it should be included within the formulation.

9.13.1.2 Alternative plots

For large amounts of data, the canonical wheels may become unwieldy as a data display. An alternative way of displaying the information as follows.

- The canonical directions for the residual comparison may be shown from left to right, instead of from 0° moving anticlockwise.
- For a canonical direction Z_i with higher variance under $\mathbf{H}_1$, $\lambda_i < 1$, we show a bar with dark shading hanging downwards from the centre. The amount of shading is proportional to $1 - \lambda_i$.
- For a canonical direction Z_i with lower variance under $\mathbf{H}_1$, $\lambda_i > 1$, we show a bar with light shading reaching upwards from the centre. The amount of shading is proportional to $1 - 1/\lambda_i$.
- The lines above and below the bars indicate c standard deviations either side of the centre, with respect to standardized residual observations under $\mathbf{H}_1$ and $\mathbf{H}_2$.
- Standardized observations corresponding to individual residual directions under some basis may be shown by different symbols. In order to display the range of the standardized observations, lines may connect the maximum and minimum observations within the same sector.
- Standardized observations of more than c standard deviations are plotted just beyond the c-standard-deviation line.

9.13.2 Example: exchangeable regressions

We illustrate the graphical comparison of exchangeable data using the exchangeable regressions data, continuing on from §9.12.2. Here, the measurements are exchangeable over experiments. We have data for three such experiments, $n = 3$, and during each such experiment there are $p = 13$ quantities measured. We decompose the comparison into a 13-dimensional mean comparison and two 13-dimensional residual comparisons based on the Helmert basis (9.63). This will allow us to make some judgements as to differences between runs, as the runs were taken in time order. Thus, we construct

$$\bar{Y} = \frac{Y_1 + Y_2 + Y_3}{3}, \tag{9.66}$$

$$\mathcal{R}_1 = \frac{1}{\sqrt{2}}(Y_2 - Y_1), \tag{9.67}$$

$$\mathcal{R}_2 = \frac{1}{\sqrt{6}}\left\{Y_3 - \frac{Y_1 + Y_2}{2}\right\}. \tag{9.68}$$

We now form the elements of the graphical comparison that we discussed at the foot of §9.13. Figure 9.8(d) shows the mean comparison, duplicating Figure 9.6(d). This contains the line plot and the residual wheel for the comparison of $\bar{Y}$, which we discussed in §9.12.2.

The comparison of specifications corresponding to the residual vector $\mathcal{R}_1 \propto Y_2 - Y_1$ is shown in Figure 9.8(a). The variance comparison reveals one direction with much higher variation under $\mathbf{H}_1$, $1/\lambda_1 = 9.02$, but the remaining directions are otherwise unexceptional: $1/\lambda_2 = 3.59$, $\lambda_{13} = 2.03$. The standardized observed residuals are shown in Table 9.10 and plotted in Figure 9.8(a). The data appear reasonably consistent with both specifications. The comparison of specifications corresponding to the residual vector $\mathcal{R}_2 \propto Y_3 - (Y_1 + Y_2)/2$ is shown in Figure 9.8(b). The variance comparison is identical to that for $\mathcal{R}_1$, by (9.62). The plot shows two standardized observed residuals beyond the three-standard-deviation boundary for the upper semicircle, suggesting that the data are inconsistent with specification $\mathbf{H}_2$ in two directions which have lower variation under $\mathbf{H}_2$. Under both specifications, the residuals are rather larger for the residual vector $\mathcal{R}_2$ than for $\mathcal{R}_1$. This adds to the suspicion that the third experiment is unusual. As the variance comparisons for the residual vectors are identical, Figure 9.8(c) shows how we may plot the $n - 1$ standardized observations on the same underlying canonical wheel for the residual variance specifications.

Figure 9.9 shows an alternative way, discussed in §9.13.1.2, of displaying the information. Standardized observations for $\mathcal{R}_1$ are shown by squares. Standardized observations for $\mathcal{R}_2$ are shown by circles, and the two are connected to emphasize the range. Standardized observations of more than $c = 3$ standard deviations are plotted just beyond the three-standard-deviation line. For this example, it is again clear that the residual vector $\mathcal{R}_2$ has some very unusual standardized observations under both specifications, suggesting that there are unanticipated differences between the third run and the average of the first two runs.

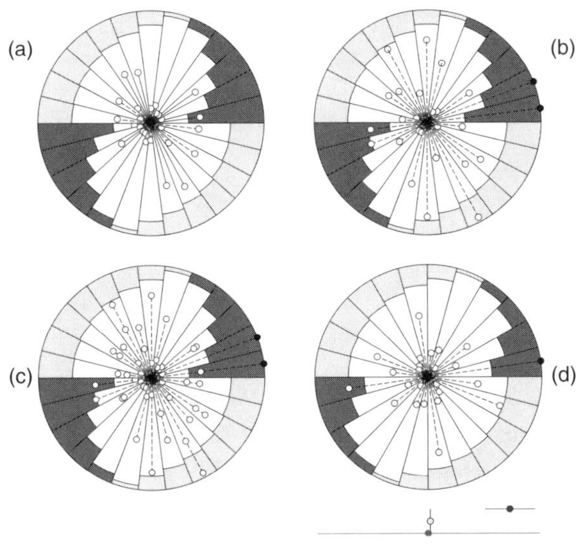

Figure 9.8 The comparison for exchangeable structures. The upper bar and semicircle correspond to $\mathbf{H}_2$, the lower to $\mathbf{H}_1$. (a) Comparison of the residual $Y_2 - Y_1$. (b) Comparison of the residual $Y_3 - (Y_1 + Y_2)/2$. (c) Superimposition of the residuals on the underlying canonical wheel for the residual variance specifications. (d) Comparison of the sample mean.

Table 9.10 Standardized residuals for the exchangeable comparison.

		$\mathcal{R}_1 \propto (Y_2 - Y_1)$		$\mathcal{R}_2 \propto \{Y_3 - (Y_1 + Y_2)/2\}$	
i	$\text{Var}_{\mathbf{H}_2}(\mathcal{R})$	r_{1i}	r_{2i}	r_{1i}	r_{2i}
1	0.1109	−0.4346	−1.3048	−1.5101	−4.5343
2	0.2789	0.1247	0.2361	−1.5891	−3.0090
3	0.4318	0.9436	1.4359	0.8892	1.3532
4	0.6038	0.1593	0.2050	−0.3042	−0.3914
5	0.8374	0.0268	0.0293	−0.2844	−0.3108
6	1.0789	−0.5074	−0.4885	−1.6825	−1.6198
7	1.3092	0.4482	0.3917	−2.5182	−2.2008
8	1.5099	1.7040	1.3867	0.9755	0.7938
9	1.6794	−1.9069	−1.4715	2.8673	2.2126
10	1.8143	0.7028	0.5218	1.4937	1.1089
11	1.9172	1.4177	1.0239	−1.7162	−1.2395
12	1.9888	−0.1052	−0.0746	0.2647	0.1877
13	2.0311	1.2756	0.8951	−0.9352	−0.6562

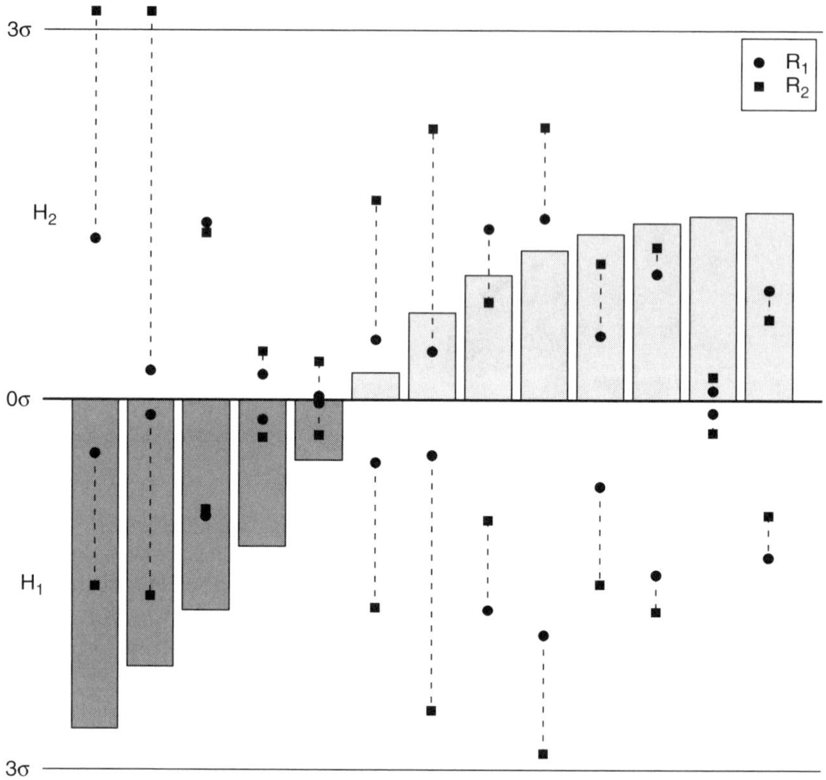

Figure 9.9 The comparison for exchangeable structures: an alternative plot for the residual vectors. Residuals beyond three standard deviations are censored.

9.14 Example: fly population dynamics

In §7.6 we described and analysed a population dynamics experiment. The model for the experiment was given in §7.6.1, with specifications $\mathbf{H}_1$ given in §7.6.2. Farrow and Goldstein (1996) considered the same model, but with slightly different specifications $\mathbf{H}_2$:

	D. melanogaster	*D. hydei*
θ_s	0.45	0.40
ψ_s	0.26	0.10
ϕ_s	−0.34	−0.18
α_s	0.2	0.2
β_s	0.7	0.7
γ_s	−0.3	−0.3
ν_s	0.04	0.02
ω_s	0.1	0.1

These were, in fact, the original specifications for the model. At a later point, we revisited this model and came to the conclusion that the priors for the local means contained too much oscillation. Further exploration of the prior means and variances led to the specifications given as $\mathbf{H}_1$ in §7.6.2, which we felt were probably more plausible. Specifically, the differences between $\mathbf{H}_1$ and $\mathbf{H}_2$ are as follows. The lower values of α_s under $\mathbf{H}_2$ allow the prior means for the local means to tend sooner to the expected equilibria. The higher values of β_s under $\mathbf{H}_2$ allow the local means to oscillate more strongly. The values of γ_s under $\mathbf{H}_2$ increase the initial cross-species inhibition, compared to $\mathbf{H}_1$. Decreased values for ω_s under $\mathbf{H}_2$ indicate a generally lower specification for the noise attached to the local means.

We now compare the beliefs generated by the two sets of specifications, and see which – if either – is supported by the data. In terms of the methodology of §9.13, we have second-order exchangeable beliefs about the quantities $Y_{ps1t}, Y_{ps2t}, \ldots, Y_{psct}$. Thus, it is appropriate to carry out the comparison suggested therein, namely (1) to summarize differences in expectation and variance for the mean part of the average, $\bar{Y}_{pst} = \frac{1}{c}\sum_{j=1}^{c} Y_{psct}$, of the exchangeable quantities; (2) to plot a residual wheel, to summarize differences in variation for the residual part of $\bar{Y}$; (3) to plot a single canonical wheel, to summarize differences in variance for the residuals of the exchangeable quantities.

We organize the comparison for each starting point separately. In §7.6.4.3 we organized the local means for starting points $p = 1$ and $p = 2$ into the 50-dimensional vectors M_1, M_2. We similarly organize the means of the data quantities as 50-dimensional vectors $\bar{Y}_1, \bar{Y}_2$, and the comparison takes place across these 50 dimensions.

9.14.1 Differences for the mean part of the average

Table 9.11 summarizes the difference in mean and variation for the mean part of the average. Specification $\mathbf{H}_1$ is clearly far superior, with small residuals. The data do not appear to support the specification $\mathbf{H}_2$: the observed residuals are over 50 standard deviations for both starting points. Recall that we are normalizing with respect to $\mathbf{H}_1$ here. We could also normalize with respect to variation under $\mathbf{H}_2$. Doing so, we find similar results.

9.14.2 Differences for the residual part of the average

The comparison for the residual part of the average is plotted in Figure 9.10. We observe that the variance specifications are mostly larger under $\mathbf{H}_1$ than under $\mathbf{H}_2$, with 38 of the 50 canonical quantities having larger variance under $\mathbf{H}_1$ (dark) and 12 having higher variance under $\mathbf{H}_2$ (light). The canonical resolution for the comparison ranges from $\lambda_1 = 0.0489$ to $\lambda_{50} = 5.37$, so that there is one direction where the variance is approximately 20 times higher under $\mathbf{H}_1$, one direction where the variance is 5.37 times higher under $\mathbf{H}_2$, and all other directions differ by amounts between these two extremes. The variance comparisons for the two starting

Table 9.11 Variance and expectation summaries for the comparison of alternative specifications for the mean part, M, for the exchangeable data of the population dynamics experiment.

	Starting point $p = 1$			
Specification	E(M)	$SD(M)$	Observed	Residual
$\mathbf{H}_1$	0	92.9775	100.5564	1.0815
$\mathbf{H}_2$	8644.8124	159.2392	100.5564	−53.6567
	Starting point $p = 2$			
Specification	E(M)	$SD(M)$	Observed	Residual
$\mathbf{H}_1$	0	93.0121	−121.7094	−1.3085
$\mathbf{H}_2$	8651.2512	159.4768	−121.7094	−55.0109

points are identical as, ignoring species, the model for $p = 1$ is identical to the model for $p = 2$.

Figure 9.10 also plots the observed residuals for each canonical direction. Triangles represent residuals of at least six standard deviations. We expect to see residuals at about one standard deviation from the centre. The upper half of this diagram compares observations to beliefs $\mathbf{H}_2$, whilst the lower half compares observations to beliefs $\mathbf{H}_1$. We observe no very large residuals under $\mathbf{H}_1$ and many for $\mathbf{H}_2$. For starting point $p = 1$, 18 of the 50 directions have large residuals under $\mathbf{H}_2$. For starting point $p = 2$, there are also 18 out of 50 directions with very large residuals under $\mathbf{H}_2$, though these are not always the same directions as for starting point $p = 1$. We conclude that, normalizing according to variation under $\mathbf{H}_1$, the data do not support $\mathbf{H}_2$. With regard to $\mathbf{H}_1$, the largest residual is 4.05 for $p = 1$ and 4.79 for $p = 2$, so there is some evidence of contradiction. If instead we make the comparison normalizing according to variation under $\mathbf{H}_2$, we find similar results but with slightly fewer very large residuals under $\mathbf{H}_2$.

9.14.3 Differences for the residual part of the average

Figure 9.11 plots the observed residuals for each canonical direction for the residual comparison. For each starting point there is a 50-dimensional residual comparison, using the Helmert basis (9.63). The two sets of specifications differ only for the local mean quantities, the $\{M_{pst}\}$, and not for the residual quantities, the $\{R_{psct}\}$. Consequently, $\mathbf{H}_1$ is identical to $\mathbf{H}_2$ for this comparison, and so there can be no canonical directions for which variances differ under the two specifications. As such, we show only the absolute residuals for each of the 50 directions, to determine whether the joint residual specification is compatible with the observations. For these plots, residuals of at least six standard deviations are censored. Circles correspond to observed residuals for differences in counts between cages 1 and 2, and squares to differences in counts between cage 3 and the average for cages 1 and 2, corresponding to the same Helmert contrasts used in §9.13.1.1.

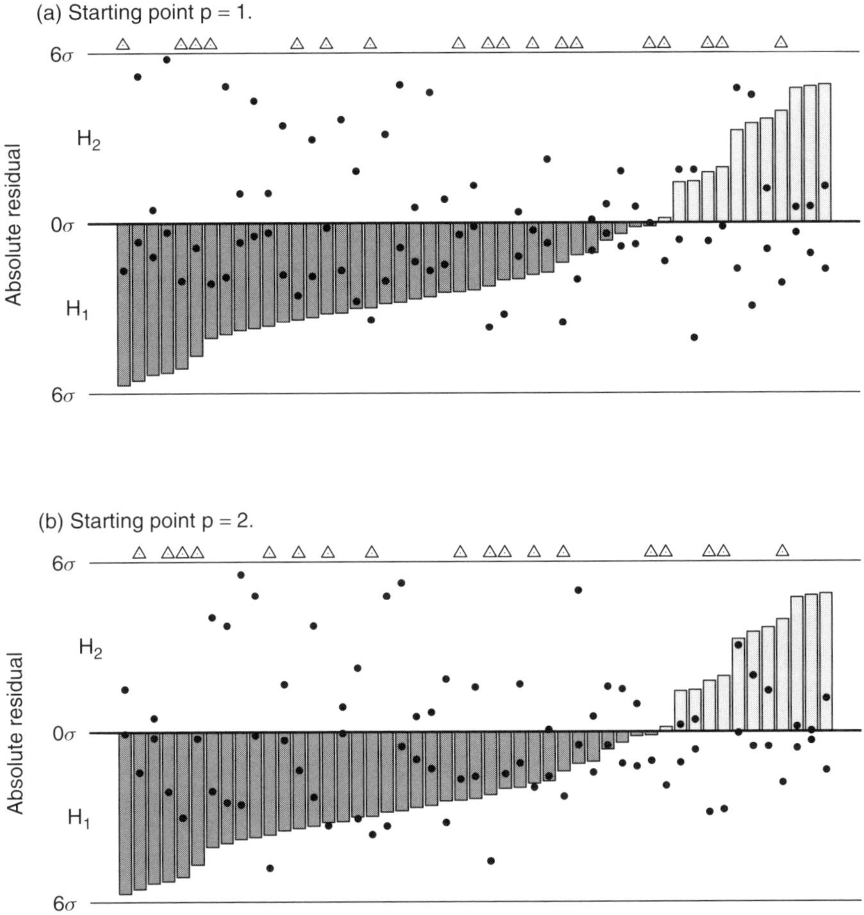

Figure 9.10 Comparison of the residual part of the average of the exchangeable quantities. Observed residuals beyond 6σ are censored.

For starting position $p = 1$, residuals are seen to be generally rather larger than expected. There are three residuals larger than six standard deviations, all corresponding to differences between cage 1 and cage 2. The conclusion is that the data do not seem to be consistent with the residual specification for starting point $p = 1$, especially in relation to differences between cage 1 and cage 2.

For starting position $p = 2$, residuals are seen to be substantially larger than expected. There are 17 residuals larger than six standard deviations, nearly all corresponding to differences between cage 3 and the average for cage 1 and cage 2. Residuals for differences between cage 1 and cage 2 appear more compatible with the belief specification. The conclusion is that the data are not consistent

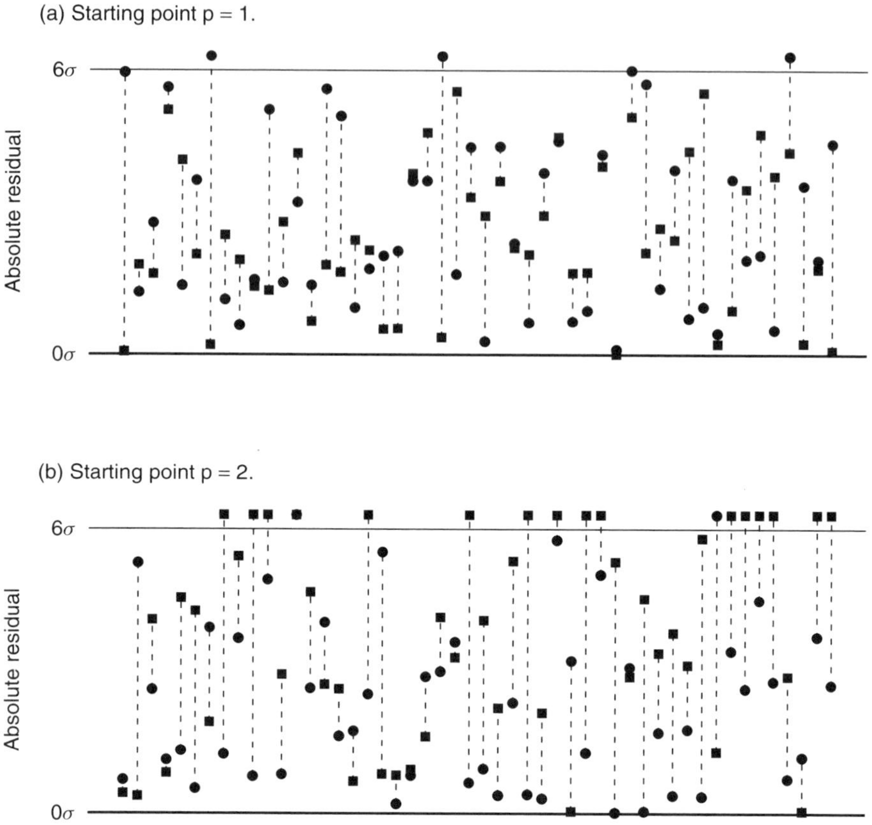

Figure 9.11 Comparison of the residual structure. Observed residuals beyond 6σ are censored.

with the residual specification for starting point $p = 2$, especially in relation to differences between cage 3 and the other two cages.

Overall, these inconsistencies tend to tally with what may be deduced from inspecting Figure 7.2 and Figure 7.3, and point at least to underestimation of differences between cages in the residual component.

9.15 Assessing robustness of specifications

In a complicated analysis, we might have to make thousands of numerical specifications. Typically, we will make such specifications by imposing pragmatic simplifications, for example treating **almost exchangeable** units as exchangeable, **almost uncorrelated** quantities as uncorrelated, and so forth. We must therefore judge the robustness of our inferences to such simplifications. In many contexts, $\mathbf{H}_1$ is fixed so that $E_{H_1}(\cdot)$ and $Var_{H_1}(\cdot)$ are the specifications that we intend to use, unless our robustness analyses force us to elaborate our modelling. The specifications $E_{H_2}(\cdot)$ and $Var_{H_2}(\cdot)$ typically vary over some class of alternative specifications. As we are interested in changes in belief over high-dimensional linear spaces, under a variety of competing specifications, measures of maximal discrepancy will often be the only practical way to identify important differences. Thus, we employ the variance and expectation comparison measures (9.2) and (9.31) described above.

With regard to assessing the robustness of prior specifications to changes in the specification, if $E_{H_1}(\cdot) = E_{H_2}(\cdot)$, then this may be addressed by evaluating $DV_{\frac{2}{1}}(\cdot)$ and $DV_{\frac{1}{2}}(\cdot)$ before observing data. If both summaries are near one, across designs, then any design choice based on reducing expected posterior variance for some weighted combinations of the quantities of interest will not be sensitive to model elaboration. If the two expectations are not equal, then we may separately compare $G_{\frac{2}{1}}$ and the residual forms for the comparison.

After sampling, we carry out two assessments. First, we can use the observed comparison to assess whether the data appear to be more or less consistent with the varied specification. Secondly, the Bayes linear adjustment implied by the sampling results in adjusted variances and expectations for the quantities of interest, and these will differ under alternative prior specifications. Consequently, with regard to assessing the robustness of posterior specifications to changes in the prior specification, we may similarly evaluate (9.2) and (9.31) for the alternative posterior beliefs.

Note that if we only vary the specification $E_{H_2}(X)$, but keep the specification $E_{H_1}(X)$ fixed, then we need only evaluate $E_{H_2}(Z_i)$ to obtain the expectation comparison (9.38). Otherwise, varying the specification $\mathbf{H}_2$ generally requires recalculation of the canonical quantities for the comparison.

9.15.1 Sensitivity analyses for expectations

When constructing a specification for a collection of quantities, one way of looking at the specification is to consider that it requires qualitative decisions as to structural relationships between the quantities of interest, followed by quantitative decisions as to the magnitude of relationship. The former concerns choice of model type; the latter concerns particular choices for uncertain quantities such as the coefficients in a linear model. For example, suppose we consider relating a response quantity Y to an explanatory quantity X. We might consider a regression equation of the form

$$Y_i = a + bX_i + \epsilon_i,$$

which constitutes the qualitative choice. We would then need to consider belief choices for the uncertain quantities a, b, and $\epsilon_1, \epsilon_2, \ldots$. Assuming that the quantities X_i are fixed, the choices are the second-order belief specifications for a, b, and the quantities $\epsilon_1, \epsilon_2, \ldots$, so that the list of possible quantitative choices for this example is $E(a)$, $E(b)$, $E(\epsilon_i)$ for all i, $Var(a)$, $Var(b)$, $Cov(a, b)$, $Cov(a, \epsilon_i)$ for all i, $Cov(b, \epsilon_i)$ for all i, and $Cov(\epsilon_i, \epsilon_j)$ for all i, j. For a sensitivity analysis, we may proceed by fixing an initial model structure, and then making a specification which we shall label as $\mathbf{H}_0$. We then explore the effects of perturbing $\mathbf{H}_0$ by altering some of the belief specifications to give another specification, and then comparing the two specifications. We repeat the process by systematically varying $\mathbf{H}_0$ and comparing to the initial specification.

Suppose we are exploring sensitivity of specifications concerning a collection X which we relate to observables under a given structural model and with a given set of belief specifications. To carry out the sensitivity analysis, we arrange the belief choices that we wish to explore as the vector $V = (V_1, \ldots, V_m)$. We now make our specification $\mathbf{H}_0$ using an initial belief choice, $V = v_0$. This leads to specifications $\mathrm{E}_{\mathbf{H}_0}(X)$ and $\mathrm{Var}_{\mathbf{H}_0}(X)$. We now choose a sequence of alternative belief choices,

$$V = v_1, V = v_2, \ldots, V = v_k, \ldots,$$

which we anticipate covers the range of plausible specifications for the problem at hand. Given the initial structural model, the alterations lead to new specifications which we label

$$\mathbf{H}_1, \mathbf{H}_2, \ldots, \mathbf{H}_k, \ldots,$$

and to alternative specifications $\mathrm{E}_{\mathbf{H}_k}(X)$ and $\mathrm{Var}_{\mathbf{H}_k}(X)$. Each such specification $\mathbf{H}_k$ can now be compared to the original specification $\mathbf{H}_0$ using the maximal summaries (9.2) and (9.31) and the bearings (9.33) for the comparisons.

For the assessment of sensitivity, we now evaluate, for any pair of values $V = v_j$, $V = v_k$, the bearings $\mathrm{G}_{\frac{j}{0}}$ and $\mathrm{G}_{\frac{k}{0}}$. These identify, relative to the initial variance $\mathrm{Var}_{\mathbf{H}_0}(X)$, the directions of maximal difference in expectations under $\mathbf{H}_j$ and $\mathbf{H}_k$, respectively. If these directions are close, we can conclude that there is little difference between $\mathbf{H}_j$ and $\mathbf{H}_k$ relative to $\mathbf{H}_0$. Consequently, we define the squared distance between the specifications $\mathbf{H}_j$ and $\mathbf{H}_k$, relative to $\mathbf{H}_0$, to be the variance of the adjusted bearing $\mathrm{G}_{\frac{j/k}{0}} = \mathrm{G}_{\frac{j}{0}} - \mathrm{G}_{\frac{k}{0}}$, so that

$$d_0(j, k) = \mathrm{Var}_{\mathbf{H}_0}(\mathrm{G}_{\frac{j/k}{0}}) = \mathrm{Var}_{\mathbf{H}_0}(\mathrm{G}_{\frac{j}{0}} - \mathrm{G}_{\frac{k}{0}}) \tag{9.69}$$

$$= \sum_{i=1}^{p} [\mathrm{E}_{\mathbf{H}_j}(Z_i) - \mathrm{E}_{\mathbf{H}_k}(Z_i)]^2. \tag{9.70}$$

As, for each $Y \in \langle B \rangle$, we have

$$\mathrm{E}_{\mathbf{H}_j}(Y) - \mathrm{E}_{\mathbf{H}_k}(Y) = \mathrm{Cov}_{\mathbf{H}_0}(Y, \mathrm{G}_{\frac{j/k}{0}}),$$

we therefore have

$$\max_{X \in \langle B \rangle} \frac{[\mathrm{E}_{\mathbf{H}_j}(X) - \mathrm{E}_{\mathbf{H}_k}(X)]^2}{\mathrm{Var}_{\mathbf{H}_0}(X)} = \mathrm{Var}_{\mathbf{H}_0}(\mathrm{G}_{\frac{j/k}{0}}). \tag{9.71}$$

Small values of $d_0(j, k)$ therefore imply no practical difference between $\mathbf{H}_j$ and $\mathbf{H}_k$, when considered as alternative specifications to $\mathbf{H}_0$, as far as the expectation specification is concerned.

For a graphical analysis, we may construct the distance matrix $\mathbf{D}_0$ with (j, k)th entry being the distance $d_0(j, k)$. We use multidimensional scaling to produce a low-dimensional graphical representation of the distance matrix, which we call a **sensitivity map**. If two points are close in the map, then the belief specification is not sensitive to the choice between them, relative to $\mathbf{H}_0$. Interpreting the major axes gives insight into the dominant features affecting sensitivity, while clusters of points identify ranges of variation amongst belief choices with similar effects.

9.15.2 Example: robustness analysis for exchangeable regressions

Consider the exchangeable regressions example discussed above and in Chapter 6. For this, quantities Y_{rt} are constructed from a number of components (a_r, b_r, t, U_{rt}, H_{rt}, E_{rt}, V_{rt}, ϕ), establishing the qualitative structure. A number of belief statements must then be provided to quantify the specification. For example, $\mathrm{E}(a_r)$ and ϕ must be chosen; the full list of uncertain quantities and their particular choices in Chapter 6 are as follows:

$$\mathrm{Var}(U_{rt}) = 0.0204, \qquad \mathrm{Var}(H_{r1}) = 0.04, \qquad \mathrm{Var}(E_{rt}) = 0.01,$$
$$\phi = 0.7, \qquad \mathrm{Var}(F_{rt}) = 0.01,$$

$$\mathrm{E}(a_r) = 1.4, \qquad \mathrm{E}(b_r) = 0.1,$$
$$\mathrm{Var}(a_r) = 0.058, \qquad \mathrm{Var}(b_r) = 0.0017,$$
$$\mathrm{Cov}(a_r, a_s) = 0.038, \qquad \mathrm{Cov}(b_r, b_s) = 0.0016,$$

for all r and $s \neq r$. We shall call this specification $\mathbf{H}_0$. Suppose that we are concerned with the sensitivity of this specification to changes in the prior means for the two regression coefficients a and b. To do this, we now consider further specifications $\mathbf{H}_1, \ldots, \mathbf{H}_{20}$ as follows.

- We consider five possible values for the prior mean for a, namely {0, 1, 1.4, 1.8, 2.8}.
- We consider three possible values for the prior mean for b, namely {0, 0.1, 0.2}.

The combination of these 5×3 possibilities yields specifications $\mathbf{H}_1, \ldots, \mathbf{H}_{15}$, of which one is identical to $\mathbf{H}_0$.

The sensitivity map comparing expectation differences for these specifications relative to $\mathbf{H}_0$ is shown in Figure 9.12. The map can be labelled to indicate which point corresponds to which specification: we show the labels for the most outlying points. One of these corresponds to $\mathbf{H}_0$, showing that there is some distance between the base specification and the remaining specifications. There is a central cluster of points corresponding to specifications with similar implications, which we might therefore describe as robust.

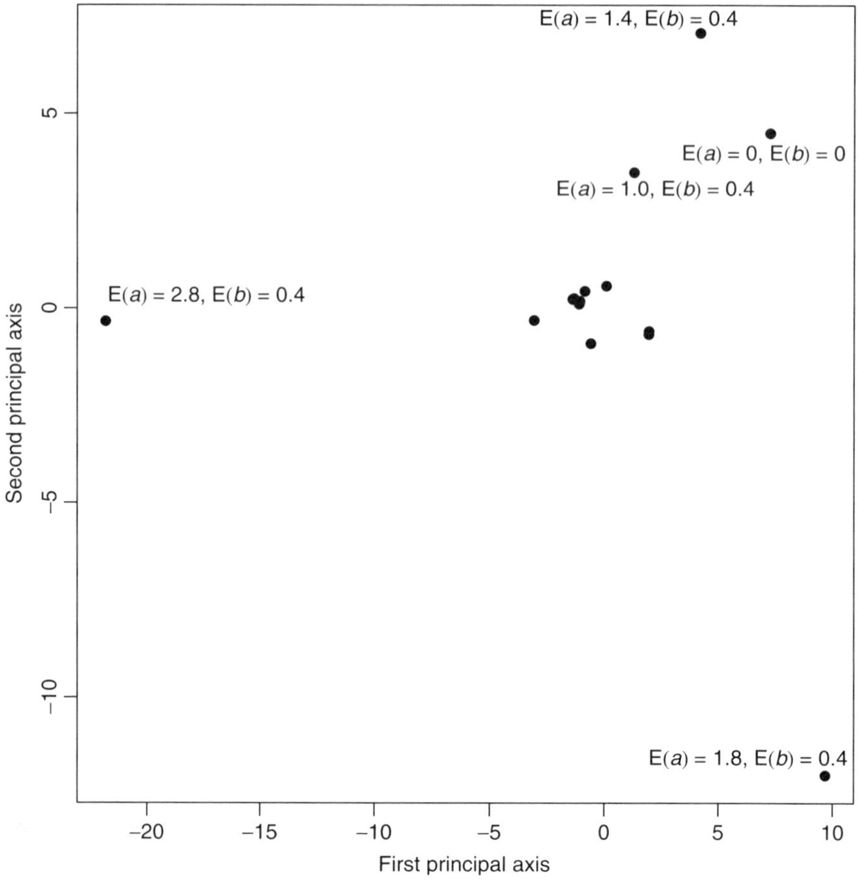

Figure 9.12 Sensitivity map for the differences of expectations. Outlying points are labelled by corresponding values of $E_{\mathbf{H}_i}(a)$ and $E_{\mathbf{H}_i}(b)$.

The most outlying points generally correspond to specifications with the highest expectation that we made for the slope quantity, $E(b) = 0.4$, and with non-zero intercept expectations. We conclude that the dominating influence for the sensitivity analysis is a large specification for the prior mean for the slope coefficient, in combination with different specifications for the prior mean for the intercept. The specification does not seem sensitive to small changes in the slope specification, whatever the intercept specification.

9.15.3 Sensitivity analyses for variances

We may similarly assess differences in the variance specification. One natural measure of distance between variance specifications under $\mathbf{H}_j$ and $\mathbf{H}_k$ is the Euclidian norm for the difference between them,

$$
\begin{aligned}
d_v(j,k) &= ||\mathrm{Var}_{\mathbf{H}_j}(X) - \mathrm{Var}_{\mathbf{H}_k}(X)|| \\
&= [\mathbf{tr}\{\{\mathrm{Var}_{\mathbf{H}_j}(X) - \mathrm{Var}_{\mathbf{H}_k}(X)\}^T\{\mathrm{Var}_{\mathbf{H}_j}(X) - \mathrm{Var}_{\mathbf{H}_k}(X)\}\}]^{\frac{1}{2}}, \qquad (9.72)
\end{aligned}
$$

which corresponds to the inner product $(A \cdot B) = \mathbf{tr}\{A^T B\}$. Small values of $d_v(j,k)$ imply no practical difference between $\mathbf{H}_j$ and $\mathbf{H}_k$, when considered as alternative variance specifications. For a graphical analysis, we may construct the distance matrix $\mathbf{D}_v$ with (j,k)th entry being the distance $d_v(j,k)$, and use multidimensional scaling to produce a low-dimensional graphical representation of the distance matrix. This can be interpreted similarly to the sensitivity map for expectation differences.

9.15.4 Example: robustness analysis for variance specifications

To return to the exchangeable regressions example, we consider 12 alternative specifications as follows.

- We consider three possible values for $\mathrm{Var}(a_r)$, namely $\{0.058, 0.068, 0.078\}$, whilst keeping $\mathrm{Cov}(a_r, a_s)$ fixed. This corresponds to weakening the relationship between an individual observation and the underlying mean component for the underlying intercept $\mathcal{M}(a)$.

- We consider two possibilities for $\mathrm{Var}(b_r)$, namely $\{0.0017, 0.0020\}$, whilst keeping $\mathrm{Cov}(b_r, b_s)$ fixed. This corresponds to weakening the relationship between an individual observation and the underlying mean component for the slope $\mathcal{M}(b)$.

Otherwise, we keep the variances for the error components fixed at their usual values. The combination of these 3×2 possibilities yields specifications $\mathbf{H}_1, \ldots, \mathbf{H}_6$.

- We keep the specifications for the regression coefficients fixed and vary each of the error component variances once: $\mathrm{Var}(E_{rt})$ at 0.05 instead of 0.01; $\mathrm{Var}(F_{rt})$ at 0.05 instead of 0.01; and $\mathrm{Var}(H_{r1})$ at 0.1 instead of 0.0204. This gives specifications $\mathbf{H}_7, \mathbf{H}_8, \mathbf{H}_9$. In each case, we are interested in seeing whether large changes for the error quantities imply large changes overall.

- We specify a model without a slope component, as in §9.12, and with a single error component E_{rt} for which we specify variances of $\{0.01, 0.04, 0.07\}$. This gives specifications $\mathbf{H}_{10}, \mathbf{H}_{11}, \mathbf{H}_{12}$. Specifications $\mathbf{H}_1$ and $\mathbf{H}_{11}$ were previously compared in §9.12.

Table 9.12 Summary of differences in variance specifications. For each specification, we fix $\text{Cov}(a_r, a_s) = 0.038$.

	$\text{Var}(a_r)$	$\text{Var}(b_r)$	$\text{Cov}(b_r, b_s)$	$\text{Var}(E_{rt})$	$\text{Var}(F_{rt})$	$\text{Var}(H_{rt})$
$\mathbf{H}_1$	0.058	0.0017	0.0016	0.01	0.01	0.0204
$\mathbf{H}_2$	0.068	0.0017	0.0016	0.01	0.01	0.0204
$\mathbf{H}_3$	0.078	0.0017	0.0016	0.01	0.01	0.0204
$\mathbf{H}_4$	0.058	0.0020	0.0016	0.01	0.01	0.0204
$\mathbf{H}_5$	0.068	0.0020	0.0016	0.01	0.01	0.0204
$\mathbf{H}_6$	0.078	0.0020	0.0016	0.01	0.01	0.0204
$\mathbf{H}_7$	0.058	0.0017	0.0016	0.05	0.01	0.0204
$\mathbf{H}_8$	0.058	0.0017	0.0016	0.01	0.05	0.0204
$\mathbf{H}_9$	0.058	0.0017	0.0016	0.01	0.01	0.1000
$\mathbf{H}_{10}$	0.058	0.0000	0.0000	0.01	0.00	0.0000
$\mathbf{H}_{11}$	0.058	0.0000	0.0000	0.04	0.00	0.0000
$\mathbf{H}_{12}$	0.058	0.0000	0.0000	0.07	0.00	0.0000

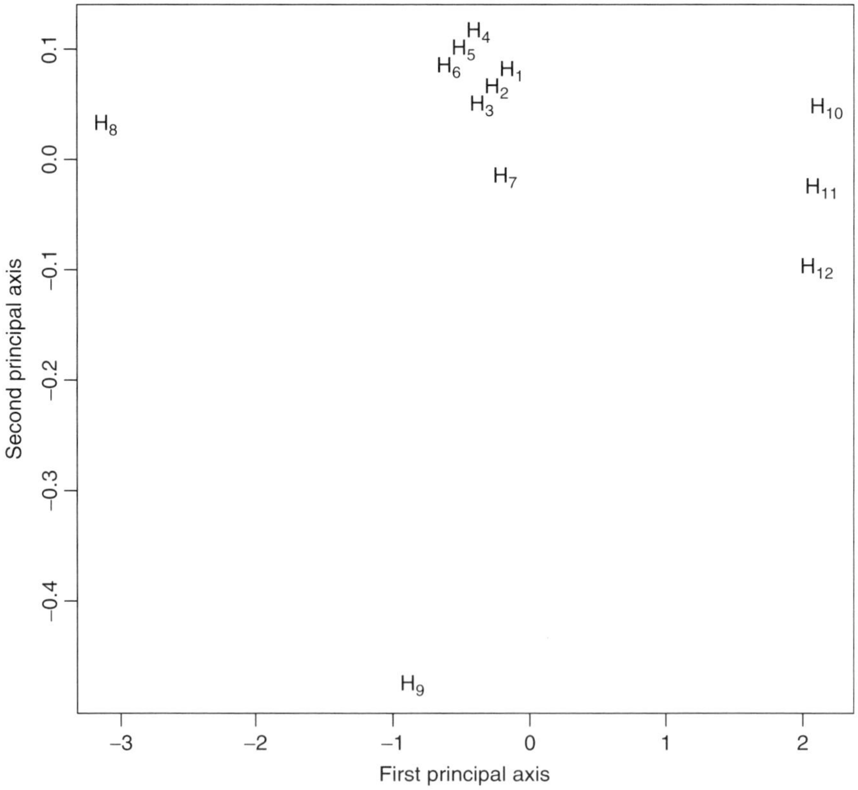

Figure 9.13 Sensitivity map for the differences in variance specifications, labelled by the specifications.

The differences between these variance specifications are summarized in Table 9.12. The sensitivity map comparing these variance matrices is shown in Figure 9.13, labelled by specification. Note that variation in the second principal axis is minor compared to variation along the first principal axis. That is, the differences between specifications can be largely expressed in the x-direction.

There is one main cluster containing the specifications $\mathbf{H}_1, \ldots, \mathbf{H}_6$, which we take to be quite similar. We conclude that the variance and covariance specifications for the slope and intercept quantities are relatively insensitive to small changes. Specification $\mathbf{H}_7$ is also close to this main cluster, suggesting insensitivity to modifying $\text{Var}(E_{rt})$ alone. Specification $\mathbf{H}_9$ is also relatively close to this cluster (as variation in the y-direction is minor) suggesting insensitivity to modifying $\text{Var}(H_{rt})$ alone. However, $\mathbf{H}_8$ is widely separated from the other specifications, suggesting that the model is sensitive to changes in specification for $\text{Var}(F_{rt})$. Finally, there is a secondary cluster corresponding to the model without a slope coefficient. This cluster is quite distant from the main cluster, indicating that the variance specifications for the models with and without the slope component are rather dissimilar. Further, the grouping in the secondary cluster shows that the specification without a slope component was not strongly sensitive to the choice for $\text{Var}(E_{rt})$.

9.16 Further reading

Basic ideas of belief comparison are described in Goldstein (1991), which includes further examples of the comparisons of specifications for the example in §9.12. Use of belief comparisons to explore issues of robustness and sensitivity is described in Goldstein and Wooff (1994).

10

Bayes linear graphical models

Graphical models offer compact pictorial representations of the qualitative structure of our beliefs. Such representations are useful for both the construction and the analysis of complex belief structures. Bayes linear graphical models perform this task for second-order specifications. The diagrams which we shall describe are closely related to Gaussian diagrams, which describe the relationships between collections of normal random quantities, as such relationships are determined by the covariance structure of the collection of quantities.

When we construct a belief structure, we often begin by forming qualitative ideas as to how the various ingredients of a problem are related. Graphical modelling helps us to visualize and then build complex structured multivariate relationships. Such pictures are particularly helpful for communicating between members of a group, and certain kinds of graphical model are sometimes termed knowledge maps to suggest their role of laying out a terrain of relationships to be explored and quantified. Further, the model reveals the most efficient way to quantify our belief specifications, by identifying which aspects of the full specification we are required to assess and which may be deduced from the form of our model.

Graphical models are helpful in two different ways in the analysis of the resulting belief specifications. First, the diagram helps us to organize the calculations which are required to solve the diagram, whereby we may break down the analysis of high-dimensional structures into a series of low-dimensional components which are sufficient to determine the whole system, and so allows us to use local computation to solve much larger problems than we could easily assess by direct computation. Secondly, the diagram provides a natural structure for displaying the results of a Bayes linear analysis in a simple graphical form.

There exists an extensive literature on graphical modelling for probabilistic systems; see, for example, Pearl (1988), Jensen (2001), Lauritzen (1996) and Cowell et al. (1999). In this chapter, we will not aim to survey this whole literature but only to provide a self-contained introduction to those elements of graphical modelling which are most relevant to the Bayes linear approach, deriving all results directly from the corresponding properties of belief adjustment.

10.1 Directed graphical models

Bayes linear graphical models represent the (linear) belief separations between collections of random quantities. Recall that separation of collections A and B by the collection C, written $\lfloor A \perp\!\!\!\perp B \rfloor / C$, is the property that C is Bayes linear sufficient for B for adjusting A. There are two basic forms for the graphical representation of belief separation, namely directed and undirected graphs.

In this section, we discuss representations of beliefs using a **directed acyclic graph**. Such a graph has nodes $B_1, \ldots B_r$, say, where each node B_i represents a collection $X_{i1}, \ldots X_{im_i}$ of random quantities. Certain nodes are joined by directed arrows, subject to the constraint that there are no directed cycles, namely directed paths which return to their starting point.

Definition 10.1 *If a directed arc goes from node A to node B, then A is termed a **parent** of B, B is termed a **child** of A, and A, B are said to be **adjacent** or **neighbour** nodes. We denote by* Pa(B) *the set of parents of B.*

The arcs express the separations of belief between the nodes by the requirement that any pair of nodes is separated by the parent nodes. We have the following definition of a directed graphical model.

Definition 10.2 *A model is a **directed (second-order) graphical model** if, for any nodes B_i and B_j, we have*

$$\lfloor B_i \perp\!\!\!\perp B_j \rfloor / (\mathrm{Pa}(B_i) \cup \mathrm{Pa}(B_j)). \tag{10.1}$$

While Definition 10.2 is a natural definition, the condition may be laborious to check for any particular graph. An alternative approach is as follows.

Definition 10.3 *We say that any ordering of the nodes with the property that any parent of a node on the graph is also a predecessor of the node in the list is a node ordering which is **consistent** with the graph.*

On any directed acyclic graph, we may construct at least one consistent ordering, by numbering at stage 1 any node with no parents as node 1, and then, at each stage, m, numbering as node m any node all of whose parents are already numbered.

This algorithm works because if, at any stage, we could not number a node, then this would imply that each unnumbered node had an unnumbered parent, which would imply that there was a cycle in the unnumbered nodes. The ordering so created must be consistent as, by construction, each node is only numbered when all parents have been numbered.

We have the following alternative definition of a directed graphical model.

Definition 10.4 *A model is a* ***directed (second-order) graphical model*** *if, when* $B_1, \ldots, B_r$ *is a consistent ordering on the nodes, then for each k, node* B_k *is separated by the parent nodes from all predecessor nodes in the list, namely*

$$\lfloor B_k \perp\!\!\!\perp B(k-1) \rfloor \,/\, \mathrm{Pa}(B_k), \tag{10.2}$$

where $B(j) = B_1 \cup \ldots \cup B_j$.

There will often be a natural ordering under which property (10.2) is relatively easy to check, so that we may use this ordering to draw the graph as, for each node in the ordering, we must only decide which sub-collection of the predecessor nodes is Bayes linear sufficient for the whole collection of predecessors. However, it is not immediately obvious whether the two properties that we have introduced are equivalent, or even whether property (10.2) is dependent on the particular choice of list ordering, for example whether there may be certain consistent orderings for a given graph for which (10.2) holds and others for which it does not. However, we will now show that the two properties are equivalent. In particular, as property (10.1) is independent of the list ordering, it follows that property (10.2) does not depend on the choice of consistent list ordering. We have the following equivalence.

Theorem 10.5 *Property* (10.2) *is equivalent to property* (10.1).

Proof. Suppose that (10.2) holds for some consistent list ordering $B_1, \ldots, B_r$. For each j, we have

$$\lfloor B_j \perp\!\!\!\perp B(j-1) \rfloor \,/\, \mathrm{Pa}(B_j),$$

so that, as the list ordering is consistent for each $i < j$, we have

$$\lfloor B_j \perp\!\!\!\perp (B_i \cup \mathrm{Pa}(B_i)) \rfloor \,/\, \mathrm{Pa}(B_j),$$

so that by Property 5.21.3 we have

$$\lfloor B_i \perp\!\!\!\perp B_j \rfloor \,/\, (\mathrm{Pa}(B_i) \cup \mathrm{Pa}(B_j)).$$

Conversely, suppose that (10.1) holds. We choose a consistent list ordering $B_1, \ldots, B_r$. Select any $s > 1$. As B_1 has no parents, we have

$$\lfloor B_s \perp\!\!\!\perp B_1 \rfloor \,/\, \mathrm{Pa}(B_s).$$

Suppose that $\lfloor B_s \perp\!\!\!\perp B_j \rfloor \,/\, \mathrm{Pa}(B_s)$, for $j = 1, \ldots, m-1$, where $m < s-1$. As the list ordering is consistent, it therefore follows that

$$\lfloor B_s \perp\!\!\!\perp \mathrm{Pa}(B_m) \rfloor \,/\, \mathrm{Pa}(B_s).$$

From (10.1), we have $\lfloor B_s \perp\!\!\!\perp B_m \rfloor \,/\, (\mathrm{Pa}(B_s) \cup \mathrm{Pa}(B_m))$. It therefore follows from Property 5.21.3 that

$$\lfloor B_s \perp\!\!\!\perp B_m \rfloor \,/\, \mathrm{Pa}(B_s).$$

Thus (10.2) follows by induction. ■

10.1.1 Construction via statistical models

It is often the case in statistical problems that our beliefs are represented through a series of models each expressing the relationship between a dependent variable and a collection of explanatory variables. In such cases, it is natural to construct the Bayes linear graphical model as follows.

We write each random quantity in the model as a node on the diagram. Optionally, fixed quantities may be added as rectangles on the diagram to aid interpretation, but have no implications for belief separation for the random quantities connected to them. We construct a list ordering for which any dependent variable comes later in the list than the explanatory quantities used to define the model for that quantity. We construct the graph by running through the elements of the list in order and draw arcs to each from a subset of elements which is jointly sufficient to separate that element from all other preceding members of the list. In particular, each dependent variable receives an arc from each explanatory quantity in the defining equation for that quantity.

10.2 Operations on directed graphs

The graphical model carries information about belief separations, which are represented by missing arcs on the graph. While we will usually want to create sparse graphs, with as few arcs as possible, sometimes we will want to introduce additional arcs or to combine nodes, to facilitate certain calculations and displays that we shall describe below. Each such operation may conceal various belief separations which were deducible on the original diagram. However, we must be careful not to carry out transformations of the diagram which introduce new belief separations which were not deducible from the original model. Therefore, it is useful to know under which conditions we may carry out such operations without creating false inferences.

Definition 10.6 *We say that graphical model $\mathcal{D}_1$* ***implies*** *model $\mathcal{D}_2$ if every belief separation for $\mathcal{D}_2$ is also a belief separation for $\mathcal{D}_1$.*

Definition 10.7 *We say that an operation on the graph is* ***allowable*** *if the resulting graph is implied by the original graph.*

Theorem 10.8 (Allowable operations on the graph)

10.8.1: *Suppose that we have a graph, and a consistent list ordering $B_1, \ldots, B_r$. Then adding a directed arc, from the lower to the higher numbered value, between any pair of nodes is an allowable operation.*

10.8.2: *Suppose that two nodes have the same children and the same parents. Then it is allowable to* ***combine*** *the two nodes into a single node with the same child and parent sets.*

10.8.3: *Suppose that, in a consistent node ordering of a graph, nodes B_i and B_j, $i < j$, have the properties that all the nodes B_k in the child set for B_i have $k \geq j$ and all the nodes B_e in the parent set for B_j have $e \leq i$. Then it is allowable to* ***combine*** *nodes B_i and B_j into a single node with child set the union of the child sets for the two nodes and parent set the union of the parent sets for the two nodes.*

In particular, if we may group the nodes on the diagram into three disjoint sets A, B, C so that B has no children in A and C has no children in A or B, then it is allowable to ***combine*** *all the nodes in B into a single node, with parents and children the union of the parent set of B and the union of the child set of B, respectively.*

Proof. For Property 10.8.1, suppose that $i < j$, and that B_i and B_j are not currently joined by an arc. If we add a new arc from B_i to B_j, then the only node for which the parent collection has changed is B_j, which has increased to $\text{Pa}(B_j) \cup B_i$. Therefore, all that we need to check is the condition that $\lfloor B_j \perp\!\!\!\perp B(j-1) \rfloor / (\text{Pa}(B_j) \cup B_i)$. This follows directly from Property 5.21.3, as $\lfloor B_j \perp\!\!\!\perp B(j-1) \rfloor / (\text{Pa}(B_j))$, and $B_i \in B(j-1)$.

Property 10.8.2 follows from (10.1), as each parent and child set is unchanged.

For Property 10.8.3, by Property 10.8.1 we may add directed arcs from each of B_i and B_j to any node in the union of the child sets for the two nodes for which such an arc is not yet present as, from the conditions imposed on nodes B_i and B_j, all nodes in the union of the child sets have higher numbers in the consistent list ordering than j. Similarly, we may add directed arcs to each of B_i and B_j from any node in the union of the parent sets for the two nodes for which such an arc is not yet present. In the resulting graph B_i and B_j have the same children and the same parents. From Property 10.8.2, we may therefore join B_i and B_j into a single node as required.

The conditions imposed on collections A, B, C ensure that, for each pair of nodes in B, these conditions are satisfied, and the result follows. ∎

Thus, we may choose to incorporate surplus arcs, if we want to track the information flow into a particularly important node, or combine nodes to clarify the structure of the graph. For example, there may be a natural time ordering on the nodes, and we may want to assess how much information about certain future events we may gain by observing various past events. We may add some extra arcs to display such information or combine nodes measured at the same time point. As long as we respect the conditions of Theorem 10.8, the graph will be valid, although some of the structure of the belief separation may be lost.

Finally, there are certain basic allowable manipulations that we can perform on the graph which follow as direct consequences of the generalized conditional independence properties of the graph. Two of the most important and well known manipulations on the graph are arc reversal and arc removal. We have the following theorems.

Theorem 10.9 (Arc reversal) *Suppose that node A is a parent of node B, and that there is no other directed path from A to B. Then it is an allowable operation to reverse the direction of the arc from node A to node B, provided that, in addition, we add arcs from each parent of A to B and add arcs from each parent of B to A.*

Proof. As there is no other directed path from A to B, no descendent of A may be an ancestor of B, so that we may construct a consistent list ordering in which each parent of A or B appears in the list ordering before A and A and B are consecutive members of the list. Therefore reversing the arc between A and B, joining the parents of A to B, and joining the parents of B to A corresponds to a list ordering which is the same as the original list ordering with A and B interchanged. There can be no directed cycles created by this operation.

Denote the collection of antecedents of A in the original list ordering as D. The original graph satisfied the properties

$$\lfloor D \perp\!\!\!\perp A \rfloor \,/\, \mathrm{Pa}(A), \quad \lfloor D \perp\!\!\!\perp B \rfloor \,/\, \mathrm{Pa}(B). \tag{10.3}$$

Denote by $\mathrm{Pa}(B_A)$ the collection of all parents of B except A. Under the new ordering the parents of A are $\mathrm{Pa}(A^*) = \mathrm{Pa}(A) \cup \mathrm{Pa}(B_A) \cup B$ and the parents of B are $\mathrm{Pa}(B^*) = \mathrm{Pa}(A) \cup \mathrm{Pa}(B_A)$. We need to show that (10.3) implies the same properties, but with $\mathrm{Pa}(A)$, $\mathrm{Pa}(B)$ replaced by $\mathrm{Pa}(A^*)$, $\mathrm{Pa}(B^*)$.

As both $\mathrm{Pa}(A) \subseteq D$ and $\mathrm{Pa}(B_A) \subseteq D$, from Property 5.21.3 we have

$$\lfloor D \perp\!\!\!\perp A \rfloor \,/\, (\mathrm{Pa}(A) \cup \mathrm{Pa}(B_A)), \quad \lfloor D \perp\!\!\!\perp B \rfloor \,/\, (\mathrm{Pa}(A) \cup \mathrm{Pa}(B_A) \cup A).$$

Again, from Property 5.21.3, we therefore have

$$\lfloor D \perp\!\!\!\perp A \cup B \rfloor \,/\, (\mathrm{Pa}(A) \cup \mathrm{Pa}(B_A)),$$

so that from Property 5.21.3

$$\lfloor D \perp\!\!\!\perp A \rfloor \,/\, \mathrm{Pa}(A^*), \quad \lfloor D \perp\!\!\!\perp B \rfloor \,/\, \mathrm{Pa}(B^*).$$

■

Theorem 10.10 (Node removal) *It is an allowable operation to remove a node A from the graph, provided that we add arcs so that each parent of A becomes a parent of each child of A. Each child of A must also be connected by an arc, the directions of the arcs between children being chosen according to a consistent node ordering, where each arc added between children of A joins the lower numbered node to the higher numbered node. Each child that receives an arc from another child in this way must also receive an arc from each parent of that child.*

This theorem may be proved most naturally by exploiting further properties of belief separation that we shall describe in §10.4, and we defer the proof of this theorem to that section.

10.3 Quantifying a directed graphical model

In many problems, we begin the task of specifying our beliefs by purely qualitative consideration of the various belief separations that we wish to impose upon our collection of beliefs. In such cases, we begin by drawing a diagram satisfying the required properties and then proceed to quantify the diagram.

We wish to specify the second-order structure expressing beliefs across all of the elements of all the collections of random quantities represented on the diagram. This may be a difficult process, for a large diagram, and so must be carried out systematically, to respect and exploit all of the coherence requirements for the joint specification. The fundamental simplification of the directed graph is that we only need to specify beliefs between neighbouring nodes in order to complete the specification over the whole graph.

Suppose that we have two nodes A and B, where A is the parent of B. We may specify directly the mean and variance for each node and the covariance between each. Alternatively, it might be more natural to specify the mean and variance for A and then to complete the belief specification by considering the adjusted expectation and variance for B given A. In particular, suppose that we may directly assess $\mathrm{E}_A(B)$, $\mathrm{Var}_A(B)$. Then, as $\mathrm{E}_A(B)$ is a linear function of A, we may deduce the corresponding mean, variance, and covariance for B as

$$\begin{aligned} \mathrm{E}(B) &= \mathrm{E}(\mathrm{E}_A(B)), \\ \mathrm{Var}(B) &= \mathrm{Var}_A(B) + \mathrm{Var}(\mathrm{E}_A(B)), \\ \mathrm{Cov}(A, B) &= \mathrm{Cov}(A, \mathrm{E}_A(B)). \end{aligned} \tag{10.4}$$

Now suppose that we have three belief structures A, B, and C, for which $\lfloor A \perp\!\!\!\perp B \rfloor / C$. By Theorem 5.23, $\lfloor A \perp\!\!\!\perp B \rfloor / C$ implies that $(A - \mathrm{E}_C(A))$ is uncorrelated with B, i.e. that

$$\mathrm{Cov}(A, B) = \mathrm{Cov}(\mathrm{E}_C(A), B). \tag{10.5}$$

Thus, to evaluate $\mathrm{Cov}(A, B)$, we assess $\mathrm{E}_C(A)$, which is determined by the covariance structure $\mathrm{Cov}(A, C)$, and then we assess $\mathrm{Cov}(\mathrm{E}_C(A), B)$, which, as $\mathrm{E}_C(A)$ is a linear form in C, is determined by $\mathrm{Cov}(C, B)$. Therefore, the covariance structure between the collections A and B is fully determined by the pair of covariance structures $\mathrm{Cov}(A, C)$ and $\mathrm{Cov}(C, B)$ and the variance matrix $\mathrm{Var}(C)$. For finite vectors A, C, B with $\lfloor A \perp\!\!\!\perp B \rfloor / C$, we have the matrix representation

$$\mathrm{Cov}(A, B) = \mathrm{Cov}(A, C)\mathrm{Var}(C)^{\dagger}\mathrm{Cov}(C, B), \tag{10.6}$$

as in Theorem 5.20. Therefore, for a general directed graphical model, the covariance structure over the full model is fully determined by the variance structure for each node, and the covariance structure between each pair of adjacent nodes. We may construct the full specification as follows. Construct a consistent list ordering

$B_1, \ldots, B_r$ of the nodes of the graph. Suppose that we have made a full second-order specification over $B(s-1)$, $1 \leq s \leq r$. We now extend this specification to the collection $B(s)$. From the belief separation, $\lfloor B_s \perp\!\!\!\perp B(s-1) \rfloor \,/\, \mathrm{Pa}(B_s)$, the covariance structure between B_s and $B(s-1)$ is fully determined by the covariance structure between B_s and $\mathrm{Pa}(B_s)$ and the covariance structure between $B(s-1)$ and $\mathrm{Pa}(B_s)$. Therefore, we may construct the full covariance structure between B_s and $B(s-1)$ from the individual covariance specifications between B_s and each member of $\mathrm{Pa}(B_s)$. Stepping through the nodes according to the list ordering, we may therefore sequentially construct the complete belief specification over the full collection of random quantities by introducing each node B_s in order and specifying the mean and variance for that node, either directly or exploiting the adjustment based on the parent nodes using (10.4), and then specifying the covariance between B_s and $\mathrm{Pa}(B_s)$.

10.4 Undirected graphs

An alternative way to represent a collection of belief separations is through an undirected graph. In such a graph, each node represents a collection of random quantities, and certain pairs of nodes are joined by undirected arcs in order to reveal various dependencies between the collections. We say that a collection of nodes C **separates** the collections A and B of nodes on such an undirected graph if every path from a node in A to a node in B passes through a node in C. We relate such a separation on the graph to a separation of beliefs as follows.

We say that an undirected graph has the **second-order version of the global Markov property** if, for any three subsets of nodes A, B, C on the graph, if C separates A from B on the graph, then $\lfloor A \perp\!\!\!\perp B \rfloor \,/\, C$.

There is a sense in which the global Markov property is a natural condition on which to base a sequence of belief adjustments, as this condition is preserved under belief adjustment. We have the following result.

Theorem 10.11 *Suppose that an undirected graphical model is second-order global Markov. Choose any node, D say, and remove node D and all arcs entering D from the diagram. The resulting diagram is second-order global Markov for the belief structure resulting from adjusting all quantities by D.*

Proof. Suppose that we remove node D and all arcs entering D from the diagram. Suppose that on the new diagram all paths from collection A to collection B pass through collection C. It follows that all paths from A to B on the original diagram pass through C or D, so that $\lfloor A \perp\!\!\!\perp B \rfloor \,/\, (C \cup D)$. From Theorem 5.25, this condition implies that $\lfloor \mathbb{A}_D(A) \perp\!\!\!\perp \mathbb{A}_D(B) \rfloor \,/\, \mathbb{A}_D(C)$, so that separation on the modified graph corresponds to the separation of adjusted beliefs as required. ■

For any directed graphical model, there is an associated undirected graphical model, termed the **moral graph**, which displays general belief separations in a direct fashion. The moral graph is constructed as follows.

Definition 10.12 *The moral graph is constructed by (i) drawing an arc between any two nodes which are parents of the same child node and which are not currently joined by an arc (i.e. 'marrying' unmarried parents), and (ii) dropping all arrows.*

Node separation on the moral graph identifies belief separation as follows.

Theorem 10.13 *For any three collections of nodes A, B, C, within a directed graphical model, construct the moral graph on A, B, C and all ancestors. If C separates A from B on this graph, then $\lfloor A \perp\!\!\!\perp B \rfloor / C$.*

Proof. The proof follows by induction on the size of the graph. That the statement is true for graphs of size three follows by checking the various cases. We now suppose that the statement is true for all graphs of size n and deduce that it is true for all graphs of size $n + 1$.

Suppose, then, that the statement is true for all graphs of size n. Now consider a directed graph of $n + 1$ nodes, with a consistent node ordering $V_1, \ldots, V_n, V_{n+1}$. Choose three collections A, B, C for which, on the moral graph on A, B, C and ancestors, there is no path from A to B except through C. We must show that $\lfloor A \perp\!\!\!\perp B \rfloor / C$.

If V_{n+1} is not a member of any of the collections A, B, C, then $\lfloor A \perp\!\!\!\perp B \rfloor / C$ from the inductive hypothesis. Suppose node $V_{n+1} \in A$; the argument is identical for B. From the inductive hypothesis, all that we need to show is that $\lfloor V_{n+1} \perp\!\!\!\perp B \rfloor / C$, as the remaining nodes in A are separated from B by C in the moral graph on the graph of size n without V_{n+1}. Let E be the nodes common to $\text{Pa}(V_{n+1})$ and to C, and let F and G be the nodes in $\text{Pa}(V_{n+1})$ and not in C, and in C but not $\text{Pa}(V_{n+1})$, respectively. From the defining property of the directed graph, we have $\lfloor V_{n+1} \perp\!\!\!\perp (G \cup B) \rfloor / \text{Pa}(V_{n+1})$. Therefore $\lfloor V_{n+1} \perp\!\!\!\perp B \rfloor / (F \cup C)$, by Property 5.21.3. Further, all paths from F to B pass through C, as otherwise there would be a path from V_{n+1} to B which does not pass through C. Therefore we have $\lfloor F \perp\!\!\!\perp B \rfloor / C$, from the inductive hypothesis, so that $\lfloor V_{n+1} \perp\!\!\!\perp B \rfloor / C$ from Property 5.21.3.

Alternatively, suppose that node V_{n+1} is in C. Now, let E denote all the remaining nodes in C except V_{n+1}. If there is a path on the moral graph from A to $\text{Pa}(V_{n+1})$ which does not pass through E, then there cannot be a path from B to $\text{Pa}(V_{n+1})$ which does not pass through E, as all the parents of V_{n+1} are joined in the moral graph, so that there would then be a path from A to B which would not pass through C. Therefore E separates $\text{Pa}(V_{n+1})$ from at least one of A and B. Suppose E separates $\text{Pa}(V_{n+1})$ from A. By the inductive hypothesis, we have $\lfloor \text{Pa}(V_{n+1}) \perp\!\!\!\perp A \rfloor / E$. However, from the definition of the graph, we have $\lfloor V_{n+1} \perp\!\!\!\perp (A \cup E) \rfloor / \text{Pa}(V_{n+1})$. Therefore, from Property 5.21.3, we have $\lfloor V_{n+1} \perp\!\!\!\perp A \rfloor / (\text{Pa}(V_{n+1}) \cup E)$. Therefore, again by Property 5.21.3, we have $\lfloor V_{n+1} \perp\!\!\!\perp A \rfloor / E$.

Further, if there is no path on the moral graph from A to B except through C, then, on the moral graph on all nodes except V_{n+1}, there is no path from A to B except through E, as any such path would still exist on the graph on $n + 1$ vertices. Therefore, from the inductive hypothesis we have $\lfloor A \perp\!\!\!\perp B \rfloor / E$.

Therefore, as $\lfloor V_{n+1} \perp\!\!\!\perp A \rfloor \,/\, E$, we have $\lfloor A \perp\!\!\!\perp (B \cup V_{n+1}) \rfloor \,/\, E$, so that, from Property 5.21.3, we have $\lfloor A \perp\!\!\!\perp B \rfloor \,/\, C$, as required. ■

Observe that the full moral graph on all of the nodes in the directed graph loses some information about belief separation. In particular, if all paths from A to B pass through C on the moral graph of A, B, C and ancestors, but there is a path from A to B which does not pass through C in the full moral graph, then this belief separation is lost when the full graph is moralized. This could happen, for example, if there was a node D which was a descendent of both A and B.

10.4.1 Node removal via the moral graph

We now use the moral graph to prove Theorem 10.10, concerning conditions for removing nodes on a directed graph.

Proof. (Theorem 10.10). Each arc added to the graph when we remove a node according to the rules stated in the theorem joins a lower numbered node to a higher numbered node, so that no directed cycles can be introduced by this procedure, and the original ordering is also a consistent node ordering for the revised graph.

Denote the children of A by $B_1, \ldots, B_r$, where B_i has the ith lowest number in the consistent node ordering. Let $\mathrm{Pa}(B_i^*)$ denote all parents of B_i in the modified graph. We must show that, for each i, $\lfloor B_i \perp\!\!\!\perp A(B_i) \rfloor \,/\, \mathrm{Pa}(B_i^*)$, where $A(B_i)$ is the collection of all nodes occurring earlier than B_i in the node listing, with A removed. Therefore, from Theorem 10.13, it is sufficient to show that, on the moral graph constructed from the original graph on $B_i \cup A(B_i) \cup A$, all paths from B_i to a member of $A(B_i)$ pass through a member of $\mathrm{Pa}(B_i^*)$.

As B_i is the highest numbered node among $B_i \cup A(B_i) \cup A$, B_i has no children, so that any path from B_i on the moral graph on this collection must pass through a member of $\mathrm{Pa}(B_i)$. Either this is a member of $\mathrm{Pa}(B_i)$ other than A, so that it is a member of $\mathrm{Pa}(B_i^*)$, or the path passes through A. All paths from A on the moral graph must

(i) pass through an arc on the original graph, i.e. a member of $\mathrm{Pa}(A)$, or one of the children $B_1, \ldots, B_{(i-1)}$, of A; or

(ii) pass through an additional arc added from A on the moral graph – these arcs connect A to each parent of $B_1, \ldots, B_{(i-1)}$.

Each of the nodes entered in (i) or (ii) is a member of $\mathrm{Pa}(B_i^*)$, proving the theorem. ■

10.5 Example

We construct a Bayes linear graphical model for the problem considered in §5.14.2. Recall that the model is

$$Y_i = a + bx_i + e_i, \quad Z_i = c + dx_i + f_i, \qquad i = 1, \ldots, 12, \tag{10.7}$$

with prior specifications as in §5.14.2.2; in particular, we have as variance matrix for $G = [a, b, c, d]$,

$$\mathrm{Var}(G) = \begin{bmatrix} 4 & -6 & -1 & 0 \\ -6 & 225 & 0 & -90 \\ -1 & 0 & 1 & -2.4 \\ 0 & -90 & -2.4 & 144 \end{bmatrix}.$$

We may construct a Bayes linear graphical model for this problem using the procedure of §10.1.1 as follows. We limit attention to pairs of quantities such as Y_i, Y_j, as the implications for the remaining terms follow in an obvious manner.

- We begin by adding nodes to the diagram for the quantities

$$Y_i, Y_j, Z_i, Z_j, a, b, c, d, e_i, e_j, f_i, f_j. \tag{10.8}$$

  We also add rectangles for the fixed quantities x_i, x_j.

- We must construct a **consistent ordering** for the quantities (10.8). As the quantities Y_i, Y_j, Z_i, Z_j are defined by the linear equations, it is natural for these to appear at the end of the list. Otherwise, any arrangement of the quantities in the sub-collection $G = \{a, b, c, d\}$ and the sub-collection $Q = \{e_i, e_j, f_i, f_j\}$ provides the basis for a consistent node ordering. We take advantage of uncorrelatedness as follows. Each pair of quantities e_i, f_i is uncorrelated with all other quantities, so we begin our list with e_i, f_i, e_j, f_j. As a is uncorrelated with d, and b with c, there may be advantages in appending a, d, b, c to our list in that order. However, we choose for the sake of illustration to retain the ordering a, b, c, d so that we use the consistent ordering

$$e_i, f_i, e_j, f_j, a, b, c, d, Y_i, Y_j, Z_i, Z_j. \tag{10.9}$$

- We now consider separations of belief for this ordering. The first node is e_i. The second is f_i, which is correlated with e_i and so needs an arc $e_i \to f_i$. The nodes e_j, f_j are similarly connected by an arc, but are separated from the first pair. The next node to consider is node a. This is uncorrelated with all nodes earlier in the list, and so needs no arcs adding. Indeed, the sub-collection $G = \{a, b, c, d\}$ is uncorrelated with the sub-collection $Q = \{e_i, e_j, f_i, f_j\}$ and so there shall be no arcs between any quantity in G and any quantity in Q for this ordering. The next node b in the ordering is correlated with a and so an arc $a \to b$ is drawn.

  We now consider the next node in the list, c, which is correlated with its predecessor a, but not its predecessor b. The question is whether both a and b need to be parents of c. To answer the question, we may check any of the properties in Theorem 5.20. We choose Property 5.20.2 as this is fairly straightforward to compute. It is simple to verify that $\lfloor c \perp\!\!\!\perp Q \rfloor \,/\, a \cup b$. However, we find that it is not true that $\lfloor c \perp\!\!\!\perp a \rfloor \,/\, b$: by Property 5.20.2, we have

$$\mathrm{Cov}(c, a) = -1 \neq \mathrm{Cov}(c, b)\mathrm{Var}(b)^{\dagger}\mathrm{Cov}(b, a) = 0.$$

Thus, we require a and b to be parents of c. Notice that b, c are not conditionally independent given a, even though they are marginally independent.

For the next node in the list, d, it is similarly trivial to check that $\lfloor d \perp\!\!\!\perp Q \rfloor \,/\, a \cup b \cup c$. However, we find that it is not true that $\lfloor d \perp\!\!\!\perp a \rfloor \,/\, c \cup b$, not true that $\lfloor d \perp\!\!\!\perp b \rfloor \,/\, a \cup c$, and not true that $\lfloor d \perp\!\!\!\perp c \rfloor \,/\, a \cup b$. Hence, we require a, b, c to be parents of d.

- Y_i is a linear function of a, b, e_i (and the fixed quantity x_i) and so these all become parents to node Y_i. Z_i is a linear function of c, d, f_i (and the fixed quantity x_i) and so these all become parents to node Z_i. Similarly, arcs connecting parent nodes to Y_j and Z_j are drawn.

The resultant Bayes linear graphical model is shown in Figure 10.1. Note that we might obtain different representations, depending on which node ordering we choose. For example, if we choose instead the ordering a, d, c, b, the resulting

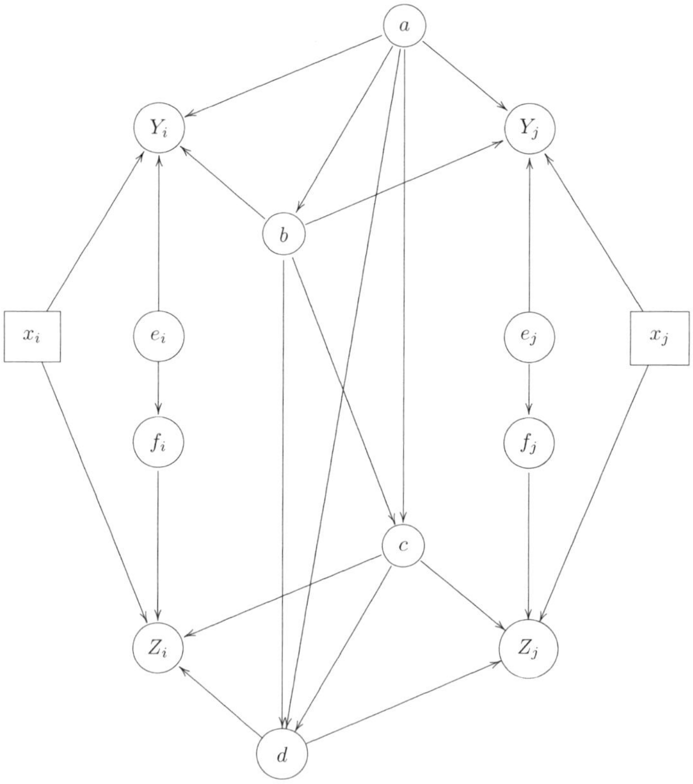

Figure 10.1 Bayes linear graphical model for model (10.7), using the consistent node ordering (10.9).

graph has

$$a \to c,\ d \to c,\ a \to b,\ d \to b,\ c \to b.$$

This is simpler, having one arc fewer, than the graph shown in Figure 10.1.

10.5.1 Plates for duplicated structures

Figure 10.1 shows that the i-subscripted quantities, $K_i = \{Y_i, e_i, f_i, Z_i, x_i\}$, are separated from the j-subscripted quantities, $K_j = \{Y_j, e_j, f_j, Z_j, x_j\}$, by the sub-collection of parameters $G = \{a, b, c, d\}$. That is, $\lfloor K_i \perp\!\!\!\perp K_j \rfloor / G$. Moreover, K_i has the same internal structure as K_j, and the arcs between K_i and G are the same as those between K_j and G. Such duplication is typical when random quantities are constructed to represent error terms and observables which are connected through an underlying model. For this example, we have 12 pairs of observables leading to collections $K_1, \ldots, K_{12}$. We can indicate such duplicated structure on the graph using the notion of a **plate**: we include a single collection of nodes K_i on the graph, draw a dashed line around the collection, and indicate how many times this plate is repeated. A plate for this example is shown in Figure 10.2.

The concept of a plate relies on fundamental properties of the graphical model. In particular, the plates such as $K_1, K_2, \ldots$ are belief separated by the parameter set such as G: this is why we only need to show a 'typical' member. Furthermore, because the whole collection is belief separated by the parameter set, we can find the covariance structure between any two plates. Indeed, one way to calculate the covariance structure between two plates is to employ the general rules for quantifying the whole model based on neighbouring nodes, developed in §10.3. Thus we may, if we wish, calculate $\text{Cov}(k_i, k_j)$ by

$$\text{Cov}(k_i, k_j) = \text{Cov}(k_i, G)\text{Var}(G)^{\dagger}\text{Cov}(G, k_j)$$

as in (10.6), for any vectors of quantities $k_i \in K_i$, $k_j \in K_j$.

10.5.2 Reading properties from the diagram

We may now read directly from the diagram some properties of the specified model. One of the most important features is that one can automatically read from the diagram that the joint specification between all the observables is determined by the specification for a, b, c, d, the specification for each e_i, f_i pair, and the covariance specification between each a, b, e_i and Y_i and between c, d, f_j and Z_j. Some belief separations are clear from the diagram and the node ordering. For example, we have

$$\lfloor Y_i \perp\!\!\!\perp Y_j \rfloor / (a, b, e_i, e_j) \quad \text{and} \quad \lfloor Y_i \perp\!\!\!\perp Z_j \rfloor / (a, b, c, d, e_i, f_j),$$

directly from Definition 10.2, as the separating set is in each case the union of all their parents.

Checking whether other belief separations hold requires a bit more effort: we must employ Theorem 10.13 and construct the **moral** graph (Definition 10.12). This is shown in Figure 10.3. For example, suppose that we wish to check further

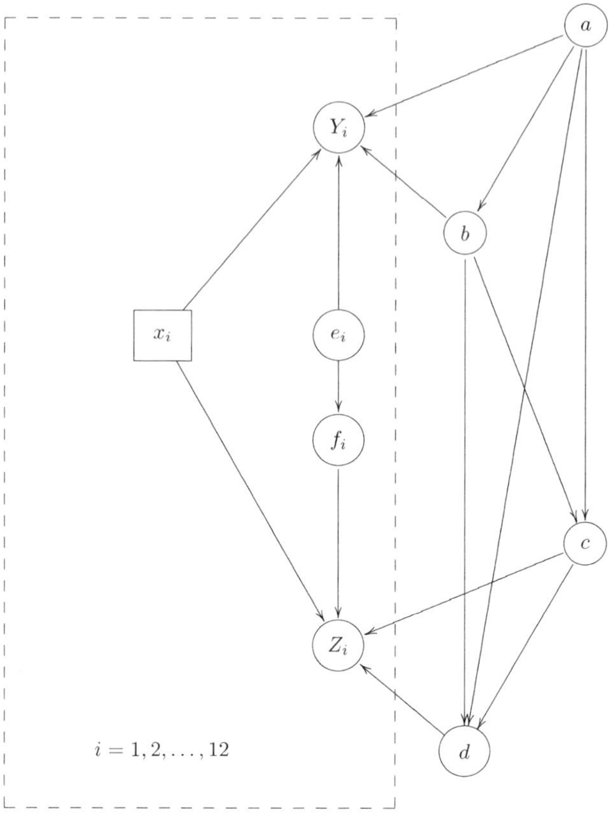

Figure 10.2 Bayes linear graphical model for model (10.7) and consistent node ordering (10.9), with a plate indicating repeated structure.

how Y_i and Y_j are separated and how Y_i and Z_j are separated. From the moral graph, we can see that

$$\lfloor Y_i \perp\!\!\!\perp Y_j \rfloor \,/\, (a, b, e_i) \quad \text{and} \quad \lfloor Y_i \perp\!\!\!\perp Z_j \rfloor \,/\, (a, b, e_i).$$

Note that identification of the separating set of nodes may not be unique. For example, we also have that

$$\lfloor Y_i \perp\!\!\!\perp Y_j \rfloor \,/\, (a, b, e_j) \quad \text{and} \quad \lfloor Y_i \perp\!\!\!\perp Z_j \rfloor \,/\, (c, d, f_j).$$

10.5.3 Alternative diagrams

Just as there is more than one possible Bayes linear graphical model for this model, so there are many possible **consistent** orderings for the nodes in this diagram. One such ordering is given by (10.9). To illustrate some properties of Theorem 10.8,

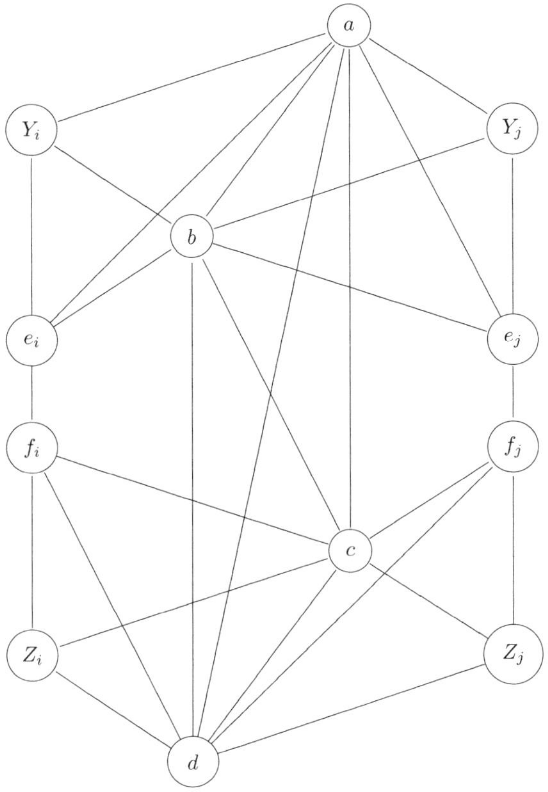

Figure 10.3 The moral graph corresponding to Figure 10.1.

we have, by Property 10.8.1, that it is an **allowable** operation to draw a directed arc between a pair of nodes such as e_i, Z_j, as e_i appears higher in the list than Z_j. The direction must be $e_i \rightarrow Z_j$. By Property 10.8.3, we can **combine** quantities a, b into a single vector node G_Y by arranging the sets of nodes as

$$\text{Group A} : e_i, e_j, f_i, f_j, Z_i, Z_j$$

$$\text{Group B} : a, b$$

$$\text{Group C} : c, d, Y_i, Y_j.$$

Having done so, we may then **combine** quantities c, d into a single node G_Z by arranging the sets of nodes as

$$\text{Group A} : G_Y, e_i, e_j, f_i, f_j, Y_i, Y_j$$

$$\text{Group B} : c, d$$

$$\text{Group C} : Z_i, Z_j.$$

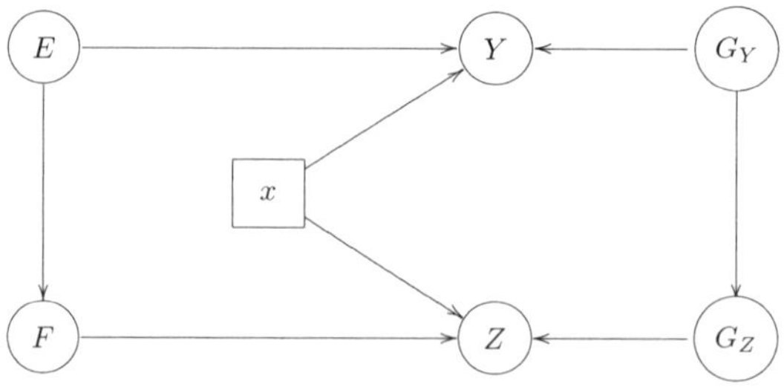

Figure 10.4 Bayes linear graphical model for model (10.7), organized into collections of interest.

Such combination shows how we may organize like quantities. For example, by Property 10.8.3, we may organize quantities as

$$E = \{e_1, \ldots, e_{12}\}, \qquad F = \{f_1, \ldots, f_{12}\},$$
$$Y = \{Y_1, \ldots, Y_{12}\}, \qquad Z = \{Z_1, \ldots, Z_{12}\}, \qquad (10.10)$$
$$G_Y = \{a, b\}, \qquad G_Z = \{c, d\}.$$

A Bayes linear graphical model for this organization is shown in Figure 10.4. This representation is natural if we wish to study the separate implications of the collections G_Y and G_Z for collections Y and Z. There are other natural organizations. We can, if we wish, similarly combine into one node the main quantities of interest, namely the coefficients $G = \{a, b, c, d\}$. In §5.14.2.5 we explored the implications of adjusting the collection of regression coefficients, G, sequentially by the pairs of measurements

$$H_i = \{Y_i, Z_i\}, \quad i = 1, 2, \ldots, 12.$$

A Bayes linear graphical model for understanding the influence of the parameters for the observables is given in Figure 10.5, where the collections are defined as

$$Q_j = \{E_j, F_j\}, \qquad Q_{[i]} = \{E_1, \ldots, E_i, F_1, \ldots, F_i\}, \qquad (10.11)$$
$$H_j = \{Y_j, Z_j\}, \qquad H_{[i]} = \{Y_1, \ldots, Y_i, Z_1, \ldots, Z_i\}. \qquad (10.12)$$
$$x_{[i]} = \{x_1, \ldots, x_i\}. \qquad (10.13)$$

This graphical model is simply extended to the full sequence of partial adjustments we may wish to perform.

10.5.4 Diagrams for inference and prediction

The diagrams we have constructed so far display the influence of parameters on observables, whereas the statistical question of interest concerns the influence of observables on parameters. To extend the example in the previous subsection, a

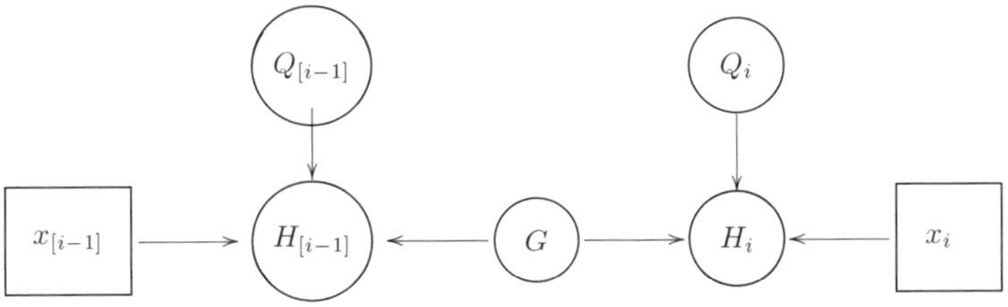

Figure 10.5 Bayes linear graphical model for model (10.7), organized for understanding the effects of sequential adjustment.

primary interest is in inference about the parameter set G given observation of all the measurements up to and including Y_{i-1}, Z_{i-1}, and, in parallel, in prediction of a future set of observables Y_i, Z_i. In order to do this we need to reverse the directions of arcs in the diagrams shown so far.

As an illustration, suppose that we have available data at time i and that we wish to adjust the parameter sets G_Y, G_Z and the next set of observables, Y_j, Z_j, by this information, and then subsequently adjust the parameter sets by Y_j, Z_j. To construct the appropriate Bayes linear graphical model, we begin with Figure 10.1 and carry out the following operations.

1. We drop the boxes representing the fixed values x_i, x_j, as these are irrelevant to our purpose.

2. We combine nodes a, b into node G_Y and nodes c, d into G_Z, so that we have an arc $G_Y \to G_Z$, arcs from Y_i, Y_j into G_Y, and arcs from Z_i, Z_j into G_Z. This is allowable by Property 10.8.2.

3. We remove from the diagram the nodes e_i, e_j, f_i, f_j, representing the unobservable nuisance quantities. Such removal is allowable by Theorem 10.10: we use the consistent node ordering (10.9), dropping in the order e_i, f_i, e_j, f_j. Following our rules for node removal, we must add arcs $Y_i \to Z_i$ and $Y_j \to Z_j$.

4. We reverse arcs from the observables Y_i, Z_i to the parameters. Such reversal is allowable by Theorem 10.9. Following our rules for node reversal, we must also add arcs $Y_i \to G_Z$ and $Z_i \to G_Y$.

5. We reverse arcs from the observables Y_j, Z_j to the parameters. In doing so, it turns out that we need to add arcs $Y_j \to G_Z$ and $Z_j \to G_Y$. We also need to add arcs from each of Y_i and Z_i to each of Y_j and Z_j.

The resultant diagram is shown in Figure 10.6. An obvious simplification is to combine nodes Y_i, Z_i into node H_i and nodes G_Y, G_Z into node G, so that the diagram has nodes H_i, H_j, G with arcs $H_i \to G$, $H_j \to G$, and $H_i \to H_j$. Similarly, we could have begun with the Bayes linear graphical model shown in Figure 10.5 (dropping the nuisance quantities $Q_{[i-1]}$, Q_i and the fixed x quantities)

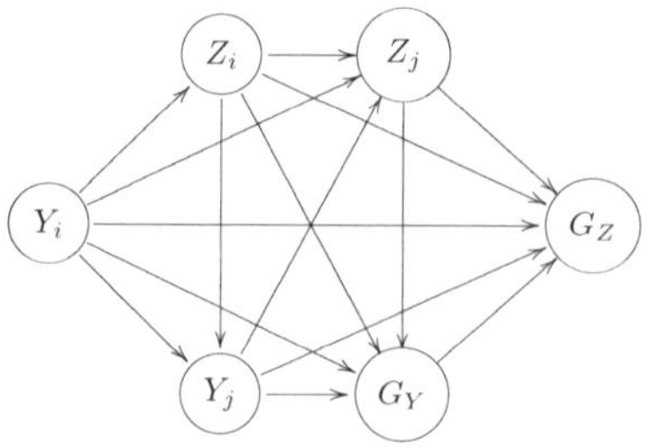

Figure 10.6 Bayes linear graphical model for model (10.7), following reversal of arcs for inference and prediction.

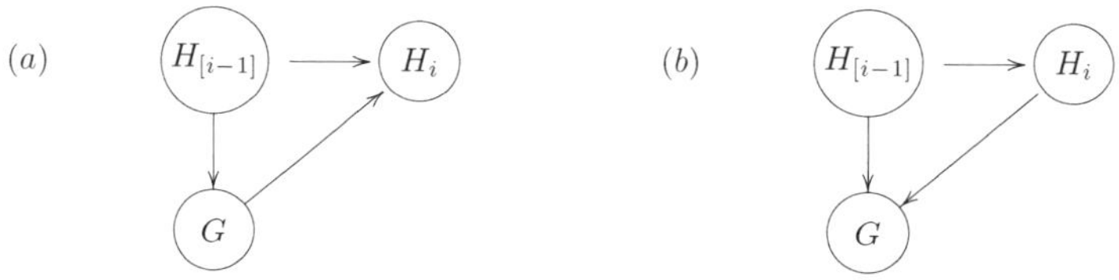

Figure 10.7 Predictive and inferential Bayes linear graphical model for model (10.7) organized into collections of interest: (a) predictive, (b) sequential.

and reversed arcs from the observables to the parameter set, following our rules for arc reversal. The resultant diagram is shown in Figure 10.7(b) and has nodes $H_{[i-1]}, H_i, G$ with arcs $H_{[i-1]} \to G$, $H_i \to G$, and $H_{[i-1]} \to H_i$. If, instead, we want only the predictive diagram, we omit stage 5 above, giving Figure 10.7(a), with arcs from the parameters and the currently observed set into the set we wish to predict.

10.6 Displaying the flow of information

Graphical models give a qualitative representation of our beliefs in a simple graphical form. These diagrams may also be used to summarize the quantitative flow of information by incorporating the various numerical measures that we have described in previous chapters. As with any system of graphical representation, a balance must be struck so as to provide large amounts of visual information to help us to understand the implications of our belief specifications and analysis, without overburdening the picture with so much information that it is difficult to learn anything at all. We shall describe the elements of a graphical toolkit which allows us to represent all of the interpretative and diagnostic aspects of a Bayes linear analysis directly upon the corresponding graph. We would only rarely be interested in seeing all of the features of the full graphical display that we will now describe.

In any particular analysis, certain features will be of particular importance, and the graphical displays should therefore be tuned to highlight these features and to de-emphasize or suppress the remainder.

To begin, suppose that we have a directed graph with consistent node ordering $B_1, \ldots, B_k$. Suppose that we want to describe the effects of the progressive adjustment of each B_i by the collection of ancestors $B(i)$, or equivalently by the parents, $\mathrm{Pa}(B_i)$. A simple quantification is the system resolution, $\mathrm{R}_{\mathrm{Pa}(B_i)}(B_i)$, namely the trace of the resolution transform for the adjustment of B_i by $\mathrm{Pa}(B_i)$ divided by the rank of $\mathrm{Var}(B_i)$. To simplify notation, we denote this by $SR_i = \mathrm{R}_{\mathrm{Pa}(B_i)}(B_i)$.

Suppose that the ordered parent nodes for B_i are $P_{i1}, \ldots, P_{ir}$. We may decompose the collection $\mathrm{Pa}(B_i)$ into the mutually uncorrelated structures $D_1, \ldots, D_r$, where $D_1 = P_{i1}$, and each D_j is the adjusted version of P_{ij} given $P_{i1}, \ldots, P_{i(j-1)}$, so that

$$D_j = \mathbb{A}_{(P_{i1} \cup \ldots \cup P_{i(j-1)})}(P_{ij}), \quad j = 2, \ldots, r.$$

Therefore, we can uniquely partition the influence of the parent nodes on B_i into the influence from the r mutually orthogonal structures D_j.

Denote the jth partial system resolution by $R_j(i) = \mathrm{R}_{D_j}(B_i)$. We may decompose the system resolution as the sum of partial resolutions, namely $SR_i = \sum_{j=1}^{r} R_j(i)$, since the partial resolutions are additive, by (5.20). We also evaluate the final portion of the resolution of variance for B_i, namely the reduction obtained from observation on B_i itself, given that we have already adjusted B_i by $B(i)$. We term this quantity the **unresolved variation** for B_i given $\mathrm{Pa}(B_i)$, denoted by

$$R_{r+1}(i) = \mathrm{R}_{\mathbb{A}_{\mathrm{Pa}(B_i)}(B_i)}(B_i) = 1 - SR_i.$$

10.6.1 Node shading

We depict this decomposition of the information on the graph by dividing the node B_i into $r + 1$ sectors so that the area of sector j is proportional to the magnitude of $R_j(i)$. There are various alternative arrangements that we may choose for the sectors. For example, we may arrange the sectors in ascending node order anticlockwise from 0 degrees. Alternatively, we may arrange the sectors radially, from the outside of the circle, with the unresolved uncertainty at the centre. The former arrangement may be better for displaying small effects and has advantages for displaying certain of the diagnostics that we will describe below. The latter arrangement has the advantage that, as the area of the sector is in units of resolved variance, the width of the sector is in units corresponding to standard deviation, which is often a more intuitive quantity for assessing the importance of the various resolutions. Further, we may use the visual image of the node shrinking to a point as we gain information to give a simple representation of the remaining uncertainty in the system as we make observations. If we wish to identify the sectors visually with the corresponding parent nodes, then we shade or colour the outer rims of the sectors appropriately; see the examples in the following section. The inner portion of each sector is reserved for displaying diagnostic information; see §10.7.

Note that this arrangement has the combination property that if we combine several numerically consecutive parent nodes, then we simply combine the corresponding adjacent sectors. If a node has a large number of parent nodes, we may therefore decide to combine qualitatively similar types of parent node, to simplify the display.

10.6.2 Arc labelling

The shading of the node depicts the amounts of information about that node successively received from each parent node. This information arrives at nodes along the arcs from parent nodes. It is useful to display the strength of the information passing along each arc, to build up a picture as to the most important aspects of the adjustment. We can express this division of information in terms of partial resolutions. First, we describe labelling for a general scenario with collections B, D, E, F, where D, E, F are parents of child node B. We represent the information leaving a parent node for a child node, and the information arriving at a child node from a parent node, as follows.

We define the information flows **leaving** D, E, and F singly to B as $\mathrm{R}_D(B)$, $\mathrm{R}_E(B)$, and $\mathrm{R}_F(B)$ respectively, representing the worth of each information source in the absence of any other. The overall resolution at the node B is $\mathrm{R}_{D\cup E\cup F}(B)$, representing the total information arriving at B.

We measure the information flow **arriving** at B from F alone as the loss in resolution at B if node F is withdrawn from the adjustment. In terms of resolutions of uncertainty, this is

$$\mathrm{R}_{D\cup E\cup F}(B) - \mathrm{R}_{D\cup E}(B). \tag{10.14}$$

We measure the information flow arriving at B from E and from D similarly, by $\mathrm{R}_{D\cup E\cup F}(B) - \mathrm{R}_{D\cup F}(B)$ and $\mathrm{R}_{D\cup E\cup F}(B) - \mathrm{R}_{E\cup F}(B)$ respectively.

The information leaving a node can be smaller than the information arriving from it. This happens typically when one node is informative for a child node only in combination with another. Sometimes, the information leaving a node is also carried partly or wholly by other parents, in which case little or none of the information will be seen to arrive at the child from this parent. The analogy here is with stepwise linear regression: the information leaving a node is akin to the predictive value of entering a single explanatory variable with no other variables fitted, while the information arriving at a node is akin to the loss in predictive power when an explanatory variable is withdrawn.

Many different kinds of labelling are possible. Here, we describe one choice. Where we wish to display such information, an arc is labelled with a rectangle as follows. The rectangle consists of a bar divided into two. The half bar nearest the parent node concerns information leaving from the parent node to the child node. The remaining half bar nearest the child node concerns information arriving at the child node from that parent.

Figure 10.8 shows the contents of the half-bar nearest the parent node. There are three regions to consider. The region $R_1^{\rightarrow} + R_2^{\rightarrow} + R_3^{\rightarrow}$ represents all the initial

← parent node $R_1^{\rightarrow}$ $\mathrm{R}_D(B)$	$R_2^{\rightarrow}$ $\mathrm{R}_{D\cup E\cup F}(B) - \mathrm{R}_D(B)$	middle of label→ $R_3^{\rightarrow}$ $1 - \mathrm{R}_{D\cup E\cup F}(B)$

Figure 10.8 Arc labels: the half-rectangle nearest the parent node summarizes information leaving the parent node.

← middle of label $R_3^{\leftarrow}$ $1 - \mathrm{R}_{D\cup E\cup F}(B)$	$R_2^{\leftarrow}$ $\mathrm{R}_{E\cup F}(B)$	child node → $R_1^{\leftarrow}$ $\mathrm{R}_{D\cup E\cup F}(B) - \mathrm{R}_{E\cup F}(B)$

Figure 10.9 Arc labels: the half-rectangle nearest the child node summarizes information arriving at the child node.

uncertainty in the child node, proportionately 1. We employ arrow notation here as a simple visual clue: arrows point the way to the centre of the label, the point dividing the half-bars. The region $R_3^{\rightarrow}$ represents the proportion of uncertainty remaining in the child after the child has been adjusted by all parent nodes, $1 - \mathrm{R}_{D\cup E\cup F}(B)$. The region $R_1^{\rightarrow} + R_2^{\rightarrow}$ thus represents the proportion of uncertainty removed in the child node B by all parents. This corresponds to the proportion of shading in the child node. The region $R_1^{\rightarrow}$ represents the resolution in uncertainty in B due solely to fitting D, i.e. $\mathrm{R}_D(B)$. Thus, when $R_1^{\rightarrow}$ is large and $R_2^{\rightarrow}$ is small, the implication is that the single source of information D is nearly sufficient for the parent nodes. When region $R_1^{\rightarrow}$ is small and $R_2^{\rightarrow}$ is large, the implication instead is that the single source of information D is not useful relative to the contributions made by the other parent nodes.

The other half of the label, nearest the child node, is similarly configured as shown in Figure 10.9: $R_3^{\leftarrow}$ is identical to $R_3^{\rightarrow}$, and $R_1^{\leftarrow} + R_2^{\leftarrow}$ is identical to $R_1^{\rightarrow} + R_2^{\rightarrow}$. However, the region $R_1^{\leftarrow}$ represents the resolution in uncertainty in B which would be lost if D were withdrawn from the adjustment. Thus, when $R_1^{\leftarrow}$ is large and $R_2^{\leftarrow}$ is small, the implication is that much of the resolution in uncertainty at B is lost if the single source of information D is withdrawn. When $R_1^{\leftarrow}$ is small and $R_2^{\leftarrow}$ is large, the implication instead is that the single source of information D contributes little extra to the information supplied already by $E \cup F$.

We typically shade the regions $R_1^{\rightarrow}$ and $R_1^{\leftarrow}$ to emphasize them. For interactive exploration on a computer, the colour shading can represent features such as parent and child node. This labelling may be modified when the actual resolution of information at a child node is small, but it is still important to visualize the information flow into the node, by de-emphasizing, or removing altogether, the empty central portion of the box which corresponds to the residual uncertainty and is given by regions $R_3^{\rightarrow}$ and $R_3^{\leftarrow}$ in Figure 10.8 and 10.9, respectively.

We apply such arc labelling to a graph on any consistent ordering of nodes. We have already recorded, on the node B_i, the cumulative effects of adding information from each parent node. On each arc, we now record the marginal effects of each parent node P_{ij}. First, we assess the information leaving the parent by measuring the effect on B_i of observing P_{ij} alone, i.e. the system resolution for the adjustment of B_i by P_{ij}, which we denote by $U_{ij}^{\rightarrow}$. Secondly, we assess the information arriving at the child along each arc using the partial system resolution for the adjustment of B_i by P_{ij} after we have adjusted B_i by all other parents of that node, which we denote by $U_{ij}^{\leftarrow}$.

The arcs carry a lot of information when, in combination, the parent nodes are highly informative about the child node. Therefore, the shading is intended to display most prominently the information flow for such nodes. If a large amount of the part of the box adjacent to P_{ij} is shaded, then the implication is that observation of this node is, by itself, strongly informative for B_i, while if a large amount of the portion of the box adjacent to B_i is shaded, then this indicates that observation of P_{ij} is important for B_i even when all other parent nodes have been observed.

While this display is very informative on a small diagram, it can become rather cluttered on a large diagram with many arcs crossing. Therefore, we may consider alternative representations, for example displaying this information using the thickness of the arcs. At the end of the arc nearest node P_{ij} the thickness of the arc is proportional to the ratio $U_{ij}^{\rightarrow}$, and at the end of the arc nearest node B_i the arc thickness is proportional to the ratio $U_{ij}^{\leftarrow}$.

10.6.3 Tracking information as it is received

So far, we have described how to display information flow when we have observed all of the quantities on the diagram according to some consistent node ordering. However, we will often wish to track information as it is received into a system. In such cases, we may still identify the effect of information on child nodes in the way that we have described above. To describe the effect of observing each node upon all of the nodes in the diagram, we shade each node as though there were an arc from the node that we have observed to each other node, and then shade the proportion of the variance resolved by the observation. As we make further observations, we may shade the additional variance resolved. We may simplify the diagram by showing only three shadings at each stage, namely variance resolved before the current observation, variance resolved by the current observation, and the unresolved variance.

10.6.4 Example

We continue our example from §10.5, using the Bayes linear graphical model shown in Figure 10.1, but aggregating like quantities such as $Y_1, \ldots, Y_{12}$ into the collection Y, and similarly for the Z_i, E_i and F_i quantities, as in (10.10). For this illustration, we will explore how observation of parents provides information about children; later we will be more interested in the implications of observing children. A consistent node ordering for this diagram is a, b, c, d, E, F, Y, Z. A Bayes linear graphical model with node shadings only is shown in Figure 10.10.

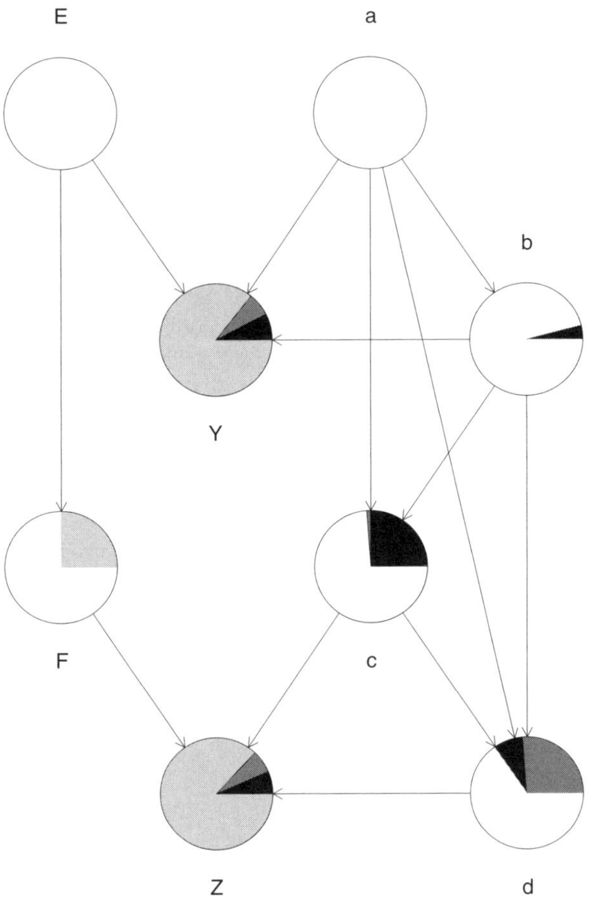

Figure 10.10 Shading nodes using resolutions and partial resolutions, from adjusting in the consistent order a, b, c, d, E, F.

Table 10.1 Resolutions used for the node shadings for Figure 10.10.

Adjustment	Overall resolution	Partial resolution
$[b/a]$	0.0400	–
$[Y/a]$	0.0738	–
$[Y/a+b]$	0.1430	0.0692
$[Y/a+b+E]$	1.0000	0.8570
$[F/E]$	0.2500	–
$[c/a]$	0.2500	–
$[c/a+b]$	0.2604	0.0104
$[d/a]$	0.0000	–
$[d/a+b]$	0.2604	0.2604
$[d/a+b+c]$	0.3462	0.0859
$[Z/c]$	0.0627	–
$[Z/c+d]$	0.1316	0.0689
$[Z/c+d+F]$	1.0000	0.8684

The shadings arise as follows. We adjust each node by its parents, according to the consistent node ordering given above.

The adjustments we carry out, together with the overall resolutions in variation for each adjustment and the partial resolution, if any, at each stage, are summarized in Table 10.1. We first adjust coefficient b by coefficient a. Observation of a is expected to resolve only 4% of the variation in quantity b. (The correlation between a and b is -0.2, and the resolution when we adjust a single quantity by another single quantity is the square of the correlation coefficient, just as for simple linear regression.) We show this proportion by shading node b by an amount $0.04 \times 360^\circ \approx 14^\circ$ anticlockwise from 0°. The node Y has three parents: we adjust first by node a, leading to a resolution of 0.0738 and an initial node shading of $0.0738 \times 360^\circ \approx 27^\circ$ anticlockwise from 0°. We then adjust the collection Y additionally by node b: this leads to a total resolution of 0.1430, of which 0.0692 is the partial resolution due to b having already taken into account a. We shade a further portion, about 25°, of node Y to reflect the partial contribution of node b. Finally, we adjust by the third parent, node E, representing individual variation terms. When we know E in addition to a and b, we know Y and so we resolve all the variance in Y and we shade the remainder of the node in consequence. Equivalently, when we observe Y any residual variation is removed.

The adjustment of node d by its parents shows only two shadings. This is because no variance is resolved by the first adjustment by parent a. This does not imply that a is not informative. The overall resolution in d given a, b, c is 34.63%, whereas the resolution in d given b, c alone is 29%: the contribution from a arises in combination with information from the other nodes.

For interactive investigation, the shadings reflect the source of the resolution. For example, we might draw the node a and arcs from it in blue, the node b and arcs from it in red, and the node E and arcs from it in green, and shade by these colours when we show the resolutions of variance in node Y. Otherwise, it is necessary to know the sequence of adjustment when evaluating the graphic, as is the case here.

The broad inferences that we draw from Figure 10.10 are: that there are relatively substantial error components attached to Y and Z; that a will not tell us much about b; similarly that c will not tell us much about d; that a resolves a quarter of the variation in c; and similarly that b resolves a quarter of the variation in d. The residual uncertainty in Y given a, b is strikingly large. This is a consequence of the magnitude of the error variances; of the number of Y_i terms in Y (each Y_i included brings another piece of error variation); and also of the correlation structure. For example, a is quite informative about each individual element Y_i but tells us almost nothing about any of the differences between the Y values, as

$$\text{Cov}(Y_i - Y_j, a) = (x_i - x_j)\text{Cov}(a, b)$$

will be typically quite small. Generally, a, b is only a two-dimensional space, and so can at most resolve variation within a two-dimensional subspace, whatever the dimension of Y.

The shadings do depend upon the order in which we assign parents for the adjustment. This is desirable when we have in mind a sequential adjustment, where there is a natural physical ordering, but less satisfactory when there is no natural ordering. Here, we could adjust node Y by node b first, and this would change the way in which node shadings are partitioned, if not the total resolution, and thus our inferences. One partial solution to this problem is to label the arcs as described in §10.6.2, which we illustrate below.

One important point to note is that when some parents of a child are aggregated, the resolution of variance in the child through observation of the aggregated parents is unchanged. For example, we can alter the diagram in Figure 10.10 to the diagram shown in Figure 10.11 by aggregating the single-quantity nodes a, b into the collection G_Y and the nodes c, d into the collection G_Z. This is an allowable operation as we described in §10.5.3. We see then that the total resolution delivered to node Y by parents a, b jointly matches the resolution delivered by their aggregate, G_Y (Table 10.2). The interpretation is similar for the nodes c, d, Z, G_Z. In cases where children with the same parents are aggregated, as here with child node G_Z and parent node G_Y, the total resolution from the adjustment of the new child by the new parent cannot be read from the first graph as it depends on the

Table 10.2 Resolutions for the node shadings for Figure 10.11.

Adjustment	Overall resolution
$[Y/G_Y]$	0.1430
$[Z/G_Z]$	0.1316
$[G_Z/G_Y]$	0.2821

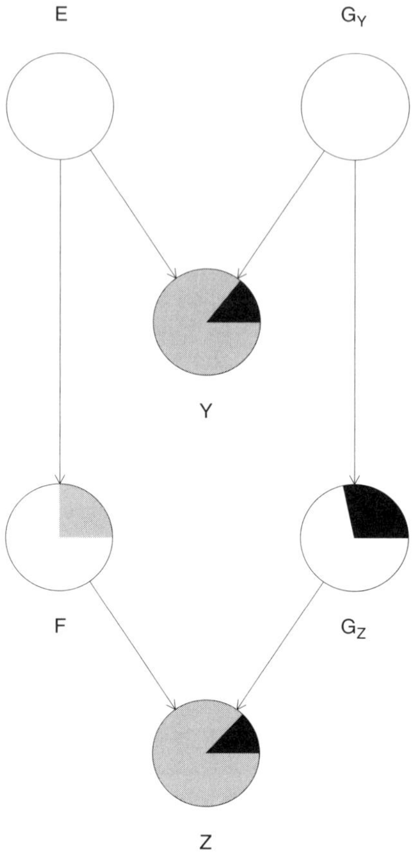

Figure 10.11 Aggregating parents leaves the total resolution in children unchanged. Adjustments are made in the order G_Y, G_Z.

nature of the correlation structure between parents and children. Altering the diagram qualitatively does not result in information loss, but can result in the display of different quantitative features, as in this case.

10.6.4.1 The heart of the transform

We noted above the large residual uncertainty in the Y collection given the parameters, largely as a consequence of the amount of variation contributed by the error quantities E. This gives a slightly misleading picture of the value of the parameters for predicting the data quantities. As an alternative, we may instead explore the implication of the parameters for the heart of the transform, $\mathbb{H}(H/G)$.

In §5.18.2 we found the heart of the transform for this problem, together with its orthogonal complement in H, $\mathbb{H}^{\perp}(H/G)$, spanned respectively by the collections

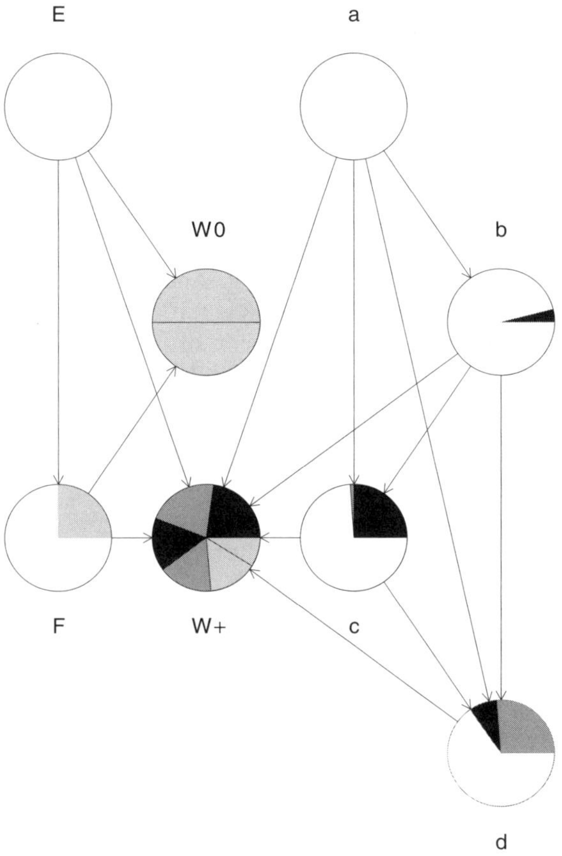

Figure 10.12 Adjusting the heart of the transform by the parameters. Adjustments are made in the consistent order a, b, c, d, E, F.

W^+ and W^0. To obtain Figure 10.12, we combined nodes for the data collections Y and Z into a single node H, this being allowable by Theorem 10.8, and then separated H into the uncorrelated collections W^+ and W^0. Arcs from the parameter set a, b, c, d into W^0 may be dropped. We then recomputed the adjustment which gave Figure 10.10, i.e. we adjusted all collections on the graph in the consistent order a, b, c, d, E, F and shaded accordingly.

We observe that the parameters are strongly informative for the heart of the transform: about 75% of the variation in W^+ is jointly resolved by them, and each of the parameters makes a substantial contribution; indeed, we can show that this is so irrespective of the order of adjustment. Only the residual collections E, F are informative for the complementary collection W^0, each resolving 50% of the variation, as we would expect as a consequence of the way in which we constructed the model.

10.6.4.2 Arc labelling

Figure 10.13 shows the labelling scheme described in §10.6.2. These labels are based on the information summarized in Table 10.3. The information leaving and arriving from a parent node is the same where a child has a single parent (for example, $a \to b$) and where a parent of a child node is uncorrelated with all other parents (for example, $E \to Y$). The arc label from node E to node Y is heavily shaded at both ends, showing that node E is an important influence on node Y. In contrast, the labels from a to Y and b to Y have little shading, showing that these carry little information about the collection Y. Almost all the information from a reaches Y: there is a small amount, $0.0738 - 0.0730 = 0.0008$, which could be carried instead by b. In this example, the sum of information arriving at a node is the total resolution for that node. However, for larger problems this will not generally be so.

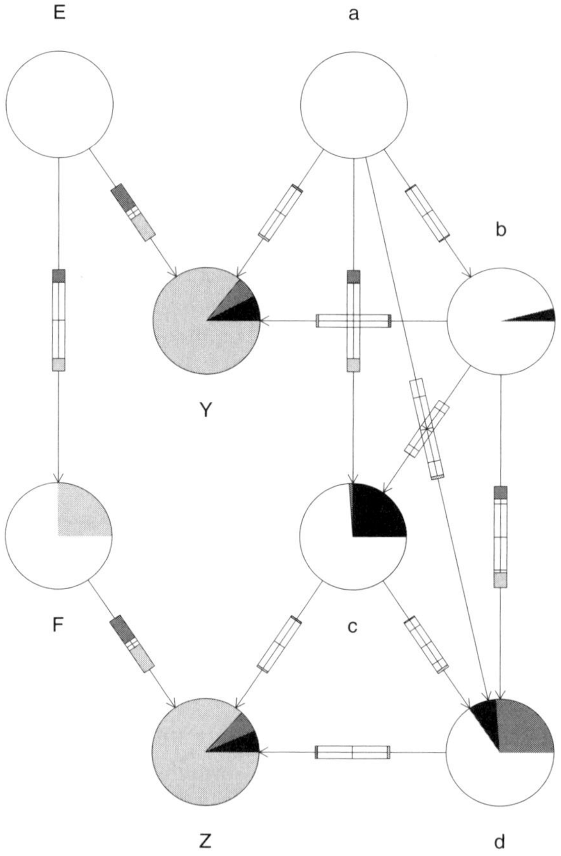

Figure 10.13 Adding labels to the diagram to show information flow.

Table 10.3 Resolutions for the arc shadings for Figure 10.13.

Parent	Information leaving	Information arriving	Child
a	0.0400	0.0400	b
a	0.2500	0.2500	c
b	0.0000	0.0104	c
a	0.0738	0.0730	Y
b	0.0700	0.0692	Y
E	0.8570	0.8570	Y
a	0.0000	0.0563	d
b	0.2500	0.2500	d
c	0.0400	0.0400	d
E	0.2500	0.2500	F
c	0.0627	0.0615	Z
d	0.0700	0.0689	Z
F	0.8684	0.8684	Z

Where the resolution in the child node is small, it can be difficult to compare the amount of information leaving and arriving. In this case, we may omit the central regions of the arc label, given as regions $R_3^{\rightarrow}$ and $R_3^{\leftarrow}$ in Figure 10.8 and 10.9, respectively. The central regions are dropped for Figure 10.14. This labelling emphasizes a parent node's relative importance given a child node's overall resolution, so that the label for $b \rightarrow d$ shows that b is relatively very important for learning about d.

One interesting feature to spot here is that the arc label from a to d reveals that no information leaves a for d, but some information – a resolution of 0.0563 – does arrive. This illustrates graphically the point made above, that two nodes can be marginally uncorrelated but that one can be informative for the other in combination with other nodes, here being b, c.

10.6.4.3 Tracking information arriving into the system

In Figure 10.15, each collection represented on the diagram has been adjusted in sequence by a, then partially by b, c, and finally by d. The colour, or style of shading, indicates the source node. We do not show arcs. Adjusting the source node by itself resolves all the variation remaining at that node. The diagram shows that a has a small influence on all the other nodes except d (with which it is uncorrelated). Observation of b then resolves some of the variation in d and a small fraction of the remaining variation in c, Y, and Z. Observation of c and d tells us nothing more about Y, but does give further information about Z.

10.7 Displaying diagnostic information

As we make observations on elements of the nodes of the graph, this allows us to carry out diagnostic criticism of the belief structure. In previous chapters, we have

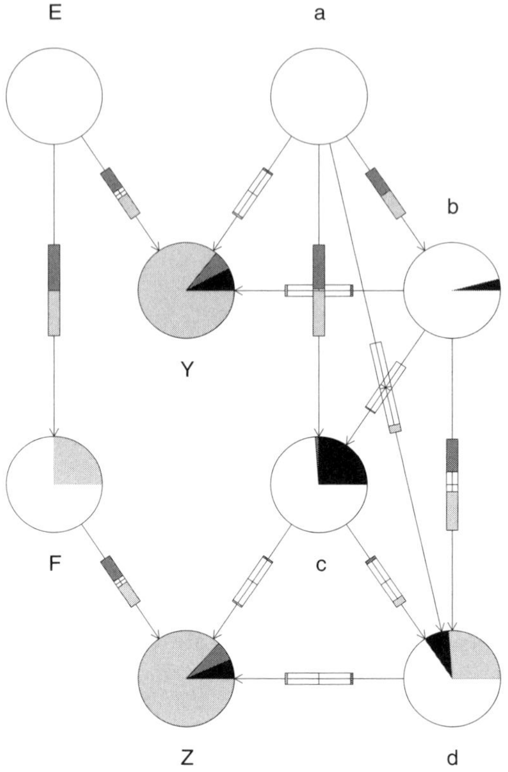

Figure 10.14 Omitting the central regions of the arc label, so as to concentrate on comparing the relative importance of the information leaving and arriving.

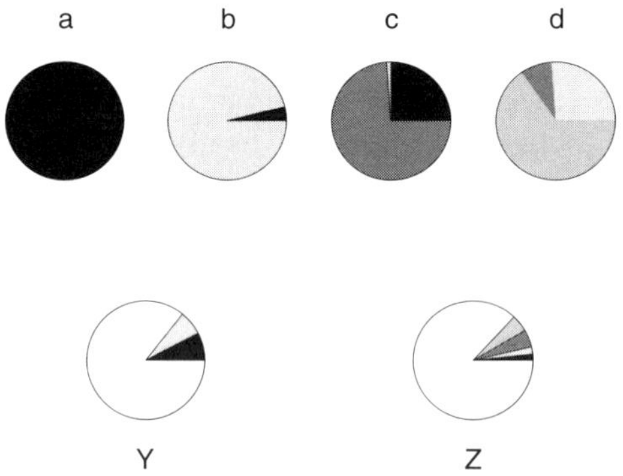

Figure 10.15 A sequential adjustment diagram, adjusting in the order a, b, c, d.

described diagnostic measures that we may evaluate for a Bayes linear analysis. We now describe how this diagnostic information may be displayed directly on the influence diagram.

Each of the shadings described in §10.6 is the trace of a full or partial belief transform. Each such trace is the expectation for the size of the bearing for the corresponding adjustment or partial adjustment. Therefore, by comparing the observed size of each adjustment with the corresponding expected size, we may make a diagnostic assessment for each arc and node shading. We discussed diagnostics for simple adjustments in §4.9 and defined the size ratio in (4.63). We discussed diagnostics for partial adjustments in §5.6 and defined the partial size ratio in (5.43).

There are two types of diagnostic information that we wish to display. First, our assessments may be wrong in that we were overconfident, which will be revealed by observed sizes for many adjustments which are far larger than their prior expectations. Secondly, we may lack confidence, in that we are actually able to predict the outcomes with greater accuracy than we have allowed in our variance specifications, which will be revealed by observed sizes for many adjustments which are far smaller than their prior expectations.

We now discuss how the diagnostics should be calculated and marked on the diagram to show the areas of agreement and conflict between prior beliefs and observations.

10.7.1 Node diagnostics

First, we consider the node diagnostics. Each sector for the node corresponds to a full or a partial adjustment. The size of the bearing, or equivalently the square of the largest standardized change in expectation for a linear combination of the elements of the node, has expectation equal to the trace of the corresponding adjusted resolution transform, which was the basis for the node shading described in §10.6. If this observed change is markedly larger or smaller than we expect, then we shade the inner portion of the corresponding sector of the node to display this. Thus, we evaluate the size ratio for the partial adjustment, namely the ratio of the observed to expected bearing size. The amount of shading corresponds to the magnitude of the diagnostic. There is no shading if the size ratio equals its expected value, namely one. Otherwise, we shade the inner sector dark or light depending on whether the size ratio is larger or smaller than one. The amount of shading may be chosen only to highlight extreme diagnostics, for example when we are trying to tune a very rough prior specification, or to highlight fairly modest discrepancies, for example when we are monitoring a diagram which has been successfully used for forecasting, over a considerable period of time. There are many different choices for the amount of shading, depending on context. Some possibilities are as follows.

- Choose some probabilistic scale for the quantity, such as a chi-squared variable with matched degrees of freedom, and mark half of the area as dark when the size ratio is in the upper 5% of the distribution, light when the size ratio is in the lower 5%, and with smooth extrapolation over the rest of the range.

- Transform the size ratio, which is a non-negative number with expectation one, to a value in (0, 1) which expresses a proportionate degree of discrepancy from one. Then, shade the corresponding sector by this proportion. We have chosen this scheme for the diagrams shown in this chapter as follows. Suppose that p is the size ratio for an adjustment. Let

$$s = \begin{cases} 1 - \sqrt{p}, & \text{if } p \leq 1, \\ 1 - \frac{1}{\sqrt{p}}, & \text{if } p > 1. \end{cases} \tag{10.15}$$

  We now shade the corresponding sector by the proportion s. The rationale behind this scheme is that p is a variance measure and so a value of $p = 4$ corresponds to two standard deviations, and so half-shading. We are free to arrange half-shading to imply lower or higher levels of discrepancy if we wish.

- Shade all the sector, but with an intensity of colour corresponding to the magnitude of the diagnostic.

While different distributional choices or schemes will give different shadings, the qualitative aspects of the display should not be greatly affected. We may show the shading in a variety of ways. In Goldstein and Wooff (1995), we showed diagnostic shadings by dividing the original sector by angle into two sectors: one shaded and one not. In this book, we show diagnostic shadings by annular sectors extending from the centre point. This has the advantage of facilitating comparison of diagnostic magnitudes using distance from centre.

We could display a variety of further diagnostic information in a similar way. For example, we could display the path correlations for the adjustment, as described in §5.9 and defined in (5.51), as follows. The path correlation in the partial adjustment from adding P_{ij} to parents $P_{i1} \cup P_{i2} \cup \ldots P_{i(j-1)}$ is a number between -1 and $+1$. We may mark this value on the radius separating sectors $j-1$ and j by a dot which is placed a distance from the centre of the circle corresponding to the path correlation, where the point is on the outer edge if the path correlation is $+1$, at the centre of the circle if -1, and so forth. While such displays may be informative, however, we must be careful to balance such information against the need to avoid overloading the graphic display with excessive visual detail.

10.7.1.1 Combining sectors

Sometimes we simplify the diagnostic picture by combining all of the sectors corresponding to the parents into a single sector. We then have two diagnostic shadings for each node. The first shading expresses the difference between the overall adjustment vector and the prior expectation. Large diagnostics for this sector will usually correspond to large diagnostics for certain of the parent nodes, and may be investigated through the various arc diagnostics that we describe below.

The shading for the second sector corresponds to the diagnostic assessment for the difference between the actual vector of values for the node and the adjusted expectation for that vector given all parents. Large diagnostics for this sector suggest that we may have been overconfident in our beliefs concerning the ability of the parent nodes to predict the child node. If, in such cases, there are further ancestral nodes which show similar large diagnostic warnings, but which are not parents of the node in question, then this may suggest modifications to the qualitative form for the graphical model.

10.7.2 Arc diagnostics

We show diagnostic information on each arc in a similar way to the nodes. For the arc labels, the regions $R_1^{\rightarrow}$ of Figure 10.8 and $R_1^{\leftarrow}$ of Figure 10.9 correspond to resolutions provided by an adjustment or a partial adjustment, and to which there are corresponding size ratios. Therefore, we shade a portion of each region according to the magnitude of the corresponding size ratio. The proportion of the area shaded is chosen just as for the node diagnostics: for the diagrams in this chapter, we choose the scheme given in (10.15). If the observed change is larger than expected, then we use dark shading, while if the change is less than expected, we use light shading. Note that if there is a single parent node, then the left and right diagnostic shading for the arc are the same and are also equal to the diagnostic shading for the first sector of the child node.

10.7.2.1 Path correlations

Where a child has more than one observed parent, we may add path correlations to the diagram. We have described above how we separate information from a parent node to a child node into the information leaving and arriving, where the latter is the information uniquely attributable to the parent node compared to other information arriving at the child node from other parents. The path correlation (§5.9) can be used as a measure of consistency between these sources of information. Thus, assuming the shading scenario presented in Figure 10.9, we calculate the path correlation (5.51) as

$$\mathrm{PC}(f \cup e, [d/f \cup e]) = \mathrm{Corr}(\mathbb{Z}_{f \cup e}(B), \mathbb{Z}_{[d/f \cup e]}(B)),$$

being the path correlation between (1) all other data sources $f \cup e$; and (2) d, having taken into account $f \cup e$. We mark the path correlation on the diagram as follows. At the end of an arc we place a small circle. We shade the circle according to the magnitude of the path correlation, with full shading when the correlation is ± 1 and no shading when the path correlation is zero. When the path correlation is negative, indicating a contradiction between the data sources, we use dark shading. Otherwise we use light shading. Our aim, as elsewhere, is to draw the eye to important diagnostics.

Note that the path correlations entering a node B from two sources, D and F, are not normally the same. This is because the path correlations attached to the two arcs are measuring different relationships. Dropping B from the notation for convenience, one path correlation is

$$\mathrm{Corr}(\mathbb{Z}_f, \mathbb{Z}_{[d/f]}) = \mathrm{Corr}(\mathbb{Z}_f, \mathbb{Z}_{d\cup f} - \mathbb{Z}_f),$$

and the other is

$$\mathrm{Corr}(\mathbb{Z}_d, \mathbb{Z}_{[f/d]}) = \mathrm{Corr}(\mathbb{Z}_d, \mathbb{Z}_{d\cup f} - \mathbb{Z}_d).$$

10.7.3 Showing implications across all nodes

So far, we have described how to display diagnostic information given that we have observed all of the quantities on the diagram. However, we will often wish to track diagnostic information as we make observations on the a system. In such cases, we may identify diagnostics for child nodes in the way that we have described above. To make diagnostic inferences over the whole diagram for each observation, we follow the corresponding procedure as described in §10.6. There, we suggested that we may shade each node as though there was an arc from the node that we have observed to each other node, and then shade the proportion of the variance resolved by the observation. As we make further observations, we shade the additional variance resolved. We may then mark a diagnostic shading for the length of the bearing for the partial adjustment corresponding to each shaded region, exactly as we have described above. The shading on the node that we have observed at each stage expresses the diagnostic warnings for the observation given all of the information gathered about the node from previous observations on the graph. The shadings on the other nodes corresponding to this partial adjustment show the diagnostic impact of the observation on the remaining nodes. In particular, if an observation is very surprising, then this picture displays the effect of the surprising observation across the whole diagram, so that we can see whether the effect is localized or whether it has important consequences across the whole collection of quantities.

10.7.4 Interpreting diagnostic warnings

Because there is much information to be displayed, and little space, we may omit some regions from the arc label. If we wish our eyes to be drawn to important diagnostics, we will want to retain all the label regions shown in Figure 10.8 and 10.9, as the diagnostics here are properly weighted by the magnitude of corresponding resolutions of variance. In other words, a large size ratio for an adjustment which carries little information for a child node may be of less interest to us than a moderate size ratio for an important adjustment. Occasionally, we may wish to present diagnostic information on the diagram without worrying whether the corresponding variance resolution is small or large: in such cases we may omit from the arc label any of the regions $R_2^{\rightarrow}$, $R_3^{\rightarrow}$, $R_3^{\leftarrow}$, $R_2^{\leftarrow}$.

There is no automatic method for assessing the implications of diagnostic warnings. Our conclusions will depend on the amount of care and detail that we have built into our prior specification, our confidence in the qualitative form and quantitative assessment of the prior specification, and the reasons that we may attribute to surprising outcomes. Qualitative assessment of distributional forms will also be relevant; for example, large diagnostics may be more revealing for unimodal distributions while small diagnostics may be more revealing for multimodal distributions.

While each diagnostic warning may be of value, in a complex diagram it will usually be the overall pattern of the diagnostic information over the whole diagram which will be of interest. The aim of the display is to enable the analyst to visualize the diagnostic performance of the belief specification over the entire system. In particular, such diagrams are of particular value for monitoring systems where similar collections of observations are made over time. For example, we may wish to forecast sales of some collection of products, in which case we might observe similar collections of diagnostic warnings on a weekly basis, or we may use this diagram as the basis of a diagnostic system for individual patients, so that patterns which might be hard to interpret in a single instance can be judged on the basis of a collection of repetitions.

10.7.5 Example: inference and prediction

In our example to this point, our concern has been the representation of uncertain quantities on a diagram, and tracking the information flow between them, supposing that we observe some and not others. In particular, we have shown how we use diagnostic labelling to display the implications of observation of parents and consequent adjustment of child nodes by parents.

Mostly, we are concerned with adjusting sets of unknowns by data (inference) and sets of future observables by data (prediction). This will often require arc reversals on the Bayes linear graphical model, as discussed in §10.5.4. The resulting Bayes linear graphical models, which have data nodes as parents of future observables and sets of unknowns, typically have quite complicated arc structures. In such cases, we may sometimes drop arcs and error quantities from the diagram in order to focus on the labelling properties.

The central questions in our example are: what do observations on Y and Z tell us about the coefficients a, b, c, d, and is what we learn consistent with what we expected? We explore these questions by adjusting these quantities by Y and then partially by Z, and by examining the adjustment and the associated diagnostics graphically. Figure 10.16 shows four styles of labelling and diagnostic shading from such a sequence of adjustments. Figure 10.16(a) shows basic node labelling without diagnostics and with arc labels as described in §10.6.2. Nodes Y, Z are fully shaded as they become known. For interactive use, colour is used to signify information source: for the node shading in Figure 10.16, dark grey signifies Y as the information source, and light grey corresponds to Z as information source. Y is the first information source fitted, so that the shading in the child nodes shows, proceeding anticlockwise, information arriving from Y, followed by partial

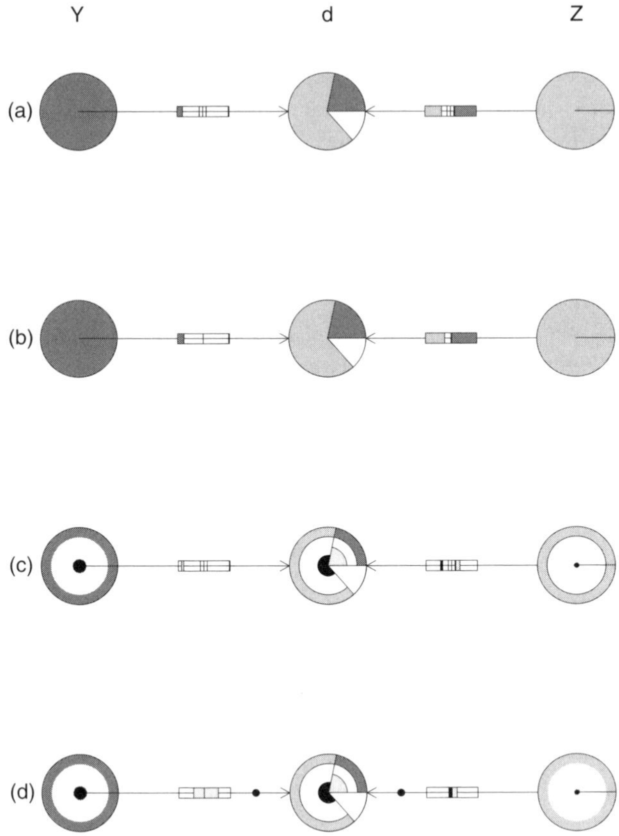

Figure 10.16 Diagnostic shading of nodes and arcs.

information arriving from Z. Figure 10.16(b) differs only in that the central regions of the arc labels have been omitted.

Figure 10.16(c) repeats Figure 10.16(a), but with diagnostics added to both nodes and arcs. Let us begin with the diagnostics for node d, recalling that the adjustment sequence is of d by Y and then partially by Z. The outer annular sectors have shadings which show the source nodes: anticlockwise, these are Y and then Z. The size ratio for the adjustment is $\text{Sr}_y(D) = 0.33$, rather smaller than expected. Thus, the inner sector for the adjustment of d by Y is partly shaded light and the amount of shading is about $1 - \sqrt{0.33} = 43\%$ of the inner sector. The size ratio for the partial adjustment by Z is $\text{Sr}_{[z/y]}(D) = 1.33$, slightly larger than expected. Thus, the inner sector for the partial adjustment of d by Z, having adjusted for Y, is partly shaded dark and the amount of shading is about $1 - 1/\sqrt{1.33} = 13\%$ of the inner sector. As we have now also observed $Y = y$ and $Z = z$, we may also form the diagnostics for these quantities, namely $\text{Sr}_y(Y)$ and $\text{Sr}_z(Z)$, and show these on the graph. The inner sector for node Y shows that the observation was about in line with what was expected: the size ratio is 1.12. Similarly, the inner sector for node Z shows a size ratio of 1.02. Recall that these nodes represent collections of 12 quantities: we could, if desired, show such diagnostics for the individual elements.

The arc diagnostic shadings in Figure 10.16(c) replace the shaded areas of the arc labels in Figure 10.16(a). Take as an example the arc label from node Z to node d. Recall that the part of the label nearest Z reflects the information leaving Z, and represents the adjustment of d by Z alone. The size ratio for this adjustment is $\mathrm{Sr}_z(d) = 0.65$. Thus, the region formerly representing resolution in the child attributable to the parent is partly shaded light and the amount of shading is about $1 - \sqrt{0.65} = 19\%$ of that part of the label. The part of the label nearest d reflects the information arriving at d from Z once other sources have been taken into account. In this case, this corresponds to the partial adjustment of d by Z having already adjusted for Y. We have seen already that the size ratio for this adjustment is $\mathrm{Sr}_{[z/y]}(D) = 1.33$. Thus, we shaded that part of the arc label dark and the amount of shading is about 13%. In summary, for this arc label the diagnostics show no major discrepancies with the prior specification, and as a consequence we see little shading.

Figure 10.16(d) repeats Figure 10.16(c), but with two differences. First, as in Figure 10.16(b), the central regions of the arc labels have been omitted. This allows us to concentrate on the diagnostics if we so wish. We note, for example, that the information both leaving and arriving from Y has a rather small size ratio, implying that the changes in expectation in d induced by observing Y are in all respects surprisingly small. Figure 10.16(d) also shows a path correlation diagnostic added at the end of each arc. In this case, the path correlation at the end of the arc from Z to d turns out to be

$$\mathrm{PC}([z/y], y) = -1,$$

and so these two sources of information are contradictory. Note that we should not read too much into such diagnostics for one-dimensional nodes: for such cases the path correlation must always be zero or ± 1.

Figure 10.17 summarizes the flow of information, without diagnostics, across all the nodes. The node shadings show that Y explains about 80% of the prior variation in a, b, and E, and about 20% of the prior variation in c, d, and F. Having already observed Y, Z appears to be essentially uninformative for a, b, E, but does then explain about a further 60% of the variation in c, d, F. Examining the arc labels, those from Y to a, b, E and from Z to c, d, F show that a lot of information leaves and arrives between these pairs of nodes. However, the arc labels between Y and c, d, F and between Z and a, b, E show that whilst some information leaves, essentially none arrives. We conclude that Y is useful for learning about a, b, E, but not useful for learning about c, d, F if we intend to observe Z, and vice versa.

In Figure 10.18 we add diagnostic information to the plot. Our eyes are immediately drawn to two features. First, many of the path correlations are fully shaded

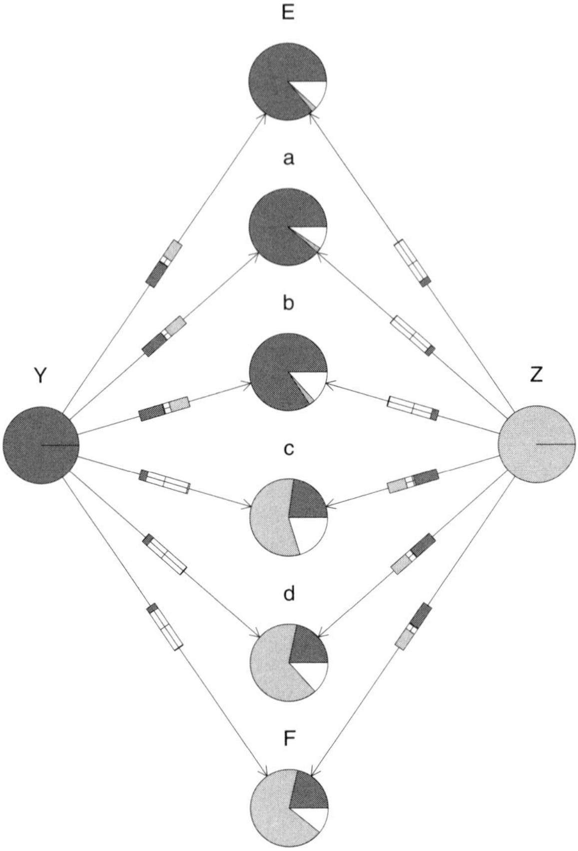

Figure 10.17 Partial adjustment by the observed quantities, without diagnostic shadings.

in black, suggesting that the two data sources are at least mildly in conflict for most of the adjustments being carried out. The path correlations for the scalar nodes a, b, c, d are not overly meaningful, as noted above. However, there are large negative path correlations for the arcs $Y \rightarrow E$ and $Z \rightarrow F$, suggesting that the information **arriving** at E from Y is at odds with the information **leaving** Z for E, and that the information **arriving** at F from Z is at odds with the information **leaving** Y for F. The other path correlations, attached to the arcs $Y \rightarrow F$ and $Z \rightarrow E$, indicate weak positive path correlations: for example, the information **arriving** at F from Y is weakly consistent with the information **leaving** Z for F. There is, in this example, a symmetry in the diagnostics. The portion of Y that is uniquely informative for its associated error quantities E is inconsistent with other information; and the portion of Z that is uniquely informative for its associated error quantities F is similarly inconsistent with other information. This suggests

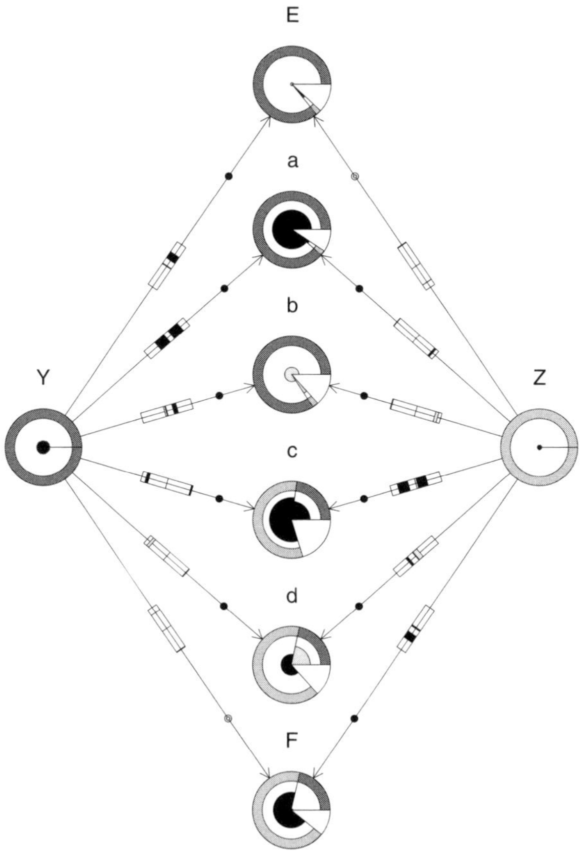

Figure 10.18 Partial adjustment by the observed quantities, with diagnostic shading of nodes and arc labels.

unwelcome features in the joint specification for the error quantities E, F. In fact, we saw in §8.12 some evidence that the correlation between each (e_i, f_i) pair should be negative, rather than positive as in our prior specification. If, instead, we take the correlation $\text{Corr}(e_i, f_i)$ to be weakly negative and recompute the adjustment and the path correlations and display these similarly to Figure 10.18, then the indications of inconsistency disappear. One obvious way to check which, if either, of the correlation models is supported by the data is to carry out the observed belief comparison described in §9.9.

Secondly, there is much diagnostic shading for nodes a and c and, to a lesser extent, F. The diagnostic shadings for a and c are large for both data sources: this is confirmed by the amount of diagnostic shading shown in the arc labels. The shadings suggest that changes in expectation for the two intercept terms in model (10.7) are rather larger than expected under both adjustments. These two features

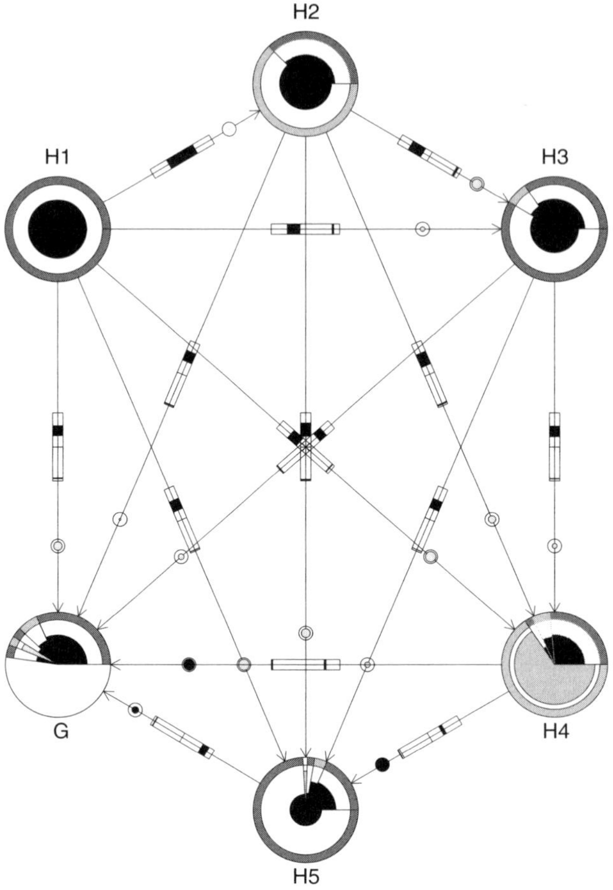

Figure 10.19 Sequential adjustment of the parameter set G and future observables by H_1, and then H_2, and so forth.

correspond to what we saw in §5.14.2.4 and in Table 5.10. That is, the adjustment of a by Y leads to a moderate positive change in expectation for a, whereas the partial adjustment by the contradictory information source Z leads to a moderate reversal.

Figure 10.19 shows the kind of Bayes linear graphical model discussed in §10.5.4. To avoid crowding the diagram, we limit attention to the six collections $H_1, \ldots, H_5, G$, where H_i is the pair of observables Y_i, Z_i. The initial Bayes linear graphical model for these quantities has arcs from G to each H_i, excluding nuisance and fixed quantities. For prediction and inference we need to reverse arcs as described in Theorem 10.9. This leads to a Bayes linear graphical model which has arcs

$$H_i \to G, \quad \forall i, \qquad H_i \to H_j, \quad \forall j > i.$$

We then carry out a sequence of partial adjustments. First, we adjust all nodes on the diagram by H_1. All the variance in H_1 becomes resolved by its observation. Its observed value is mildly surprising, as shown by near-half shading of the inner portion of the node: the corresponding size ratio is 3.04. For predicting future observables, about one-third of the variation in H_2 is resolved by observation of H_1, as indicated by outer node shading moving anticlockwise from 0° to about 120° for node H_2. The proportion of variation explained for more distant observations $H_3, \ldots$ falls slightly at each point, dropping to about 20% for H_5. Observed overall changes in expectation for these nodes are more or less the same as for H_1. The inferential adjustment of node G by node H_1 shows an explanation of variance of again about 30%, with an overall change in expectation similar in magnitude to those for the observables.

Secondly, we adjust all the nodes (except H_1) partially by H_2. This resolves about an extra 10% of the remaining variation in H_3, and similar amounts for the other observables and for the parameter set G. The magnitude of observed changes in expectation are similar to those for the previous adjustment. We continue making partial adjustments as time progresses. As far as the parameter set G is concerned, we receive less and less information from subsequent observations.

The outstanding feature on the diagram concerns node H_4. It has very full light shading of the inner part of the node corresponding to its actual observed value, suggesting that its forecast given H_1, H_2, H_3 was much closer to its actual value than would have been expected, given the amount of variation remaining. Parts of nodes G and H_5 are shaded to show the partial adjustment by H_4: these too show quite full light shading, corresponding also to surprisingly small changes in expectation relative to variance explained.

As far as the arc labels are concerned, there is generally rather more shading at the sending end than at the receiving end, suggesting that the information carried by a single information source can be largely replaced by information from other information sources. The path correlations between information sources are displayed as shaded roundels near the ends of arcs. Most are shaded light to some degree, indicating that the information sent uniquely by the source node is in general agreement with the information sent by the other nodes. The exception concerns information sent uniquely from node H_4 to nodes H_5 and G: the full dark shading indicates that this information source contradicts the other remaining information sources. Closer inspection reveals, for example, that the implication of data H_4 is to increase the expectations across the components of G, whilst the implication of the other data is to reduce them. The magnitude of the changes in expectation are small, so that in this case we should not be too concerned about such contradiction.

10.8 Local computation: directed trees

Graphical structure helps us to quantify beliefs over a graphical model, by restricting prior specification to neighbouring nodes. We now discuss how to use the local graphical structure, in a similar way, to simplify the computation for large graphical models given observations. Local computation for a general graph may be complex because of complicated interrelationships resulting in there being many paths between a pair of nodes in the graph. Propagation is much simpler if we may reduce the diagram to a connected tree.

Definition 10.14 *A* ***connected tree*** *is a graph for which every pair of nodes is connected by one and only one path.*

Note that we may often remove undirected cycles from general directed graphs by introducing new nodes and combining nodes, for example following the conditions of Theorem 10.8.

In this section, we describe how to propagate beliefs around a directed tree. Suppose first that we have three belief structures A, B, C for which $\lfloor A \perp\!\!\!\perp B \rfloor / C$. We have noted, in Theorem 5.20, that the covariance structure between the collections A and B is determined by the pair of covariance structures $\text{Cov}(A, C)$ and $\text{Cov}(C, B)$. In §5.17, we showed how to exploit the belief separation $\lfloor A \perp\!\!\!\perp B \rfloor / C$, in order to evaluate the adjustment of $C \cup B$ by A based strictly on the pairwise evaluations of C by A and of B by C.

Now consider a general directed connected tree. First, note that for any tree, any node A separates the graph into two parts. Let $A^{\leftarrow}$ be the collection of all nodes on the graph which are connected to A by a path which passes through a child of A, and ${}^{\rightarrow}A$ be all nodes joined to A by a path through a parent of A. The node sets ${}^{\rightarrow}A$, $A^{\leftarrow}$ must be disjoint, and contain all nodes except A as the tree is connected. Further, there can be no path on the moral graph from ${}^{\rightarrow}A$ to $A^{\leftarrow}$ which does not pass through A, as otherwise there would be two paths between a pair of nodes on the tree. Therefore, for each node A, we have $\lfloor {}^{\rightarrow}A \perp\!\!\!\perp A^{\leftarrow} \rfloor / A$.

Further, if A and C are both parents of a node B, then $A \in C^{\leftarrow}$ and $C \in A^{\leftarrow}$. Therefore, apart from nodes A and C, sets $A^{\leftarrow}$ and $C^{\leftarrow}$ are identical. Therefore, if A and C are both parents of B, then we can divide the nodes of the diagram into three collections, U consisting of all nodes connected to A and C through parent nodes, V consisting of A and C, and W consisting of all nodes connected to A and C through child nodes. There are no directed arrows from V to U or from W to U or V. Therefore, from Theorem 10.8, we may merge the A and C into a single node. The resulting structure will again be a tree, as there are no paths between A and C on the original tree except for the path through B.

To summarize our discussion above, we have the following theorem.

Theorem 10.15 *In a connected directed tree,*

10.15.1: *for any node A, we have* $\lfloor {}^{\rightarrow}A \perp\!\!\!\perp A^{\leftarrow} \rfloor / A$*;*

10.15.2: *it is allowable to join the parents of any node into a single node which has as parents the union of the parent set of the joined nodes, and has as children the union of the child set of the joined nodes. The resulting graph is again a directed tree.*

We will now consider how information propagates from the observation of a particular node A through the tree.

10.8.1 Propagation

The first stage in the propagation is to split the tree. Suppose that A has k children, labelled $A_{(1)}, \ldots, A_{(k)}$. We may divide the nodes in the tree into A and $k+1$ disjoint groups. The jth group is all nodes B on the tree for which the unique path from node A to B passes through $A_{(j)}$, for $j = 1, \ldots, k$. Group $k+1$ is the collection of nodes for which the path passes through an ancestor of A. The split is shown in Figure 10.20. From Theorem 10.13, these $k+1$ groups of nodes are all separated from each other by node A. Therefore, adjustment by A separates the tree into $k+1$ separate diagrams, through which we may separately propagate information.

Now, we consider how to propagate information forward, i.e. through any one of the first k groups of nodes. Observe first, from Theorems 5.25 and 10.15, that the directed graph for the adjusted beliefs within this collection is exactly as for the original subgraph, and so is itself a tree, where we replace each node B by the corresponding node $\mathbb{A}_A(B)$, or equivalently we replace each $\mathrm{E}(B)$, $\mathrm{Var}(B)$, and each $\mathrm{Cov}(B, C)$ for neighbouring nodes by $\mathrm{E}_A(B)$, $\mathrm{Var}_A(B)$, $\mathrm{Cov}_A(B, C)$. Note in particular that for any node B for which there is not a directed path from A, we have $A \perp B$ so that beliefs are not adjusted for any such node. Each node along each such directed path separates the children from the ancestors, from Theorem 10.13. The propagation along each directed path from A therefore follows by the rules laid out at the beginning of this section.

Finally, we propagate information backwards through subgraph $k+1$. We may propagate beliefs back through the tree in exactly the same way that we propagated forwards, to determine beliefs within individual nodes. However, adjustment by A induces dependencies between all of the direct ancestors of A, which removes the tree structure in this subgraph. From Theorem 10.13, we may retrieve the tree

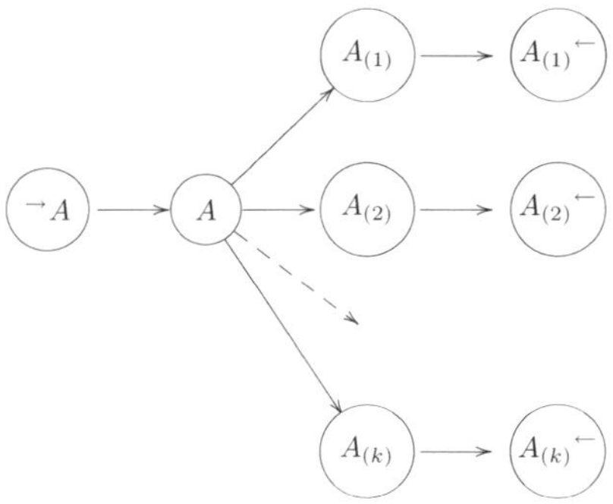

Figure 10.20 Splitting a directed connected tree.

structure by working back through ancestors, and at each node, combining all of the direct ancestors of that node into a single node, which is an allowable operation from Theorem 10.15. This will give us a properly updated new tree.

If we only observe a subset A^* of the elements of A, then propagation is as above, but the tree is not separated at A, and, just as for the other nodes, A is replaced by $\mathbb{A}_{A^*}(A)$.

10.8.2 Example

In order to construct Figure 10.19 we carried out a sequence of partial adjustments over all the quantities remaining at each stage. That is, we made the global adjustment of the collection $\{G, H_1, H_2, H_3, H_4, H_5\}$ by H_1 at step one, and so forth. Local computation over the directed tree provides an alternative as follows. We begin with the directed tree shown in Figure 10.21(a). It is an allowable operation to reverse the arc between H_1 and G. We now observe the collection of

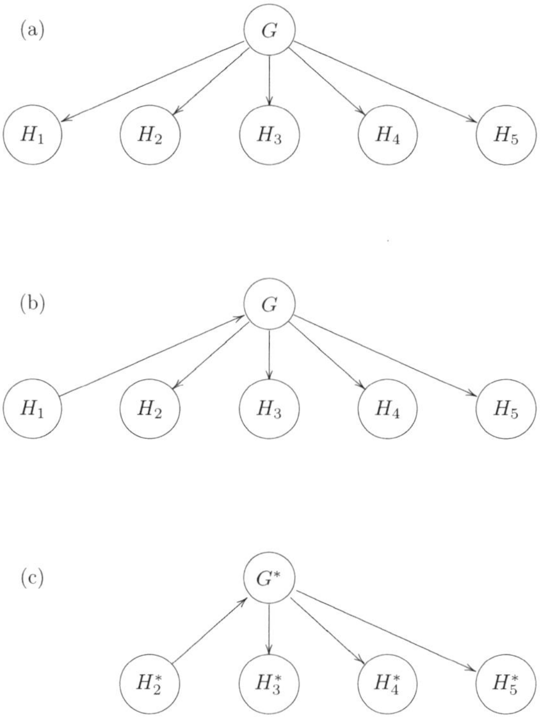

Figure 10.21 Local computation over a directed tree: (a) initial tree; (b) tree with the first observable placed at the head of the tree; (c) tree remaining after the evidence from H_1 has been propagated, and the arc from G to H_2 reversed. G^* is an abbreviation for $\mathbb{A}_{H_1}(G)$, and similarly for H_i^*.

quantities $H_1 = h_1$ and propagate the evidence over the remaining nodes. To do so, we construct the adjusted versions

$$G^* = \mathbb{A}_{H_1}(G),\ H_2^* = \mathbb{A}_{H_1}(H_2), \ldots, H_5^* = \mathbb{A}_{H_1}(H_5),$$

and replace the prior variance structure by variances and covariances calculated over these adjusted versions. In parallel, we update expectations to

$$\mathrm{E}(G^*) = \mathrm{E}_{h_1}(G), \quad \mathrm{E}(H_2^*) = \mathrm{E}_{h_1}(H_2),$$

and so forth. This completes the update for the evidence $H_1 = h_1$. Node H_1 may now be dropped from the tree. The algorithm may now continue as above, by reversing the arc from G to H_2, giving Figure 10.21(c), and propagating the evidence $H_2 = h_2$.

10.9 Junction trees

While propagation around directed trees is straightforward, it may be difficult to reduce a complex graph to a suitable directed tree structure. It is easier to describe procedures to reduce a directed graph to an undirected Markov tree. Therefore, we now give a general algorithm to reduce any graph to an undirected model, which will give a general approach to belief propagation. Note that the two approaches may often be combined, as we may reduce our graph to a directed tree in which each of the nodes is itself a large subgraph of the original graph.

We now describe how to construct the **junction tree**, which forms the basis for local computation in many kinds of graphical model. Proof that the algorithm does produce the junction tree, with properties as described, is given in, for example, Lauritzen (1996). Note that various of the steps in the following algorithm are not uniquely defined. In each case an arbitrary choice may be made.

1. Create the **moral graph**, by joining all parents and dropping arrows.

2. Triangulate the graph, by adding sufficient edges to ensure that there are no cycles of length four or more without a chord.

3. Carry out a **maximum cardinality search**. We may arbitrarily label any node as node 1. At each stage k, we label as node k the node on the graph with the largest number of labelled neighbours. (If and only if the graph is triangulated, at each stage when we label node k, all labelled neighbours of this node will be neighbours of each other.)

4. Order the **cliques**. The cliques are the maximal sets of nodes which are all joined to each other. For each clique, note the highest labelled node, and label the cliques in the order of these values.

5. Create the junction tree. The nodes of the tree are the cliques. Each clique is joined to at most one of the lower numbered cliques as follows. From the

above construction, it turns out that the intersection of the nodes in a clique and the nodes in all lower numbered cliques will be contained in at least one of the lower numbered cliques. Place a link between the clique and one of the lower numbered cliques which contain the intersection.

The basic properties of the junction tree are as follows.

Property 10.16 (Properties of the junction tree)

10.16.1: *If the original directed graph represents a second-order graphical model, then the junction tree is an undirected graph with the second-order global Markov property.*

10.16.2: *There is at most one path between any two nodes on the graph.*

10.16.3: *If a node of the original graph is contained in two nodes on the junction tree, then it is contained in all nodes on the unique path between these nodes.*

10.16.4: *Suppose that nodes A, B are adjacent on the junction tree. Let Z be the collection of nodes from the original graph which are in the intersection of A and B. Let U be the collection of nodes in A but not in Z, and let V be the nodes in B but not in Z. Then we must have $\lfloor U \perp\!\!\!\perp V \rfloor / Z$.*

10.16.5: *As a consequence of Property 10.16.4, the covariance between adjacent nodes on the junction tree may be derived from the covariances within each node, as from* (10.6) *(or, equivalently, Property 5.20.2),*

$$\text{Cov}(U, V) = \text{Cov}(U, Z)\text{Var}(Z)^{\dagger}\text{Cov}(Z, V).$$

10.10 Sequential local computation on the junction tree

Beliefs may be propagated around the junction tree as follows. Suppose that D is a node of the original graph, and that we observe $D = d$; if we are doing this analysis at the design stage, before making any observations, to see which nodes are worth observing, then we proceed as below but we do not pass around actual values for d.

All that matters for adjusting beliefs across the graph are the observed values d and the covariance between D and all the other nodes. So, we first pass the covariance function $\text{Cov}(D, \cdot)$ around the graph. This proceeds as follows.

D is contained in each of some connected sequence of nodes on the junction tree, so that $\text{Cov}(D, \cdot)$ is already determined for these nodes. For each other node in turn proceed as follows.

Suppose that we have already assessed $\text{Cov}(D, A)$ for node A, and we wish to pass the covariance to adjacent node B. If node A has subsets U, S and node B

has V, S, where U, V, S are disjoint, then $\lfloor U \perp\!\!\!\perp V \rfloor / S$. Therefore, having found $\mathrm{Cov}(D, A)$, we may find $\mathrm{Cov}(D, B)$ using (10.6) by

$$\mathrm{Cov}(D, V) = \mathrm{Cov}(D, S)\mathrm{Var}(S)^{\dagger}\mathrm{Cov}(S, V).$$

Within each node on the junction tree, we therefore compute adjusted means, variances, and covariances via (3.21), (3.30), and (3.31), i.e.

$$\mathrm{E}_D(B) = \mathrm{E}(B) + \mathrm{Cov}(B, D)\mathrm{Var}(D)^{\dagger}(d - \mathrm{E}(D)),$$

$$\mathrm{Var}_D(B) = \mathrm{Var}(B) - \mathrm{Cov}(B, D)\mathrm{Var}(D)^{\dagger}\mathrm{Cov}(D, B),$$

$$\mathrm{RVar}_D(B) = \mathrm{Cov}(B, D)\mathrm{Var}(D)^{\dagger}\mathrm{Cov}(D, B).$$

To find the belief transform, for each node B on the original graph, we evaluate the resolution transform matrix as $\mathbb{T}_{B:D} = \mathrm{Var}(B)^{\dagger}\mathrm{RVar}_D(B)$. Now observe the following.

First, adjustment by a node on the original graph preserves separations in the junction tree, by Theorem 5.25. Therefore, the new junction tree, with D removed from all nodes where it occurs on the original junction tree, is a valid junction tree for the structure where all beliefs are adjusted by D and corresponds to the second-order global Markov undirected graph on which we remove D and all arcs into D from the original moral graph by Theorem 10.11. Secondly, for any B, D_1, D_2 we have

$$\mathrm{Cov}(B, \mathbb{A}_{D_1}(D_2)) = \mathrm{Cov}(\mathbb{A}_{D_1}(B), \mathbb{A}_{D_1}(D_2)),$$

so that

$$\mathrm{RVar}_{[D_2/D_1]}(B) = \mathrm{RVar}_{[D_2/D_1]}(\mathbb{A}_{D_1}(B)).$$

Therefore, we can adjust every node on the diagram by D_1, then 'forget' the fact that we have adjusted B by D_1 and combine the resolved variance for the adjustment of B by D_1 with the resolved variance of $\mathbb{A}_{D_1}(B)$ by $\mathbb{A}_{D_1}(D_2)$ and still get the overall resolved variance for B by $D_1 \cup D_2$, namely we can move between stepwise adjustment and overall adjustment, using relations (5.4) and (5.10), i.e.

$$\mathrm{E}_{D_1 \cup D_2}(B) = \mathrm{E}_{D_1}(B) + \mathrm{E}_{[D_2/D_1]}(B), \tag{10.16}$$

$$\mathrm{RVar}_{D_1 \cup D_2}(B) = \mathrm{RVar}_{D_1}(B) + \mathrm{RVar}_{[D_2/D_1]}(B). \tag{10.17}$$

Thus, when we adjust by D_2, and subsequently $D_3, D_4, \ldots$, we may locally compute adjusted means, variances, and covariances by repeating the above steps but using the current adjusted means, variances, and covariances in precisely the same way as above.

The only difference comes when we evaluate the resolution transform. This is because $\mathbb{T}_{\mathbb{A}_{D_1}(B):[D_2/D_1]}$ is not the same as $\mathbb{T}_{B:[D_2/D_1]}$. Therefore, having found $\mathrm{RVar}_{[D_2/D_1]}(\mathbb{A}_{D_1}(B))$ in the adjustment stage of the algorithm, we would assess $\mathbb{T}_{B:[D_2/D_1]}$ as

$$\mathbb{T}_{B:[D_2/D_1]} = \mathrm{Var}(B)^{\dagger}\mathrm{RVar}_{[D_2/D_1]}(\mathbb{A}_{D_1}(B)).$$

It follows that if we want to evaluate these transforms, then we need to hold the inverse of the original variance matrix for each node on the original graph.

10.11 Example: correlated regressions

We continue the example of §10.5. For illustration, we explore the sequential adjustment of all other quantities by the data quantities. Organize the parameter quantities as $G = \{a, b, c, d\}$ and the error and data quantities as Q_i and H_i, as defined in (10.11) and (10.12). The graphical model for this organization is similar to Figure 10.5, but with all the Q_i, H_i explicitly represented, $i = 1, \ldots, 12$. The moral graph is obtained by marrying all nodes Q_i to node G, and by dropping arcs. Triangulation is trivial as there are no chord cycles of length four or more. The cliques comprise the 12 collections $J_i = \{G, Q_i, H_i\}$, $i = 1, \ldots, 12$, each containing G and each such that $\lfloor J_i \perp\!\!\!\perp J_j \rfloor \,/\, G$. There are many alternative choices of junction tree possible. For example, we may take J_1 as the root of the tree and link the other cliques to it. Alternatively, we may link J_1 to J_2, J_2 to J_3, and so forth. Note that this model has a simple plate representation. A high-level junction tree for all such plate representations is obtained trivially in this fashion, and will contain as many nodes in the junction tree as there are plates.

10.12 Example: problems of prediction in a large brewery

This example is part of an analysis of problems of prediction for a large brewery. The example arises from a project to develop a user-friendly computer-based decision support tool for use by managers, and is described in Spiropoulos (1995). We are very grateful to Takis Spiropoulos and Malcolm Farrow for describing the problem, and making available the specifications and the data. There are many quantities which have a bearing on decision making at the brewery – for example, the brewery produces different kinds of beers, demand for beer varies over time, production targets may be set or not, and may be met or not, depot stocks may be low or high, and so forth. The brewery staff have some expertise in judging the relationships between the variables, and assessing the implications of, say, a drop in depot orders for production volumes. There is some data concerning previous sales, depot orders, productions, and so forth. Examples of the kinds of question that need to be answered are: how much beer of each type must be produced next week, and what depot stocks are likely to be at Christmas.

We will not go into specific details of the elicitation and modelling process for this problem. Instead, we will present a broad picture of how such a problem might be tackled, using a Bayes linear approach. By doing so, we may incorporate valuable prior expert information, without requiring the specification of a joint probability distribution over all the quantities of interest – something certainly beyond the capabilities of industrial managers.

10.12.1 Problem summary

The brewery produces four beers. In the packaging plant, beer is drawn from vessels and packaged into kegs. Filled kegs become part of the stock held at the brewery until they are delivered to depots. Individual depots issue orders in advance to the brewery. These orders are used as information in planning the number of kegs of each kind of beer to be produced, but the deliveries need not exactly match the orders. The depots then sell the beer to the retail trade, subject to demand for each kind of beer. The orders that the depot makes to the brewery depend on the sales from the depots to the retail trade, as there is a limit to the volume of stock held at any given depot, and also a limit on the shelf life of a given kind of beer. In addition, to help predict the depot orders, the following information is available: previous order figures; the brewery's demand forecasts; the latest available depot stocks; and the depots' demand forecasts. One concern of the packaging plant is to meet demand from the depots, at the same time avoiding holding too much stock at the depots, or holding stock for too long.

10.12.2 Identifying the quantities of interest

The general problem was organized by focusing on certain quantities viewed by brewery staff as important. For simplicity, our analysis deals with weekly totals, with quantities totalled over depots, so that the analysis proceeds as though there is only one depot. The main quantities of interest are as follows: (1) the total volume of each beer sold each week, totalled over all the depots; (2) the previous week's forecast for this total, and (3) the forecast for this total made 2 weeks earlier; (4) the total volume of stock of each beer currently held at all depots; (5) deliveries made of each kind of beer from the packaging plant to the depots; (6) the total orders for each kind of beer requested by the depots; (7) the 2-weeks-ahead forecast made by the depots for the total volume of sales for each kind of beer. Each week, these seven quantities are observed for each of the four kinds of beer: 28 quantities in all. We use the following notation for these quantities, where $b = 1, 2, 3, 4$ indexes beer type. In fact, beer 1 represents a brand of bitter-style beer; whilst the other three beers are lager styles, of which beers 2 and 3 are brands which compete in the same market, and beer 4 has more of a niche market.

- V_{bt} is the total volume of beer b sold by the depots in week t.
- $V^{(1)}_{b,t-1}$ is the 1-week-ahead forecast (i.e. made at time $t-1$) for V_{bt}.
- $V^{(2)}_{b,t-2}$ is the 2-weeks-ahead forecast (i.e. made at time $t-2$) for V_{bt}.
- H_{bt} is the total volume of stock of beer b, above a fixed target level, held in depots at the end of week t.
- D_{bt} is the total volume of deliveries of beer b sent by the brewery to the depots during week t.
- F_{bt} is the latest available depot forecast of total demand for beer b. Depot demand forecasts are made 2 weeks ahead, so that the forecast for week t is prepared in week $t-2$.

- O_{bt} is the total order of beer b made by the depots in week t for the next week $t+1$.

For each kind of quantity we gather the four beers into a collection, or vector, for which we use the same notation, dropping the b subscript, so that:

- V_t is the collection of total volumes of beer sold by the depots in week t.
- $V_{t-1}^{(1)}$ is the 1-week-ahead forecast for V_t.
- $V_{t-2}^{(2)}$ is the 2-weeks-ahead forecast for V_t.
- H_t is the collection of total volumes of stock of beer held in depots at the end of week t.
- D_t is the collection of total volumes of deliveries of beer sent by the brewery to the depots during week t.
- F_t is the collection of latest available depot forecasts of total demand in week t for beer. These forecasts are prepared in week $t-2$.
- O_t is the collection of total orders of beer made by the depots in week t for the next week $t+1$.

The principal interest for the brewery in this analysis is to reduce its uncertainties about the orders quantities $O_t, O_{t+1}, \ldots,$ as the role of the brewery is to fulfil depot orders.

10.12.3 Modelling

We now describe our modelling for this problem. Any complex problem is amenable to many different modelling strategies, and we could postulate a variety of alternative models for the brewery. Our aim is to illustrate, with graphical models, the Bayes linear analysis of a particular collection of beliefs, and so we leave aside discussion of the appropriateness of this modelling strategy, except in so far as problems are highlighted by our diagnostics.

The relationships that we use derive from the beliefs and practices of the brewery staff. Previous forecasting strategies tended to be based on examining sales figures over recent weeks and sales figures of corresponding periods in the previous year. We use a Box–Jenkins style approach, assuming that differencing sales volumes V_{bt} once seasonally and once non-seasonally can be expected to remove seasonal effects and any linear trend. Then, if the resulting differenced series evinces stationarity with zero mean, a long-term forecast function should project trends linearly. In fact, previous data did suggest, in the notation of Box and Jenkins (1970), as appropriate the multiplicative $(0, 1, 1) \times (0, 1, 1)_{52}$ model:

$$(1-B)(1-B^{52})V_{bt} = (1-\theta_{b,1})(1-\theta_{b,52})\epsilon_{bt}, \tag{10.18}$$

for some $\theta_{b,1}$ and $\theta_{b,52}$, where $\mathrm{E}(\epsilon_{bt}) = 0$ and $\mathrm{Cov}(\epsilon_{bt}, \epsilon_{bs}) = 0$, for $t \neq s$, and B is the backshift operator such that $BV_t = V_{t-1}$. We gather the four quantities ϵ_{bt} into the vector ϵ_t, and gather the θs into diagonal matrices:

$$\theta_1 = \mathbf{diag}\{\theta_{1,1}, \theta_{2,1}, \theta_{3,1}, \theta_{4,1}\},$$

$$\theta_{52} = \mathbf{diag}\{\theta_{1,52}, \theta_{2,52}, \theta_{3,52}, \theta_{4,52}\}.$$

The qualitative features of the prior beliefs over the second-order structure may be expressed through a series of linear relations. Using our vector and matrix quantities, we model V_t and its 1-week-ahead forecast $V_{t-1}^{(1)}$ as

$$V_t = V_{t-1}^{(1)} + \epsilon_t, \tag{10.19}$$

$$V_{t-1}^{(1)} = V_{t-1} - \theta_1 \epsilon_{t-1} + c_{t-1}^{(1)}, \tag{10.20}$$

$$c_{t-1}^{(1)} = V_{t-52} - V_{t-53} - \theta_{52}\epsilon_{t-52} + \theta_1\theta_{52}\epsilon_{t-53}, \tag{10.21}$$

where $c_{t-1}^{(1)}$ is a modifying vector of values calculated from the corresponding values of V_t in previous years. We have from (10.19) that, up to time t, all of the information useful for linear prediction of V_t is carried by $V_{t-1}^{(1)}$; on the graphical model, $V_{t-1}^{(1)}$ is the only parent node of V_t. Thus V_t is constructed from its one-step-ahead forecast, perturbed by ϵ_t, to which we return below.

This model is a simplification to the extent that years are not exactly 52 weeks long, and so the model will drift out of phase over time unless corrected. Also, no account is taken of exceptional effects such as public holidays, major sporting fixtures, and so forth, where these do not occur in the same week in each year. (The model could be modified to take these factors into account.) In practice, the 1-week-ahead forecasts are not known with certainty beforehand as a result of such irregular exceptional circumstances, and because there is only a limited amount of historic data available.

The collections $\{\epsilon_t\}$ represent what we hope to be stationary mean-zero time series, one for each kind of beer. Hence we model ϵ_t as

$$\epsilon_t = M\eta_t, \tag{10.22}$$

where η_t is a vector of four uncorrelated random quantities, and M is a 4×4 matrix of constants which expresses the relationships between variation across beers.

The 2-weeks-ahead forecasts are modelled as

$$V_{bt}^{(2)} = V_{bt}^{(1)} + c_{bt}^{(2)} \tag{10.23}$$

$$c_{bt}^{(2)} = V_{b,t-50} - V_{b,t-51} - \theta_{b,52}\epsilon_{b,t-50} + \theta_{b,1}\theta_{b,52}\epsilon_{b,t-51}, \tag{10.24}$$

so that $c_t^{(2)}$ is a vector of values calculated from the corresponding values of V_t in previous years.

Prior beliefs about the remaining quantities are expressed through the following relationships:

$$F_t = V_t^{(2)} + \phi F_{t-1} - \phi V_{t-1}^{(2)} + \eta_t^{(F)}, \tag{10.25}$$

$$D_t = 0.5 O_{t-1} + 0.5 V_{t-1}^{(1)} - 0.5 H_{t-1} + \eta_t^{(D)}, \tag{10.26}$$

$$H_t = H_{t-1} + D_t - V_t, \tag{10.27}$$

$$O_t = F_{t-1} - H_{t-1} + F_{t-2} - O_{t-1} + \eta_t^{(O)}, \tag{10.28}$$

where $\eta_t^{(F)}, \eta_t^{(D)}, \eta_t^{(O)}$ are vectors of uncorrelated quantities and ϕ is a constant. The specification is completed by quantifying the various prior means and variances which generate the structure.

Figure 10.22, which is adapted from Goldstein et al. (1993), shows a Bayes linear graphical model summarizing the basic relationships amongst the quantities of interest. Each node on the diagram represents a collection, or vector, containing four beer quantities. For example the node labelled O_t represents the collection of total orders of beer made by the depots in week t for delivery in week $t+1$. Directed arcs on the diagram from one node (a parent) to another (a child) represent the potential capacity of the quantities in the parent collection to help explain the variance in the quantities in the child collection. The existence of an arc implies potentially relevant predictive information but does not imply a causal relationship. For example, from Figure 10.22 observe that depot orders in any given week (the node O_t) are influenced by quantities in four other collections: F_{t-2}, F_{t-1}, H_{t-1}, and O_{t-1}, and are conditionally independent, given these four parent collections, of any other collections up to and including week t. On the other hand, the depot orders are themselves predictive not only for next week's orders, but also for next week's deliveries, D_t.

10.12.4 Initialization values and specifications

We now quantify the model. First, we specify values for the constants that we have introduced: ϕ, the λ values used to construct M, expressing the relationships between variation across beers, and the θ values parameterizing (10.18):

$$\phi = 0.4, \tag{10.29}$$

$$\lambda_1 = 0.87, \quad \lambda_2 = 0.78, \quad \lambda_3 = 0.58, \tag{10.30}$$

$$\theta_1 = \mathbf{diag}\{0.95, 0.9, 0.8, 0.8\}, \tag{10.31}$$

$$\theta_{52} = \mathbf{diag}\{0.2, 0.7, 0.7, 0.7\}, \tag{10.32}$$

$$M = \begin{bmatrix} \lambda_1 & -1 & 0 & 0 \\ \lambda_1(1-\lambda_2) & 1-\lambda_2 & -1 & 0 \\ \lambda_1\lambda_2(1-\lambda_3) & \lambda_2(1-\lambda_3) & 1-\lambda_3 & -1 \\ \lambda_1\lambda_2\lambda_3 & \lambda_2\lambda_3 & \lambda_3 & 1 \end{bmatrix}. \tag{10.33}$$

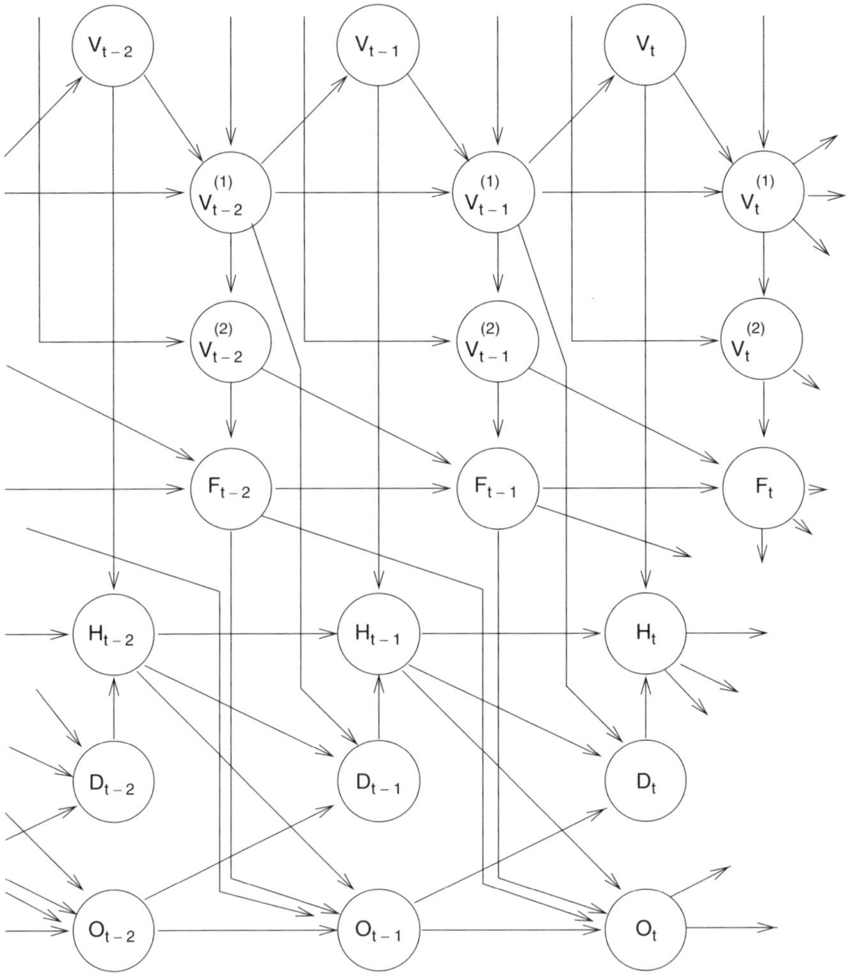

Figure 10.22 Bayes linear graphical model for the brewery problem.

Secondly, we specify the second-order structure over the collections of error components: η_t, $\eta_t^{(F)}$, $\eta_t^{(D)}$, and $\eta_t^{(O)}$. All such quantities are taken to have expectation zero, are uncorrelated, and uncorrelated over time and across beers. Thus all expectation and covariance specifications for these quantities are zero except for the variance specifications summarized in Table 10.4, which are constant over time t.

Some historic data are available, beginning at what we shall take to be week $t = 1$. The data consists of observed values of all 28 quantities in the seven collections: V_t, $V_t^{(1)}$, $V_t^{(2)}$, F_t, D_t, O_t, H_t. Checking the requirements of our model, and

Table 10.4 Error component variances.

Beer, b	$\text{Var}(\eta_{bt})$	$\text{Var}(\eta_{bt}^{(F)})$	$\text{Var}(\eta_{bt}^{(O)})$	$\text{Var}(\eta_{bt}^{(D)})$
1	1,840,000	22,500	27,000	25,600
2	17,000	90,000	119,000	136,900
3	38,000	302,500	416,000	184,900
4	18,000	90,000	119,000	78,400

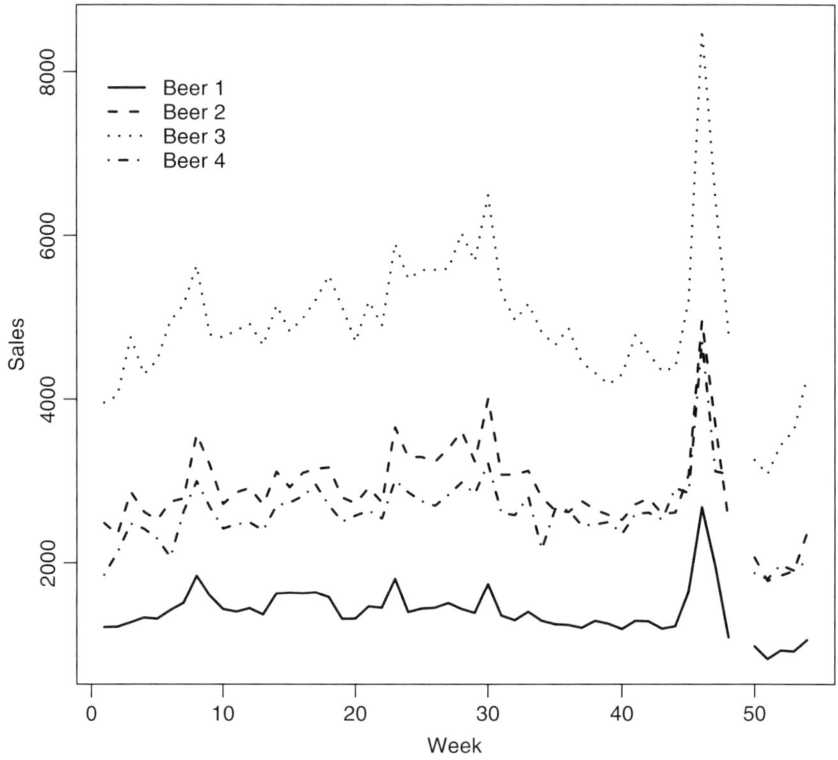

Figure 10.23 Sales of four kinds of beer over 54 weeks.

recalling that we take into account events in the same week of the preceding year, the earliest time t for which the requisite historic data are available to generate the full model is week $t = 55$. Beer sales for the first 54 weeks are given in Table 10.5, these being the observed values of the quantities in the collections $V_1, \ldots, V_{54}$. A simple way of handling the observed quantities, in linear equations such as (10.27), is to treat them as variables with zero variance and expectation equal to their observed value. The data are plotted in Figure 10.23: beer 1 has the lowest

Table 10.5 Sales of four beers, $V_1, \ldots, V_4$, at weeks $t = 1, 2, \ldots, 54$. No information was available for week $t = 49$.

t	V_{1t}	V_{2t}	V_{3t}	V_{4t}	t	V_{1t}	V_{2t}	V_{3t}	V_{4t}
1	1212	2484	3956	1852	28	1434	3608	6027	2980
2	1216	2343	4039	2124	29	1386	3236	5686	2843
3	1270	2864	4752	2477	30	1739	4003	6491	3213
4	1328	2619	4300	2414	31	1354	3074	5299	2606
5	1314	2521	4475	2293	32	1297	3076	4962	2583
6	1420	2746	4959	2067	33	1400	3122	5168	2805
7	1508	2783	5152	2626	34	1290	2795	4809	2165
8	1841	3572	5632	2994	35	1248	2637	4658	2656
9	1598	3192	4787	2677	36	1239	2619	4866	2643
10	1433	2717	4749	2417	37	1204	2752	4444	2454
11	1401	2859	4839	2470	38	1289	2646	4317	2468
12	1443	2912	4914	2490	39	1253	2583	4179	2497
13	1366	2708	4654	2396	40	1189	2524	4303	2354
14	1624	3110	5148	2699	41	1289	2714	4783	2599
15	1633	2920	4825	2734	42	1282	2786	4554	2610
16	1627	3095	4982	2813	43	1193	2595	4336	2528
17	1637	3146	5215	2952	44	1223	2613	4390	2914
18	1579	3162	5511	2699	45	1641	3070	5203	2850
19	1314	2796	5119	2503	46	2677	4954	8512	4705
20	1317	2725	4697	2576	47	1960	3680	6459	3115
21	1466	2917	5198	2636	48	1087	2497	4791	3080
22	1450	2735	4901	2541	49				
23	1805	3655	5900	3007	50	978	2064	3253	1873
24	1396	3290	5470	2860	51	823	1779	3080	1803
25	1439	3290	5578	2751	52	927	1844	3439	1975
26	1449	3232	5581	2696	53	916	1902	3623	1907
27	1506	3396	5594	2830	54	1056	2371	4280	2029

sales, and beer 3 the highest. Sales reach a peak in week 46. Sales figures are absent for week 49. The observed values of ϵ_t, the difference between the 1-week-ahead forecast and the actual sales figure, are given in Table 10.6 and plotted in Figure 10.24. The relationship between the four beers can be clearly seen in the way that movements in sales of one beer type are tracked quite closely by sales of other beer types. With regard to seasonal or periodic components, as there is only just over 1 years' data, it is impossible to verify the assumption of a strong yearly component. However, visually there is some evidence of a monthly or 4-weekly cycle which is not modelled. Such cyclic behaviour is more strongly evident when the sales forecasts errors are examined in Figure 10.24, which also shows the very large positive error in week 46 (sales were much higher than the forecast) followed by large negative errors (much smaller sales than forecast) in the following two weeks. Such swings could be exacerbated by reporting lags and the granularity of the data.

Table 10.6 Forecast errors for beer sales, $\epsilon_1, \ldots, \epsilon_4$, at weeks $t = 1, 2, \ldots, 54$. No information was available for week $t = 49$.

t	ϵ_{1t}	ϵ_{2t}	ϵ_{3t}	ϵ_{4t}	t	ϵ_{1t}	ϵ_{2t}	ϵ_{3t}	ϵ_{4t}
1	139	276	94	−97	28	−95	−657	−392	−103
2	72	−34	68	130	29	−194	−661	−445	−22
3	117	239	469	304	30	427	425	290	402
4	−84	−18	−426	−41	31	−256	−507	−412	−198
5	−50	−109	−129	−117	32	−200	−299	−860	−240
6	22	−59	72	−323	33	132	105	248	160
7	−43	−117	162	−131	34	75	182	520	−256
8	363	587	515	368	35	79	260	394	419
9	−31	−258	−612	−272	36	136	212	359	87
10	−41	−568	−512	−546	37	5	483	−133	−215
11	−387	−682	−725	−652	38	3	114	−296	−32
12	−99	−228	−116	−62	39	−18	163	−259	107
13	−57.	−29	−345	−250	40	−142	66	−111	−196
14	217	490	−107	−128	41	76	220	404	238
15	−153	−350	−1094	−193	42	58	355	188	−32
16	−79	−467	−905	−222	43	15	50	−78	−120
17	134	−93	−274	157	44	−11	229	74	132
18	6	−146	−124	−278	45	361	500	1020	26
19	−177	−774	−624	−531	46	775	780	1757	986
20	−156	−634	−729	−282	47	−583	−35	62	−458
21	92	−133	129	−93	48	−840	−1638	−2114	−410
22	70	32	−9	136	49				
23	417	591	877	333	50	55	−499	−1045	−382
24	−348	−297	−280	7	51	142	225	−30	−306
25	65	238	468	118	52	391	339	660	137
26	−91	−155	−417	−279	53	161	512	916	315
27	−63	−341	−840	−181	54	232	723	999	266

Construction of the model at time $t = 55$ also requires past observations on depot forecasts, F; depot orders, O; and depot stocks, H. Reported depot stocks are given in relation to certain fixed targets, so that positive (negative) values indicate a surplus (deficit) in stock from a fixed target. For example, for beer 3 we have:

- $F_{3,54} = 3785$, meaning that the forecast made in week 54 for the depot demand in week 56 is 3785 kegs;

- $O_{3,54} = 3639$, meaning that the depot ordered 3639 kegs in week 54, to be delivered in week 55;

- $D_{3.54} = 4127$, meaning that 4127 kegs were delivered to the brewery in week 54;

- $H_{3,54} = -505$, meaning that the total depot stock was 505 kegs short of target in week 54.

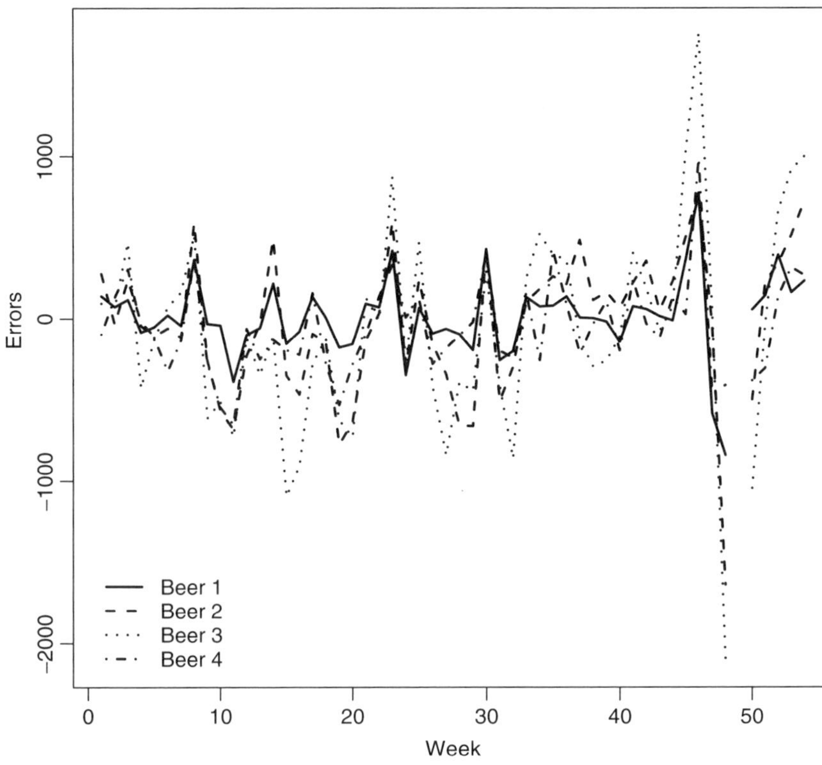

Figure 10.24 Sales forecast errors for four kinds of beer over 54 weeks.

Table 10.7 Depot forecasts, orders and deliveries, weeks 53–54.

	Beer style			
	1	2	3	4
$F_{b,53}$	1121	2274	3811	2075
$F_{b,54}$	1128	2236	3785	1858
$O_{b,54}$	788	1530	3639	1481
$D_{b,54}$	897	2165	4127	1641
$H_{b,54}$	−181	840	−505	−473

The remaining inputs for these quantities are shown in Table 10.7. This completes the inputs necessary to construct the model, so we now proceed to construct the model from the relations (10.19), (10.20), (10.22), (10.23), and (10.25)–(10.28) for weeks $t = 55$ onwards, for as far into the future as we wish.

10.12.5 Examining the generated model

Suppose that we generate the model for several weeks, beginning at week $t = 55$. The prior structure so generated is detailed and has very many specifications. For example, Table 10.8 shows prior expectations and standard deviations as they are constructed for beer sales and depot orders for the first three weeks, for each beer style. Thus, the expected total sales of beer 3 for the first three weeks for the constructed model are 879.5, 976.66, and 956.72, respectively, with a standard deviation of about 220 in each case that varies little from week to week. Standard deviations for the order quantities rise substantially from week 55 to week 56, but less substantially thereafter: this is an artefact of the initialization of the model.

There are very many specifications generated from the model, even though we are considering only the second-order structure. As examples of the features we may want to explore we have the following. Table 10.9 shows the prior correlation matrices for beer sales and beer orders at weeks $t = 55, 56, 57$. The correlations at any given time point between sales for different beer styles are generally large and positive, as one would expect. However, observe that the correlation structure

Table 10.8 Some expectations and standard deviations for beer sales and orders, rounded to the nearest integer.

$t = 55$	E(·)	SD(·)	$t = 56$	E(·)	SD(·)	$t = 57$	E(·)	SD(·)
V_1	880	219	V_1	977	220	V_1	957	220
V_2	2051	326	V_2	1970	328	V_2	1937	329
V_3	3903	420	V_3	4013	428	V_3	4039	436
V_4	2029	565	V_4	2166	576	V_4	2104	587
O_1	1642	164	O_1	640	394	O_1	1186	425
O_2	3340	345	O_2	1380	770	O_2	2376	828
O_3	4462	645	O_3	3656	1254	O_3	4623	1334
O_4	2925	345	O_4	1424	931	O_4	2422	971

Table 10.9 Vector correlations at fixed times for beer sales and orders.

	Week $t = 55$				Week $t = 56$				Week $t = 57$		
	V_{1t}	V_{2t}	V_{3t}		V_{1t}	V_{2t}	V_{3t}		V_{1t}	V_{2t}	V_{3t}
V_{2t}	0.59			V_{2t}	0.59			V_{2t}	0.59		
V_{3t}	0.68	0.63		V_{3t}	0.67	0.62		V_{3t}	0.67	0.62	
V_{4t}	0.70	0.64	0.84	V_{4t}	0.69	0.64	0.84	V_{4t}	0.68	0.64	0.84
	O_{1t}	O_{2t}	O_{3t}		O_{1t}	O_{2t}	O_{3t}		O_{1t}	O_{2t}	O_{3t}
O_{2t}	0.00			O_{2t}	0.16			O_{2t}	0.16		
O_{3t}	0.00	0.00		O_{3t}	0.16	0.12		O_{3t}	0.15	0.11	
O_{4t}	0.00	0.00	0.00	O_{4t}	0.30	0.22	0.25	O_{4t}	0.28	0.21	0.23

Table 10.10 Correlation matrices between beer sales and orders at times t and $r = t + 1$.

	Week $t = 55$ with week 56					Week $t = 56$ with week 57			
	$V_{1,r}$	$V_{2,r}$	$V_{3,r}$	$V_{4,r}$		$V_{1,r}$	$V_{2,r}$	$V_{3,r}$	$V_{4,r}$
V_{1t}	0.05	0.06	0.13	0.14	V_{1t}	0.05	0.06	0.14	0.14
V_{2t}	0.03	0.10	0.12	0.13	V_{2t}	0.03	0.11	0.13	0.14
V_{3t}	0.03	0.06	0.20	0.17	V_{3t}	0.04	0.07	0.23	0.19
V_{4t}	0.04	0.06	0.17	0.20	V_{4t}	0.04	0.08	0.19	0.23
	$O_{1,r}$	$O_{2,r}$	$O_{3,r}$	$O_{4,r}$		$O_{1,r}$	$O_{2,r}$	$O_{3,r}$	$O_{4,r}$
O_{1t}	−0.42	0.00	0.00	0.00	O_{1t}	−0.39	−0.05	−0.03	−0.05
O_{2t}	0.00	−0.45	0.00	0.00	O_{2t}	−0.06	−0.40	−0.02	−0.04
O_{3t}	0.00	0.00	−0.52	0.00	O_{3t}	−0.06	−0.04	−0.38	−0.04
O_{4t}	0.00	0.00	0.00	−0.37	O_{4t}	−0.11	−0.06	−0.04	−0.29

across beer sales weakens quite slowly over time. For example, the correlation between sales of beers 1 and 4 drops from 0.698 in week 55 to 0.690 in week 56, and to 0.683 in week 57. Examining the prior correlation structure for the beer orders, we see that these are uncorrelated in week 55 – an artefact of the model initialization – and are fairly weakly correlated in later weeks. For orders too, there is an apparent weakening of the correlation structure over time. The correlation structure for later weeks turns out to be similar to that for week 57.

Table 10.10 shows some prior correlation matrices between vector collections 1 week apart. For example, the correlation between sales of beer 1 in weeks 55 and 56 is only 0.050, rising slightly for weeks 56 and 57. Correlations for the other kinds of beer are somewhat stronger. These positive correlations summarize the beliefs that sales increases one week are likely to be weakly associated with sales increases the next week. The prior correlation matrices for beer orders show negative correlations, in particular for beers of the same kind, showing that high orders one week are associated with lower orders the next week. As time progresses, weak negative correlations are induced between orders for different beers. The correlation structure for later weeks is about the same for the sales quantities, whilst for the order quantities the one-week correlations remain negative and are slightly larger in magnitude.

Table 10.11 similarly shows some prior correlation matrices between vector collections 2 weeks apart. For example, for the beer sales quantities, V_t has a slightly weaker correlation with V_{t+2} than it has with V_{t+1}. On the other hand, for the beer orders, because there is a negative correlation between the quantities 1 week apart, $\text{Corr}(O_{bt}, O_{b,t+1}) < 0$, there is necessarily a weaker but positive correlation between quantities 2 weeks apart, $\text{Corr}(O_{bt}, O_{b,t+2}) > 0$. The correlation structure for later weeks turns out to be slightly larger in magnitude, but otherwise similar.

Table 10.11 Correlation matrices between beer sales and orders at times t and $r = t + 2$.

	Week $t = 55$ with week 57					Week $t = 56$ with week 58			
	$V_{1,r}$	$V_{2,r}$	$V_{3,r}$	$V_{4,r}$		$V_{1,r}$	$V_{2,r}$	$V_{3,r}$	$V_{4,r}$
V_{1t}	0.05	0.06	0.13	0.13	V_{1t}	0.05	0.06	0.14	0.14
V_{2t}	0.03	0.10	0.12	0.12	V_{2t}	0.03	0.11	0.13	0.13
V_{3t}	0.03	0.06	0.19	0.16	V_{3t}	0.04	0.07	0.22	0.19
V_{4t}	0.04	0.06	0.16	0.19	V_{4t}	0.04	0.07	0.19	0.22
	$O_{1,r}$	$O_{2,r}$	$O_{3,r}$	$O_{4,r}$		$O_{1,r}$	$O_{2,r}$	$O_{3,r}$	$O_{4,r}$
O_{1t}	0.19	0.00	0.00	0.00	O_{1t}	0.20	0.04	0.05	0.09
O_{2t}	0.00	0.21	0.00	0.00	O_{2t}	0.04	0.20	0.04	0.07
O_{3t}	0.00	0.00	0.24	0.00	O_{3t}	0.04	0.03	0.20	0.08
O_{4t}	0.00	0.00	0.00	0.18	O_{4t}	0.07	0.06	0.08	0.25

There are clearly many such prior correlations that we could display and comment upon: for vectors of one kind of quantity such as sales in a given week; across vectors of one kind of quantity during different weeks and for different lags; and of course across different kinds of quantities such as beer sales and orders at a given time point, or at separated time points. However, even for this problem, which has had many complicating features removed, there are so many specifications that it is very time-consuming to go beyond a cursory examination. In the remainder of this example, we see how labelled graphical models can help us to focus on the key features, and on diagnostic exploration.

10.12.6 Basic adjustment

There are many adjustments of potential interest for this problem. For illustration we will concentrate on one very limited aspect, namely the value of the information available in one week for predicting beer orders in a given future week.

We may use local computation to simplify our calculations. In particular, we propagate beliefs using the directed tree algorithm, where we aggregate all nodes for a particular week as a single node W_t. The tree then becomes W_t to W_{t+1} to W_{t+2} and so forth. In order to impose this structure, we must deal with the arc from F_{t-2} to O_t, which we can handle by adding a deterministic node to W_t of the form $G_t = F_{t-1}$.

Suppose that our aim is to predict beer orders in week 62, so that the collection of interest is O_{62}. We will adjust beer orders in week 62 by the information available in previous weeks. Note from Figure 10.22 that the collections $H_{t-1}, F_{t-1}, G_{t-1} = F_{t-2}, O_{t-1}$, are sufficient for all quantities up to week $t-1$ inclusive for predicting beer orders for week t. Therefore, instead of adjusting by the full collection W_t, we may restrict attention to the adjustment by this sufficient collection.

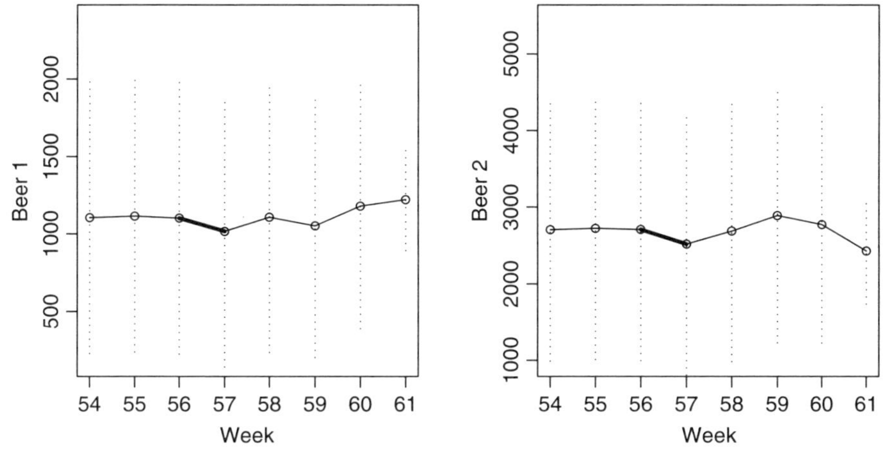

Figure 10.25 Adjusted expectations and variances for orders as we accumulate evidence. Plotted are the adjusted expectations with two-standard-deviation bounds for $O_{b,62}$, given all the information available up to that week. Week 54 information is prior. Successive forecasts are connected with a line, emphasized when the change is surprising.

Figure 10.25 shows a simple graphical summary for the sequence of forecasts made for orders. Under this model we have essentially prior information at week 54, and then information on depot stocks and so forth for each subsequent week. To assess the value of the information available weeks earlier, we adjust beer orders at week 62 by the information available at week 55, and then partially by the extra information available at week 56, and so forth. In each case we may calculate an updated adjusted expectation and an adjusted variance, and we may calculate the standardized change in adjustment. For each beer, we plot the adjusted expectation at that point, with two-standard-deviation bounds to convey the uncertainty remaining. We connect successive forecasts. Where the standardized change in adjustment is surprising, we emphasize the connection. The plots show that, for all four beers, the orders were roughly in agreement with the prior expectation and prior variance. Evidently, there is little information carried by information more than 1 week away, as the adjusted variances reduce only marginally from week to week, except for the final forecast made at week 61. All of the adjustments moving from week 56 to week 57 are somewhat surprising: all are between two and three standard deviations.

10.12.7 Exploration via graphical models

There is a vast number of adjustments, resolutions, expectations, variances, size diagnostics, and so forth that we might look at for this example. We show in this section how we can focus on the principal features by examining the labelled graphical model. Our interest is in three main areas.

- We want to see the implications of the model for reducing uncertainty in observables such as future beer orders.
- We want to know whether the data are consistent with the prior specification, and, if not, whether any discrepancies are important.
- We want to track information flow and diagnostics as they combine over time.

By way of illustration, we begin by looking at diagrams representing all the quantities over the 3-week period starting at week 55, the first time for which the model is fully constructed. We show the quantities at each week as seven collections of four-dimensional vectors. We may employ the separations shown below when making adjustments:

$$\lfloor V_{57} \perp\!\!\!\perp V_{56} \cup V_{56}^{(2)} \cup F_{56} \cup H_{56} \cup D_{56} \cup O_{56} \rfloor \;/\; V_{56}^{(1)},$$

$$\lfloor V_{57}^{(1)} \perp\!\!\!\perp V_{56} \cup V_{56}^{(2)} \cup F_{56} \cup H_{56} \cup D_{56} \cup O_{56} \rfloor \;/\; V_{56}^{(1)},$$

$$\lfloor V_{57}^{(2)} \perp\!\!\!\perp V_{56} \cup V_{56}^{(2)} \cup F_{56} \cup H_{56} \cup D_{56} \cup O_{56} \rfloor \;/\; V_{56}^{(1)},$$

$$\lfloor F_{57} \perp\!\!\!\perp V_{56} \cup V_{56}^{(1)} \cup H_{56} \cup D_{56} \cup O_{56} \rfloor \;/\; V_{56}^{(2)} \cup F_{56},$$

$$\lfloor H_{57} \perp\!\!\!\perp V_{56} \cup V_{56}^{(2)} \cup F_{56} \cup D_{56} \rfloor \;/\; H_{56} \cup V_{56}^{(1)} \cup O_{56},$$

$$\lfloor D_{57} \perp\!\!\!\perp V_{56} \cup V_{56}^{(2)} \cup F_{56} \cup D_{56} \rfloor \;/\; H_{56} \cup V_{56}^{(1)} \cup O_{56},$$

$$\lfloor O_{57} \perp\!\!\!\perp V_{56} \cup V_{56}^{(1)} \cup V_{56}^{(2)} \cup D_{56} \rfloor \;/\; H_{56} \cup F_{56} \cup O_{56} \cup F_{55}.$$

Our diagrams only show the arcs for adjustments of child nodes by the parents given by these separations. In our diagrams, we show the observed quantities at week 55;

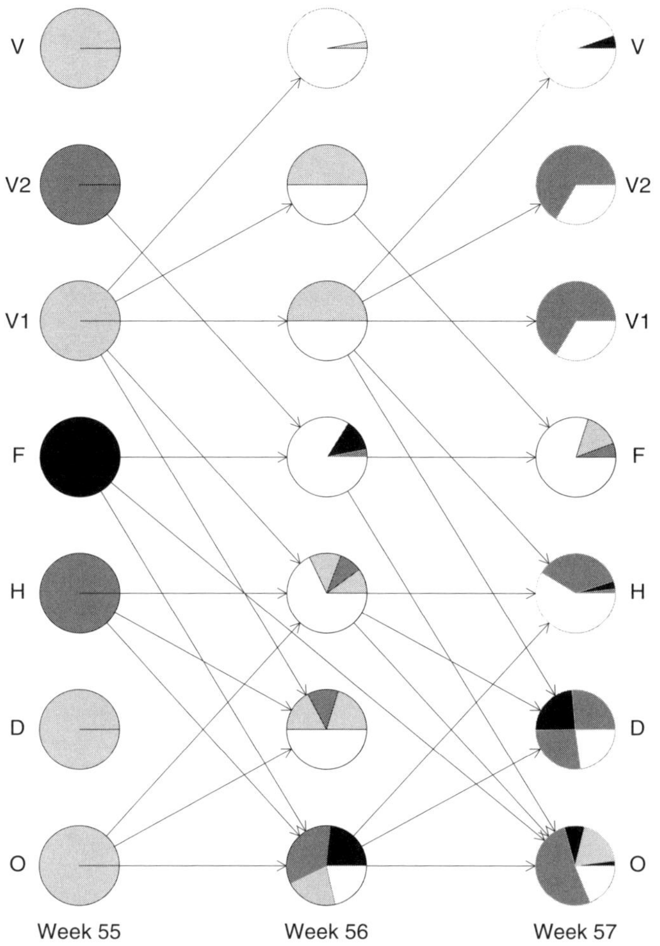

Figure 10.26 Node influence. Node shadings show the variance resolutions and partial resolutions as information arrives from a sequence of parents.

the quantities at week 56 adjusted by those at week 55; and the quantities at week 57 adjusted by those at week 56. For interactive use we use colour to track influence; in monochrome we have to distinguish between information sources using shading, where possible. Each node represents a collection of four quantities, so that each column constitutes the 28 quantities of interest for one week.

Figure 10.26 shows node resolutions as we adjust quantities and track their implications over time. To help aid interpretation, if a node is adjusted by more than one parent node, the adjustments take place in left-to-right and vertically descending order. For example, node O_{57} is adjusted by F_{55}, F_{56}, H_{56}, and O_{56}, in that order. The resolutions and partial resolutions contributed by the parent

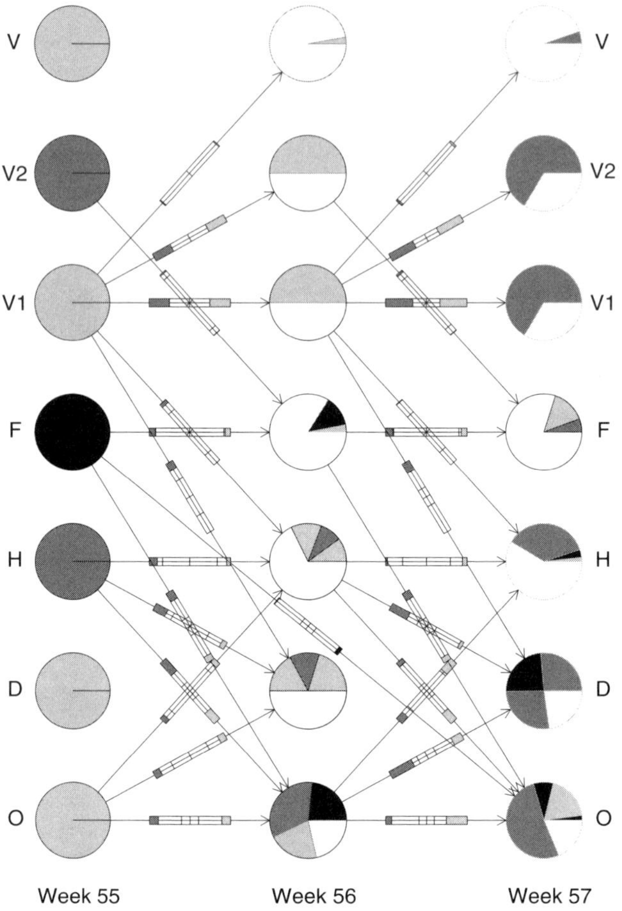

Figure 10.27 Arc influence. Arc shadings show the amount of information leaving a parent node, and the amount arriving at a child node.

node are shown sequentially and anticlockwise from 0°. The interpretation for this node is that F_{55} delivers little resolution, F_{56} partially resolves about a fifth of the remaining variation, H_{56} slightly less than this, and finally the partial adjustment by O_{56} resolves most of the remaining variation – and is obviously an important source of information.

In Figure 10.27 we add arc labels to show the amount of information leaving parent nodes and arriving at child nodes. For example, for adjusting node D_{57}, relatively a lot of information leaves O_{56} and about half of it arrives. Quite a lot of information leaves H_{56} for D_{57}, but little arrives. Relatively little information leaves $V_{56}^{(1)}$ for D_{57}, and – visually – none appears to arrive. The implication is that O_{56} is very important for learning about D_{57}; that H_{56} is less important, but

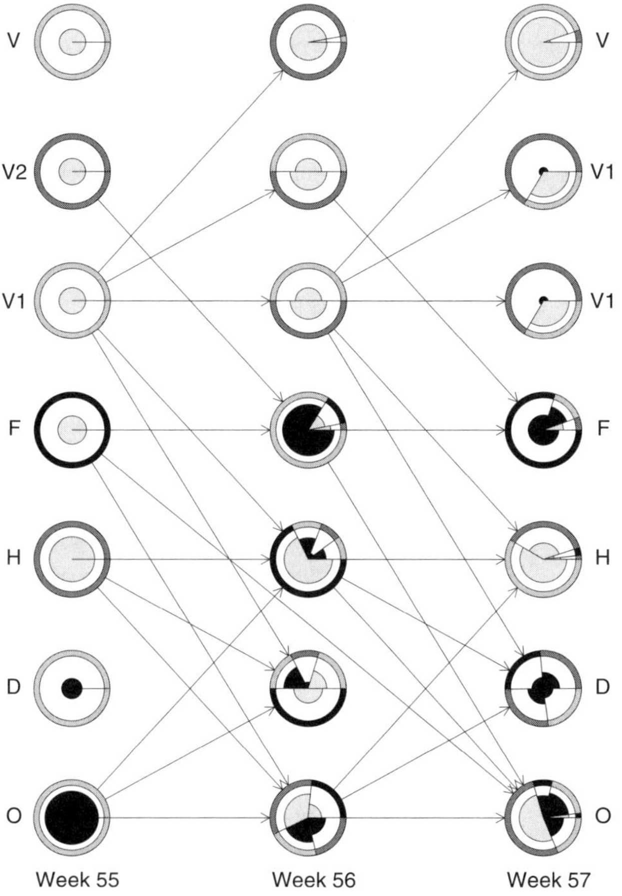

Figure 10.28 Node diagnostics. Inner portions of sectors show diagnostic ratios: dark shading for surprisingly large changes in expectation, light shading for surprisingly small changes in expectation.

probably still useful; and that $V_{56}^{(1)}$ is unimportant. It is simple, if desired, to calculate resolutions and partial resolutions for an alternative sequence of adjustments such as this. Here we find that the variance resolution in D_{57} due to O_{56} alone is 0.6156; the partial resolution given by adjusting also by H_{56} is 0.1556; and the partial resolution given by adjusting next by $V_{56}^{(1)}$ is 0.0009, confirming the interpretation we drew from the diagram.

In Figure 10.28, we add node diagnostics to the plot by shading inner parts of a sector corresponding to an adjustment, and we take into account actual observation. Observation of a node resolves any remaining variation in the node. Here, we will also introduce the notion that we can track influence via colour or shading.

We attach shadings to parent nodes via the final part of the outer annular sector corresponding to the residual variation removed when the parent node is actually observed. To return to the adjustment of O_{57}, its four parent nodes were, in order, allocated the shadings black (F_{55}), light grey (F_{56}), black (H_{56}), and dark grey (O_{56}). Therefore, the resolutions shown in the outer annular sector for node O_{57} are, from 0° and moving anticlockwise, shaded black, light grey, black, and dark grey to represent the resolutions and partial resolutions attributable to these parent nodes. The fifth and final sector carries information and diagnostics for the observed value of the collection O_{57}. The interpretation for this node as follows. Recall that the inner sector shading, described in §10.7.1, has heavy (light) shading for surprisingly large (small) squared changes in expectation relative to prior variance. In these diagrams, half shading of an area corresponds to a change of two standard deviations, or equivalently a size ratio of 4 or $\frac{1}{4}$ (10.15), and more shading to larger changes.

- The adjustment by parent node F_{55} was only very weakly informative.
- Two of the adjustments (by F_{56} and H_{56}) led to minor resolutions in variances and also to quite surprisingly large changes in squared expectation.
- The partial adjustment by parent O_{56} resolved a lot more of the variance, but led to surprisingly small changes in expectation.
- The final observation of O_{57} was relatively quite far from its adjusted expectation given its forecast from its parent nodes.

The actual resolutions and partial resolutions for the adjustment of node O_{57}, and the corresponding size ratios, are shown in Table 10.12. It is notable that the changes in expectation are all mildly surprising, even for the apparently uninformative parent F_{55}.

In Figure 10.29 we add diagnostics to the arc labels. Recall that the arc labels portray the resolutions and size ratios for the adjustments which correspond to information leaving a parent node, and information arriving at a child node. We saw in Figure 10.27 that the arc resolutions in this example are mostly quite small. As such, in order to visualize arc diagnostics we expand the area allocated for

Table 10.12 Resolutions and partial resolutions for the adjustment of node O_{57}, together with the corresponding size ratios.

Parent node	Resolution	Size ratio	Shading of inner sector (%)
F_{55}	0.0199	0.14	62
F_{56} (partial)	0.1922	7.53	64
H_{56} (partial)	0.0813	3.68	48
O_{56} (partial)	0.5203	0.21	54
O_{57} (observed)	0.1863	2.78	40

diagnostic shading to the full label. Thus, an unshaded half-label corresponds to a size ratio of one and unsurprising changes in expectation. A fully shaded half-label corresponds to a very large size ratio and a highly surprising change in expectation: dark shading for aberrantly large changes, light shading for unusually small changes. Note that we calculate arc labels and their diagnostics before we adjust a node by its observed value: the information arriving from a parent node **after** we have observed the child node is necessarily zero.

Figure 10.29 shows, for example, that the information leaving $V_{56}^{(2)}$ for F_{57} corresponded to an adjustment with a size ratio of slightly less than one: there is a small amount of light shading for the half-label nearest the parent. However,

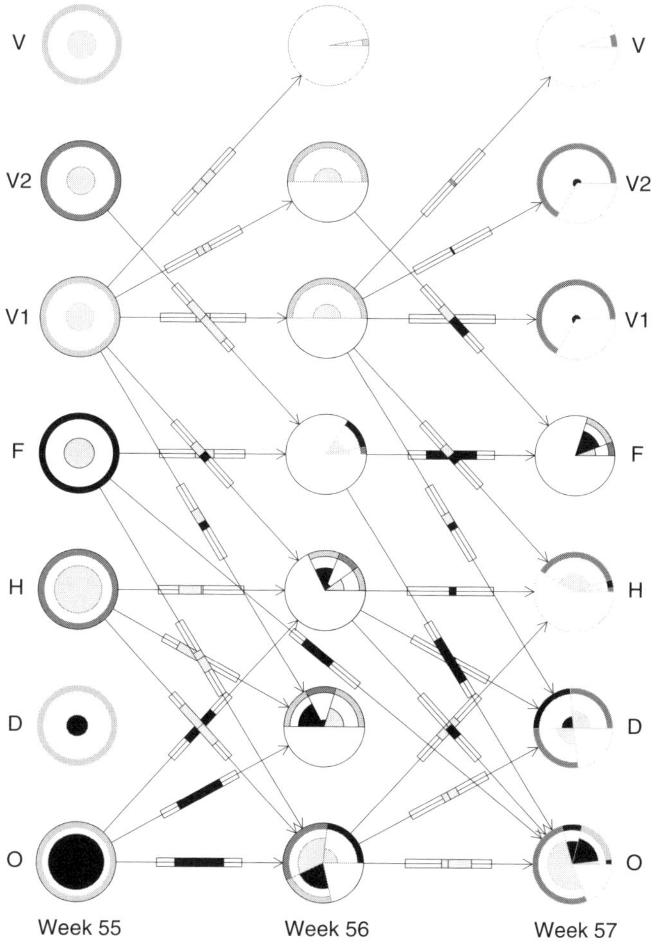

Figure 10.29 Diagnostics for nodes and arcs. Arc diagnostics are expanded to cover the label.

the information arriving at F_{57} from $V_{56}^{(2)}$ corresponded to an adjustment with a size ratio of about four, as there is roughly half dark shading for the half-label nearest the child. The partial information arriving from the other parent, F_{56}, is also heavily and darkly shaded. The implication is that the prediction for F_{57} given its parents is rather different than its prior expectation, relative to its prior variance.

Figure 10.30 provides a summary of overall influences over the three-week period. Each node is adjusted by all the information available beforehand, and then partially by its observation, as in §10.7.1.1. We also add **path correlations** to the arcs as described in §10.7.2.1. Lightly shaded circles at the ends of arcs show that the information arriving from a parent is strongly consistent with the

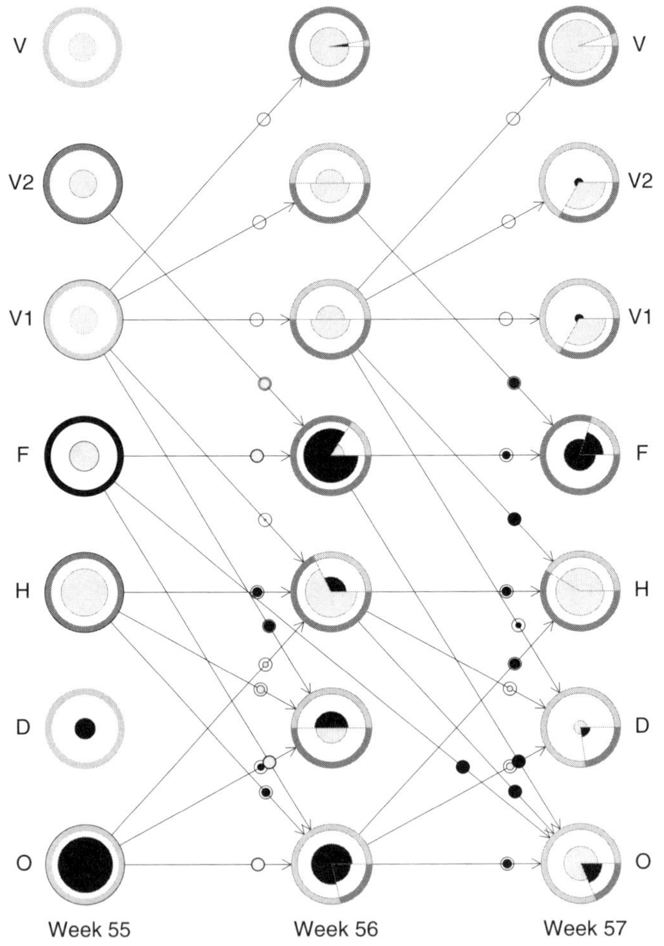

Figure 10.30 Combined and observed adjustments, with path correlations indicating consistency of parent contributions.

information arriving from other parents combined. Darkly shaded circles show that the information arriving from a parent is contradictory to the information arriving from other parents combined. No shading shows that sources of information are uncorrelated as far as the joint adjustment is concerned. Note that for this shading, sources are contradictory if the changes in expectation implied by the two sources of information are opposite in direction. In Figure 10.30, we see a balance of compatibilities. Parent F_{56}, for example, sends information which is contradictory to other data sources, as both its arcs possess darkly shaded small circles. Given that the observed value for F_{56} is rather unusual (there is heavy dark shading for the diagnostic portion corresponding to its observation), we might wish to explore such discrepancies further by examining individual adjustments and influences for the elements in the collection.

Figure 10.31 summarizes the essential features of the brewery model for an 8-week period, and shows the arrival of evidence, together with summaries of the combination of evidence with prior information to generate or revise predictions, and diagnostic measures useful for comparing expected to actual behaviour. Starting from week 56, each collection in Figure 10.31 is adjusted by all the relevant information available up to the preceding week. This information gives rise to adjusted expectations for the quantities in the collection, together with a diagnostic summarizing the surprisingness of the observed prediction. Finally, each week the quantities themselves become observed. It is now straightforward to identify the important characteristics of the system.

- The previous week's data are only weakly informative for sales volumes, V, and depot forecasts, F, though this does improve as we get later in the series. The remaining collections are quite well predicted.
- We see a lot of light shading, indicating that the changes in expectation were generally rather smaller than expected, suggesting that we exaggerated prior variability in the model.
- We see some dark shading for the first few weeks, indicating surprisingly large changes in expectation: these gradually die out, and may be the artefacts of initialization conditions.
- The dark shading in the node V_{60}, representing beer volumes at week 60, indicates a potentially serious anomaly which deserves investigation.
- The value of accumulated information varies with the kind of collection.

As time progresses, we gradually learn more and more about beer volumes – comparing nodes V_{56} and V_{62}, we see that one week's information is almost worthless, whereas several weeks' information explains roughly 20% of the uncertainty. However, comparing nodes D_{57} and D_{62}, we see that two weeks' information for the delivery nodes is about as useful as seven weeks' information. Notice that these summaries of behaviour over the system can also be readily interpreted by interested non-technical users.

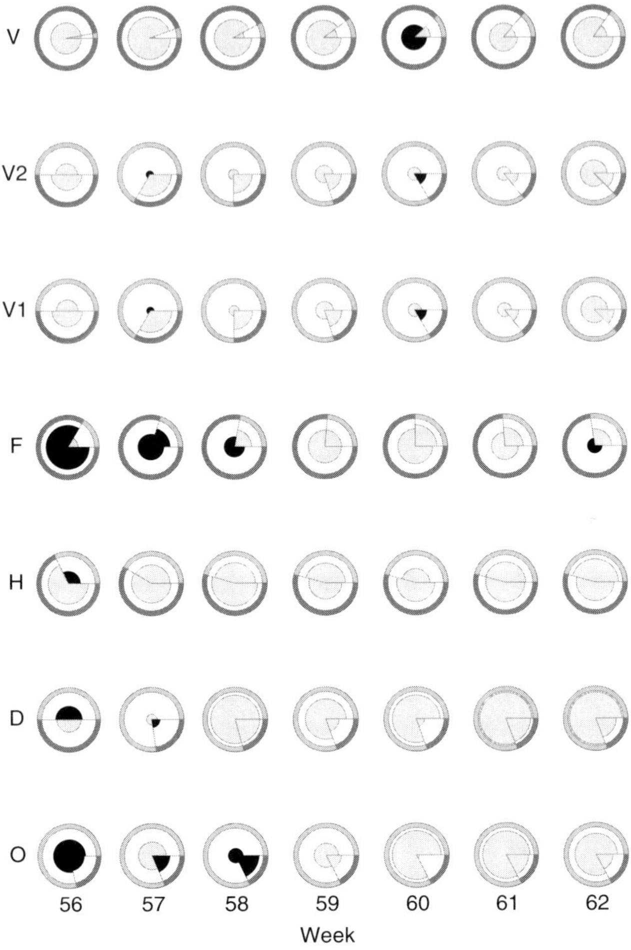

Figure 10.31 Variance resolutions and diagnostics over time for multiple collections.

10.13 Local computation for global adjustment of the junction tree

The algorithm in §10.10 is appropriate when adjustments are carried out in a natural sequence. Often, however, data is associated with many different clique tree nodes, with no natural ordering to the data, and no interest in the sequential effect of the introduction of evidence. We now describe an algorithm which is generally more efficient for such cases, requiring only two passes through the junction tree for the global incorporation of evidence at arbitrarily many nodes. However, each pass is computationally much more intensive than for the previous algorithm, and so the sequential algorithm may be preferable if evidence is only to be introduced at a handful of nodes.

10.13.1 Merging separate adjustments

The sequential algorithm is simple because propagation of a single piece of evidence is a natural geometric process, as described in §10.8. The merging of messages required by batch algorithms is less natural from a geometric viewpoint. We require certain additional results in order to separate out the effects of the various partial adjustments.

In particular, we require the ability to combine information from different parts of the graph. The crucial step in the algorithm is as follows. Suppose that $\lfloor A \perp\!\!\!\perp C \rfloor \,/\, B$. Suppose that we have assessed the adjustment of B by A and C separately. We will now show how these two separate adjustments may be merged in order to assess the combined adjustment of B by $A \cup C$. This is equivalent to assessing the adjustment of B by$[C/A]$, for then we can construct the combined adjustment using properties such as (10.16). The additional results that we require are given in the following theorem.

Theorem 10.17 *For $\lfloor A \perp\!\!\!\perp C \rfloor \,/\, B$ we have, for each $X \in [B]$,*

$$\mathbb{T}_{B:[C/A]}(X) = \mathbb{S}_{B:A}(\mathbb{S}_{B:A} + \mathbb{S}_{B:C} - \mathbb{S}_{B:C}\mathbb{S}_{B:A})^{\dagger}\mathbb{T}_{B:C}\mathbb{S}_{B:A}(X), \tag{10.34}$$

$$\mathrm{E}_{[C/A]}(X) = (\mathrm{E}_C - \mathrm{E}_A\mathbb{T}_{B:C})(\mathbb{S}_{B:A} + \mathbb{S}_{B:C} - \mathbb{S}_{B:A}\mathbb{S}_{B:C})^{\dagger}\mathbb{S}_{B:A}(X), \tag{10.35}$$

$$\mathbb{T}_{B:C}(X) = (\mathbb{S}_{B:A}\mathbb{T}^{\dagger}_{B:[C/A]}\mathbb{S}_{B:A} + \mathbb{T}_{B:A})^{\dagger}(X), \tag{10.36}$$

$$\mathrm{E}_C(X) = (\mathrm{E}_{[C/A]}(\mathbb{T}_{B:C} + \mathbb{S}^{\dagger}_{B:A}\mathbb{S}_{B:C}) + \mathrm{E}_A\mathbb{T}_{B:C})(X). \tag{10.37}$$

Proof. Each element of $[C/A]$ is of the form $Y - \mathrm{E}_A(Y) = (\mathrm{I} - \mathrm{E}_A)(Y)$ for some $Y \in [C]$. Now, for any $X \in [B]$, let $c_A(X)$ be the element of $[C]$ for which $\mathrm{E}_{[C/A]}(X) = (\mathrm{I} - \mathrm{E}_A)c_A(X)$.

For all $X \in [B]$, $(X - \mathrm{E}_{[C/A]}(X)) = (\mathrm{I} - \mathrm{E}_{[C/A]})(X) \perp [C/A]$ and so $(X - \mathrm{E}_{[C/A]}(X)) \perp \mathrm{E}_{[C/A]}(Y) = (\mathrm{I} - \mathrm{E}_A)(Y)$, for all $Y \in [C]$. Now since $(\mathrm{I} - \mathrm{E}_A)$ is a projection, this gives

$$\begin{aligned}
&(\mathrm{I} - \mathrm{E}_A)(X - \mathrm{E}_{[C/A]}(X)) \perp Y \\
\Rightarrow\ &(\mathrm{I} - \mathrm{E}_A)(X - \mathrm{E}_{[C/A]}(X)) \perp [C] \\
\Rightarrow\ &(\mathrm{I} - \mathrm{E}_A)(X - c_A(X)) \perp [C] \\
\Rightarrow\ &\mathrm{E}_C((\mathrm{I} - \mathrm{E}_A)(X - c_A(X))) = 0 \\
\Rightarrow\ &\mathrm{E}_C((\mathrm{I} - \mathrm{E}_A)(X)) = \mathrm{E}_C((\mathrm{I} - \mathrm{E}_A)c_A(X)) \\
\Rightarrow\ &\mathrm{E}_B(\mathrm{E}_C((\mathrm{I} - \mathrm{E}_A)(X))) = \mathrm{E}_B(\mathrm{E}_C((\mathrm{I} - \mathrm{E}_A)c_A(X))) \\
\Rightarrow\ &\mathrm{E}_B(\mathrm{E}_C((\mathrm{I} - \mathrm{E}_A)(X))) = \mathrm{E}_B(\mathrm{E}_C(\mathrm{E}_{[C/A]}(X))).
\end{aligned} \tag{10.38}$$

As $\lfloor A \perp\!\!\!\perp C \rfloor / B)$, we also have

$$\begin{aligned} \mathrm{E}_B(\mathrm{E}_C((\mathrm{I}-\mathrm{E}_A)(X))) &= \mathrm{E}_B(\mathrm{E}_C((\mathrm{I}-\mathrm{E}_B\mathrm{E}_A)(X))) \\ &= \mathbb{T}_{B:C}(\mathrm{I}-\mathbb{T}_{B:A})(X) \\ &= \mathbb{T}_{B:C}\mathbb{S}_{B:A}(X), \end{aligned} \tag{10.39}$$

and so from (10.38) we have

$$\mathbb{T}_{B:C}\mathbb{S}_{B:A}(X) = \mathrm{E}_B(\mathrm{E}_C(\mathrm{E}_{[C/A]}(X))). \tag{10.40}$$

Now

$$\begin{aligned} \mathrm{E}_B(\mathrm{E}_C(\mathrm{E}_{[C/A]}(X))) &= \mathrm{E}_B(\mathrm{E}_C((\mathrm{I}-\mathrm{E}_A)c_A(X))) \\ &= \mathrm{E}_B((\mathrm{I}-\mathrm{E}_C\mathrm{E}_A)c_A(X)) \\ &= \mathrm{E}_B((\mathrm{I}-\mathrm{E}_C\mathrm{E}_A)\mathrm{E}_B c_A(X)) \qquad (10.41) \\ &= (\mathrm{I}-\mathrm{E}_B\mathrm{E}_C\mathrm{E}_A)\mathrm{E}_B(c_A(X)). \qquad (10.42) \end{aligned}$$

Multiplying (10.42) through by $(\mathrm{I}-\mathrm{E}_B\mathrm{E}_A)(I-\mathrm{E}_B\mathrm{E}_C\mathrm{E}_A)^{\dagger}$ gives

$$\begin{aligned} &(\mathrm{I}-\mathrm{E}_B\mathrm{E}_A)(I-\mathrm{E}_B\mathrm{E}_C\mathrm{E}_A)^{\dagger}\mathrm{E}_B(\mathrm{E}_C(\mathrm{E}_{[C/A]}(X))) \\ &= (\mathrm{I}-\mathrm{E}_B\mathrm{E}_A)(I-\mathrm{E}_B\mathrm{E}_C\mathrm{E}_A)^{\dagger}(\mathrm{I}-\mathrm{E}_B\mathrm{E}_C\mathrm{E}_A)\mathrm{E}_B(c_A(X)) \\ &= (\mathrm{I}-\mathrm{E}_B\mathrm{E}_A)\mathrm{E}_B(c_A(X)). \end{aligned}$$

This last step is clear when the operator $(\mathrm{I}-\mathrm{E}_B\mathrm{E}_C\mathrm{E}_A)$ is invertible. In fact, it is also valid when singular as the null space of the operator is equal to the null space of $\mathrm{E}_{[C/A]}$. And so

$$\begin{aligned} &(\mathrm{I}-\mathrm{E}_B\mathrm{E}_A)(\mathrm{I}-\mathrm{E}_B\mathrm{E}_C\mathrm{E}_A)^{\dagger}\mathrm{E}_B(\mathrm{E}_C(\mathrm{E}_{[C/A]}(X))) \\ &= \mathrm{E}_B(\mathrm{I}-\mathrm{E}_B\mathrm{E}_A)c_A(X) \\ &= \mathrm{E}_B(\mathrm{I}-\mathrm{E}_A)c_A(X) \\ &= \mathrm{E}_B\mathrm{E}_{[C/A]}(X) \\ &= \mathbb{T}_{B:[C/A]}(X). \end{aligned} \tag{10.43}$$

Therefore (10.40) and (10.43) together imply that for all $X \in [B]$, we have

$$\begin{aligned} \mathbb{T}_{B:[C/A]}(X) &= (\mathrm{I}-\mathrm{E}_B\mathrm{E}_A)(I-\mathrm{E}_B\mathrm{E}_C\mathrm{E}_A)^{\dagger}\mathbb{T}_{B:C}\mathbb{S}_{B:A}(X) \\ &= \mathbb{S}_{B:A}(\mathrm{I}-\mathbb{T}_{B:C}\mathbb{T}_{B:A})^{\dagger}\mathbb{T}_{B:C}\mathbb{S}_{B:A}(X) \end{aligned}$$

(as $\lfloor A \perp\!\!\!\perp C \rfloor / B)$, and this gives (10.34). Equation (10.35) may be derived similarly. Equations (10.36) and (10.37) are obtained by inverting (10.34) and (10.35), respectively. ■

10.13.2 The global adjustment algorithm

We want to adjust beliefs over the whole graph by the collection of observations $D_{(1)}, \ldots, D_{(k)}$, where each $D_{(i)}$ is contained in some node D_j on the moral graph. At each node J_r of the junction tree, $\mathrm{E}(J_r)$ and $\mathrm{Var}(J_r)$ are stored. A current value for the resolution transform $\mathbb{T}(J_r) = \mathbb{T}_{J_r:D_*}$ and the adjusted expectation $A(J_r) = \mathrm{E}_{D_*}(J_r)$ are also stored, and these are initialized to zero (D_* denotes some subset of $D_{(1)}, \ldots, D_{(k)}$).

10.13.3 Absorption of evidence

The fundamental operation on the junction tree is that of absorption of a piece of evidence. Suppose that the evidence $\{\mathrm{E}_{D_{(i)}}(J_r), \mathbb{T}_{J_r:D_{(i)}}\}$ is obtained by J_r, which has current adjustment information $A(J_r)$ and $\mathbb{T}(J_r)$. The current adjustment information represents the effect of some evidence D, where $\lfloor D_{(i)} \perp\!\!\perp D \rfloor \,/\, J_r$. Using (10.34) and (10.35), $\mathbb{T}_{J_r:\mathbb{A}_D(D_{(i)})}$ and $\mathrm{E}_{\mathbb{A}_D(D(i))}(J_r)$ may be computed. Then using (5.15) and (5.4), $\mathbb{T}_{J_r:(D_{(i)}+D)}$ and $\mathrm{E}_{D(i)+D}(J_r)$ can be computed and these then replace the old values of $\mathbb{T}(J_r)$ and $A(J_r)$. The new evidence $D_{(i)}$ is said to have been **absorbed** into the adjustment information. Our use of this term is somewhat different from that of many similar local computation algorithms, in that the prior expectation and variance specifications at the node are not replaced by the adjusted expectation and variance. Instead, the information is absorbed into the current belief transform and not into the prior belief specifications.

10.13.3.1 Entering evidence

If the collection $D_{(i)}$ is to be observed, and this is contained in node D_j on the moral graph, then the evidence $\{\mathrm{E}_{D_{(i)}}(J_r), \mathbb{T}_{J_r:D_{(i)}}\}$ is computed for each node J_r containing D_j, and these pieces of evidence are **absorbed** by each J_r respectively.

10.13.3.2 Message-passing

When requested for a message, a node J_r will compute

$$R(J_r) = \mathrm{RVar}_{D_*}(J_r) = \mathrm{Var}(J_r)\mathbb{T}(J_r)$$

and then return the message $\{A(J_r), R(J_r)\}$ to the caller (which, unless J_r is the root node, is an adjacent node on the junction tree).

10.13.3.3 Processing a collect-phase message

When a node J_s receives the message $\{\mathrm{E}_D(J_r), \mathrm{RVar}_D(J_r)\}$ from J_r, J_s first extracts the marginals $\mathrm{E}_D(W_{rs})$ and $\mathrm{RVar}_D(W_{rs})$ where W_{rs} represents the quantities that J_r and J_s have in common and D is the evidence represented by the message. It can then compute $\mathbb{T}_{W_{rs}:D} = \mathrm{Var}(W_{rs})^{\dagger}\mathrm{RVar}_D(W_{rs})$. Now on the collect phase we have $\lfloor D \perp\!\!\perp J_s \rfloor \,/\, W_{rs}$, so we can use Properties 5.23.1 and 5.24.2 to compute the message $\{\mathrm{E}_D(J_s), \mathbb{T}_{J_s:D}\}$ ready for absorption by J_s.

10.13.3.4 Processing a distribute-phase message

On the distribute phase, the node J_r receives from J_s the message comprising $\{\mathrm{E}_{D+D'}(J_s), \mathrm{RVar}_{D+D'}(J_s)\}$, where D represents the information already absorbed by J_r, and D' represents extra evidence to be absorbed by J_r. J_r first extracts the marginals $\mathrm{E}_{D+D'}(W_{rs})$ and $\mathrm{RVar}_{D+D'}(W_{rs})$, and then computes the value of $\mathbb{T}_{W_{rs}:D+D'} = \mathrm{Var}(W_{rs})^{\dagger}\mathrm{RVar}_{D+D'}(W_{rs})$. Using (5.4) and(5.15), both $\mathrm{E}_{[D'/D]}(W_{rs})$ and $\mathbb{T}_{[D'/D]}(W_{rs})$ may be formed. In the distribute phase, we have $\lfloor D \perp\!\!\!\perp D' \rfloor \,/\, W_{rs}$ and so (10.37) and (10.36) can be used to compute $\mathrm{E}_{D'}(W_{rs})$ and $\mathbb{T}_{W_{rs}:D'}$. Finally, since $\lfloor D' \perp\!\!\!\perp J_r \rfloor \,/\, W_{rs}$ we can use Properties 5.23.1 and 5.24.2 to form the message $\{\mathrm{E}_{D'}(J_r), \mathbb{T}_{J_r:D'}\}$ ready for absorption by J_r.

10.13.3.5 Collection and distribution of evidence

We enter the data collection into the graph by carrying out a collect operation, followed by a distribute operation, as follows.

Collecting evidence Pick an arbitrary root node and send it the message `CollectEvidence`. When a node, J_r, receives this message, it sends the message to each of its other neighbours and processes and absorbs each message in turn. It then returns the message $\{A(J_r), R(J_r)\}$ to the caller.

When the collection phase is complete, the adjustment information at each node represents the adjustment by all evidence lower than it in the junction tree (with respect to the chosen root node). We now carry out the distribute operation, as follows.

Distributing evidence Send the message `DistributeEvidence` to the root node. On receipt of this message, the node, J_r, should process and absorb any message, and then pass the message `DistributeEvidence` $\{A(J_r), R(J_r)\}$ to all other neighbours.

When the distribution phase is completed, the adjustment information at each node represents the global adjustment of that node by all evidence in the junction tree.

The distribute phase of this algorithm could be simplified if the prior expectation and variance specification for each node were replaced by their adjusted values during the collect phase. This would lead to an algorithm more directly comparable with other commonly used local computation routines. However, for a full diagnostic analysis of the belief revision process it is useful to preserve the prior structure together with the full transforms for the global adjustment. In particular, the above algorithm allows direct construction of the diagnostic graphics that we have described. Further, since each absorption corresponds to a partial belief update, the associated diagnostics may be computed and displayed on a version of the partial adjustment graph. These diagnostic graphics can act as a monitoring process for each of the various calculations, and should help to highlight possible problems, such as data contamination, problems with prior specifications or computational problems.

10.14 Further reading

Basic properties of Bayes linear graphical models are covered in Goldstein (1990). Further properties of such models, with emphasis on graphical diagnostics and local computation, are developed in Wilkinson (1998) and Goldstein and Wilkinson (2000), giving examples of the application of the algorithm described in §10.13, using the software BAYES-LIN; see Appendix C. Graphics for the Bayes linear graphical model in the context of the brewery example are developed in Goldstein et al. (1993), and more details covering the development of the Bayes linear decision support system for this problem are given in Farrow et al. (1997). The canonical and residual wheels shown in this chapter were produced using [B/D]; see Appendix C and Goldstein and Wooff (1995) for details.

11

Matrix algebra for implementing the theory

In this chapter we present some algebra and definitions germane to the implementation of Bayes linear statistics, which requires some matrix theory as preamble. In particular, we are careful to implement the methodology so that any degeneracies in variance–covariance specifications are handled routinely rather than forbidden. We regard the specification of a zero variance, whether intended or not, for a given linear combination as an interesting feature of a problem, not one that will break its analysis. Therefore, we need to work with generalized matrix inverses rather than simple inverses, and we need to find generalized inverses of partitioned matrices. Many of the results in this chapter, therefore, are to do with careful handling of possibly singular matrices. We also concentrate on providing solutions to the generalized eigenvalue problem, as this is the kernel of linear statistical analysis.

11.1 Basic definitions

Definition 11.1 *The rank of a matrix A,* $\mathbf{rk}\{A\}$, *is the dimension of its column space, and so also the dimension of its row space.*

Definition 11.2 *The trace of a symmetric matrix A,* $\mathbf{tr}\{A\}$, *is the sum of its diagonal values.*

11.2 Covariance matrices and quadratic forms

Many of the matrices involved in this book are variance–covariance matrices, and as such they are required to possess certain properties. In particular, they are required to be **non-negative definite**. Suppose that the $n \times n$ variance matrix $A = \text{Var}(B)$ is specified over a vector of quantities $B = [B_1\ B_2\ \ldots\ B_n]^T$.

Bayes Linear Statistics: Theory and Methods M. Goldstein and D.A. Wooff

Definition 11.3 *We call a matrix A non-negative definite if it is real and symmetric, and if $y^T Ay \geq 0$ for all vectors y. Such a matrix has non-negative eigenvalues and rank* $\mathbf{rk}\{A\} \leq n$.

Definition 11.4 *We call a matrix A positive semi-definite if it is real and symmetric, and if $y^T Ay \geq 0$ for all vectors y, and $y^T Ay = 0$ for some non-null vectors y. Such a matrix has at least one eigenvalue equal to zero, the remaining eigenvalues positive, and rank* $\mathbf{rk}\{A\} < n$.

Definition 11.5 *We call a matrix A positive definite if it is real and symmetric, and if $y^T Ay > 0$ for all vectors y. Such a matrix has positive eigenvalues and rank* $\mathbf{rk}\{A\} = n$.

In common with Searle (1982), Wilkinson (1965), and many others, we distinguish between **positive definite** and **positive semi-definite** matrices. Let us see why this is useful, with reference to the variance matrix A. In the former case, the matrix A being positive definite expresses the fact that every linear combination of the B_is has a positive variance; whereas in the latter case, the matrix A being positive semi-definite expresses that fact that **at least one** linear combination has a variance equal to zero.

11.3 Generalized inverses

11.3.1 Basic properties

Definition 11.6 *The Moore–Penrose generalized inverse for any real matrix A is the unique matrix $A^\dagger$ satisfying all four properties below. A reflexive generalized inverse A_r^- need satisfy only Properties 11.6.1 and 11.6.2; and a simple generalized inverse A^- need satisfy only Property 11.6.1.*

11.6.1: $AA^-A = A$.

11.6.2: $A^-AA^- = A^-$.

11.6.3: $(AA^-)^T = AA^-$.

11.6.4: $(A^-A)^T = A^-A$.

For most problems we encounter, we will employ the Moore–Penrose generalized inverse. This is the most highly demanding in terms of properties, but possesses some nice features. There are occasions where we need to resort to some of the weaker generalized inverses.

11.3.2 Computing the Moore–Penrose inverse

Lemma 11.7 *The Moore–Penrose generalized inverse may be calculated as $A^\dagger = A^T(AA^T)^-A(A^TA)^-A^T$, where A^- is any generalized inverse of A.*

Lemma 11.8 *If N is non-negative definite, then $N^\dagger = A(A^T A)^{-2} A^T$ for any A such that $N = AA^T$ subject to A having full column rank. Further, such a matrix A exists.*

Lemma 11.9 *Let A be a non-negative definite matrix of dimension n and rank r at least unity. Suppose that A has r, $1 \le r \le n$, positive eigenvalues $\lambda_1 \ge \lambda_2 \ge \ldots \ge \lambda_r > 0$ and $n - r$ eigenvalues equal to zero. Suppose that the $n \times 1$ orthonormal eigenvectors corresponding to positive eigenvalues are $x_1, x_2, \ldots, x_r$, arranged as the columns of the matrix X. Represent by Λ the $r \times r$ diagonal matrix whose ith entry is λ_i. Then*

$$A = X \Lambda X^T, \tag{11.1}$$

$$A^\dagger = X \Lambda^{-1} X^T. \tag{11.2}$$

Lemma 11.10 *Lemma 11.9 can be stated slightly differently to include redundant structure, if any. Let A be a non-negative definite matrix of dimension n and rank at least unity. Suppose that A has eigenvalues $\lambda_1 \ge \lambda_2 \ge \ldots \ge \lambda_n \ge 0$. Suppose that the $n \times 1$ orthonormal eigenvectors corresponding to positive eigenvalues are $x_1, x_2, \ldots, x_n$, arranged as the columns of the matrix X. Represent by Λ the $n \times n$ diagonal matrix whose ith entry is λ_i. Then*

$$A = X \Lambda X^T, \tag{11.3}$$

$$A^\dagger = X \Lambda^\dagger X^T. \tag{11.4}$$

$\Lambda^\dagger$ is diagonal and has values equal to λ_i^{-1} if $\lambda_i > 0$ and zero otherwise.

11.3.3 Other properties of generalized inverses

Lemma 11.11 *$A^T A (A^T A)^- A^T = A^T$ for any generalized inverse A^-.*

Lemma 11.12 *$A(A^T A)^- A^T = AA^\dagger$ for any generalized inverse A^-. This follows from Lemmas 11.7 and 11.11.*

Lemma 11.13 *Suppose that A is an $m \times n$ matrix. If A has rank m, then $A^\dagger = A^T(AA^T)^{-1}$. Similarly, if A has rank n then $A^\dagger = (A^T A)^{-1} A^T$. See, for example, §20.2 of Harville (1997).*

Lemma 11.14 $\mathbf{rk}\{A_r^-\} = \mathbf{rk}\{A\}$ *for any reflexive generalized inverse A_r^- of A.*

Lemma 11.15 (Invariance property) *Let k be any suitably dimensioned vector. If $k^T A^* A = k^T$ for the particular choice of generalized inverse A^* of A, then*

$$k^T A^- A = k^T$$

for every other generalized inverse of A. (See, for example, Searle 1982, p. 285.)

Lemma 11.16 *If A^- is a generalized inverse of A, then all possible generalized inverses A^* of A can be constructed from*

$$A^* = A^- + U - A^- A U A A^-,$$

where U is arbitrary. This restates Theorem 2.4.1a of Rao and Mitra (1971).

Lemma 11.17 *If A^- is a generalized inverse of A, then all possible generalized inverses A^* of A can be constructed from*

$$A^* = A^- + V(I - AA^-) + (I - A^- A)W$$

where V, W are arbitrary. This restates Theorem 2.4.1b of Rao and Mitra (1971) and equation 23 of Searle (1982).

Lemma 11.18 *If A^- is any generalized inverse of A, then*

$$\mathbf{rk}\{A\} = \mathbf{rk}\{A^- A\} = \mathbf{rk}\{AA^-\} = \mathbf{tr}\{A^- A\} = \mathbf{tr}\{AA^-\}.$$

See, for example, Lemma 10.2.5 of Harville (1997).

Lemma 11.19 *For any matrix A,*

$$\mathbf{rk}\{A^\dagger\} = \mathbf{rk}\{A\}.$$

Definition 11.20 *We define for convenience*

$$A^\perp = I - AA^-,$$

where $A^\perp A = 0$ for any matrix A.

11.4 Multiplication laws

Lemma 11.21 *Suppose that A, B, C are any conformable matrices. Then we have $A^T AB = A^T AC$ if and only if $AB = AC$. This is known elsewhere as the star cancellation law; see Proposition 0.2.2 of Campbell and Meyer (1991).*

Lemma 11.22 *Suppose that a matrix equation involves a matrix A preceded (followed) throughout by A^T. Then if A^T premultiplies (postmultiplies) both sides of the equation, it may be cancelled out. For example,*

$$A^T ABA^T A = A^T AC$$

$$\Rightarrow \quad ABA^T A = AC.$$

See Rayner and Livingstone (1965), Basilevsky (1983), and Appendix A4.2 of Guttman (1982).

11.5 Range and null space of a matrix

Definition 11.23 *The range of any matrix A,* **range**$\{A\}$*, is the linear span of the columns of A, also known as the column space of A. Denote the null space of the matrix A by* **null**$\{A\}$.

Definition 11.24 *A vector b is in the range of a matrix A, $b \in$* **range**$\{A\}$*, if b can be expressed as a linear combination of a basis for the columns of A.*

Definition 11.25 *If every column c of a matrix C is such that $c \in$* **range**$\{A\}$ *then we shall write $C \in$* **range**$\{A\}$*. Similarly, if every column c of a matrix C is such that $c \in$* **null**$\{A\}$ *then we shall write $C \in$* **null**$\{A\}$.

Lemma 11.26 *For any matrix A,* **range**$\{A^\dagger\} =$ **range**$\{A^T\}$*. See, for example, Theorem 20.5.1 of Harville (1997).*

Lemma 11.27 *If $b \in$* **range**$\{A\}$ *then $AA^-b = b$ for any generalized inverse A^- of A. Conversely, if $AA^-b = b$ then $b \in$* **range**$\{A\}$*. See, for example, Lemma 2.2.4 of Rao and Mitra (1971).*

Lemma 11.28 *If C is any matrix all of whose columns are in* **range**$\{A\}$*, then $AA^-C = C$ for any generalized inverse A^- of A. Conversely, if we have $AA^-C = C$, then $C \in$* **range**$\{A\}$.

Lemma 11.29 *Suppose that A is any $n \times r$ matrix whose columns a_i are orthonormal and provide a basis for the $n \times 1$ vector b. Then $AA^Tb = b$. This follows as $b \in$* **range**$\{A\}$ *and because $A^\dagger = A^T$, and by Lemma 11.27.*

Lemma 11.30 *Suppose that A is any $n \times r$ matrix and that b is any $r \times 1$ vector. If $b \in$* **null**$\{A\}$ *then $A^Tb = 0$. Conversely, if $A^Tb = 0$ then we have $b \in$* **null**$\{A^T\}$*. (See, for example, Searle 1982, p. 246.)*

Theorem 11.31 *Let A be any matrix and suppose that b is any vector such that $b \in$* **range**$\{A\}$*. Then $b^TA^-b = b^TA^\dagger b$ is the same for any choice of generalized inverse of A.*

Proof. For some V, W,

$$
\begin{aligned}
b^TA^-b &= b^T[A^\dagger + V(I - AA^\dagger) + (I - A^\dagger A)W]b, \qquad \text{by Lemma 11.17,} \\
&= b^TA^\dagger b + b^TV(I - AA^\dagger)b + b^T(I - A^\dagger A)Wb \\
&= b^TA^\dagger b
\end{aligned}
$$

as $AA^\dagger b = b$ and $b^TA^\dagger A = b^T$ by Lemma 11.27. ■

Lemma 11.32 *Let A be any $m \times n$ matrix and let B be any $m \times p$ matrix. If* **range**$\{A\} \in$ **range**$\{B\}$ *and* **rk**$\{A\} =$ **rk**$\{B\}$ *then* **range**$\{A\} =$ **range**$\{B\}$*. See, for example, Theorem 4.4.6 of Harville (1997).*

11.6 Rank conditions

Lemma 11.33 *Let A be any $m \times n$ matrix and B be any $n \times p$ matrix. Then* $\mathbf{rk}\{AB\} = \mathbf{rk}\{B\}$ *if and only if* $\mathbf{range}\{B\} \perp \mathbf{null}\{A\}$, *or equivalently if* $\mathbf{range}\{B\} \in \mathbf{range}\{A\}$. *See, for example, Theorem 17.5.4 of Harville (1997).*

Lemma 11.34 *Let A be any $m \times n$ matrix and B be any $m \times m$ non-negative definite matrix. Then* $\mathbf{rk}\{A^T BA\} = \mathbf{rk}\{BA\} = \mathbf{rk}\{A^T B\}$. *See, for example, Theorem 14.11.2 of Harville (1997).*

11.7 Partitioned matrices

11.7.1 Definiteness for a partitioned real symmetric matrix

Consider the real symmetric matrix

$$M = \begin{bmatrix} E & F \\ F^T & H \end{bmatrix}.$$

where E and H are square real symmetric matrices. The following crucial result was shown in Marsaglia and Styan (1974).

Theorem 11.35 *M is non-negative definite if and only if the following three properties hold:*

11.35.1: *E is non-negative definite;*

11.35.2: $F \in \mathbf{range}\{E\}$;

11.35.3: *$H - F^T E^- F$ is non-negative definite for any choice of generalized inverse for E.*

Proof. First assume that all three conditions hold. Then it is simple to show that M is non-negative definite as follows. By Property 11.35.1, E is non-negative definite, and so affords the representation $E = Q\Psi Q^T$, where the columns of Q are the orthonormal eigenvectors $\{q_i\}$ corresponding to the positive eigenvalues $\{\psi_i\}$ of E, and Ψ is the diagonal matrix of these ordered positive eigenvalues. By Property 11.35.2, the columns of Q form a basis for each column in F, so that we must be able to write $F = QG$ for some G. Suppose we write $K = \Psi^{\frac{1}{2}} Q^T$, so that we may write $E = K^T K$ and $F = K^T T$ for some T. Then, for any choice of generalized inverse E^- for E, we have

$$F^T E^- F = T^T K(K^T K)^- K^T T = KK^\dagger$$

by Lemma 11.12. Hence $H - F^T E^- F$ is independent of the choice of generalized inverse E^-.

Now suppose that $c^T = [a^T \; b^T]$ is any vector partitioned so as to be conformable with the partition of M. We have

$$\begin{aligned} c^T M c &= a^T K^T K a + 2a^T K^T T b + b^T H b \\ &= a^T K^T K a + 2a^T K^T T b + b^T (H - F^T E^- F) b + b^T T^T T b \\ &= h^T h + b^T (H - F^T E^- F) b \\ &\geq 0, \end{aligned}$$

by Property 11.35.3, where $h^T = [a^T K^T \ b^T T^T]$. ■

Notice that by symmetry an alternative set of conditions, each implying the other, could be stated. Property 11.35.1 is essential as all principal submatrices of a non-negative definite matrix must be non-negative definite. Property 11.35.2 is essential as the following demonstrates. Suppose that F is any conformable matrix, and form the matrix quadratic

$$\begin{bmatrix} I & 0 \\ -F^T E^- & I \end{bmatrix} \begin{bmatrix} E & F \\ F^T & H \end{bmatrix} \begin{bmatrix} I & -E^- F \\ 0 & I \end{bmatrix} = \begin{bmatrix} E & (I - EE^-)F \\ F^T(I - E^- E) & H - F^T E^- F \end{bmatrix}. \tag{11.5}$$

Note that in calculating the bottom right submatrix on the right-hand side of (11.5), we have *en passant*

$$F^T(E^- - E^- E E^-)F = 0,$$

as we can write any $F = F_1 + F_2$, where F_1 and F_2 are constructed from **range**$\{E\}$ and **null**$\{E\}$, respectively. F_2 is orthogonal to $E^- - E^- E E^-$, and $F_1^T(E^- - E^- E E^-)F_1 = 0$ by invariance. The resulting matrix on the right-hand side of (11.5) cannot be non-negative definite unless we have $(I - EE^-)F = 0$, which is the case only if $F \in$ **range**$\{E\}$. Property 11.35.3 is clearly essential as the submatrix $H - F^T E^- F$ must here be non-negative definite. For details, see Marsaglia and Styan (1974).

11.7.2 Generalized inverses for partitioned non-negative definite matrices

Consider the $(n + m) \times (n + m)$ non-negative definite matrix

$$M = \begin{bmatrix} E & F \\ F^T & H \end{bmatrix}$$

where E is $n \times n$ non-negative definite, H is $m \times m$ non-negative definite, and F is $n \times m$. We must have that $H - F^T E^\dagger F$ is non-negative definite and that $F \in$ **range**$\{E\}$ to satisfy the conditions of Theorem 11.35.

Definition 11.36 *The* ***Schur complement of*** E ***in*** M *is uniquely*

$$S = H - F^T E^- F$$

for any choice of generalized inverse E^-. *The Schur complement of* H *in* M *is uniquely* $T = E - FH^- F^T$ *for any choice of generalized inverse* H^-.

Lemma 11.37

$$\mathbf{rk}\{M\} = \mathbf{rk}\{E\} + \mathbf{rk}\{H - F^T E^- F\} = \mathbf{rk}\{E\} + \mathbf{rk}\{S\}.$$

This follows by equation 10 of Marsaglia and Styan (1974) and Lemma 11.28.

Lemma 11.38 *For any choices of generalized inverses E^- and S^-,*

$$G = \begin{bmatrix} E^- + E^- F S^- F^T E^- & -E^- F S^- \\ -S^- F^T E^- & S^- \end{bmatrix}$$

is a generalized inverse for M.

This follows via equations (12) and (13) of Marsaglia and Styan (1974) and Lemma 11.37.

Lemma 11.39 *The generalized inverse $M^- = G$ given in Lemma 11.38 is the Moore–Penrose generalized inverse $M^\dagger$ if and only if the following three conditions given by equation (26) of Marsaglia and Styan (1974) are satisfied:*

11.39.1: *we choose $E^- = E^\dagger$;*

11.39.2: *we choose $S^- = S^\dagger$;*

11.39.3: $\mathbf{rk}\{H\} = \mathbf{rk}\{H - F^T E^\dagger F\} = \mathbf{rk}\{S\}$.

Note that as $\mathbf{rk}\{M\} = \mathbf{rk}\{E\} + \mathbf{rk}\{S\}$, the third condition will be satisfied trivially whenever we have $\mathbf{rk}\{M\} = \mathbf{rk}\{E\} + \mathbf{rk}\{H\}$.

Lemma 11.40 *The generalized inverse $M^- = G$ given in Lemma 11.38 is a reflexive generalized inverse M_r^- if and only if the following two conditions are satisfied:*

11.40.1: *we choose $E^- = E_r^-$;*

11.40.2: *we choose $S^- = S_r^-$.*

The choices $E^- = E^\dagger$ and $S^- = S^\dagger$ suffice.

11.8 Solving linear equations

Definition 11.41 *A system of linear equations $Ax = b$ is **consistent**, and so may be solved, if $b \in$ **range**$\{A^T\}$.*

Lemma 11.42 *If A is non-negative definite, then the system of linear equations $Ax = b$ is consistent if $b \in$ **range**$\{A\}$.*

Consider a consistent system of linear equations $Ax = b$, where $A \neq 0$ is a non-negative definite matrix of dimension r and rank r'; b is some $r \times 1$ vector such that $b \in$ **range**$\{A\}$; and x is some $r \times 1$ solution vector.

Lemma 11.43 *If A is full rank, we have uniquely*

$$x = A^{-1}b = A^{\dagger}b.$$

If A is not full rank, there are an infinite number of solutions:

$$x = A^{-}b + (I - A^{-}A)t$$

for an arbitrary r-dimensional vector t, and for any A^{-} satisfying at least Property 11.6.1 for generalized inverses. All possible solutions can be generated by taking any one generalized inverse and varying the arbitrary vector t.

Lemma 11.44 (Invariance properties) *Suppose that x_i and x_j are any two solutions. Then $k^T x_i = k^T x_j$ if $k^T A^{-} A = k^T$.*

Lemma 11.45 *There is one unique solution x having minimum Euclidian norm, and it is given by taking any A^{-} possessing Properties 11.6.1 and 11.6.4 for generalized inverses. Consequently, the solution $A^{\dagger}b$ is the unique minimum norm solution.*

Lemma 11.46 *Consider a consistent system of linear equations $AX = B$, where $A \neq 0$ is a non-negative definite matrix of dimension r and rank r'; B is some $r \times m$ matrix whose columns are contained in* **range**$\{A\}$*; and X is some $r \times m$ solution matrix. Then,*

$$X = A^{-}B + (I - A^{-}A)T$$

for an arbitrary $r \times m$ matrix T, and for any A^{-} satisfying at least Property 11.6.1 for generalized inverses. All possible solutions can be generated by taking any one generalized inverse and varying the arbitrary matrix T.

11.9 Eigensolutions to related matrices

The following results are useful in solving for the eigenstructure of the resolution transform. This is asymmetric by definition, but for computational purposes it is better to work with a symmetrized version. Suppose that H is any $n \times m$ matrix, and suppose that

$$T = HH^T \quad \text{and} \quad S = H^T H,$$

so that T is $n \times n$ non-negative definite and S is $m \times m$ non-negative definite. Both T and S have rank at most $\min(m, n)$. Suppose that we need to find the non-degenerate eigenstructure of T. Suppose that T has eigenvectors $x_1, x_2, \ldots, x_n$ corresponding to eigenvalues $\lambda_1 \geq \lambda_2 \geq \ldots \lambda_n \geq 0$. Suppose that $n > m$. Suppose that S has r (where $r \leq m$) positive eigenvalues $\lambda_1, \ldots, \lambda_r$ represented as the diagonal entries of the $r \times r$ matrix Λ corresponding to orthonormal eigenvectors $y_1, \ldots, y_r$ represented as the columns of the $n \times r$ matrix Y.

Lemma 11.47 *The r positive eigenvalues λ_i of S are also eigenvalues of T, and the corresponding eigenvectors of T are transforms $x_i \propto Hy_i$ of those of S. Additionally, T has a further $n-r$ eigenvalues equal to zero, with corresponding constructible eigenvectors. For an orthonormal collection of eigenvectors corresponding to positive eigenvalues for T we choose the r vectors*

$$x_i = \lambda_i^{-\frac{1}{2}} Hy_i,$$

and can represent these eigenvectors as columns of the $n \times r$ matrix X, where $X = HY\Lambda^{-\frac{1}{2}}$.

This follows straightforwardly as

$$Sy_i = \lambda_i y_i \quad \Rightarrow HSy_i = \lambda_i Hy_i \quad \Rightarrow HH^T Hy_i = \lambda_i Hy_i \quad \Rightarrow THy_i = \lambda_i Hy_i.$$

11.10 Maximizing a ratio of quadratic forms

Suppose that A and B are $n \times n$ non-negative definite matrices where, for $r \leq n$, $\mathbf{rk}\{B\} = r$, and where $\mathbf{null}\{B\} \subseteq \mathbf{null}\{A\}$. Suppose that B has positive eigenvalues $\psi_1, \dots, \psi_r$ collected into the diagonal matrix Ψ, and corresponding orthonormal eigenvectors $q_1, \dots, q_r$ organized as the columns of the $n \times r$ matrix Q. Suppose also that there are further orthonormal eigenvectors $q_{r+1}, \dots, q_n$ to correspond to zero eigenvalues. Suppose that we construct the $r \times r$ non-negative definite matrix

$$C = \Psi^{-\frac{1}{2}} Q^T A Q \Psi^{-\frac{1}{2}},$$

and suppose that C has eigenvalues $\lambda_1 \geq \lambda_2 \geq \dots \geq \lambda_r \geq 0$ corresponding to orthonormal eigenvectors $x_1, \dots, x_r$ organized as the columns of the $r \times r$ matrix X.

Theorem 11.48

$$\max_h \left\{ \frac{h^T A h}{h^T B h} \right\} = \lambda_1 \geq 0,$$

and the maximum is attained for $h \propto Q\Psi^{-\frac{1}{2}} x_1$.

Proof. For $h \in \mathbf{null}\{B\}$, $h \in \mathbf{null}\{A\}$ also, and we must have $\frac{h^T Ah}{h^T Bh} = 0$. For $h \in \mathbf{range}\{B\}$ we can write $h = Q\Psi^{-\frac{1}{2}} a$ for some vector $a \in \mathbf{range}\{B\}$. Now our problem reduces to finding

$$\max_a \left\{ \frac{a^T Ca}{a^T a} \right\}. \tag{11.6}$$

The solution to the reformulated problem is well known. Expression (11.6) is maximized by choosing $a \propto x_1$, with

$$\max_a \left\{ \frac{a^T Ca}{a^T a} \right\} = \lambda_1.$$

■

Note that the requirement **null**$\{B\} \subseteq$ **null**$\{A\}$ for Theorem 11.48 is satisfied when $B - A$ is non-negative definite, but otherwise to require $B - A$ non-negative definite is unnecessarily stringent.

Theorem 11.49 *For any vector* $g \in$ **range**$\{B\}$, *where* B *is non-negative definite,*

$$\max_h \left\{ \frac{[h^T g]^2}{h^T B h} \right\} = g^T B^- g \geq 0,$$

and the maximum is attained for any $h \propto B^- g$.

Proof. First we establish the theorem when we use the Moore–Penrose generalized inverse. From Theorem 11.48, we have in this case that

$$C = \Psi^{-\frac{1}{2}} Q^T g g^T Q \Psi^{-\frac{1}{2}},$$

which has one positive eigenvalue $\lambda = g^T B^\dagger g$ corresponding to an eigenvector proportional to $\Psi^{-\frac{1}{2}} Q^T g$. Hence we find that

$$\max_h \left\{ \frac{[h^T g]^2}{h^T B h} \right\} = g^T B^\dagger g,$$

and the maximum is attained for $h \propto B^\dagger g$. To extend the result to any generalized inverse, we have by Theorem 11.31 that $g^T B^- g = g^T B^\dagger g$ for $g \in$ **range**$\{B\}$ and any choice of generalized inverse for B. Finally, suppose we take $h \propto B^- g$. It follows from Lemma 11.17 that, for some V, W, we can generate all B^- from

$$B^- = B^\dagger + V(I - BB^\dagger) + (I - B^\dagger B)W$$

so that $h = B^- g = B^\dagger g + (I - B^\dagger B)Wg$ as $g \in$ **range**$\{B\}$. It follows that $g^T h = g^T B^\dagger g$ and

$$h^T B h = [g^T B^\dagger + g^T W^T (I - BB^\dagger)] B [B^\dagger g + (I - B^\dagger B) W g] = g^T B^\dagger g,$$

by Lemma 11.27 and Definition 11.6. Thus, for any $h \propto B^- g$, we have

$$\frac{[h^T g]^2}{h^T B h} = g^T B^\dagger g.$$

■

11.11 The generalized eigenvalue problem

11.11.1 Introduction

Suppose that A and B are $n \times n$ non-negative definite matrices. The generalized eigenvalue problem is to find generalized eigenvalues λ and generalized eigenvectors z to solve

$$Az = \lambda B z. \tag{11.7}$$

In the context of Bayes linear methods, this problem arises in two areas. First, computation of the resolution transform involves solving (11.7) where it is known that $B - A$ is non-negative definite. Secondly, the comparison of belief hypotheses requires solving (11.7) more generally. We do, however, require solutions which allow degeneracy within either A or B or both. Whenever A, B are non-negative definite, various deficiencies can arise in (11.7).

Definition 11.50 (Generalized eigenvalue problem deficiencies)

11.50.1: *The pencil $A - \lambda B$ may be singular. If so, a subspace of the null spaces of A and B will be common to both, and there will exist vectors v such that $v^T Av = v^T Bv = 0$. Such an eventuality corresponds to specifications of* $\text{Var}_A(v^T X) = 0 = \text{Var}_B(v^T X)$.

11.50.2: *There may exist vectors v such that $v^T Av > 0$ and $v^T Bv = 0$, corresponding to specifications of* $\text{Var}_A(v^T X) > 0$ *and* $\text{Var}_B(v^T X) = 0$. *If so, part of the null space of B is in* $\mathbf{range}\{A\}$.

11.50.3: *There may exist vectors v such that $v^T Av = 0$ and $v^T Bv > 0$, corresponding to specifications of* $\text{Var}_A(v^T X) = 0$ *and* $\text{Var}_B(v^T X) > 0$. *If so, part of the null space of A is in* $\mathbf{range}\{B\}$.

The remaining case of interest concerns, of course, the existence of vectors v such that both $v^T Av > 0$ and $v^T Bv > 0$, corresponding to specifications of $\text{Var}_A(v^T X) > 0$ and $\text{Var}_B(v^T X) > 0$.

11.11.2 The QZ algorithm

The QZ algorithm, presented in Moler and Stewart (1973) and extended in Ward (1975), represents an efficient way of determining the eigenstructure of the generalized eigenvalue problem (11.7). The algorithm simultaneously diagonalizes A, B, whence scalar pairs a_i, b_i are determined from each, where the generalized eigenvalues are computed as the ratios $\lambda_i = a_i/b_i$. It is simple to handle cases where either a_i or b_i is zero. However, in Wilkinson (1979) it is demonstrated that when both $a_i = b_i = 0$ (for example, this will be the case when A, B share eigenvectors with common eigenvalue zero) this leads to an arbitrariness in the non-zero values of $\lambda_j = a_j/b_j$. It is suggested therein that if this is a possibility, then the singular part of the pencil should be extracted before application of the QZ algorithm.

11.11.3 An alternative algorithm

First, we form the compound matrix $A + B$ and determine its eigenstructure. Suppose that there are m eigenvalues $(\theta_1, \ldots, \theta_m)$ corresponding to eigenvectors $(r_1, \ldots, r_m)$, and that the remaining eigenvectors $(r_{m+1}, \ldots, r_n)$ correspond to zero eigenvalues. We will assume that $\mathbf{rk}\{A + B\} \geq 1$, so that $n \geq m \geq 1$. Arrange the eigenvectors $(r_1, \ldots, r_m)$ as the columns of the $n \times m$ matrix W, the eigenvectors $(r_{m+1}, \ldots, r_n)$ as the columns of the $n \times (n-m)$ matrix R, and the positive eigenvalues as the entries of the diagonal $m \times m$ matrix Θ. Suppose that the n eigenvectors $(r_1, \ldots, r_n)$ are constructed so as to be orthonormal.

Property 11.51 *The following are elementary properties of this decomposition.*

> **11.51.1:** *The columns of W are orthonormal and span $[A]$ and $[B]$, so that we must have $WW^TA = A$ and $WW^TB = B$, by Lemma 11.27.*
>
> **11.51.2:** *$R^TA = R^TB = 0$ as $R^T(A+B)R = 0$, and because A and B are non-negative definite.*

Now we form $K = \Theta^{-\frac{1}{2}}W^TAW\Theta^{-\frac{1}{2}}$ and $G = \Theta^{-\frac{1}{2}}W^TBW\Theta^{-\frac{1}{2}}$. Then, excepting the null space which is common to both A and B, which we now consider extracted, we can rewrite (11.7) as

$$Ky = \lambda Gy, \tag{11.8}$$

where y is such that $z = W^T\Theta^{-\frac{1}{2}}y$. This follows as

$$\begin{aligned}
& Ky = \lambda Gy \\
\Rightarrow\quad & \Theta^{-\frac{1}{2}}W^TAW\Theta^{-\frac{1}{2}}y = \lambda\Theta^{-\frac{1}{2}}W^TBW\Theta^{-\frac{1}{2}}y \\
\Rightarrow\quad & W^TAW\Theta^{-\frac{1}{2}}y = \lambda W^TBW\Theta^{-\frac{1}{2}}y \\
\Rightarrow\quad & WW^TAW\Theta^{-\frac{1}{2}}y = \lambda WW^TBW\Theta^{-\frac{1}{2}}y \\
\Rightarrow\quad & AW\Theta^{-\frac{1}{2}}y = \lambda BW\Theta^{-\frac{1}{2}}y \\
\Rightarrow\quad & Az = \lambda Bz.
\end{aligned}$$

Hence, having excluded the null space common to both A and B, we can solve the generalized eigenvalue problem (11.8), and thus solve (11.7). The new generalized eigenvalue problem (11.8) does not have deficiency 11.50.1, but may well have deficiency 11.50.2 or 11.50.3. As such, it would now normally be appropriate to apply the QZ algorithm as the pencil $K - \lambda G$ is regular. However, for implementing Bayes linear theory, it remains key to deal with deficiencies 11.50.2 and 11.50.3, which are the cases where the columns of G do not necessarily span those of K, or vice versa. As an alternative to the QZ algorithm, notice that

$$\begin{aligned}
K + G &= \Theta^{-\frac{1}{2}}W^TAW\Theta^{-\frac{1}{2}} + \Theta^{-\frac{1}{2}}W^TBW\Theta^{-\frac{1}{2}} \\
&= \Theta^{-\frac{1}{2}}W^T(A+B)W\Theta^{-\frac{1}{2}} \\
&= \Theta^{-\frac{1}{2}}W^TW\Theta W^TW\Theta^{-\frac{1}{2}} \\
&= I_m.
\end{aligned}$$

Hence, (11.8) can itself be transformed into two alternative simple eigenvalue problems:

$$G = \delta y, \qquad \text{with } \lambda = \frac{1}{\delta} - 1, \tag{11.9}$$

or

$$K = \mu y, \qquad \text{with } \lambda = \frac{\mu}{1-\mu}.$$

We will solve (11.9). Both G and K are non-negative definite, and we have $K + G = I_m$. Thus the eigenvalues δ of G must satisfy $1 \geq \delta \geq 0$. Suppose that we obtain t positive ordered eigenvalues $(\delta_1, \ldots, \delta_t)$ and $m - t$ eigenvalues equal to zero. Arrange the positive eigenvalues as the entries of the $t \times t$ diagonal matrix Δ. Suppose that we construct corresponding orthonormal eigenvectors $(y_1, \ldots, y_m)$, where we gather those corresponding to the positive eigenvalues as the columns of the $n \times t$ matrix Y, and the remainder as the columns of the $n \times (m - t)$ matrix V. We will normalize the first t eigenvectors further by postmultiplying Y by $\Delta^{-\frac{1}{2}}$, for reasons which will become clear.

Now, $[Y\Delta^{-\frac{1}{2}} \,:\, V]$ is a matrix of generalized eigenvectors for problem (11.8). Consequently, a matrix of eigenvectors for the original problem (11.7) (excluding the shared null space) is $W\Theta^{-\frac{1}{2}}[Y\Delta^{-\frac{1}{2}} \,:\, V]$. Now, explicitly including the shared null space eigenvectors yields a full generalized eigenvector matrix

$$Z = [W\Theta^{-\frac{1}{2}}Y\Delta^{-\frac{1}{2}} \,:\, W\Theta^{-\frac{1}{2}}V \,:\, R] \tag{11.10}$$

for the original problem (11.7). The eigenvalues we deal with as follows. We have already extracted the common null space corresponding to which are eigenvalues of zero in both A and B. We have remaining the eigenvalues δ_i for the simple problem (11.9). These we interpret as follows. Suppose that there are s values of $\delta_i = 1$, each of which yields $\lambda_i = 0$. These correspond to cases of deficiency 11.50.3. There are $m - t$ values of $\delta_i = 0$, each of which yields indeterminate λ_i, corresponding to a case of deficiency 11.50.2. Values of $0 < \delta_i < 1$ yield $0 < \lambda_i = \delta_i^{-1} - 1 < \infty$.

We can show easily that the generalized eigenvectors are orthogonal under both belief specifications. They have been normalized so that they have variances unity in B and λ_i in A, or zero in B and either unity or zero in A. The eigenstructure is best reported in terms of paired quadratic forms, as in Table 11.1. The interpretation of the eigenvectors given in the first two rows of the table is given in Goldstein (1991). A full example comparing two variance specifications is given in §9.2.

11.11.4 An algorithm for $B - A$ non-negative definite

Now we consider the case where $B - A$ is known to be non-negative definite, so that deficiency 11.50.2 cannot occur. In this case, there is an alternative approach.

Suppose that the matrix B has eigenstructure as follows. B has r (where $r \leq n$) positive eigenvalues $\psi_1 \geq \ldots \geq \psi_r > 0$ which we collect into the diagonal matrix Ψ. Suppose that corresponding to these eigenvalues are orthonormal eigenvectors

Table 11.1 Generalized eigenstructure under potential rank deficiencies.

Cases	Eigenvector z_i	$\mathrm{Var}_A(z_i^T X)$	$\mathrm{Var}_B(z_i^T X)$
$i = 1, \ldots, s$	$W\Theta^{-\frac{1}{2}} y_i$	0	1
$i = s+1, \ldots, t$	$\frac{1}{\sqrt{\delta_i}} W\Theta^{-\frac{1}{2}} y_i$	$\lambda_i = \frac{1}{\delta_i} - 1$	1
$i = t+1, \ldots, m$	$W\Theta^{-\frac{1}{2}} y_i$	1	0
$i = m+1, \ldots, n$	r_i	0	0

$q_1, \ldots, q_r$ collected as the columns of the $n \times r$ matrix Q, where $Q^T Q = I_r$. Suppose that we construct $n - r$ orthonormal eigenvectors $q_{r+1}, \ldots, q_n$ corresponding to the zero eigenvalues of B. Suppose that A has rank $m \leq r$.

Theorem 11.52 *Construct the non-negative definite $r \times r$ matrix*

$$K = \Psi^{-\frac{1}{2}} Q^T A Q \Psi^{-\frac{1}{2}},$$

and suppose that K has eigenvalues $\lambda_1, \ldots, \lambda_r$ corresponding to orthonormal eigenvectors $y_1, \ldots, y_r$. Then these eigenvalues are also generalized eigenvalues of problem (11.7), *corresponding to generalized eigenvectors*

$$x = Q\Psi^{-\frac{1}{2}} y.$$

Proof. B has rank $r \leq n$. Any of its eigenvectors q_i such that $Bq_i = 0$ must also be a generalized eigenvector corresponding to a zero generalized eigenvalue for the problem $Ax = \lambda Bx$. This follows as $B - A$ is non-negative definite, so that $q_i^T B q_i \geq q_i^T A q_i$, so that $Bq_i = 0 \Rightarrow Aq_i = 0$ as A is non-negative definite. (Notice that $B - A$ non-negative definite is more stringent a requirement than necessary in that the result can be obtained under the weaker requirement that for every x for which $x^T Bx = 0$ it is also the case that $x^T Ax = 0$.) It follows that the eigenvectors $q_{r+1}, \ldots, q_n$ of B corresponding to zero eigenvalues of B are also eigenvectors within the generalized formulation corresponding to zero eigenvectors. For the generalized eigenvectors x corresponding to the remaining generalized eigenvalues, these must be such that $x \in \mathbf{range}\{Q\}$ for every such x. As such, by Lemma 11.29, we must have $QQ^T x = x$. The proof now follows as

$$\begin{aligned}
Ax &= \lambda Bx \\
&= \lambda Q\Psi Q^T x \\
\Rightarrow \quad \Psi^{-\frac{1}{2}} Q^T A Q \Psi^{-\frac{1}{2}} \Psi^{\frac{1}{2}} Q^T x &= \lambda \Psi^{\frac{1}{2}} Q^T x \\
\Rightarrow \quad Ky &= \lambda y.
\end{aligned}$$

■

Note that $m < r$ implies that some of the generalized eigenvalues $\lambda_1, \ldots, \lambda_r$ will be zero.

Lemma 11.53 *If we arrange the generalized eigenvectors $x_1, \ldots, x_r$ as the columns of the $n \times r$ matrix X, then the generalized eigenvector matrix X simultaneously diagonalizes both A and B.*

This follows as

$$X^T A X = Y^T \Psi^{-\frac{1}{2}} Q^T A Q \Psi^{-\frac{1}{2}} Y = Y^T K Y = \Lambda,$$

$$\text{and} \quad X^T B X = Y^T \Psi^{-\frac{1}{2}} Q^T B Q \Psi^{-\frac{1}{2}} Y = Y^T Y \quad = I_r.$$

Lemma 11.54 *Suppose that A, B, and $B - A$ are non-negative definite. Then the generalized eigenvalues satisfy*

$$1 \geq \lambda_1 \geq \ldots \lambda_r \geq \lambda_{r+1} \geq \ldots \lambda_n \geq 0.$$

This follows because

$$A = Q \Psi^{\frac{1}{2}} Y \Lambda Y^T \Psi^{\frac{1}{2}} Q^T$$

$$\text{and} \quad B - A = Q \Psi^{\frac{1}{2}} Y (I_r - \Lambda) Y^T \Psi^{\frac{1}{2}} Q^T,$$

where $QQ^T A = A$ by Lemma 11.29. Thus Λ must be non-negative as A is non-negative definite, and $I_r - \Lambda$ must be non-negative as $B - A$ is non-negative definite. When B is not full rank, the generalized eigenvalues

$$\lambda_{r+1} = \psi_{r+1} = 0, \ldots, \lambda_n = \psi_n = 0.$$

Theorem 11.55 *$X^\dagger = Y^T \Psi^{\frac{1}{2}} Q^T$ is the unique Moore–Penrose generalized inverse of X.*

Proof.

$$\begin{aligned} XX^\dagger &= Q\Psi^{-\frac{1}{2}} Y Y^T \Psi^{\frac{1}{2}} Q^T = QQ^T && \text{is symmetric,} \\ X^\dagger X &= Y^T \Psi^{\frac{1}{2}} Q^T Q \Psi^{-\frac{1}{2}} Y = I_r && \text{is symmetric,} \\ XX^\dagger X &= I_r X = X, && \text{and} \\ X^\dagger X X^\dagger &= X^\dagger I_r = X^\dagger, \end{aligned}$$

so that $X^\dagger$ satisfies all the conditions required under Definition 11.6 for it to be the unique Moore–Penrose generalized inverse of X. ■

Lemma 11.56 *We have the following representations:*

$$A = (X^\dagger)^T \Lambda X^\dagger, \tag{11.11}$$

$$B = (X^\dagger)^T X^\dagger, \tag{11.12}$$

$$B^\dagger = XX^T, \tag{11.13}$$

$$X^\dagger B^\dagger (X^\dagger)^T = I_r. \tag{11.14}$$

Thus, $X^\dagger$ diagonalizes $B^\dagger$. There is not an equivalent result for $A^\dagger$ unless $\mathbf{rk}\{A\} = \mathbf{rk}\{B\}$.

Theorem 11.57 *Let*

$$G = \alpha A + \beta B, \quad \alpha \geq 0,\ \beta > 0.$$

Then

$$G^\dagger = X(\beta I_r + \alpha\Lambda)^{-1} X^T.$$

Proof. We need to satisfy the conditions of Definition 11.6. We have

$$\begin{aligned} G &= \alpha A + \beta B \\ &= (X^\dagger)^T(\alpha\Lambda)X^\dagger + (X^\dagger)^T(\beta I_r)X^\dagger, \qquad \text{by Lemma 11.56,} \\ &= (X^\dagger)^T(\beta I_r + \alpha\Lambda)X^\dagger. \end{aligned}$$

Hence,

$$\begin{aligned} GG^\dagger &= (X^\dagger)^T(\beta I_r + \alpha\Lambda)X^\dagger X(\beta I_r + \alpha\Lambda)^{-1}X^T \\ &= (X^\dagger)^T X^T = QQ^T \quad \text{is symmetric;} \\ G^\dagger G &= X(\beta I_r + \alpha\Lambda)^{-1}X^T(X^\dagger)^T(\beta I_r + \alpha\Lambda)X^\dagger \\ &= XX^\dagger = QQ^T \quad \text{is symmetric;} \\ GG^\dagger G &= QQ^T G = G; \\ G^\dagger G G^\dagger &= G^\dagger QQ^T = G^\dagger, \end{aligned}$$

so that $G^\dagger$ satisfies all the conditions required for it to be the unique Moore–Penrose generalized inverse of G. ■

Note that when $\beta = 0$ the rank of G reduces from $\mathbf{rk}\{B\} = r$ to $\mathbf{rk}\{A\}$, and the above representations may not be used.

11.12 Direct products of matrices

Let $\mathbf{1}_m$ be the $m \times 1$ vector of ones. Let $\mathbf{I}_m$ be the $m \times m$ identity matrix. Let $\mathbf{J}_m = \mathbf{1}_m\mathbf{1}_m^T$ be the $m \times m$ matrix of ones. Let $\mathbf{P}_m$ be the $m \times m$ matrix with $(1, 1)$th entry unity and all other entries equal to zero.

11.12.1 The Helmert matrix

Definition 11.58 *We will write $\mathbf{H}_m$ to represent the $m \times m$ transpose of the Helmert matrix of order m. The first column of $\mathbf{H}_m$ is $\frac{1}{\sqrt{m}}\mathbf{1}_m$. The ith, $i > 1$, column is*

$$\frac{1}{\sqrt{i(i-1)}}\left[-\mathbf{1}_{i-1}^T \quad (i-1) \quad 0 \quad \ldots \quad 0\right]^T.$$

The Helmert matrix is helpful when writing replicated eigensolutions in higher dimensions. The orthonormal columns of the $\mathbf{H}_m$ matrix are useful in representing m linear combinations, the first being an average and the remainder being $m - 1$ orthogonal contrasts. Note that $\mathbf{J}_m \mathbf{H}_m = m \mathbf{H}_m \mathbf{P}_m$.

11.12.2 Direct products

The notation $A \otimes B$ is used for the direct product of A and B. That is, if A is any $p \times q$ matrix with (i, j)th element a_{ij}, and B is any $s \times t$ matrix, then $A \otimes B$ is the $ps \times qt$ matrix

$$A \otimes B = \begin{bmatrix} a_{11}B & a_{12}B & \dots & a_{1q}B \\ a_{21}B & a_{22}B & \dots & a_{2q}B \\ \vdots & \vdots & \ddots & \vdots \\ a_{p1}B & a_{p2}B & \dots & a_{pq}B \end{bmatrix}.$$

The notation $A \oplus B$ is used for the direct sum of A and B. For matrices A and B of any dimensions, this is defined as the block matrix

$$A \oplus B = \begin{bmatrix} A & 0 \\ 0 & B \end{bmatrix}.$$

A discussion of the properties of direct sums and products can be found in Searle (1982) or Harville (1997). Some properties are as follows.

Lemma 11.59 *For any matrices A, B, $(A \otimes B)^T = A^T \otimes B^T$.*

Lemma 11.60 *For conformable matrices A, C and B, D we have*

$$(A \otimes B)(C \otimes D) = AC \otimes BD.$$

Lemma 11.61 *Let $A - B$ be a positive definite $r \times r$ matrix and let B be a non-negative definite $r \times r$ matrix. Then*

$$[\mathbf{I}_n \otimes (A - B) + \mathbf{J}_n \otimes B]^{-1} = \mathbf{I}_n \otimes [A - B]^{-1} - \mathbf{J}_n \otimes [A + (n-1)B]^{-1} B [A - B]^{-1}. \quad (11.15)$$

The proof follows trivially by verifying that the product of the original with the inverse equals $\mathbf{I}_n \otimes \mathbf{I}_r$.

Lemma 11.62 *Let A and B be $r \times r$ matrices. Suppose the matrix $A + (n - 1)B$ has eigenvalues $\lambda_1 \geq \lambda_2 \geq \dots \geq \lambda_r$ corresponding to eigenvectors $v_1, \dots, v_r$. Arrange the eigenvalues as the diagonal elements of the matrix Λ and the eigenvectors as vectors of the matrix V, so that $[A + (n - 1)B]V = V\Lambda$. Suppose also that the matrix $A - B$ has eigenvalues $\delta_1 \geq \delta_2 \geq \dots \geq \delta_r$ corresponding to eigenvectors $w_1, \dots, w_r$. Arrange the eigenvalues as the diagonal elements of the matrix Δ*

and the eigenvectors as vectors of the matrix W, so that $(A - B)W = W\Delta$. *Then the matrix*

$$\mathbf{I}_n \otimes (A - B) + \mathbf{J}_n \otimes B$$

has eigenvalue matrix

$$\tilde{\lambda} = \Lambda \oplus \Delta \oplus \ldots \oplus \Delta,$$

where there are $r - 1$ *terms* Δ, *so that each eigenvalue* δ_i *is of multiplicity* $r - 1$. *The corresponding eigenvector matrix is*

$$\tilde{V} = \begin{bmatrix} h_1 \otimes V & h_2 \otimes W & h_3 \otimes W & \ldots & h_n \otimes W \end{bmatrix},$$

where h_i *is the* ith *column of the Helmert matrix* $\mathbf{H}_n$.

The result can be shown trivially by multiplying out the terms, and noting that $\mathbf{J}_n h_1 = n h_1$ and that $\mathbf{J}_n h_i = 0,\ i > 1$. If A, B, $A - B$ are non-negative definite then the eigenvector matrices V, W can be chosen to be orthonormal, and if so $\tilde{V}$ is orthonormal.

12

Implementing Bayes linear statistics

12.1 Introduction

In this chapter we deal with the technical aspects of implementing Bayes linear methodology. This includes checking the coherence of belief specifications, checking whether data are consistent with belief specifications, finding matrix representations of the resolution transform, and so forth. We assume the results and definitions given in Chapter 11. We begin by establishing some general notation. Suppose that there are three collections of unknown quantities which we organize into vectors as

$$B = [B_1\ B_2\ \ldots\ B_{n_B}]^T, \quad D = [D_1\ D_2\ \ldots\ D_{n_D}]^T, \quad F = [F_1\ F_2\ \ldots\ F_{n_F}]^T.$$

We work with three collections as we wish to consider adjusting a collection B by a data collection D, and then a further partial adjustment by a data collection F. We distinguish between the collections and vectors only as necessary. We write the variance matrices as

$$\Sigma_B = \text{Var}(B), \quad \Sigma_D = \text{Var}(D), \quad \Sigma_F = \text{Var}(F),$$

and the covariance matrices as

$$\Sigma_{BD} = \text{Cov}(B, D) = \Sigma_{DB}^T, \quad \Sigma_{BF} = \text{Cov}(B, F) = \Sigma_{FB}^T,$$
$$\Sigma_{DF} = \text{Cov}(D, F) = \Sigma_{FD}^T.$$

Observed values are represented by lower case, so that d_j is the observed value of D_j, and d is the observed value of the vector D.

Bayes Linear Statistics: Theory and Methods M. Goldstein and D.A. Wooff

12.2 Coherence of belief specifications

12.2.1 Coherence for a single collection

Definition 12.1 *Second-order belief specifications over any collection B are* ***finite*** *and* ***coherent*** *if and only if*

> **12.1.1:** $\mathrm{E}(B_i)$ *is finite for each* B_i,
>
> **12.1.2:** $\mathrm{Var}(B_i)$ *is finite for each* B_i,
>
> **12.1.3:** *the joint variance–covariance matrix* $\mathrm{Var}(B)$ *is non-negative definite.*

For the remainder of this chapter we assume that all expectations and variances are finite, so that coherence is essentially identified with non-negative definiteness of the joint variance–covariance matrix, i.e. the condition that we do not assign negative variance to any linear combination of the elements of B.

Our Bayes linear approach is quite general in not insisting that all the variance matrices at every point be positive definite. Indeed, it is extremely useful to be able to work with structures which may contain some degree of linear degeneracy (and to detect such degeneracy via canonical analysis). By requiring that $\mathrm{Var}(B)$ be non-negative definite rather than positive definite, we allow, for convenience, some linear combinations to have variance zero, and we make this allowance at all stages of the analysis. However, we restrict attention to $\mathrm{Var}(B) \neq 0$, $\mathrm{Var}(D) \neq 0$, and $\mathrm{Var}(F) \neq 0$.

12.2.2 Coherence for two collections

Definition 12.2 *Second-order belief specifications over two collections B, D are* ***finite*** *and* ***coherent*** *if and only if*

> **12.2.1:** $\mathrm{E}(B_i)$ *is finite for each* B_i, *and* $\mathrm{E}(D_j)$ *is finite for each* D_j,
>
> **12.2.2:** $\mathrm{Var}(B_i)$ *is finite for each* B_i, *and* $\mathrm{Var}(D_j)$ *is finite for each* D_j
>
> **12.2.3:** *the joint variance–covariance matrix*
>
> $$\mathrm{Var}\left(\begin{bmatrix} B \\ D \end{bmatrix}\right) = \begin{bmatrix} \Sigma_B & \Sigma_{BD} \\ \Sigma_{BD}^T & \Sigma_D \end{bmatrix}$$
>
> *is non-negative definite.*

Lemma 12.3 *The matrix*

$$\begin{bmatrix} \Sigma_B & \Sigma_{BD} \\ \Sigma_{BD}^T & \Sigma_D \end{bmatrix}$$

is non-negative definite if and only if the following three conditions are met:

> **12.3.1:** Σ_D *is non-negative definite;*

12.3.2: $\Sigma_{DB} \in$ **range**$\{\Sigma_D\}$ *(or, equivalently,* $\Sigma_D^{\perp}\Sigma_{DB} = 0$*);*

12.3.3: $\Sigma_B - \Sigma_{BD}\Sigma_D^{\dagger}\Sigma_{DB}$ *is non-negative definite.*

These conditions straightforwardly follow from Theorem 11.35. The matrix which appears in Property 12.3.3 is uniquely defined irrespective of the choice of generalized inverse indicated in Theorem 11.35, so the if-and-only-if part of this lemma holds if we adopt the Moore–Penrose generalized inverse at this point.

Remark. Coherence over a pair of collections may also be deduced via properties of the resolution transform; see Theorem 12.37. Note also that the matrix in Property 12.3.3 is the adjusted variance matrix for B given D, $\mathrm{Var}_D(B)$, as defined in (3.30); see §12.5.

We may state an alternative set of coherence requirements to Lemma 12.3 as follows.

Lemma 12.4 *The matrix*

$$\begin{bmatrix} \Sigma_B & \Sigma_{BD} \\ \Sigma_{BD}^T & \Sigma_D \end{bmatrix}$$

is non-negative definite if and only if the following three conditions are met:

12.4.1: Σ_B *is non-negative definite;*

12.4.2: $\Sigma_{BD} \in$ **range**$\{\Sigma_B\}$ *(or, equivalently,* $\Sigma_B^{\perp}\Sigma_{BD} = 0$*);*

12.4.3: $\Sigma_D - \Sigma_{DB}\Sigma_B^{\dagger}\Sigma_{DB}$ *is non-negative definite.*

12.2.3 Coherence for three collections

In this section we deal with coherence for three collections. This is necessary when we want to implement partial adjustment.

Definition 12.5 *Second-order belief specifications over three collections B, D, F are **finite** and **coherent** if and only if*

12.5.1: $\mathrm{E}(B_i)$ *is finite for each* B_i, $\mathrm{E}(D_j)$ *is finite for each* D_j, *and* $\mathrm{E}(F_k)$ *is finite for each* F_k,

12.5.2: $\mathrm{Var}(B_i)$ *is finite for each* B_i, $\mathrm{Var}(D_j)$ *is finite for each* D_j, *and* $\mathrm{Var}(F_k)$ *is finite for each* F_k

12.5.3: *the joint variance–covariance matrix*

$$\mathrm{Var}\left(\begin{bmatrix} B \\ D \\ F \end{bmatrix}\right) = \begin{bmatrix} \Sigma_B & \Sigma_{BD} & \Sigma_{BF} \\ \Sigma_{BD}^T & \Sigma_D & \Sigma_{DF} \\ \Sigma_{BF}^T & \Sigma_{DF}^T & \Sigma_F \end{bmatrix} \tag{12.1}$$

is non-negative definite.

For purposes of computation, it is often not appropriate to check the definiteness of the entire matrix (12.1), and instead we require alternative conditions on various of its submatrices. Such is the purpose of the following theorem. First, define for convenience the matrices

$$S = \Sigma_F - \Sigma_{FD}\Sigma_D^{\dagger}\Sigma_{DF}, \tag{12.2}$$

$$K^T = \Sigma_{FB} - \Sigma_{FD}\Sigma_D^{\dagger}\Sigma_{DB}. \tag{12.3}$$

Theorem 12.6 *The matrix* (12.1) *is non-negative definite if and only if Properties 12.3.1–12.3.3 are met, together with the following conditions.*

12.6.1: $\Sigma_{DF} \in$ **range**$\{\Sigma_D\}$ *(or, equivalently,* $\Sigma_D^{\perp}\Sigma_{DF} = 0$*);*

12.6.2: S *is non-negative definite;*

12.6.3: $K^T = \in$ **range**$\{S\}$ *(or, equivalently,* $S^{\perp}K^T = 0$*);*

12.6.4: $\Sigma_B - \Sigma_{BD}\Sigma_D^{\dagger}\Sigma_{DB} - KS^{\dagger}K^T$ *is non-negative definite.*

Proof. By Theorem 11.35, (12.1) is non-negative definite if and only if the following conditions are met:

1. the matrix
$$\begin{bmatrix} \Sigma_D & \Sigma_{DF} \\ \Sigma_{DF}^T & \Sigma_F \end{bmatrix}$$
is non-negative definite;

2. the matrix
$$\begin{bmatrix} \Sigma_{DB} \\ \Sigma_{FB}^T \end{bmatrix} \in \textbf{range}\left\{\begin{bmatrix} \Sigma_D & \Sigma_{DF} \\ \Sigma_{FD}^T & \Sigma_F \end{bmatrix}\right\};$$

3. the matrix
$$\Sigma_B - \begin{bmatrix} \Sigma_{BD} & \Sigma_{BF} \end{bmatrix} \begin{bmatrix} \Sigma_D & \Sigma_{DF} \\ \Sigma_{DF}^T & \Sigma_F \end{bmatrix}^{-} \begin{bmatrix} \Sigma_{BD}^T \\ \Sigma_{BF}^T \end{bmatrix}$$
is non-negative definite.

The first of these conditions requires, by Lemma 12.3, that Σ_D be non-negative definite, that $\Sigma_{DF} \in$ **range**$\{\Sigma_D\}$, and that S be non-negative definite. The first of these is guaranteed by Property 12.3.1; the latter two must be satisfied and appear as conditions (Property 12.6.1 and Property 12.6.2) in the theorem. Property 12.6.3 is necessary to guarantee the second of the conditions because

$$\left(I - \begin{bmatrix} \Sigma_D & \Sigma_{DF} \\ \Sigma_{FD}^T & \Sigma_F \end{bmatrix}\begin{bmatrix} \Sigma_D & \Sigma_{DF} \\ \Sigma_{FD}^T & \Sigma_F \end{bmatrix}^{\dagger}\right)\begin{bmatrix} \Sigma_{DB} \\ \Sigma_{FB} \end{bmatrix} = \begin{bmatrix} \Sigma_{DB} \\ \Sigma_{FB} - S^{\perp}K^T \end{bmatrix} = \begin{bmatrix} \Sigma_{DB} \\ \Sigma_{FB} \end{bmatrix}$$

if and only if $S^{\perp}K^T = 0$, i.e. $K^T \in$ **range**$\{S\}$. Property 12.6.4 follows by simplifying the matrix expression in the third condition, noticing that this expression is (12.25), addressed in Theorem 12.52. ■

Remark. The matrices K and S can be interpreted as adjusted covariance and adjusted variance matrices respectively:

$$S = \text{Var}(\mathbb{A}_D(F)) = \text{Var}_D(F),$$

$$K^T = \text{Cov}(\mathbb{A}_D(F), \mathbb{A}_D(B)).$$

See §12.11.

12.3 Consistency of data with beliefs

12.3.1 Consistency for a single collection

Definition 12.7 *For any collection D such that belief specifications over D are finite and coherent according to Definition 12.1, the observed value d of D is* ***consistent with the belief specifications*** *if and only if we have $[d - \text{E}(D)] \in \textbf{range}\{\Sigma_D\}$. Equivalently, $\Sigma_D^{\perp}[d - \text{E}(D)] = 0$.*

If Σ_D is full rank, then finite observations d are automatically consistent with the belief specifications. Otherwise, if Σ_D is not full rank, suppose that $a \in \textbf{null}\{\Sigma_D\}$; then $\text{Var}(a^T D) = 0$ and this implies that $a^T D$ is known so that $a^T[d - \text{E}(D)] = 0$ for any possible observation d and every vector $a \in \textbf{null}\{\Sigma_D\}$.

Theorem 12.8 *If data d are consistent with their belief specifications then all linear transformations of the data are consistent.*

Proof. Suppose that $V = GD$ is a collection of m linear transformations $(V_1, \ldots, V_m)$ of the quantities $(D_1, \ldots, D_n)$, where G is some $m \times n$ real matrix, and suppose that $v = Gd$ are the corresponding observations. Now the data v are consistent if $a^T[v - \text{E}(V)] = 0$ for all $a \in \textbf{null}\{\text{Var}(V)\}$. For such a, $a^T\text{Var}(V)a = a^T G\Sigma_D G^T a = 0$. Hence $G^T a \in \textbf{null}\{\Sigma_D\}$, and

$$a^T[v - \text{E}(V)] = a^T G[d - \text{E}(D)] = 0$$

as the data d are consistent. ■

Theorem 12.9 *If data d are consistent with their belief specifications then the value of*

$$[d - \text{E}(D)]^T \Sigma_D^{-}[d - \text{E}(D)]$$

is the same for every choice of generalized inverse Σ_D^{-} of Σ_D.

Proof. If the data are consistent then $[d - \text{E}(D)] \in \textbf{range}\{\Sigma_D\}$, so that

$$\Sigma_D^{\perp}[d - \text{E}(D)] = 0$$

and

$$[d - \text{E}(D)]^T \Sigma_D^{\perp} = 0.$$

Now, by Lemma 11.17, all generalized inverses Σ_D^- of Σ_D can be constructed from

$$\Sigma_D^- = \Sigma_D^\dagger + V\Sigma_D^\perp + (I - \Sigma_D^\dagger \Sigma_D)W$$

for arbitrary V, W. Hence all possible values of

$$[d - \mathrm{E}(D)]^T \Sigma_D^- [d - \mathrm{E}(D)]$$

can be constructed, for arbitrary V, W, as

$$\begin{aligned}[d - \mathrm{E}(D)]^T \Sigma_D^- [d - \mathrm{E}(D)] =& [d - \mathrm{E}(D)]^T (\Sigma_D^\dagger + V\Sigma_D^\perp \\ &+ (I - \Sigma_D^\dagger \Sigma_D)W)[d - \mathrm{E}(D)] \qquad (12.4)\\ =& [d - \mathrm{E}(D)]^T \Sigma_D^\dagger [d - \mathrm{E}(D))]. \qquad (12.5)\end{aligned}$$

■

12.3.2 Consistency for a partitioned collection

By Definition 12.7, observed data $D = d$ and $F = f$ are consistent with their belief specifications if and only if, for any generalized inverse,

$$\left(I - \begin{bmatrix}\Sigma_D & \Sigma_{DF}\\ \Sigma_{DF}^T & \Sigma_F\end{bmatrix}\begin{bmatrix}\Sigma_D & \Sigma_{DF}\\ \Sigma_{DF}^T & \Sigma_F\end{bmatrix}^-\right)\begin{bmatrix}d - \mathrm{E}(D)\\ f - \mathrm{E}(F)\end{bmatrix} = 0. \qquad (12.6)$$

This necessary consistency condition can be re-expressed as in the following theorem. Suppose that S is as defined in (12.2).

Theorem 12.10 *If the beliefs specified over the collections D and F are jointly coherent, and if the data d are consistent with the beliefs specified over the collection D, then the data f are consistent if and only if*

$$S^\perp[\Sigma_{FD}\Sigma_D^\dagger(d - \mathrm{E}(D)) - (f - \mathrm{E}(F))] = 0, \qquad (12.7)$$

or, equivalently,

$$\Sigma_{FD}\Sigma_D^\dagger[d - \mathrm{E}(D)] - [f - \mathrm{E}(F)] \in \mathbf{range}\{S\}.$$

Proof.

$$\begin{aligned}&\begin{bmatrix}\Sigma_D & \Sigma_{DF}\\ \Sigma_{DF}^T & \Sigma_F\end{bmatrix}\begin{bmatrix}\Sigma_D & \Sigma_{DF}\\ \Sigma_{DF}^T & \Sigma_F\end{bmatrix}^\dagger\begin{bmatrix}d - \mathrm{E}(D)\\ f - \mathrm{E}(F)\end{bmatrix}\\ &= \begin{bmatrix}\Sigma_D\Sigma_D^\dagger[d - \mathrm{E}(D)] - \Sigma_D^\perp\Sigma_{DF}S^\dagger[\Sigma_{FD}\Sigma_D^\dagger(d - \mathrm{E}(D)) - (f - \mathrm{E}(F))]\\ (I - SS^\dagger)[\Sigma_{FD}\Sigma_D^\dagger(d - \mathrm{E}(D))] + SS^\dagger(f - \mathrm{E}(F))\end{bmatrix}\\ &= \begin{bmatrix}(d - \mathrm{E}(D))\\ (I - SS^\dagger)[\Sigma_{FD}\Sigma_D^\dagger(d - \mathrm{E}(D))] + SS^\dagger(f - \mathrm{E}(F))\end{bmatrix}\end{aligned}$$

as $\Sigma_D \Sigma_D^\dagger (d - \mathrm{E}(D)) = d - \mathrm{E}(D)$, because the data d are consistent, and because $\Sigma_D \Sigma_D^\dagger \Sigma_{DF} = \Sigma_{DF}$ by coherence of the joint variance matrix over D and F. The proof now follows after some rearrangement. ■

In some circumstances, the eigenstructure of S may be available, in which case the following alternative checks may be made, with obvious proof.

Theorem 12.11 *If the joint variance–covariance matrix specified over D and F is coherent, and if the data d are consistent with the variance–covariance matrix Σ_D, then the data f are consistent (1) if S is full rank, (2) when S is not full rank, but*

$$g^T(\mathrm{E}_d(F) - f) = g^T[\Sigma_{FD}\Sigma_D^\dagger(d - \mathrm{E}(D)) - (f - \mathrm{E}(F))] = 0 \qquad (12.8)$$

for every eigenvector g of S corresponding to a zero eigenvalue: $Sg = 0$.

12.4 Adjusted expectation

Suppose that $c \neq 0$ is any $n_B \times 1$ vector, and let a be some $n_D \times 1$ vector. Let $c^T B$ be any linear combination of the elements of B.

Theorem 12.12 $\mathrm{Var}(c^T B - a^T D)$ *is minimized for some* $a \in$ **range**$\{\Sigma_D\}$ *which satisfies* $\Sigma_{DB}c = \Sigma_D a$.

Proof. To show that any solution a must be in **range**$\{\Sigma_D\}$, decompose a into $a = g + f$, where $g \in$ **range**$\{\Sigma_D\}$ and $f \in$ **null**$\{\Sigma_D\}$. Then

$$\begin{aligned}\mathrm{Var}(c^T B - a^T D) &= \mathrm{Var}(c^T B - g^T D) + f^T \Sigma_D f - 2c^T \Sigma_{BD} f + 2g_T \Sigma_D f \\ &= \mathrm{Var}(c^T B - g^T D),\end{aligned}$$

where, by Lemma 11.30,

$$f^T \Sigma_D f = 0, \quad g^T \Sigma_D f = 0, \quad c^T \Sigma_{BD} f = 0,$$

as $f \in$ **null**$\{\Sigma_D\}$ and $\Sigma_{BD}^T \in$ **range**$\{\Sigma_D\}$.

To show that we must have $\Sigma_{DB}c = \Sigma_D a$, suppose that $\Sigma_{DB}c = g + f$, for some $g \in$ **range**$\{\Sigma_D\}$ and $f \in$ **null**$\{\Sigma_D\}$. We have $f^T \Sigma_{DB} c = f^T g + f^T f$, where $f^T \Sigma_{DB} c = 0$, as $\Sigma_{DB} \in$ **range**$\{\Sigma_D\}$; and $f^T g = 0$; so that $f^T f = 0$, implying that we must have $f = 0$. As such, we must have $\Sigma_{DB} c = g$ for some $g \in$ **range**$\{\Sigma_D\}$. Finally, all $a \in$ **range**$\{\Sigma_D\}$ can be constructed from $a = \Sigma_D^\dagger g$, as $\Sigma_D \Sigma_D^\dagger a = a$ by Lemma 11.27. ■

Theorem 12.13 *The adjusted expectation for $c^T B$ is unique, and can be calculated as*

$$\mathrm{E}_D(c^T B) = c^T \mathrm{E}(B) + c^T \Sigma_{BD} \Sigma_D^\dagger (D - \mathrm{E}(D)).$$

Proof. From §3.1, the adjusted expectation for $c^T B$ is the linear combination $\mathrm{E}_D(c^T B) = a_0 + a^T D$ which minimizes

$$\mathrm{E}([c^T B - a_0 - a^T D]^2) = \mathrm{Var}(c^T B - a^T D) + [\mathrm{E}(c^T B) - a_0 - a^T \mathrm{E}(D)]^2 \quad (12.9)$$

over all collections $a = (a_0, a_1, \ldots, a_{n_D})$. The second term in (12.9) is minimized by taking $a_0 = \mathrm{E}(c^T B) - a^T \mathrm{E}(D)$. By Theorem 12.12, the first term is minimized by taking any a such that $\Sigma_{DB} c = \Sigma_D a$. This system of linear equations is consistent by Lemma 11.42 as $\Sigma_{DB} c \in \mathbf{range}\{\Sigma_D\}$, and has general solution

$$a = \Sigma_D^- \Sigma_{DB} c + (I - \Sigma_D^- \Sigma_D)t \quad (12.10)$$

for an arbitrary conformable vector t, by Lemma 11.43. The first term in (12.10) is in $\mathbf{range}\{\Sigma_D\}$ as required. However, the second is in $\mathbf{null}\{\Sigma_D\}$, by Lemma 11.30, as $\Sigma_D(I - \Sigma_D^- \Sigma_D)t = 0$ for any generalized inverse. Therefore, as we require our solution to be in $\mathbf{range}\{\Sigma_D\}$, we may drop the second term in (12.10). The adjusted expectation is unique whatever generalized inverse we take: we use the Moore–Penrose inverse. ■

By Lemma 11.45, the Moore–Penrose inverse, together with reflexive generalized inverses, gives a solution for a that has minimum norm compared to the solutions for a given by other choices of generalized inverse. For vector calculations we have directly from Theorem 12.13:

Corollary 12.14 *The adjusted expectation for the vector B is unique, and can be calculated as*

$$\mathrm{E}_D(B) = \mathrm{E}(B) + \Sigma_{BD} \Sigma_D^\dagger (D - \mathrm{E}(D)).$$

Corollary 12.15 *The observed adjusted expectation for the vector B, given observation of consistent data $D = d$, is unique, and can be calculated as*

$$\mathrm{E}_d(B) = \mathrm{E}(B) + \Sigma_{BD} \Sigma_D^\dagger (d - \mathrm{E}(D)).$$

12.5 Adjusted and resolved variance

The adjusted expectation $\mathrm{E}_D(B)$ has variance

$$\begin{aligned}
\mathrm{Var}(\mathrm{E}_D(B)) &= \mathrm{Var}(\mathrm{E}(B) + \Sigma_{BD} \Sigma_D^\dagger (D - \mathrm{E}(D))) \\
&= \Sigma_{BD} \Sigma_D^\dagger \Sigma_D \Sigma_D^\dagger \Sigma_{DB} \\
&= \Sigma_{BD} \Sigma_D^\dagger \Sigma_{DB}, \quad \text{by Definition 11.6,} \\
&= \mathrm{RVar}_D(B),
\end{aligned}$$

which is termed the resolved variance matrix as in (3.31). The residual vector, i.e. the adjusted version (3.22), has variance

$$\mathrm{Var}(\mathbb{A}_D(B)) = \mathrm{Var}(B - \mathrm{E}_D(B)) = \mathrm{Var}(B) - \mathrm{RVar}_D(B) = \mathrm{Var}_D(B),$$

which is termed the adjusted variance matrix.

Theorem 12.16 *The adjusted and resolved variance matrices are unique and can be calculated using any generalized inverse of Σ_D. That is, for any generalized inverse,*

$$\begin{aligned}\text{Var}_D(B) &= \Sigma_B - \Sigma_{BD}\Sigma_D^{-}\Sigma_{BD} = \Sigma_B - \Sigma_{BD}\Sigma_D^{\dagger}\Sigma_{BD},\\ \text{RVar}_D(B) &= \Sigma_{BD}\Sigma_D^{-}\Sigma_{BD} = \Sigma_{BD}\Sigma_D^{\dagger}\Sigma_{BD}.\end{aligned}$$

Proof. This follows directly by Theorem 11.31. ■

12.6 The resolved variance matrix

Following §3.7, the resolved variance matrix is defined to be

$$\text{RVar}_D(B) = \text{Var}(\text{E}_D(B)) = \text{RVar}_D(B) = \Sigma_{BD}\Sigma_D^{\dagger}\Sigma_{DB}.$$

This matrix has a number of properties as follows.

Theorem 12.17 *The matrix*

$$\begin{bmatrix}\text{RVar}_D(B) & \Sigma_{BD}\\ \Sigma_{DB}^T & \Sigma_D\end{bmatrix} \tag{12.11}$$

is non-negative definite.

Proof. $\text{Var}([B^T\ D^T]^T)$ is non-negative definite, so that by Theorem 11.35, Σ_D is non-negative definite and $\Sigma_{DB} \in \textbf{range}\{\Sigma_D\}$. Equivalently,

$$\Sigma_D\Sigma_D^{-}\Sigma_{DB} = \Sigma_{DB}$$

for any generalized inverse of Σ_D. Now consider the matrix

$$\text{Var}\left(\begin{bmatrix}\Sigma_{BD}\Sigma_D^{\dagger}D\\ D\end{bmatrix}\right) = \begin{bmatrix}\Sigma_{BD}\Sigma_D^{\dagger}\\ I\end{bmatrix}\Sigma_D\begin{bmatrix}\Sigma_{BD}\Sigma_D^{\dagger}\\ I\end{bmatrix}^T.$$

This matrix is non-negative definite because Σ_D is non-negative definite, Therefore,

$$\begin{bmatrix}\Sigma_{BD}\Sigma_D^{\dagger}\Sigma_D\Sigma_D^{\dagger}\Sigma_{DB} & \Sigma_{BD}\Sigma_D^{\dagger}\Sigma_D\\ \Sigma_D\Sigma_D^{\dagger}\Sigma_{DB} & \Sigma_D\end{bmatrix} = \begin{bmatrix}\text{RVar}_D(B) & \Sigma_{BD}\\ \Sigma_{DB} & \Sigma_D\end{bmatrix}$$

is non-negative definite, as $\Sigma_D\Sigma_D^{\dagger}\Sigma_{DB} = \Sigma_{DB}$, and $\Sigma_D^{\dagger}\Sigma_D\Sigma_D^{\dagger} = \Sigma_D^{\dagger}$ for Moore–Penrose generalized inverses. ■

Theorem 12.18 *The matrix $\Sigma_D - \Sigma_{DB}\text{RVar}_D(B)^{\dagger}\Sigma_{BD}$ is non-negative definite.*

Proof. This follows by applying Theorem 11.35 to the non-negative definite matrix of Theorem 12.17. ■

Theorem 12.19 $\mathbf{rk}\{\mathrm{RVar}_D(B)\} = r_{\mathbb{T}} = \mathbf{rk}\{\Sigma_{DB}\}$.

Proof.

$$\begin{aligned}\mathbf{rk}\{\mathrm{RVar}_D(B)\} &= \mathbf{rk}\{\Sigma_{BD}\Sigma_D^{\dagger}\Sigma_{DB}\}\\ &= \mathbf{rk}\{\Sigma_D^{\dagger}\Sigma_{DB}\}, \qquad \text{by Lemma 11.34,}\\ &\qquad\qquad\qquad\qquad \text{as } \Sigma_D \text{ is non-negative definite,}\\ &= \mathbf{rk}\{\Sigma_{DB}\}\end{aligned}$$

by Lemma 11.33, as $\mathbf{range}\{\Sigma_{DB}\} \in \mathbf{range}\{\Sigma_D^{\dagger}\} = \mathbf{range}\{\Sigma_D\}$. ■

Theorem 12.20 $\mathbf{range}\{\mathrm{RVar}_D(B)\} = \mathbf{range}\{\Sigma_{BD}\}$.

Proof. We have $\Sigma_{BD} \in \mathbf{range}\{\mathrm{RVar}_D(B)\}$ by applying Theorem 11.35 directly to the non-negative definite matrix (12.11). Alternatively, we have Theorem 12.19, so that we may apply Lemma 11.32 and the proof follows. ■

12.7 Matrix representations of the resolution transform

In §3.9.1 we gave a definition (3.65) for the resolution transform matrix. We now generalize the definition and establish properties. We use the following notation for convenience:

$$\mathrm{r}_B = \mathbf{rk}\{\mathrm{Var}(B)\}, \tag{12.12}$$

$$\mathrm{Var}(B)\mathrm{Q}_B = \mathrm{Q}_B\Psi_B, \tag{12.13}$$

where Ψ_B is the $\mathrm{r}_B \times \mathrm{r}_B$ diagonal matrix with values $\Psi_{B_1}, \ldots, \Psi_{B_{\mathrm{r}_B}} > 0$ being the positive eigenvalues of the variance matrix $\mathrm{Var}(B)$, with corresponding orthonormal eigenvectors collected as the columns of the $n_B \times \mathrm{r}_B$ matrix Q_B.

Lemma 12.21 *A matrix representation of the resolution transform for the adjustment of B by D is any matrix $\mathbb{T}_{B:D}$ in the class of matrices satisfying*

$$\Sigma_B\mathbb{T}_{B:D} = \Sigma_{BD}P,$$

where P is any matrix in the class of matrices satisfying

$$\Sigma_D P = \Sigma_{DB}.$$

Theorem 12.22 *All matrix representations of the resolution transform for the adjustment of B by D are of the form*

$$\mathbb{T}_{B:D} = \Sigma_B^{\dagger}\Sigma_{BD}\Sigma_D^{\dagger}\Sigma_{DB} + (I - \Sigma_B^{\dagger}\Sigma_B)H_2,$$

where H_2 is any conformable arbitrary matrix. If $\mathrm{r}_B = n_B$ then the arbitrary part vanishes.

Proof. By Lemma 11.46, the linear equations $\Sigma_D P = \Sigma_{DB}$ have solutions of the form

$$P = \Sigma_D^{\dagger}\Sigma_{DB} + (I - \Sigma_D^{\dagger}\Sigma_D)H_1,$$

where H_1 is arbitrary. These linear equations are consistent if the belief specifications over B, D are coherent by Lemma 11.42 as $\Sigma_{DB} \in \textbf{range}\{\Sigma_D\}$. Similarly, the linear equations $\Sigma_B \mathbb{T}_{B:D} = \Sigma_{BD} P$ have solutions of the form

$$\mathbb{T}_{B:D} = \Sigma_B^{\dagger}\Sigma_{BD} P + (I - \Sigma_B^{\dagger}\Sigma_B)H_2,$$

where H_2 is arbitrary. These linear equations are consistent by Lemma 11.42 as $\Sigma_{BD} \in \textbf{range}\{\Sigma_B\}$. Consequently, all matrix representations of the resolution transform can be generated from

$$\begin{aligned}\mathbb{T}_{B:D} &= \Sigma_B^{\dagger}\Sigma_{BD}\Sigma_D^{\dagger}\Sigma_{DB} + \Sigma_B^{\dagger}\Sigma_{BD}(I - \Sigma_D^{\dagger}\Sigma_D)H_1 + (I - \Sigma_B^{\dagger}\Sigma_B)H_2,\\ &= \Sigma_B^{\dagger}\Sigma_{BD}\Sigma_D^{\dagger}\Sigma_{DB} + (I - \Sigma_B^{\dagger}\Sigma_B)H_2,\end{aligned}$$

as $\Sigma_{BD}\Sigma_D^{\dagger}\Sigma_D = \Sigma_{BD}$ by Lemma 11.28. ∎

12.7.1 The symmetrized resolution transform matrix

The resolution transform matrix, $\mathbb{T}_{B:D}$, is often asymmetric. For the purposes of calculation of its eigenstructure, it can be helpful to work with the symmetrized resolution transform matrix, defined as follows.

Definition 12.23 *The symmetrized resolution transform matrix is the non-negative definite* $\mathrm{r}_B \times \mathrm{r}_B$ *matrix*

$$\begin{aligned}\tilde{\mathbb{T}}_{B:D} =& \Psi_B^{\frac{1}{2}} \mathrm{Q}_B^T \mathbb{T}_{B:D} \mathrm{Q}_B \Psi_B^{-\frac{1}{2}}\\ =& \Psi_B^{\frac{1}{2}} \mathrm{Q}_B^T (\Sigma_B^{\dagger}\Sigma_{BD}\Sigma_D^{\dagger}\Sigma_{DB} + (I - \Sigma_B^{\dagger}\Sigma_B)H_2)\mathrm{Q}_B \Psi_B^{-\frac{1}{2}}, \quad \textit{for arbitrary } H_2,\\ =& \Psi_B^{\frac{1}{2}} \mathrm{Q}_B^T \Sigma_B^{\dagger}\Sigma_{BD}\Sigma_D^{\dagger}\Sigma_{DB}\mathrm{Q}_B \Psi_B^{-\frac{1}{2}}, \quad \textit{as } \mathrm{Q}_B^T(I - \Sigma_B^{\dagger}\Sigma_B) = 0,\\ =& \Psi_B^{-\frac{1}{2}} \mathrm{Q}_B^T \Sigma_{BD}\Sigma_D^{\dagger}\Sigma_{DB}\mathrm{Q}_B \Psi_B^{-\frac{1}{2}}.\end{aligned}$$

Theorem 12.24 $\textbf{rk}\{\tilde{\mathbb{T}}_{B:D}\} = r_{\mathbb{T}} = \textbf{rk}\{\Sigma_{DB}\} = \textbf{rk}\{\Sigma_{BD}\}$.

Proof.

$$\begin{aligned}\textbf{rk}\{\tilde{\mathbb{T}}_{B:D}\} =& \textbf{rk}\{\Psi_B^{-\frac{1}{2}} \mathrm{Q}_B^T \Sigma_{BD}\Sigma_D^{\dagger}\Sigma_{DB}\mathrm{Q}_B \Psi_B^{-\frac{1}{2}}\}\\ =& \textbf{rk}\{\Psi_B^{-\frac{1}{2}} \mathrm{Q}_B^T \Sigma_{BD}\Sigma_D^{\dagger}\Sigma_{DB}\}, \quad \text{by Lemma 11.34},\\ =& \textbf{rk}\{\Sigma_{BD}\Sigma_D^{\dagger}\Sigma_{DB}\}, \quad \text{by Lemma 11.33},\\ & \qquad \text{as } \textbf{range}\{\Sigma_{BD}\} \in \textbf{range}\{\Psi_B^{-\frac{1}{2}}\mathrm{Q}_B^T\} = \textbf{range}\{B\},\\ =& \textbf{rk}\{\Sigma_{DB}\}, \quad \text{by Theorem 12.19}.\end{aligned}$$

∎

Corollary 12.25 *The resolution transform matrix has rank*

$$\mathbf{rk}\{\mathbb{T}_{B:D}\} = r_{\mathbb{T}} = \mathbf{rk}\{\mathrm{Cov}(B, D)\}.$$

Proof. This follows similarly to Theorem 12.24. ∎

Definition 12.26 *Suppose that the symmetrized resolution transform has eigenvalues $\lambda_1 \geq \lambda_2 \geq \ldots \geq \lambda_{\mathrm{r}_B} \geq 0$ corresponding to eigenvectors $g_1, \ldots, g_{\mathrm{r}_B}$. We gather the eigenvalues into the $\mathrm{r}_B \times \mathrm{r}_B$ diagonal matrix Λ, and we gather the eigenvectors as the columns of the matrix $\tilde{\mathrm{Z}}$. We may write $\tilde{\mathbb{T}}_{B:D} = \tilde{\mathrm{Z}}\Lambda\tilde{\mathrm{Z}}^T$ and $\tilde{\mathbb{T}}^{\dagger}_{B:D} = \tilde{\mathrm{Z}}\Lambda^{\dagger}\tilde{\mathrm{Z}}^T$. The eigenvector matrix $\tilde{\mathrm{Z}}$ has the property that $\tilde{\mathrm{Z}}\tilde{\mathrm{Z}}^T = \tilde{\mathrm{Z}}^T\tilde{\mathrm{Z}} = I_{\mathrm{r}_B}$. By Theorem 12.24, the number of positive eigenvalues λ_i is equal to $r_{\mathbb{T}} = \mathbf{rk}\{\Sigma_{BD}\}$, where*

$$0 \leq \mathbf{rk}\{\Sigma_{BD}\} = r_{\mathbb{T}} \leq \min(\mathrm{r}_B, \mathrm{r}_D).$$

Remark. The resolution transform $\mathbb{T}_{B:D}$ and its symmetrized version $\tilde{\mathbb{T}}_{B:D}$ are equivalent in the sense of §11.9, in that they have the same positive eigenvalues and have related eigenvectors. It is computationally advantageous, for numerical stability, to compute first the eigensolution to the symmetrized version. Where there are subsets of non-distinct eigenvalues, the corresponding eigenvectors are not uniquely defined. Eigenvalues may only be computed to machine accuracy, and so computer implementations for Bayes linear methods may have to make automated judgements as to the rank of the resolution transform, depending on the magnitude of the smallest computed eigenvalue which is deemed to be positive.

Theorem 12.27 *If $\mathrm{r}_B = n_B$, the matrix of right eigenvectors of the resolution transform matrix is $\mathrm{Z} = \mathrm{Q}_B\Psi_B^{-\frac{1}{2}}\tilde{\mathrm{Z}}$ corresponding to diagonal eigenvalue matrix Λ.*

Proof. By Theorem 12.22, if $\mathrm{r}_B = n_B$, there is no arbitrary element in $\mathbb{T}_{B:D}$. Hence

$$\begin{aligned}
\mathbb{T}_{B:D}\mathrm{Z} &= \Sigma_B^{\dagger}\Sigma_{BD}\Sigma_D^{\dagger}\Sigma_{DB}\mathrm{Q}_B\Psi_B^{-\frac{1}{2}}\tilde{\mathrm{Z}} \\
&= \mathrm{Q}_B\Psi_B^{-\frac{1}{2}}\tilde{\mathbb{T}}_{B:D}\tilde{\mathrm{Z}} \\
&= \mathrm{Q}_B\Psi_B^{-\frac{1}{2}}\tilde{\mathrm{Z}}\Lambda \\
&= \mathrm{Z}\Lambda.
\end{aligned}$$

∎

Theorem 12.28 *If $\mathrm{r}_B < n_B$ then the matrix representation of $\mathbb{T}_{B:D}$ is not uniquely defined. However, if we take the arbitrary part H_2 in Theorem 12.22 to be zero, the matrix representation given by $\mathbb{T}_{B:D} = \Sigma_B^{\dagger}\Sigma_{BD}\Sigma_D^{\dagger}\Sigma_{DB}$ has right eigenvector matrix $[\mathrm{Z}\ \mathrm{Q}_B^0]$, where Q_B^0 is a matrix whose columns are eigenvectors $\mathrm{q}_{B\mathrm{r}_B+1}, \ldots, \mathrm{q}_{Bn_B}$ of Σ_B corresponding to zero eigenvalues of Σ_B. The corresponding eigenvalues are $\lambda_1, \ldots, \lambda_{\mathrm{r}_B}$ corresponding to right eigenvectors $Z_1, \ldots Z_{\mathrm{r}_B}$ and $\lambda_{\mathrm{r}_B+1} = 0, \ldots, \lambda_{n_B} = 0$, corresponding to right eigenvectors $\mathrm{q}_{B\mathrm{r}_B+1}, \ldots, \mathrm{q}_{Bn_B}$.*

Proof. As above, $\mathbb{T}_{B:D}\mathrm{Z} = \mathrm{Z}\Lambda$, establishing the columns of Z as eigenvectors corresponding to eigenvalues λ_i. Finally, for $\mathrm{r}_B < i \leq n_B$,

$$\mathbb{T}_{B:D}\mathrm{q}_{Bi} = \Sigma_B^{\dagger}\Sigma_{BD}\Sigma_D^{\dagger}\Sigma_{DB}\mathrm{q}_{Bi} = 0$$

as $\mathrm{q}_{Bi} \in \mathbf{null}\{\Sigma_B\}$ and $\Sigma_{BD}^T \in \mathbf{range}\{\Sigma_B\}$. ∎

Theorem 12.29 *The matrix of eigenvectors* Z *simultaneously diagonalizes the prior, resolved, and adjusted variance matrices, i.e.* $\mathrm{Var}(B)$, $\mathrm{RVar}_D(B)$, *and* $\mathrm{Var}_D(B)$, *to*

$$\mathrm{Z}^T\mathrm{Var}(B)\mathrm{Z} = I_{\mathrm{r}_B},$$
$$\mathrm{Z}^T\mathrm{RVar}_D(B)\mathrm{Z} = \Lambda$$
$$\mathrm{Z}^T\mathrm{Var}_D(B)\mathrm{Z} = I_{\mathrm{r}_B} - \Lambda.$$

Proof. We have

$$\mathrm{Z}^T\mathrm{Var}(B)\mathrm{Z} = \tilde{\mathrm{Z}}^T\Psi_B^{-\frac{1}{2}}\mathrm{Q}_B^T\Sigma_B\mathrm{Q}_B\Psi_B^{-\frac{1}{2}}\tilde{\mathrm{Z}} = I_{\mathrm{r}_B}$$

and

$$\mathrm{Z}^T\mathrm{RVar}_D(B)\mathrm{Z} = \tilde{\mathrm{Z}}^T\Psi_B^{-\frac{1}{2}}\mathrm{Q}_B^T\Sigma_{BD}\Sigma_D^{\dagger}\Sigma_{DB}\mathrm{Q}_B\Psi_B^{-\frac{1}{2}}\tilde{\mathrm{Z}} = \tilde{\mathrm{Z}}^T\tilde{\mathbb{T}}_{B:D}\tilde{\mathrm{Z}} = \Lambda.$$

The final part follows trivially. ∎

12.7.2 The transform for the reverse adjustment

The following results concern the implementation of the resolution transform for the reverse adjustment, corresponding to (3.76) and (3.77).

Theorem 12.30 *If* $\mathrm{r}_D = n_D$, *the resolution transform matrix* $\mathbb{T}_{D:B}$ *for the adjustment of D by B has the representation*

$$\mathbb{T}_{D:B} = \Sigma_D^{\dagger}\Sigma_{DB}\Sigma_B^{\dagger}\Sigma_{BD} \tag{12.14}$$

with right eigenvector matrix $\Sigma_D^{\dagger}\Sigma_{DB}\mathrm{Z}(\Lambda^{\dagger})^{\frac{1}{2}}$ *corresponding to eigenvalues* Λ.

Proof. The basic representation follows directly from Theorem 12.22. The canonical structure follows as

$$\begin{aligned}\mathbb{T}_{D:B}(\Sigma_D^{\dagger}\Sigma_{DB}\mathrm{Z}(\Lambda^{\dagger})^{\frac{1}{2}}) &= \Sigma_D^{\dagger}\Sigma_{DB}\Sigma_B^{\dagger}\Sigma_{BD}(\Sigma_D^{\dagger}\Sigma_{DB}\mathrm{Z}(\Lambda^{\dagger})^{\frac{1}{2}}) \\ &= \Sigma_D^{\dagger}\Sigma_{DB}\mathbb{T}_{B:D}\mathrm{Z}(\Lambda^{\dagger})^{\frac{1}{2}} \\ &= (\Sigma_D^{\dagger}\Sigma_{DB}\mathrm{Z}(\Lambda^{\dagger})^{\frac{1}{2}})\Lambda.\end{aligned}$$

∎

Theorem 12.31 *If* $\mathrm{r}_D < n_D$, *the resolution transform matrix* $\mathbb{T}_{D:B}$ *for the adjustment of* D *by* B *has the representation*

$$\mathbb{T}_{D:B} = \Sigma_D^{\dagger}\Sigma_{DB}\Sigma_B^{\dagger}\Sigma_{BD} + (I - \Sigma_D^{\dagger}\Sigma_D)H_3 \tag{12.15}$$

where H_3 *is arbitrary. If we take the arbitrary part* $H_3 = 0$ *then a subset of the right eigenvector matrix for the representation is given by the columns of the matrix* $\Sigma_D^{\dagger}\Sigma_{BD}\mathrm{Z}(\Lambda^{\dagger})^{\frac{1}{2}}$, *with corresponding eigenvalues* Λ.

Proof. This follows similarly to Theorems 12.30 and 12.28. ■

Further right eigenvectors, all of which correspond to eigenvalue zero, can be constructed from $\mathbf{null}\{\Sigma_D\}$ if $\mathrm{r}_D \le \mathrm{r}_B$, and from $\mathbf{null}\{\Sigma_D\}$ and $\mathbf{null}\{\Sigma_B\}$ if $\mathrm{r}_D > \mathrm{r}_B$.

12.7.3 Inverses for the resolved variance matrix

Theorem 12.32

$$\mathrm{RVar}_D(B)_r^- = \mathrm{Q}_B\Psi_B^{-\frac{1}{2}}\tilde{\mathrm{Z}}\Lambda^{\dagger}\tilde{\mathrm{Z}}^T\Psi_B^{-\frac{1}{2}}\mathrm{Q}_B^T$$

is a reflexive generalized inverse for $\mathrm{RVar}_D(B)$, *but not necessarily the Moore–Penrose generalized inverse.*

Proof.

$$\begin{aligned}\mathrm{RVar}_D(B) &= \Sigma_{BD}\Sigma_D^{\dagger}\Sigma_{DB} \\ &= \mathrm{Q}_B\Psi_B^{\frac{1}{2}}\tilde{\mathbb{T}}_{B:D}\Psi_B^{\frac{1}{2}}\mathrm{Q}_B^T\end{aligned}$$

by Definition 12.23 and because $\mathrm{Q}_B\mathrm{Q}_B^T\Sigma_{BD} = \Sigma_{BD}$, as

$$\Sigma_{BD} \in \mathbf{range}\{\mathrm{Q}_B\} = \mathbf{range}\{\Sigma_B\}.$$

Hence we have

$$\begin{aligned}\mathrm{RVar}_D(B)\mathrm{RVar}_D(B)_r^- &= \mathrm{Q}_B\Psi_B^{\frac{1}{2}}\tilde{\mathrm{Z}}\Lambda\tilde{\mathrm{Z}}^T\Psi_B^{\frac{1}{2}}\mathrm{Q}_B^T\mathrm{Q}_B\Psi_B^{-\frac{1}{2}}\tilde{\mathrm{Z}}\Lambda^{\dagger}\tilde{\mathrm{Z}}^T\Psi_B^{-\frac{1}{2}}\mathrm{Q}_B^T \\ &= \mathrm{Q}_B\Psi_B^{\frac{1}{2}}\tilde{\mathrm{Z}}\Lambda\Lambda^{\dagger}\tilde{\mathrm{Z}}^T\Psi_B^{-\frac{1}{2}}\mathrm{Q}_B^T,\end{aligned}$$

whence

$$\mathrm{RVar}_D(B)\mathrm{RVar}_D(B)_r^-\mathrm{RVar}_D(B) = \mathrm{RVar}_D(B)$$

and

$$\mathrm{RVar}_D(B)_r^-\mathrm{RVar}_D(B)\mathrm{RVar}_D(B)_r^- = \mathrm{RVar}_D(B)_r^-,$$

so that $\mathrm{RVar}_D(B)_r^-$ is a reflexive generalized inverse of $\mathrm{RVar}_D(B)$ by Definition 11.6. However, $\mathrm{RVar}_D(B)\mathrm{RVar}_D(B)_r^-$ and $\mathrm{RVar}_D(B)_r^-\mathrm{RVar}_D(B)$ are not necessarily symmetric, and therefore $\mathrm{RVar}_D(B)_r^-$ is not necessarily the Moore–Penrose inverse. ■

Theorem 12.33 *When $\tilde{\mathbb{T}}_{B:D}$ is full rank,*

$$\text{RVar}_D(B)^\dagger = \text{RVar}_D(B)_r^- = \text{Q}_B \Psi_B^{-\frac{1}{2}} \tilde{\text{Z}} \Lambda^\dagger \tilde{\text{Z}}^T \Psi_B^{-\frac{1}{2}} \text{Q}_B^T.$$

Proof. If $\tilde{\mathbb{T}}_{B:D}$ is full rank, then $\Lambda^\dagger = \Lambda^{-1}$ and $\tilde{\text{Z}}\tilde{\text{Z}}^T = I$. We then obtain that

$$\begin{aligned}
\text{RVar}_D(B)\text{RVar}_D(B)_r^- &= \text{Q}_B \Psi_B^{\frac{1}{2}} \tilde{\text{Z}} \Lambda \Lambda^\dagger \tilde{\text{Z}}^T \Psi_B^{-\frac{1}{2}} \text{Q}_B^T \\
&= \text{Q}_B \Psi_B^{\frac{1}{2}} \tilde{\text{Z}} \tilde{\text{Z}}^T \Psi_B^{-\frac{1}{2}} \text{Q}_B^T \\
&= \text{Q}_B \text{Q}_B^T
\end{aligned}$$

is symmetric. Similarly, $\text{RVar}_D(B)_r^- \text{RVar}_D(B)$ is symmetric, and so by Definition 11.6, $\text{RVar}_D(B)_r^- = \text{RVar}_D(B)^\dagger$ is the Moore–Penrose generalized inverse. ■

12.7.4 Canonical quantities

Following §3.9.1, the canonical quantities and canonical resolutions can be calculated as follows.

Definition 12.34 *The canonical quantities for the adjustment of B by D are the linear combinations implied by the columns $\text{Z}_1, \ldots, \text{Z}_{\text{r}_B}$ of the $n_B \times \text{r}_B$ matrix Z with corresponding canonical resolutions $\lambda_1, \ldots, \lambda_{\text{r}_B}$. That is, the ith canonical quantity is $\text{Y}_i = \text{Z}_i^T(B - \text{E}(B))$.*

Of these r_B canonical quantities, some may correspond to canonical resolutions equal to zero. There are, by Theorem 12.24, exactly $r_{\mathbb{T}} = \mathbf{rk}\{\Sigma_{DB}\}$ canonical quantities with positive resolution. Thus, the structure of the canonical quantities is as follows.

- If $r_{\mathbb{T}} > 0$, then there are $r_{\mathbb{T}}$ canonical quantities

  $$\text{Y}_i = \text{Z}_i^T(B - \text{E}(B)), \quad i = 1, \ldots, r_{\mathbb{T}},$$

  corresponding to positive canonical resolutions λ_i. These are quantities in $[B]$ about which the data quantities in $[D]$ are informative.

- If $\text{r}_B > r_{\mathbb{T}}$, then there are $\text{r}_B - r_{\mathbb{T}}$ further canonical quantities

  $$\text{Y}_i = \text{Z}_i^T(B - \text{E}(B)), \quad i = r_{\mathbb{T}} + 1, \ldots, \text{r}_B,$$

  corresponding to zero canonical resolutions λ_i. These are quantities in $[B]$ about which the data quantities in $[D]$ are not informative, but for which a different set of data quantities might be informative.

- If $r_B < n_B$, then there are $n_B - r_B$ further canonical quantities

$$Y_j = q_{Bj}^T(B - E(B)), \quad j = r_B + 1, \ldots, n_B,$$

corresponding to zero canonical resolutions. These degenerate canonical quantities are uncorrelated with all other canonical quantities, and have expectation and variance zero.

We gather the first r_B canonical quantities into the vector $Y = [Y_1 \ldots Y_{r_B}]$.

Definition 12.35 *The vector* Y *of canonical quantities has prior expectation, adjusted expectation, prior variance matrix, resolved variance matrix, and adjusted variance matrix as follows:*

$$\begin{aligned}
E(Y) &= E(Z^T(B - E(B))) & & = 0,\\
E_D(Y) &= E_D(Z^T(B - E(B))) & & = 0,\\
\mathrm{Var}(Y) &= \mathrm{Var}(Z^T(B - E(B))Z) & = Z^T\Sigma_B Z & = I_{r_B},\\
\mathrm{RVar}_D(Y) &= \mathrm{RVar}_D(Z^T(B - E(B)) & = Z^T\mathrm{RVar}_D(B)Z & = \Lambda,\\
\mathrm{Var}_D(Y) &= \mathrm{Var}_D(Z^T(B - E(B)) & = Z^T\mathrm{Var}_D(B)Z & = I_{r_B} - \Lambda.
\end{aligned}$$

These follow from basic definitions for adjusted expectations and variances and by Theorem 12.29.

Definition 12.36 *The canonical quantities for the adjustment of D by B are the quantities* $Y_1^*, \ldots, Y_{r_B}^*$ *gathered into the vector* Y^*, *where* Y^* *is constructed from the right eigenvectors of the resolution transform matrix* $\mathbb{T}_{D:B}$:

$$\begin{aligned}
Y^* &= (\Sigma_D^\dagger \Sigma_{DB} Z(\Lambda^\dagger)^{\frac{1}{2}})^T(D - E(D))\\
&= (\Lambda^\dagger)^{\frac{1}{2}} Z^T(E_D(B) - E(B))\\
&= (\Lambda^\dagger)^{\frac{1}{2}} E_D(Y),
\end{aligned}$$

establishing (3.78), *where* Y *are the canonical quantities for the adjustment of B by D.*

There may be further degenerate canonical quantities corresponding to canonical resolutions equal to zero.

12.7.5 Coherence via the resolution transform matrix

The following theorem shows how the eigenstructure for $\tilde{\mathbb{T}}_{B:D}$ can be used to check a necessary coherence requirement.

Theorem 12.37 *Property 12.3.3 is satisfied if the following three conditions are satisfied:*

12.37.1: Σ_B *is non-negative definite;*

12.37.2: $\Sigma_{BD} \in$ **range**$\{\Sigma_B\}$*;*

12.37.3: *all the eigenvalues* λ *of* $\tilde{\mathbb{T}}_{B:D}$ *are contained in the interval* $1 \geq \lambda \geq 0$.

Proof. By Property 12.37.1 we have the representation $\Sigma_B = \mathrm{Q}_B \Psi_B \mathrm{Q}_B^T$. By Property 12.37.2, $\mathrm{Q}_B \mathrm{Q}_B^T \Sigma_{BD} = \Sigma_{BD}$. Consequently, we can write the matrix in Property 12.3.3 as

$$\Sigma_B - \Sigma_{BD} \Sigma_D^{\dagger} \Sigma_{DB} = \mathrm{Q}_B \Psi_B^{\frac{1}{2}} (I - \tilde{\mathbb{T}}_{B:D}) \Psi_B^{\frac{1}{2}} \mathrm{Q}_B^T.$$

By Property 12.37.3 the matrix $I - \tilde{\mathbb{T}}_{B:D}$ is non-negative definite as all its eigenvalues lie in [0, 1], and thus it follows that $\mathrm{Q}_B \Psi_B^{\frac{1}{2}} (I - \tilde{\mathbb{T}}_{B:D}) \Psi_B^{\frac{1}{2}} \mathrm{Q}_B^T$ must also be non-negative definite. ■

12.8 Assessing discrepant data

Let h be any $n_D \times 1$ vector. Following §4.1, the standardized observation and discrepancy of $h^T d$ are, for $\mathrm{Var}(h^T D) > 0$,

$$\mathrm{S}(h^T d) = \frac{h^T (d - \mathrm{E}(D))}{\sqrt{\mathrm{Var}(h^T (D - \mathrm{E}(D)))}},$$

$$\mathrm{Dis}(h^T d) = \frac{[h^T (d - \mathrm{E}(D))]^2}{\mathrm{Var}(h^T (D - \mathrm{E}(D)))} = \frac{[h^T (d - \mathrm{E}(D))]^2}{h^T \Sigma_D h}.$$

Theorem 12.38 *If data d are consistent with their corresponding belief specifications, the linear combination* $h^T d$ *having maximal discrepancy is given by taking* $h \propto \Sigma^{\dagger}(d - \mathrm{E}(D))$. *The maximal discrepancy is*

$$\begin{aligned}
\max_h \{\mathrm{Dis}(h^T d)\} &= \max_h \left\{ \frac{[h^T (d - \mathrm{E}(D))]^2}{h^T \Sigma_D h} \right\} \\
&= (d - \mathrm{E}(D))^T \Sigma_D^{-} (d - \mathrm{E}(D)) \\
&= (d - \mathrm{E}(D))^T \Sigma_D^{\dagger} (d - \mathrm{E}(D)).
\end{aligned} \tag{12.16}$$

Proof. This follows by Corollary 11.49. ■

Theorem 12.39 *The maximal discrepancy* (12.16) *has its prior expectation equal to* **rk**$\{\Sigma_D\}$.

Proof.

$$\begin{aligned}\mathrm{E}((D-\mathrm{E}(D))^T\Sigma_D^{\dagger}(D-\mathrm{E}(D))) &= \mathrm{E}(\mathbf{tr}\{[D-\mathrm{E}(D)]^T\Sigma_D^{\dagger}[D-\mathrm{E}(D)]\})\\ &= \mathrm{E}(\mathbf{tr}\{\Sigma_D^{\dagger}[D-\mathrm{E}(D)][D-\mathrm{E}(D)]^T\})\\ &= \mathbf{tr}\{\Sigma_D^{\dagger}\mathrm{E}([D-\mathrm{E}(D)][D-\mathrm{E}(D)]^T)\}\\ &= \mathbf{tr}\{\Sigma_D^{\dagger}\Sigma_D\}\\ &= \mathbf{rk}\{\Sigma_D\}, \qquad \text{by Lemma 11.18.}\end{aligned}$$

■

In §4.2, the linear combination having maximal discrepancy is termed the discrepancy vector,

$$\dot{w}_d = [d-\mathrm{E}(D)]^T\Sigma^{\dagger}[D-\mathrm{E}(D)].$$

Its properties summarized there are easily established.

12.9 Consistency of observed adjustments

Theorem 12.40 *For a Bayes linear adjustment for a finite coherent specification, if data d are consistent in the sense of Definition 12.7 then the observed adjusted expectations are consistent.*

Proof. This follows directly from Theorem 12.8. ■

Lemma 12.41 *The adjustment discrepancy vector* (4.26) *is calculated using any generalized inverse of* $\mathrm{RVar}_D(B)$ *as*

$$\ddot{a}_d = \mathrm{RVar}_D(B)^{-}(\mathrm{E}_d(B)-\mathrm{E}(B)). \tag{12.17}$$

Lemma 12.42 *As a linear combination of the B_is, the induced discrepancy vector can be expressed using the reflexive generalized inverse given in Theorem 12.32 as*

$$\begin{aligned}\ddot{\mathbb{Y}}_d(B) &= \ddot{a}_d{}^T(B-\mathrm{E}(B))\\ &= [\mathrm{E}_d(B)-\mathrm{E}(B)]^T \mathrm{Q}_B\Psi_B^{-\frac{1}{2}}\tilde{\mathrm{Z}}\Lambda^{\dagger}\tilde{\mathrm{Z}}^T\Psi_B^{-\frac{1}{2}}\mathrm{Q}_B^T(B-\mathrm{E}(B))\\ &= [\mathrm{E}_d(B)-\mathrm{E}(B)]^T\mathrm{Z}\Lambda^{\dagger}\mathrm{Z}^T(B-\mathrm{E}(B))\\ &= (\mathrm{E}_d(Y))^T\Lambda^{\dagger}Y, \end{aligned}\tag{12.18}$$

where Z *is the matrix whose columns are the eigenvectors of the resolution matrix, and* $Y = \mathrm{Z}^T(B-\mathrm{E}(B))$ *are the canonical quantities for the adjustment, with* $\mathrm{E}(Y)=0$. $\ddot{\mathbb{Y}}_d(B)$ *is the discrepancy vector for the adjustment* (4.29), (4.36). *Similarly,* $\ddot{W}_d$ (4.24), (4.35), *may be constructed as*

$$\ddot{W}_d = \ddot{a}_d{}^T(\mathrm{E}_D(B)-\mathrm{E}(B)) \tag{12.19}$$

$$= \mathrm{E}_d(Y)\Lambda^{\dagger}\mathrm{E}_D(Y). \tag{12.20}$$

Lemma 12.43 *We may calculate the adjustment discrepancy* (4.23) *as*

$$\begin{aligned}\mathrm{Dis}_d(B) &= \mathrm{Dis}(\mathrm{E}_d(B)) &(12.21)\\ &= (\mathrm{E}_d(B) - \mathrm{E}(B))^T \mathrm{RVar}_D(B)^- (\mathrm{E}_d(B) - \mathrm{E}(B)) &(12.22)\end{aligned}$$

for any generalized inverse of $\mathrm{RVar}_D(B)$. *It is convenient to use a reflexive or Moore–Penrose generalized inverse.*

Theorem 12.44 *For any finite coherent specification with consistent data d, the value of*

$$\mathrm{Dis}_d(B) = \mathrm{Dis}(\mathrm{E}_d(B))$$

is the same for any choice of generalized inverse of $\mathrm{RVar}_D(B) = \mathrm{Var}(\mathrm{E}_d(B))$.

Proof. The observed adjustments are consistent with the beliefs specified about them, by Theorem 12.40. Applying Theorem 12.9 proves the result. ■

Theorem 12.45 *The prior expectation of*

$$\mathrm{Dis}_D(B) = (\mathrm{E}_D(B) - \mathrm{E}(B))^T \mathrm{RVar}_D(B)^- (\mathrm{E}_D(B) - \mathrm{E}(B))$$

is $\mathrm{E}(\mathrm{Dis}_D(B)) = r_{\mathbb{T}} = \mathbf{rk}\{\Sigma_{DB}\}$.

Proof.

$$\begin{aligned}\mathrm{E}(\mathrm{Dis}_D(B)) &= \mathrm{E}((\mathrm{E}_D(B) - \mathrm{E}(B))^T \mathrm{RVar}_D(B)^- (\mathrm{E}_D(B) - \mathrm{E}(B)))\\ &= \mathrm{E}((\mathrm{E}_D(B) - \mathrm{E}(B))^T \mathrm{RVar}_D(B)^\dagger (\mathrm{E}_D(B) - \mathrm{E}(B)))\\ &= \mathbf{tr}\{\mathrm{E}(\mathrm{RVar}_D(B)^\dagger (\mathrm{E}_D(B) - \mathrm{E}(B))(\mathrm{E}_D(B) - \mathrm{E}(B))^T)\}\\ &= \mathbf{tr}\{\mathrm{RVar}_D(B)^\dagger \mathrm{E}((\mathrm{E}_D(B) - \mathrm{E}(B))(\mathrm{E}_D(B) - \mathrm{E}(B))^T)\}\\ &= \mathbf{tr}\{\mathrm{RVar}_D(B)^\dagger \mathrm{RVar}_D(B)\}\\ &= \mathbf{rk}\{\mathrm{RVar}_D(B)\} = r_{\mathbb{T}}.\end{aligned}$$

■

12.9.1 Partitioning the discrepancy

The overall data discrepancy can be partitioned as described in §4.4.5 into parts relevant to the adjustment of B by D, and a residual part.

Theorem 12.46 *Let* $G = D - \mathrm{E}_{\mathrm{E}_D(B)}(D) = \mathbb{A}_{\mathrm{E}_D(B)}(D)$ *with observed value g. Then, for any consistent observations d,*

$$\mathrm{Dis}(d) = \mathrm{Dis}_d(B) + \mathrm{Dis}(g).$$

Proof. We have by Definition 12.43 that

$$\begin{aligned}\text{Dis}_d(B) &= \text{Dis}(\text{E}_d(B)) \\ &= (\text{E}_d(B) - \text{E}(B))^T \text{RVar}_D(B)^\dagger (\text{E}_d(B) - \text{E}(B)) \\ &= (d - \text{E}(D))^T \Sigma_D^\dagger \Sigma_{DB} \text{RVar}_D(B)^\dagger \Sigma_{BD} \Sigma_D^\dagger (d - \text{E}(D)).\end{aligned}$$

Now, $\text{E}(G) = 0$ and $\text{Var}(G) = \Sigma_D - \Sigma_{DB}\text{RVar}_D(B)^\dagger \Sigma_{BD}$, so that

$$\text{Dis}(g) = g^T(\Sigma_D - \Sigma_{DB}\text{RVar}_D(B)^\dagger \Sigma_{BD})^\dagger g.$$

g is the observed value of

$$\begin{aligned}G &= D - \text{E}_{\text{E}_D(B)}(D) \\ &= D - \{\text{E}(D) + \text{Cov}(D, \text{E}_D(B))\text{Var}(\text{E}_D(B))^\dagger [\text{E}_D(B) - \text{E}(\text{E}_D(B))]\} \\ &= D - \text{E}(D) - \Sigma_{DB}\text{RVar}_D(B)^\dagger \Sigma_{BD}\Sigma_D^\dagger (D - \text{E}(D)) \\ &= \Sigma_D \Sigma_D^\dagger (D - \text{E}(D)) - \Sigma_{DB}\text{RVar}_D(B)^\dagger \Sigma_{BD}\Sigma_D^\dagger (D - \text{E}(D)) \\ &= (\Sigma_D - \Sigma_{DB}\text{RVar}_D(B)^\dagger \Sigma_{BD})\Sigma_D^\dagger (D - \text{E}(D)) \\ &= \text{Var}(G)\Sigma_D^\dagger (D - \text{E}(D))\end{aligned}$$

so that $g = \text{Var}(G)\Sigma_D^\dagger (d - \text{E}(D))$.

Note that we can write

$$\Sigma_D \Sigma_D^\dagger (D - \text{E}(D)) = D - \text{E}(D)$$

above because $[D - \text{E}(D)] \in \textbf{range}\{\Sigma_D\}$). It follows that

$$\begin{aligned}\text{Dis}(g) =& g^T \text{Var}(G)^\dagger g \\ =& (d - \text{E}(D))^T \Sigma_D^\dagger \text{Var}(G)\text{Var}(G)^\dagger \text{Var}(G)\Sigma_D^\dagger (d - \text{E}(D)) \\ =& (d - \text{E}(D))^T \Sigma_D^\dagger (\Sigma_D - \Sigma_{DB}\text{RVar}_D(B)^\dagger \Sigma_{BD})\Sigma_D^\dagger (d - \text{E}(D)) \\ =& (d - \text{E}(D))^T \Sigma_D^\dagger (d - \text{E}(D)) \\ & - (d - \text{E}(D))^T \Sigma_D^\dagger \Sigma_{DB}\text{RVar}_D(B)^\dagger \Sigma_{BD}\Sigma_D^\dagger (d - \text{E}(D)) \\ =& \text{Dis}(d) - \text{Dis}_d(B).\end{aligned}$$

■

Theorem 12.47 *For any consistent observations d, $\text{Dis}(d) \geq \text{Dis}_d(B)$, with equality if and only if $\textbf{rk}\{\Sigma_{DB}\} = \textbf{rk}\{\Sigma_D\}$.*

Proof. The inequality follows directly from Theorem 12.46 as

$$\text{Dis}(d) = \text{Dis}_d(B) + g^T \text{Var}(G)^\dagger g \geq \text{Dis}_d(B).$$

Equality is attained only for $\text{Dis}(g) = 0$. We may write

$$\text{Dis}(g) = (d - \text{E}(D))^T \Sigma_D^\dagger (\Sigma_D - \Sigma_{DB} \text{RVar}_D(B)^\dagger \Sigma_{BD}) \Sigma_D^\dagger (d - \text{E}(D)),$$

so that $\text{Dis}(g) = 0$ for any consistent g if and only if

$$\Sigma_D^\dagger (\Sigma_D - \Sigma_{DB} \text{RVar}_D(B)^\dagger \Sigma_{BD}) \Sigma_D^\dagger = 0.$$

Now,

$$\begin{aligned} &\Sigma_D^\dagger (\Sigma_D - \Sigma_{DB} \text{RVar}_D(B)^\dagger \Sigma_{BD}) \Sigma_D^\dagger \\ &\quad = \text{Q}_D \Psi_D^{-\frac{1}{2}} (I_{\text{r}_D} - A(A^T A)^\dagger A^T) \Psi_D^{-\frac{1}{2}} \text{Q}_D^T, \end{aligned}$$

where

$$A = \Psi_D^{-\frac{1}{2}} \text{Q}_D^T \Sigma_{DB},$$

which is equal to zero if and only if $I_{\text{r}_D} = A(A^T A)^\dagger A^T$, i.e. if and only if $I_{\text{r}_D} = AA^\dagger$ by Lemma 11.12. For $I_{\text{r}_D} = AA^\dagger$ we must have

$$\begin{aligned} \text{r}_D &= \mathbf{rk}\{AA^\dagger\} \\ &= \mathbf{rk}\{A\}, \quad \text{by Lemma 11.18,} \\ &= \mathbf{rk}\{\Psi_D^{-\frac{1}{2}} \text{Q}_D^T \Sigma_{DB}\} \\ &= \mathbf{rk}\{\Sigma_{DB}\}, \quad \text{by Lemma 11.34, as } \Sigma_{DB} \in \mathbf{range}\{\Sigma_D\}. \end{aligned}$$

Hence it is a necessary condition for equality that $\mathbf{rk}\{\Sigma_{BD}\} = \mathbf{rk}\{\Sigma_D\}$. Alternatively, if $\mathbf{rk}\{\Sigma_{BD}\} < \mathbf{rk}\{\Sigma_D\}$, then $AA^\dagger$ is an idempotent non-negative definite matrix with rank less than r_D, and the condition $I_{\text{r}_D} = AA^\dagger$ clearly cannot be met. Suppose we meet the condition that $\mathbf{rk}\{\Sigma_{BD}\} = \mathbf{rk}\{\Sigma_D\} = \text{r}_D$. Then, as $\mathbf{rk}\{A\} = \text{r}_D$ and as A is an $\text{r}_D \times n_B$ matrix, we have that $A^\dagger = A^T(AA^T)^{-1}$ by Lemma 11.13, so that

$$AA^\dagger = AA^T(AA^T)^{-1} = I_{\text{r}_D}.$$

Hence, this rank condition is both necessary and sufficient. ■

Theorem 12.48 *The discrepancy for the observed value of $\mathbb{A}_{\text{E}_D(B)}(D)$ has prior expectation*

$$\text{E}(\text{Dis}(\mathbb{A}_{\text{E}_D(B)}(D))) = \mathbf{rk}\{\Sigma_D\} - \mathbf{rk}\{\Sigma_{BD}\}.$$

Proof. Follows from Theorems 12.46, 12.39, and 12.45. ■

12.10 The bearing and size of adjustment

Let h be any $n_B \times 1$ vector. Following §4.6, the size of the adjustment for the quantity $h^T B$ is

$$\text{Size}_d(h^T B) = \frac{[h^T(\text{E}_d(B) - \text{E}(B))]^2}{h^T\text{Var}(B)h}.$$

Theorem 12.49 *For finite and coherent belief specifications and data consistent with them, the linear combination $h^T B$ having maximal size of adjustment is given by taking*

$$h \propto \Sigma_B^\dagger(\text{E}_d(B) - \text{E}(B)) = \Sigma_B^\dagger \Sigma_{BD} \Sigma_D^\dagger (d - \text{E}(D)).$$

The maximal discrepancy is, for any choice of generalized inverse,

$$\text{Size}_d(B) = \max_h\{\text{Size}_d(h^T B)\} = \text{E}_d(Y)^T\text{E}_d(Y),$$

where $Y = \text{Z}^T(B - \text{E}(B))$ are the canonical quantities.

Proof. Σ_B is non-negative definite and

$$\text{E}_d(B) - \text{E}(B) = \Sigma_{BD}\Sigma_D^\dagger(d - \text{E}(D)) \in \mathbf{range}\{\Sigma_B\}$$

so that Theorem 11.49 applies directly. For the size,

$$\begin{aligned}
\text{Size}_d(B) &= \max_h \left\{ \frac{[h^T(\text{E}_d(B) - \text{E}(B))]^2}{h^T\Sigma_B h} \right\} \\
&= (\text{E}_d(B) - \text{E}(B))^T \Sigma_B^\dagger (\text{E}_d(B) - \text{E}(B)) \\
&= (\text{E}_d(B) - \text{E}(B))^T \Sigma_B^- (\text{E}_d(B) - \text{E}(B)) \\
&= (\text{E}_d(B) - \text{E}(B))^T \text{Z}\text{Z}^T (\text{E}_d(B) - \text{E}(B)), \quad \text{as } \text{Z}\text{Z}^T = \Sigma_B^\dagger \\
&= \text{E}_d(Y)^T\text{E}_d(Y).
\end{aligned}$$

■

Definition 12.50 *The quantity corresponding to the maximum* (4.49) *is denoted the bearing, $\mathbb{Z}_d(B)$. That is, let*

$$\dot{h}_d = \Sigma_B^\dagger(\text{E}_d(B) - \text{E}(B)) = \Sigma_B^\dagger\Sigma_{BD}\Sigma_D^\dagger(d - \text{E}(D)).$$

Then

$$\begin{aligned}
\mathbb{Z}_d(B) &= \dot{h}_d^T(B - \text{E}(B)) \\
&= (B - \text{E}(B))^T \Sigma_B^\dagger (\text{E}_d(B) - \text{E}(B)) \\
&= (B - \text{E}(B))^T \Sigma_B^\dagger \Sigma_{BD} \Sigma_D^\dagger (d - \text{E}(D)). \qquad (12.23)
\end{aligned}$$

Theorem 12.51

$$\mathbb{Z}_d(B) = Y^T \mathrm{E}_d(Y).$$

Proof.

$$\begin{aligned}\mathbb{Z}_d(B) &= (B - \mathrm{E}(B))^T \Sigma_B^{\dagger}(\mathrm{E}_d(B) - \mathrm{E}(B)) \\ &= (B - \mathrm{E}(B))^T \mathrm{Z}\mathrm{Z}^T(\mathrm{E}_d(B) - \mathrm{E}(B)) \text{ as } \Sigma_B^{\dagger} = \mathrm{Z}\mathrm{Z}^T \\ &= [\mathrm{Z}^T(B - \mathrm{E}(B))]^T[\mathrm{E}_d(\mathrm{Z}^T(B - \mathrm{E}(B)))] = Y^T \mathrm{E}_d(Y).\end{aligned}$$

■

This restates (4.55). It is easily shown that

$$\mathrm{Var}(\mathbb{Z}_d(B)) = \mathrm{Size}_d(B) = \mathrm{E}_d(Y)^T \mathrm{E}_d(Y).$$

12.11 Partial adjustments

Suppose that we have collections of uncertain quantities B, D, F with coherent beliefs and consistent data as described in §12.2.3 and §12.3.2, respectively. We assume that the initial adjustment of B by D has already taken place, so that we know already the adjusted expectations $\mathrm{E}_D(B)$, possibly with observed values $\mathrm{E}_d(B)$; the adjusted variance matrix $\mathrm{Var}_D(B)$; and the resolution matrix $\mathbb{T}_{B:D}$. We are now concerned with obtaining $\mathrm{E}_{D\cup F}(B)$, $\mathrm{Var}_{D\cup F}(B)$, and $\mathbb{T}_{B:D\cup F}$ in terms of $\mathrm{E}_D(B)$, $\mathrm{Var}_D(B)$, and $\mathbb{T}_{B:D}$ and further quantities reflecting the change in adjustment. Recall that the adjusted versions of B and F, having fitted each on D, are

$$\begin{aligned}\mathbb{A}_D(B) &= B - \mathrm{E}_D(B), \\ \mathbb{A}_D(F) &= F - \mathrm{E}_D(F).\end{aligned}$$

It is helpful to employ S as defined in (12.2) and K as defined in (12.3), and to note that these can be identified as

$$K = \mathrm{Cov}_D(B, F) = \mathrm{Cov}(\mathbb{A}_D(B), \mathbb{A}_D(F)), \quad S = \mathrm{Var}(\mathbb{A}_D(F)) = \mathrm{Var}_D(F).$$

Theorem 12.52 *The partial resolved variance, i.e. the reduction in variance in B due to fitting on F as well as D (5.9), is*

$$\begin{aligned}\mathrm{Var}_D(B) - \mathrm{Var}_{D\cup F}(B) &= K S^{\dagger} K^T \qquad (12.24) \\ &= \mathrm{RVar}_{\mathbb{A}_D(F)}(\mathbb{A}_D(B))\end{aligned}$$

Proof. By Theorem 12.16 we have uniquely, and independently of choice of generalized inverse,

$$\mathrm{Var}_{D\cup F}(B) = \Sigma_B - \begin{bmatrix}\Sigma_{BD} & \Sigma_{BF}\end{bmatrix}\begin{bmatrix}\Sigma_D & \Sigma_{DF} \\ \Sigma_{DF}^T & \Sigma_F\end{bmatrix}^{-}\begin{bmatrix}\Sigma_{BD}^T \\ \Sigma_{BF}^T\end{bmatrix}. \qquad (12.25)$$

We use the following generalized inverse given in (11.38):

$$\begin{bmatrix} \Sigma_D & \Sigma_{DF} \\ \Sigma_{DF}^T & \Sigma_F \end{bmatrix}^{-} = \begin{bmatrix} \Sigma_D^{\dagger} + \Sigma_D^{\dagger}\Sigma_{DF}S^{\dagger}\Sigma_{DF}^T\Sigma_D^{\dagger} & -\Sigma_D^{\dagger}\Sigma_{DF}S^{\dagger} \\ -S^{\dagger}\Sigma_{DF}^T\Sigma_D^{\dagger} & S^{\dagger} \end{bmatrix} \tag{12.26}$$

Inserting this choice of generalized inverse into (12.25) yields the result (12.24) after some rearrangement. ■

Notice that S is uniquely the Schur complement of Σ_D in

$$M = \begin{bmatrix} \Sigma_D & \Sigma_{DF} \\ \Sigma_{DF}^T & \Sigma_F \end{bmatrix},$$

and is independent of the choice of generalized inverse of Σ_F; see (11.36).

We may write $K = \mathrm{Cov}_D(B, F)$ as in (12.3), and $S = \mathrm{Var}_D(F)$ in (12.2) as we have

$$\mathrm{Var}_D\left(\begin{bmatrix} B \\ F \end{bmatrix}\right) = \begin{bmatrix} \Sigma_B - \Sigma_{BD}\Sigma_D^{\dagger}\Sigma_{BD}^T & \Sigma_{BF} - \Sigma_{BD}\Sigma_D^{\dagger}\Sigma_{FD}^T \\ \Sigma_{BF}^T - \Sigma_{FD}\Sigma_D^{\dagger}\Sigma_{BD}^T & \Sigma_F - \Sigma_{BF}\Sigma_D^{\dagger}\Sigma_{BF}^T \end{bmatrix} \tag{12.27}$$

$$= \begin{bmatrix} \mathrm{Var}_D(B) & \mathrm{Cov}_D(B, F) \\ \mathrm{Cov}_D(F, B) & \mathrm{Var}_D(F) \end{bmatrix} \tag{12.28}$$

$$= \begin{bmatrix} \mathrm{Var}_D(B) & K \\ K^T & S \end{bmatrix}. \tag{12.29}$$

Remark. Note that 12.6.2–12.6.4 are necessary and sufficient conditions for the matrix (12.28) to be non-negative definite.

Corollary 12.53 *The partial adjusted expectation, i.e. the change in adjusted expectations* (5.4)*, may be obtained as follows, using the generalized inverse* (12.26)*:*

$$\begin{aligned} \mathrm{E}_{[F/D]}(B) &= \mathrm{E}_{D\cup F}(B) - \mathrm{E}_D(B) \\ &= KS^{\dagger}[(F - \mathrm{E}(F)) - \Sigma_{DF}^T\Sigma_D^{\dagger}(D - \mathrm{E}(D))] \qquad (12.30) \\ &= KS^{\dagger}\mathbb{A}_D(F). \end{aligned}$$

We can obtain the observed partial adjusted expectations by replacing F by f throughout (12.30).

Remark. As

$$\Sigma_{FD}\Sigma_D^{\dagger}(d - \mathrm{E}(D)) - (f - \mathrm{E}(F)) = \mathrm{E}_d(F) - f,$$

condition (12.7) of Theorem 12.10 is equivalent to requiring that $\mathrm{E}_d(F) - f$ be in the null space of $\mathrm{Var}_D(F)$.

12.11.1 Partial and relative adjustment transforms

Corollary 12.54 *The partial resolution transform matrix* (5.15) *can be calculated as follows.*

$$\mathbb{T}_{B:[F/D]} = \mathbb{T}_{B:D\cup F} - \mathbb{T}_{B:D} = \Sigma_B^{\dagger} K S^{\dagger} K^T. \tag{12.31}$$

It is important to note that this partial resolution transform matrix is not in general equal to the relative adjusted belief transform (5.26),

$$\mathbb{T}_{B:F(D)} = \mathbb{T}_{\mathbb{A}_D(B):\mathbb{A}_D(F)}, \tag{12.32}$$

the resolution matrix for the adjustment of B given F having already accounted for D in the sense that this resolution matrix relates to the space spanned by $\mathbb{A}_D(B)$ rather than the space spanned by B. The two matrices are related multiplicatively via

$$\mathbb{T}_{B:[F/D]} = (I - \mathbb{T}_{B:D})\mathbb{T}_{\mathbb{A}_D(B):\mathbb{A}_D(F)},$$

corresponding to (5.27) after some arrangement.

Theorem 12.55 *The partial resolution transform matrix* (5.15), (12.31) *and the relative adjusted belief transform matrix* (5.26), (12.32) *have the same rank, $r_{\mathbb{P}}$, defined as*

$$\mathbf{rk}\{\mathbb{T}_{B:F(D)}\} = \mathbf{rk}\{\mathbb{T}_{B:[F/D]}\} = \mathbf{rk}\{K\} = r_{\mathbb{P}}.$$

Proof. We have by Corollary 12.25 that

$$\mathbf{rk}\{\mathbb{T}_{B:[F/D]}\} = \mathbf{rk}\{\text{Cov}(B, \mathbb{A}_D(F))\} = \mathbf{rk}\{\text{Cov}_D(B, F)\} = \mathbf{rk}\{K\},$$

and similarly that

$$\begin{aligned}\mathbf{rk}\{\mathbb{T}_{B:F(D)}\} &= \mathbf{rk}\{\mathbb{T}_{\mathbb{A}_D(B):\mathbb{A}_D(F)}\} = \mathbf{rk}\{\text{Cov}(\mathbb{A}_D(B), \mathbb{A}_D(F))\} \\ &= \mathbf{rk}\{\text{Cov}_D(B, F)\} = \mathbf{rk}\{K\}.\end{aligned}$$

■

12.11.2 Calculating the partial bearing

Theorem 12.56 *The partial bearing is given by the change in bearing* (5.39), *given by*

$$\begin{aligned}\mathbb{Z}_{[f/d]}(B) &= \mathbb{Z}_{d\cup f}(B) - \mathbb{Z}_d(B) &(12.33)\\ &= (B - \text{E}(B))^T \Sigma_B^{\dagger} K S^{\dagger}[(f - \text{E}(F)) - \Sigma_{DF}^T \Sigma_D^{\dagger}(d - \text{E}(D))] &(12.34)\\ &= (B - \text{E}(B))^T \Sigma_B^{\dagger} K S^{\dagger} \mathbb{A}_d(f). &(12.35)\end{aligned}$$

Proof. By (12.23),

$$\mathbb{Z}_{d\cup f}(B) - \mathbb{Z}_d(B) = (B - \mathrm{E}(B))^T \Sigma_B^{\dagger}[\mathrm{E}_{d\cup f}(B) - \mathrm{E}_d(B)],$$

and $\mathrm{E}_{d\cup f}(B) - \mathrm{E}_d(B)$ is given by (12.30) with observations d, f inserted. ■

Corollary 12.57 *The partial bearing change is the quantity $X \in \langle B\rangle$ maximizing the size of the partial adjustment* (5.36)*:*

$$\frac{[\mathrm{E}_{d\cup f}(X) - \mathrm{E}_d(X)]^2}{\mathrm{Var}(X)}, \tag{12.36}$$

with maximum given by

$$(\mathrm{E}_{d\cup f}(B) - \mathrm{E}_d(B))^T \Sigma_B^{\dagger}(\mathrm{E}_{d\cup f}(B) - \mathrm{E}_d(B)) = \mathrm{Var}(\mathbb{Z}_{[f/d]}(B)). \tag{12.37}$$

Corollary 12.58 *The expected value of* (12.37)*, evaluated prior to observing $D = d$, $F = f$, is given by*

$$\mathbf{tr}\{\mathbb{T}_{B:D\cup F} - \mathbb{T}_{B:D}\}. \tag{12.38}$$

Corollary 12.59 *The covariance between the initial bearing and the partial bearing is*

$$\mathrm{Cov}(\mathbb{Z}_d(B), \mathbb{Z}_{d\cup f}(B) - \mathbb{Z}_d(B)) = (\mathrm{E}_d(B) - \mathrm{E}(B))^T \Sigma_B^{\dagger}(\mathrm{E}_{d\cup f}(B) - \mathrm{E}_d(B)),$$

with prior expected value

$$\mathrm{E}(\mathrm{Cov}(\mathbb{Z}_D(B), \mathbb{Z}_{D\cup F}(B) - \mathbb{Z}_D(B))) = 0.$$

12.12 Exchangeable adjustments

12.12.1 Notation

Suppose that we intend to adjust B by $\mathcal{S}_n(D)$, where $\mathcal{S}_n(D)$ is the mean of n exchangeable vectors $D_1, \ldots, D_n$. Suppose that B respects exchangeability with this sequence of vectors (see Definition 7.1). We will use the notation of §6.4 for the mean-component and residual-component variance matrices. Thus, suppose that each D_i has the representation

$$D_i = \mathcal{M}(D) + \mathcal{R}_i(D),$$

with variance specifications

$$\mathrm{Var}(\mathcal{M}(D)) = \Gamma, \qquad \mathrm{Var}(\mathcal{R}_i(D)) = [\Sigma - \Gamma], \quad i = 1, 2, \ldots, n,$$

so that $\mathcal{S}_n(D)$ has prior variance

$$\mathrm{Var}(\mathcal{S}_n(D)) = \Sigma_D = \Gamma + \frac{1}{n}[\Sigma - \Gamma].$$

Suppose also that $\Sigma_B = \text{Var}(B)$, $\Sigma_{BD} = \text{Cov}(B, D_i) = \text{Cov}(B, \mathcal{S}_n(D))$. The joint variance matrix over $B, \mathcal{S}_n(D)$ is thus

$$\begin{bmatrix} \Sigma_B & \Sigma_{BD} \\ \Sigma_{DB} & \Gamma + \frac{1}{n}[\Sigma - \Gamma] \end{bmatrix}. \tag{12.39}$$

Suppose that each vector D_i has observation d_i.

12.12.2 Coherence requirements for exchangeable adjustments

Coherence requirements for general exchangeable adjustments are as follows. We require the matrix (12.39) to be non-negative definite. However, a rather stronger condition is required: (12.39) must continue to remain non-negative definite as $n \to \infty$, and so we require the following property.

Definition 12.60 *Beliefs for an exchangeable adjustment are coherent if and only if* $[\Sigma - \Gamma]$ *is non-negative definite and*

$$\begin{bmatrix} \Sigma_B & \Sigma_{BD} \\ \Sigma_{DB} & \Gamma \end{bmatrix} \text{ is non-negative definite.} \tag{12.40}$$

Equivalently, we require $\text{Var}([B \;\; \mathcal{M}(D)]^T)$ to be non-negative definite. Coherence for (12.40) can be assessed via Lemma 12.3.

12.12.3 Data consistency

Definition 12.61 *By Definition 12.7, data for an exchangeable adjustment are consistent if and only if*

$$d_i \in \textbf{range}\{\Sigma\}, \quad i = 1, 2, \ldots, n.$$

Note that the condition must be met for every observation d_i, in that the fact of the mean data vector $\mathcal{S}_n(d)$ being consistent, $\mathcal{S}_n(d) \in \textbf{range}\{\Sigma\}$, does not imply that the individual observations are consistent.

12.12.4 Pure exchangeable adjustments

Pure exchangeable adjustments arise when $B = \mathcal{M}(D)$, so that interest is in learning about the underlying mean of a sequence of exchangeable vectors. Pure predictive exchangeable adjustments arise when $B = D_f$, $f > n$, for a future observation in the exchangeable sequence. We consider these two cases jointly as they share many features. We will suppose that Γ and $[\Sigma - \Gamma]$ are $k \times k$ non-negative definite matrices, so that D_i is a collection of k quantities. We will suppose that $\textbf{rk}\{\Gamma\} = m$ and $\textbf{rk}\{\Sigma\} = r$, where $m \le r \le k$.

12.12.4.1 Mean components

For adjusting mean components, we let $B = \mathcal{M}(D)$. Belief specifications are as follows:

$$\Sigma_B = \Gamma, \tag{12.41}$$

$$\Sigma_{BD} = \Sigma_{DB} = \Gamma, \tag{12.42}$$

$$\Sigma_D = \Gamma + \frac{1}{n}[\Sigma - \Gamma]. \tag{12.43}$$

By Definition 12.60, these specifications are coherent if $\Gamma, [\Sigma - \Gamma]$ are non-negative definite. For convenience, define

$$\mathbb{T}_n = \mathbb{T}_{\mathcal{M}(D):\mathcal{S}_n(D)} \tag{12.44}$$

as a matrix representation of the belief transform for this adjustment.

12.12.4.2 Predictive components

For adjusting the predictive components, let $B = D_f$, $f > n$. Belief specifications are as follows:

$$\Sigma_B = \Sigma, \tag{12.45}$$

$$\Sigma_{BD} = \Sigma_{DB} = \Gamma, \tag{12.46}$$

$$\Sigma_D = \Gamma + \frac{1}{n}[\Sigma - \Gamma]. \tag{12.47}$$

By Definition 12.60, these specifications are coherent if $\Gamma, [\Sigma - \Gamma]$ are non-negative definite. For convenience, define

$$\mathbb{T}_n^* = \mathbb{T}_{D_f:\mathcal{S}_n(D)} \tag{12.48}$$

as a matrix representation of the belief transform for this adjustment.

12.12.4.3 Computing the resolution transforms

The resolution transform matrices $\mathbb{T}_n$ and $\mathbb{T}_n^*$ can be computed as follows. Solve the generalized eigenvalue problem

$$\Gamma x = \lambda \Sigma x$$

as in §11.11.4, where we use the same notation with $A = \Gamma$ and $B = \Sigma$, and where we meet the requirement that $B - A = [\Sigma - \Gamma]$ is non-negative definite. In summary, suppose that Σ has $r \leq k$ positive eigenvalues $\psi_1 \geq \ldots \geq \psi_r > 0$ which we collect into the diagonal matrix Ψ. Suppose that corresponding to these eigenvalues are orthonormal eigenvectors $q_1, \ldots, q_r$ collected as the columns of the $k \times r$ matrix Q. Suppose that we construct $k - r$ orthonormal eigenvectors $q_{r+1}, \ldots, q_k$ corresponding to the zero eigenvalues of Σ. Let C be the $r \times r$ non-negative definite matrix

$$C = \Psi^{-\frac{1}{2}} Q^T \Gamma Q \Psi^{-\frac{1}{2}}.$$

Suppose that C has r orthonormal eigenvectors $y_1, y_2, \ldots, y_r$, arranged as columns of the $k \times r$ matrix Y, corresponding to eigenvalues

$$\lambda_1 \geq \lambda_2 \geq \ldots \geq \lambda_m > 0 \quad \text{and} \quad \lambda_{m+1} = \ldots = \lambda_r = 0,$$

and write

$$x_i = Q\Psi^{-\frac{1}{2}} y_i, \quad X = Q\Psi^{-\frac{1}{2}} Y. \tag{12.49}$$

Arrange $x_1, x_2, \ldots, x_r$ as the columns of the $k \times r$ matrix X. Note that the eigenvectors $y_1, \ldots, y_m$ form a basis for Γ. Arrange the ordered eigenvalues $\lambda_1, \lambda_2, \ldots, \lambda_r$ as the values of the diagonal matrix Λ, and the m positive eigenvalues as the values of the diagonal matrix Λ^*.

Lemma 12.62 *For $n > 0$,*

$$\left(\Gamma + \frac{1}{n}[\Sigma - \Gamma]\right)^{\dagger} = X \Delta_n X^T, \tag{12.50}$$

where

$$\Delta_n = \begin{bmatrix} n(I_m + (n-1)\Lambda^*)^{-1} & 0 \\ 0 & nI_{r-m} \end{bmatrix}. \tag{12.51}$$

This follows by Theorem 11.57.

Theorem 12.63 *A matrix representation for the resolution transform T_n^* is given by*

$$T_n^* = X \Lambda \Delta_n X^T \Gamma, \tag{12.52}$$

and T_n^* has eigenvector matrix X corresponding to eigenvalues as the elements of the diagonal matrix $\Lambda^2 \Delta_n$.

Proof. By Theorem 12.22, a matrix representation for the belief transform T_n^* for this pure exchangeable case is given by

$$T_n^* = \Sigma^{\dagger}\Gamma \left(\Gamma + \frac{1}{n}[\Sigma - \Gamma]\right)^{\dagger} \Gamma + [I - \Sigma^{\dagger}\Sigma]H_2, \tag{12.53}$$

where H_2 is an arbitrary $k \times k$ matrix. We will use the representation offered by taking $H_2 = 0$, and take

$$T_n^* = \Sigma^{\dagger}\Gamma \left(\Gamma + \frac{1}{n}[\Sigma - \Gamma]\right)^{\dagger} \Gamma \tag{12.54}$$

$$= XX^T \Gamma X \Delta_n X^T \Gamma \tag{12.55}$$

$$= X \Lambda \Delta_n X^T \Gamma, \tag{12.56}$$

as $X^T \Gamma X = \Lambda$. This provides the representation. For the eigenstructure we have

$$T_n^* X = X \Lambda \Delta_n X^T \Gamma X \tag{12.57}$$

$$= X \Lambda \Delta_n \Lambda. \tag{12.58}$$

Consequently, $x_1, \ldots, x_m$ are eigenvectors of T_n^* corresponding to eigenvalues

$$\frac{n\lambda_i^2}{1 + (n-1)\lambda_i};$$

and $x_{m+1}, \ldots, x_r$ are eigenvectors of T_n^* corresponding to zero eigenvalues. ■

Theorem 12.64 *Two alternative matrix representations of the resolution transform* $\mathbb{T}_n$ *are*

$$\mathbb{T}_n^a = \Gamma^\dagger \Gamma X \Delta_n X^T \Gamma, \tag{12.59}$$

$$\mathbb{T}_n^b = X \Delta_n X^T \Gamma. \tag{12.60}$$

These two representations are identical if $\mathbf{rk}\{\Gamma\} = \mathbf{rk}\{\Sigma\}$. *Both representations have eigenvalues as the elements of the diagonal matrix* $\Delta_n \Lambda$. $\mathbb{T}_n^a$ *has eigenvector matrix* $\Gamma^\dagger \Gamma X$, *whilst* $\mathbb{T}_n^b$ *has eigenvector matrix* X.

Proof. By Theorem 12.22, a matrix representation for the belief transform $\mathbb{T}_n$ for this pure exchangeable case is given by

$$\mathbb{T}_n = \Gamma^\dagger \Gamma \left(\Gamma + \frac{1}{n}[\Sigma - \Gamma] \right)^\dagger \Gamma + (I - \Gamma^\dagger \Gamma) H_3,$$

where H_3 is an arbitrary $k \times k$ matrix. We obtain T_n^a by taking $H_3 = 0$ and we obtain T_n^b by taking $H_3 = X \Delta_n X^T \Gamma$, employing Lemma 12.62 in each case. That these representations are the same if $\mathbf{rk}\{\Gamma\} = \mathbf{rk}\{\Sigma\}$ follows by Lemma 11.28 and because $\mathbf{range}\{X\} = \mathbf{range}\{\Gamma\}$ in this case. For the eigenstructure, we have

$$\mathbb{T}_n^b X = X \Delta_n X^T \Gamma X \tag{12.61}$$

$$= X \Delta_n \Lambda, \tag{12.62}$$

$$\text{and } \mathbb{T}_n^a X = \Gamma^\dagger \Gamma X \Delta_n X^T \Gamma X \tag{12.63}$$

$$= \Gamma^\dagger \Gamma X \Delta_n \Lambda, \tag{12.64}$$

$$\text{so that } \mathbb{T}_n^a \Gamma^\dagger \Gamma X = \Gamma^\dagger \Gamma X \Delta_n \Lambda, \tag{12.65}$$

as $\mathbb{T}_n^a \Gamma^\dagger \Gamma = \mathbb{T}_n^a$ because $\Gamma \Gamma^\dagger \Gamma = \Gamma$. Thus, $x_1, x_2, \ldots, x_m$ are eigenvectors of $\mathbb{T}_n^b$, and they correspond to eigenvalues

$$\frac{n\lambda_i}{1 + (n-1)\lambda_i}.$$

Additionally, $x_{m+1}, \ldots, x_r$ are eigenvectors of $\mathbb{T}_n^b$ corresponding to zero eigenvalues. $\mathbb{T}_n^a$ has the same eigenvalues, but transformed eigenvectors. ■

For the mean-component adjustment, to obtain eigenvectors with prior variance unity, we scale X appropriately by using instead the eigenvectors $X\Lambda^{-\frac{1}{2}}$. For the predictive component adjustment, the eigenvectors already have prior variance unity. For exchangeable adjustments it may be more natural to choose the representation $\mathbb{T}_n^b$ as it has the same algebraic eigenvectors as $\mathbb{T}_n^*$ when $\mathbf{rk}\{\Gamma\} < \mathbf{rk}\{\Sigma\}$.

12.12.5 General exchangeable adjustments

Theorem 12.65 *For a general second-order exchangeable adjustment the adjusted expectation, adjusted variance, and resolution transform matrix can be calculated as*

$$\mathrm{E}_n(B) = \mathrm{E}(B) + \Sigma_{BD} X \Delta_n X^T [\mathcal{S}_n(D) - \mathrm{E}(D)], \tag{12.66}$$

$$\mathrm{Var}_n(B) = \Sigma_B - \Sigma_{BD} X \Delta_n X^T \Sigma_{DB}, \tag{12.67}$$

$$\mathbb{T}_{B:\mathcal{S}_n(D)} = \Sigma_B^{\dagger} \Sigma_{BD} X \Delta_n X^T \Sigma_{DB}, \tag{12.68}$$

where X, Δ_n are as given in (12.49) *and Lemma 12.62.*

Thus, all the quantities of interest depend on the sample size n only through the values of the averages $\mathcal{S}_n(D)$ and the eigenvalue-type quantities Δ_n defined in (12.51) in Lemma 12.62.

Lemma 12.66

$$\mathbf{tr}\{\mathbb{T}_n\} = \sum_{i=1}^{m} \delta_{in} t_i^*, \tag{12.69}$$

where $t_1^*, \ldots, t_m^*$ are the diagonal elements of $X^T \Sigma_{DB} \Sigma_B^{\dagger} \Sigma_{BD} X$.

Lemma 12.67 *The maximum value of* (12.69), *given by taking an infinite sample size, is*

$$\mathbf{tr}\{\mathbb{T}_\infty\} = \lim_{n\to\infty} \mathbb{T}_n = \sum_{i=1}^{m} \frac{t_i^*}{\lambda_i} = \phi. \tag{12.70}$$

Theorem 12.68 *For a general second-order exchangeable adjustment with resolution transform $\mathbb{T}_n$,*

$$\mathbf{tr}\{\mathbb{T}_{n_{\min}}\} \le \beta \le \mathbf{tr}\{\mathbb{T}_{n_{\max}}\},$$

where β is any value such that

$$0 < \beta < \phi = \sum_{i=1}^{m} \frac{t_i^*}{\lambda_i},$$

$$n_{\min} = \mathbf{ceiling}\left\{\frac{1-\lambda_1}{\lambda_1}\frac{\beta}{\phi-\beta}\right\},$$

$$n_{\max} = \mathbf{floor}\left\{\frac{1-\lambda_m}{\lambda_m}\frac{\beta}{\phi-\beta}\right\}.$$

Proof.

$$\mathbf{tr}\{\mathbb{T}_n\} = \sum_{i=1}^{m} \lambda_{i(n)} \frac{t_i^*}{\lambda_i},$$

where we have the ordering

$$1 \geq \lambda_{1(n)} \geq \ldots \geq \lambda_{m(n)} > 0.$$

Thus

$$\sum_{i=1}^{m} \lambda_{m(n)} \frac{t_i^*}{\lambda_i} \leq \mathbf{tr}\{\mathbb{T}_n\} \leq \sum_{i=1}^{m} \lambda_{1(n)} \frac{t_i^*}{\lambda_i}$$

$$\Rightarrow \quad \phi\lambda_{m(n)} \leq \mathbf{tr}\{\mathbb{T}_n\} \leq \phi\lambda_{1(n)},$$

as the values t_i^* are non-negative because they are the diagonal values of a non-negative definite matrix. We thus require $n_{\min}$ such that $\beta \geq \phi\lambda_{1(n_{\min})}$ and $n_{\max}$ such that $\beta \leq \phi\lambda_{m(n_{\max})}$. Applying (6.57) in each case now gives the result. ■

Corollary 12.69 *The smallest sample size n guaranteeing* $\mathbf{tr}\{\mathbb{T}_n\} \geq \beta$ *is bounded by* $n_{\min} \leq n \leq n_{\max}$.

Proof. This follows directly from Theorem 12.68, and because $\mathbf{tr}\{\mathbb{T}_n\}$ is monotone non-decreasing in n. ■

Corollary 6.6 provides sample sizes required for a specified resolution of variance for a single canonical quantity.

Lemma 12.70 *Let* $L = \Sigma_{BD}X$. *For each element* $B_j \in B$, *the resolved variance is*

$$\text{RVar}_n(B_j) = \sum_{i=1}^{m} \delta_{in} L_{ij}^2. \tag{12.71}$$

The resolved variance thus depends on sample size n only through the quantities Δ_n (12.51).

Lemma 12.71 *The maximum value of* (12.71), *given by taking an infinite sample size, is*

$$\text{RVar}_\infty(B_j) = \lim_{n\to\infty} \text{RVar}_n(B_j) = \sum_{i=1}^{m} \frac{L_{ij}^2}{\lambda_i} = \phi_j. \tag{12.72}$$

Theorem 12.72 *For a general second-order exchangeable adjustment,*

$$\text{RVar}_{n_{j\min}}(B_j) \leq \beta_j \leq \text{RVar}_{n_{j\max}}(B_j),$$

where β_j *is any value such that*

$$0 < \beta_j < \phi_j = \sum_{i=1}^{m} \frac{L_{ij}^2}{\lambda_i},$$

$$n_{j\min} = \mathbf{ceiling}\left\{\frac{1-\lambda_1}{\lambda_1}\frac{\beta_j}{\phi_j - \beta_j}\right\},$$

$$n_{j\max} = \mathbf{floor}\left\{\frac{1-\lambda_m}{\lambda_m}\frac{\beta_j}{\phi_j - \beta_j}\right\}.$$

Proof. This follows as in the proof of Theorem 12.69. ■

Corollary 12.73 *The sample size guaranteeing* $\mathrm{RVar}_n(B_j) \geq \beta_j$ *is bounded by*

$$n_{j\min} \leq n \leq n_{j\max}.$$

12.13 Implementing comparisons of belief

12.13.1 Expectation comparisons

Here we address briefly the algebraic and geometric constructions for the belief comparison bearings discussed in §9.6. Using the notation therein, we wish to maximize (9.31) and (9.32), under the constraint that $\mathrm{Var}_{\mathbf{H}_1}(X)$ and $\mathrm{Var}_{\mathbf{H}_2}(X)$ are positive definite. Suppose we let $B = \mathrm{Var}_{\mathbf{H}_1}(X)$ and $A = \mathrm{Var}_{\mathbf{H}_2}(X)$. Then, via §11.11.4, we can form generalized eigenvectors W corresponding to eigenvalues Λ to solve the generalized eigenvalue problem $AW = BW\Lambda$, where the eigenvalues are all positive and where the eigenvectors are all invertible and normalized such that $W^TAW = \Lambda$ and $W^TBW = I$. Further, writing $G = W^{-1}$ for convenience, we have $B = G^TG$ and $A = G^T\Lambda G$. Thus, writing $c = Gh$ (i.e. $h = Wc$), we have

$$DE_{12}(X) = \max_h \frac{[h^T[\mathrm{E}_{\mathbf{H}_2}(X) - \mathrm{E}_{\mathbf{H}_1}(X)]]^2}{h^T\mathrm{Var}_{\mathbf{H}_1}(X)h} \tag{12.73}$$

$$= \max_c \frac{[c^T[\mathrm{E}_{\mathbf{H}_2}(W^TX) - \mathrm{E}_{\mathbf{H}_1}(W^TX)]]^2}{c^Tc} \tag{12.74}$$

$$= [\mathrm{E}_{\mathbf{H}_2}(W^TX) - \mathrm{E}_{\mathbf{H}_1}(W^TX)]^T[\mathrm{E}_{\mathbf{H}_2}(W^TX) - \mathrm{E}_{\mathbf{H}_1}(W^TX)] \tag{12.75}$$

by Theorem 11.49, so that W^TX corresponds to Z of §9.6 in (9.38). Similarly,

$$DE_{21}(X) = \max_h \frac{[h^T[\mathrm{E}_{\mathbf{H}_2}(X) - \mathrm{E}_{\mathbf{H}_1}(X)]]^2}{h^T\mathrm{Var}_{\mathbf{H}_2}(X)h} \tag{12.76}$$

$$= \max_c \frac{[c^T[\mathrm{E}_{\mathbf{H}_2}(W^TX) - \mathrm{E}_{\mathbf{H}_1}(W^TX)]]^2}{c^T\Lambda c} \tag{12.77}$$

$$= [\mathrm{E}_{\mathbf{H}_2}(W^TX) - \mathrm{E}_{\mathbf{H}_1}(W^TX)]^T\Lambda^{-1}[\mathrm{E}_{\mathbf{H}_2}(W^TX) - \mathrm{E}_{\mathbf{H}_1}(W^TX)] \tag{12.78}$$

by Theorem 11.49, giving the result which corresponds to (9.39). Clearly we have that $\mathrm{Var}_{\mathbf{H}_1}(W^TX) = I$ and $\mathrm{Var}_{\mathbf{H}_2}(W^TX) = \Lambda$, showing that W^TX satisfies the normalizations of §9.6.

12.13.2 Comparison of exchangeable beliefs

Consider two sets of belief specifications for a sample of n exchangeable vectors $X = X_1, \ldots, X_n$, where we have the representation

$$X_i = \mathcal{M}(X) + \mathcal{R}_i(X), \quad i = 1, 2, \ldots,$$

and where each X_i contains r quantities $X_{i1}, \ldots, X_{ir}$. Suppose that we specify

$$\text{Var}_{\mathbf{H}_1}(\mathcal{M}(X)) = \Gamma_1, \qquad \text{Var}_{\mathbf{H}_1}(\mathcal{R}_i(X_i)) = \Sigma_1 - \Gamma_1, \quad \forall i, \tag{12.79}$$

$$\text{Var}_{\mathbf{H}_2}(\mathcal{M}(X)) = \Gamma_2, \qquad \text{Var}_{\mathbf{H}_2}(\mathcal{R}_i(X_i)) = \Sigma_2 - \Gamma_2, \quad \forall i, \tag{12.80}$$

where $\Gamma_1, \Gamma_2, \Sigma_1, \Sigma_2, \Sigma_1 - \Gamma_1, \Sigma_2 - \Gamma_2$, are all non-negative definite $r \times r$ matrices. To simplify the construction, we will require $\Sigma_2 - \Gamma_2$ to be positive definite; if this is not the case, the construction is more difficult but can be circumvented, for example, by transforming the quantities into $\mathbf{range}\{\Sigma_2 - \Gamma_2\}$. Gather the quantities into the vector $X = [X_1^T \; X_2^T \; \ldots \; X_n^T]^T$. The two specifications lead to variance matrices

$$\text{Var}_{\mathbf{H}_1}(X) = \mathbf{I}_n \otimes (\Sigma_1 - \Gamma_1) + \mathbf{J}_n \otimes \Gamma_1, \tag{12.81}$$

$$\text{Var}_{\mathbf{H}_2}(X) = \mathbf{I}_n \otimes (\Sigma_2 - \Gamma_2) + \mathbf{J}_n \otimes \Gamma_2, \tag{12.82}$$

using the direct product notation of §11.12.2. To compare the belief specifications, we must solve the generalized eigenvalue problem

$$\text{Var}_{\mathbf{H}_1}(X)U = \text{Var}_{\mathbf{H}_2}(X)U\Phi,$$

where U are the generalized eigenvectors and Φ the corresponding eigenvalues. We can obtain the eigenstructure as follows, provided that $\Sigma_2 - \Gamma_2$ is positive definite:

$$\text{Var}_{\mathbf{H}_1}(X)U = \text{Var}_{\mathbf{H}_2}(X)U\Phi, \tag{12.83}$$

$$\Rightarrow \text{Var}_{\mathbf{H}_2}(X)^{-1}\text{Var}_{\mathbf{H}_1}(X)U = U\Phi, \tag{12.84}$$

$$\Rightarrow [\mathbf{I}_n \otimes G - \mathbf{J}_n \otimes L][\mathbf{I}_n \otimes (\Sigma_1 - \Gamma_1) + \mathbf{J}_n \otimes \Gamma_1]U = U\Phi, \tag{12.85}$$

by Lemma 11.61, where

$$G = (\Sigma_2 - \Gamma_2)^{-1}, \tag{12.86}$$

$$L = (\Sigma_2 + (n-1)\Gamma_2)^{-1}\Gamma_2(\Sigma_2 - \Gamma_2)^{-1}, \tag{12.87}$$

$$\Rightarrow [\mathbf{I}_n \otimes G(\Sigma_1 - \Gamma_1) + \mathbf{J}_n \otimes [G\Gamma_1 - L(\Sigma_1 + (n-1)\Gamma_1)]]U = U\Phi, \tag{12.88}$$

$$\Rightarrow [\mathbf{I}_n \otimes (A - B) + \mathbf{J}_n \otimes B]U = U\Phi, \tag{12.89}$$

where

$$A - B = G(\Sigma_1 - \Gamma_1), \tag{12.90}$$

$$A = G\Sigma_1 - L(\Sigma_1 + (n-1)\Gamma_1), \tag{12.91}$$

$$B = G\Gamma_1 - L(\Sigma_1 + (n-1)\Gamma_1). \tag{12.92}$$

Now it follows from Lemma 11.62 that the eigenstructure of (12.89) can be obtained from the eigenstructure of $A - B$ and $A + (n-1)B$, where

$$A - B = G(\Sigma_1 - \Gamma_1) \tag{12.93}$$

$$= (\Sigma_2 - \Gamma_2)^{-1}(\Sigma_1 - \Gamma_1), \tag{12.94}$$

$$A + (n-1)B = (\Sigma_2 + (n-1)\Gamma_2)^{-1}(\Sigma_1 + (n-1)\Gamma_1). \tag{12.95}$$

This leads us to the following result.

Theorem 12.74 *Suppose that we obtain the generalized eigenstructure*

$$(\Sigma_1 - \Gamma_1)W = (\Sigma_2 - \Gamma_2)W\Delta \tag{12.96}$$

$$\textit{i.e. } \mathrm{Var}_{\mathbf{H}_1}(\mathcal{R}_i(X))W = \mathrm{Var}_{\mathbf{H}_2}(\mathcal{R}_i(X))W\Delta \tag{12.97}$$

$$\textit{and } (\Sigma_1 + (n-1)\Gamma_1)U = (\Sigma_2 + (n-1)\Gamma_2)U\Lambda \tag{12.98}$$

$$\textit{i.e. } \mathrm{Var}_{\mathbf{H}_1}(\mathcal{M}(X))U = \mathrm{Var}_{\mathbf{H}_2}(\mathcal{M}(X))U\Lambda, \tag{12.99}$$

where we can choose U, W to be orthonormal. Then the generalized eigenstructure (12.83) *is given by*

$$\Phi = \Lambda \oplus \Delta \oplus \ldots \oplus \Delta,$$

where there are $r-1$ terms Δ, so that each eigenvalue δ_i is of multiplicity $r-1$. The corresponding eigenvector matrix is

$$U = \begin{bmatrix} h_1 \otimes V & h_2 \otimes W & h_3 \otimes W & \ldots & h_n \otimes W \end{bmatrix},$$

where h_i is the ith *column of the Helmert matrix $\mathbf{H}_n$ (see Definition 11.58).*

Notice that (12.96) provides the canonical structure for the comparison of variance specifications for the sample averages, whilst (12.98) provides the canonical structure for the comparison of variance specifications for the residual structures. Therefore, to obtain the canonical structure for the comparison of exchangeable beliefs, it is necessary only to make the comparison for the sample averages (which does depend on the sample size) and the comparison for one residual structure (which does not depend on the sample size). Further, the two comparisons can be made separately.

A

Notation

Notation used in the book is briefly as follows, with page numbers showing the first, or main, definition.

$A^{\perp}$	The orthogonal part of a matrix A.	434
$A^{\dagger}$	The Moore–Penrose generalized inverse of matrix A.	432
A^{-}	A generalized inverse of the matrix A.	432
$\dot{a}_d$	Coefficients for the discrepancy vector for data d.	99
$\ddot{a}_d$	Coefficients for the adjustment discrepancy vector for data d.	106
$\lfloor A \perp\!\!\!\perp B \rfloor / C$	Collection A is separated from collection B by collection C.	167
$\mathbb{A}_D(B)$	Adjusted version of B given D.	57
$\langle C \rangle$	The collection of linear combinations of elements of the collection C.	75
$\{C\}$	The base of a collection.	82
$[C]$	The belief structure over the collection C.	82
$\mathrm{c}_i(X)$	Resolution in X contributed by the ith canonical direction.	79
$\mathrm{CF}(A)$	The correlation matrix derived from the variance matrix A.	285
$\mathrm{Corr}(X, Y)$	The correlation matrix for vectors X, Y.	8
$\mathrm{Corr}_D(X, Y)$	The correlation between X, Y in the adjusted variance matrix given by adjusting by D.	130
$\mathrm{Cov}(X, Y)$	The covariance matrix for vectors X, Y.	8
$\mathrm{Cov}_D(X, Y)$	The adjusted covariance between X, Y given by adjusting by D.	58
$\mathrm{CR}(\mathcal{M}(V))$	The prior correlation matrix for a vector $\mathcal{M}(V)$ of population residual variances.	285

Bayes Linear Statistics: Theory and Methods M. Goldstein and D.A. Wooff

$\mathrm{CR}_n(\mathcal{M}(V))$	The updated correlation matrix for a vector $\mathcal{M}(V)$ of population residual variances.	285
$\mathrm{DE}_{\frac{2}{1}}(B)$	Maximum squared difference in expectation over collection B between two specifications $\mathbf{H}_1$ and $\mathbf{H}_2$, relative to variances under $\mathbf{H}_1$.	312
$\mathbf{diag}\{\cdot\}$	The diagonal matrix with diagonal values as given.	234
$\mathrm{Dis}(d)$	Discrepancy in a collection $D = d$.	96
$\mathrm{Dis}_d(B)$	Adjustment discrepancy for B given adjustment by $D = d$.	105
$\mathrm{Dr}(d)$	Discrepancy ratio for a collection $D = d$.	97
$\mathrm{Dr}_d(B)$	Adjustment discrepancy ratio for B given adjustment by $D = d$.	106
$\mathrm{DV}_{\frac{2}{1}}(B)$	Maximal variance ratio $\mathbf{H}_2$:$\mathbf{H}_1$ for comparing variance specifications over collection B.	294
$\mathrm{E}(X)$	The expectation vector for the vector X.	8
$\mathrm{E}_D(B)$	Adjusted expectation for the vector B adjusted by the vector D.	64
$\mathrm{E}_d(B)$	Observed adjusted expectation for the vector B adjusted by the vector $D = d$.	104
$\mathrm{E}_{[F/D]}(B)$	The partial adjustment of B by F given D.	126
$\mathrm{E}_{(n)}(\mathcal{M}(V))$	The semi-adjusted residual variance matrix.	286
$\mathrm{G}_{\frac{2}{1}}$	The bearings for the belief comparison of $\mathbf{H}_1$ and $\mathbf{H}_2$, norming according to variances under $\mathbf{H}_1$.	312
$\dot{h}_d$	Coefficients for the bearing for data d.	113
$\mathbb{H}(D/B)$	The heart of the transform for the adjustment of B by D.	81
$\mathbb{I}$	The identity operator or identity matrix.	83
$\mathrm{Kur}(X)$	The kurtosis for the random quantity X.	268
$\mathcal{M}(X)$	The mean component for the exchangeable sequence $X_1, X_2, \ldots$.	185
$\mathbf{null}\{A\}$	Null space of a matrix or transform.	435
$\mathrm{P}(X)$	Prevision (probability or expectation) for the random quantity X.	34
$\mathrm{Pa}(B)$	The parents of node B.	356
$\mathrm{PC}(d, f)$	The path correlation between two data sources d, f, for adjusting a third collection.	138
$\mathrm{RA}_F(B/D)$	The relative adjustment ratio for B by F given prior adjustment by D.	129
$\mathbf{range}\{A\}$	The range of a matrix.	435
r_B	The rank of the variance matrix for the collection B.	76
$\mathrm{RCorr}_D(X, Y)$	The correlation between X, Y in the resolved variance matrix given by adjusting by D.	107

$\mathrm{RCov}_D(X, Y)$	The resolved covariance between X, Y given by adjusting by D.	58
$\mathrm{R}_d(X)$	Resolution in the quantity X given adjustment by $D = d$.	58
$\mathcal{R}_i(X)$	The residual component for the ith individual of the exchangeable sequence $X_1, X_2, \ldots$.	185
$\mathbf{rk}\{A\}$	Column rank of a matrix or transform A.	431
$\mathrm{R}_{[F/D]}(X)$	The partial resolution in X given by adjusting by F in addition to D.	128
$r_{\mathbb{P}}$	The rank of the partial resolution transform matrix.	135
$r_{\mathbb{T}}$	The rank of the resolution transform matrix.	76
$\mathrm{RU}_D(B)$	Residual uncertainty in the collection B given adjustment by D.	79
$\mathrm{RVar}_D(X)$	The resolved variance matrix for B given by adjusting by D.	57
$\mathcal{S}_n(\cdot)$	An average of n terms in an exchangeable sequence.	206
S_n^2	The sample variance matrix.	282
$\mathrm{S}(X)$	Standardized version of X.	8
$\mathrm{S}(x)$	Standardized version of the observed value x.	95
$\mathrm{S}_d(X)$	Standardized value of the adjustment of X by $D = d$.	105
$\mathrm{Size}_d(B)$	Size of the adjustment of the collection B by $D = d$.	114
$\mathrm{Sr}_d(B)$	Size ratio for the adjustment of the collection B by $D = d$.	116
$\mathbb{S}_{B:D}$	The adjusted belief transform matrix for the adjustment of B by D.	81
$\mathbb{S}_{B:F(D)}$	The relative adjusted belief transform for $[B/D]$ given F.	130
$\mathbb{T}_{B:D}$	The resolution transform matrix for the adjustment of B by D.	77
$\mathbb{T}_{B:[F/D]}$	The partial resolution transform matrix for the adjustment of B by F in addition to D.	128
$\mathbb{T}_{B:F(D)}$	The relative resolution transform of B by F given D.	130
$\mathbf{tr}\{A\}$	The trace of a matrix or transform A.	431
$\mathrm{Var}(X)$	The variance–covariance matrix for the vector X.	8
$\mathrm{Var}_D(B)$	The adjusted variance matrix for B given by adjusting by D.	57
$\dot{W}_d$	Discrepancy vector for data d.	99
$\ddot{W}_d$	Adjustment discrepancy vector for data d.	106
$\dot{w}_d$	Observed discrepancy vector for data d.	99
$\ddot{\mathbb{Y}}_d(B)$	The discrepancy vector in B induced by the adjustment by $D = d$.	106
$\mathbb{Z}_d(B)$	The bearing for collection B given data d.	114

$\mathbb{Z}_{[f/d]}(B)$	The partial bearing for B given initial adjustment by data d and then partial adjustment by data f.	135
$\mathbb{Z}(B, D)$	The joint bearing for collections B and D.	114
ζ_i	Partial canonical resolution.	128

B

Index of examples

Oral glucose tolerance test

Simple one-dimensional problem

Algebraic example

Regression with correlated bivariate responses

Analysing exchangeable regressions

A population dynamics experiment

Comparison of degenerate variance matrices

Prediction in a large brewery

Anscombe data sets

Other examples

C

Software for Bayes linear computation

C.1 [B/D]

[B/D] is the computer implementation of the Bayes linear methodology developed by us, initially at the University of Hull and thereafter at the University of Durham. The package is freely available as Wooff and Goldstein (2000b). Manuals and software guides are available as Wooff (2000b), Goldstein (2000), Wooff (2000a), and Wooff and Goldstein (2000a). A brief description of the language, with examples, may also be found in Goldstein and Wooff (1995). Note that this package provides a programming language and does not have a graphical user interface.

Most of the calculations in this book were carried out using [B/D]. Some of the graphics shown in Chapter 10 are produced directly from [B/D]. The remaining graphics in the book were produced by the statistical package R (R Development Core Team 2006) using computations imported from [B/D].

C.2 BAYES-LIN

BAYES-LIN, written by Darren Wilkinson, is an object-oriented environment for Bayes linear local computation. It is intended for people who already know about Bayes linear methods, graphical modelling and local computation, and want a collection of object-oriented programming tools for carrying out computations.

BAYES-LIN is a set of modules for the XLISP-STAT statistical programming environment, and hence assumes some familiarity with that system and with the basic concepts of object-oriented programming. Local computation in BAYES-LIN is achieved via message-passing between objects representing clique-tree nodes. It

Bayes Linear Statistics: Theory and Methods M. Goldstein and D.A. Wooff

uses directed acyclic graph nodes as the basis for model specification, and automatically constructs an appropriate junction tree for computation. The underlying theory is given in Goldstein and Wilkinson (2000), and Wilkinson (1998) describes the object-oriented approach to Bayes linear local computation. The package is available as Wilkinson (2000).

References

Anscombe FJ (1973) Graphs in statistical analysis. *American Statistician* **27**, 17–21.

Arthur W and Farrow M (1987) On detecting interactions between species in population dynamics. *Biological Journal of the Linnean Society* **32**, 271–279.

Bachman G and Narici L (1966) *Functional Analysis*. Academic Press, New York.

Basilevsky A (1983) *Applied Matrix Algebra in the Statistical Sciences*. North-Holland, New York.

Bassett EE, Bremner JM, Jolliffe IT, Jones B, Morgan BJT and North PM (2000) *Statistics: Problems and Solutions*, 2nd edn. World Scientific, Singapore.

Bernardo JM and Smith AFM (1994) *Bayesian theory*. John Wiley & Sons, Ltd, Chichester.

Box GEP and Jenkins GM (1970) *Time Series Analysis: Forecasting and Control*. Holden-Day, San Francisco.

Box GEP and Tiao GC (1973) *Bayesian Inference in Statistical Analysis*. Addison-Wesley, Reading, MA.

Campbell SL and Meyer CD (1991) *Generalized Inverses of Linear Transformations*. Dover, New York.

Coolen FPA, Goldstein M and Munro M (2001) Generalized partition testing via Bayes linear methods. *Information and Software Technology* **43**, 783–793.

Cowell RG, Dawid AP, Lauritzen SL and Spiegelhalter DJ (1999) *Probabilistic Networks and Expert Systems*. Springer-Verlag, Berlin.

Craig PS, Goldstein M, Seheult AH and Smith JA (1996) Bayes linear strategies for history matching of hydrocarbon reservoirs. In *Bayesian Statistics 5* (ed. Bernardo JM *et al.*), pp. 69–98. Oxford University Press, Oxford.

Craig PS, Goldstein M, Seheult AH and Smith JA (1997) Pressure matching for hydrocarbon reservoirs: a case study in the use of Bayes linear strategies for large computer experiments (with discussion). In *Case Studies in Bayesian Statistics Volume III* (ed. Gatsonis C, Hodges JS, Kass RE, McCulloch R, Rossi P and Singpurwalla ND), pp. 37–93. Springer-Verlag, New York.

Craig PS, Goldstein M, Seheult AH and Smith JA (1998) Constructing partial prior specifications for complex physical models (with discussion). *The Statistician* **47**, 37–68.

Craig PS, Goldstein M, Rougier JC and Seheult AH (2001) Bayesian forecasting using large computer models. *Journal of the American Statistical Association* **96**, 717–729.

De Finetti B (1937) Foresight: its logical laws, its subjective sources. *Annales de l'Institut Henri Poincaré*.

Bayes Linear Statistics: Theory and Methods M. Goldstein and D.A. Wooff

De Finetti B (1974) *Theory of Probability, vol. 1*. John Wiley & Sons, Inc., New York.

De Finetti B (1975) *Theory of Probability, vol. 2*. John Wiley & Sons, Inc., New York.

Draper NR and Smith H (1998) *Applied Regression Analysis*, 3rd edn. John Wiley & Sons, Inc., New York.

Farrow M (2003) Practical building of subjective covariance structures for large complicated systems. *The Statistician* **52**(4), 553–574.

Farrow M and Goldstein M (1992) Reconciling costs and benefits in experimental design. In *Bayesian Statistics 4* (ed. Bernardo JM *et al.*). Oxford University Press, Oxford.

Farrow M and Goldstein M (1993) Bayes linear methods for grouped multivariate repeated measurement studies with application to crossover trials. *Biometrika* **80**(1), 39–59.

Farrow M and Goldstein M (1996) Diagnostic geometry for Bayes linear prediction systems. In *Bayesian Statistics 5* (ed. Bernardo JM *et al.*), pp. 561–568. University Press, Oxford.

Farrow M and Goldstein M (2006) Trade-off sensitive experimental design: a multicriterion, decision theoretic, Bayes linear approach. *Journal of Statistical Planning and Inference* **136**, 498–526.

Farrow M and Leyland AH (1991) Interpretation of oral glucose tolerance test results. In *Statistics in medicine* (ed. Dunstan F and Pickles J), pp. 249–266. Oxford University Press, Oxford.

Farrow M, Goldstein M and Spiropoulos T (1997) Developing a Bayes linear decision support system for a brewery. In *The Practice of Bayesian analysis* (ed. French S and Smith JQ), pp. 71–106. Edward Arnold, London.

Garthwaite PH, Kadane JB and O'Hagan A (2005) Statistical methods for eliciting prior distributions. *Journal of the American Statistical Association* **100**, 680–701.

Goldstein M (1974) Approximate Bayesian inference with incompletely specified prior distributions. *Biometrika* **61**, 629–631.

Goldstein M (1975a) Approximate Bayesian solutions to some nonparametric problems. *Annals of Statistics* **3**, 512–517.

Goldstein M (1975b) A note on some Bayesian nonparametric estimates. *Annals of Statistics* **3**, 736–740.

Goldstein M (1976) Bayesian analysis of regression problems. *Biometrika* **63**, 51–58.

Goldstein M (1979) The variance modified linear Bayes estimator. *Journal of the Royal Statistical Society, Series B* **41**, 96–100.

Goldstein M (1980) The linear Bayes regression estimator under weak prior assumptions. *Biometrika* **67**, 621–628.

Goldstein M (1981) Revising previsions: a geometric interpretation. *Journal of the Royal Statistical Society, Series B* **43**, 105–130.

Goldstein M (1983a) General variance modifications for linear Bayes estimators. *Journal of the American Statistical Association* **78**, 616–618.

Goldstein M (1983b) The prevision of a prevision. *Journal of the American Statistical Association* **78**, 817–819.

Goldstein M (1984) Turning probabilities into expectations. *Annals of Statistics* **12**, 1551–1557.

Goldstein M (1985) Temporal coherence. In *Bayesian Statistics 2* (ed. Bernardo JM *et al.*), pp. 231–248. North-Holland, Amersterdam and Valencia University Press, Valencia.

Goldstein M (1986a) Exchangeable belief structures. *Journal of the American Statistical Association* **81**, 971–976.

Goldstein M (1986b) Separating beliefs. In *Bayesian Inference and Decision Techniques* (ed. Goel P and Zellner A). North-Holland, Amsterdam.

Goldstein M (1987a) Can we build a subjectivist statistical package? In *Probability and Bayesian Statistics* (ed. Viertl R), pp. 203–217. Plenum, New York.

Goldstein M (1987b) Systematic analysis of limited belief specifications. *The Statistician* **36**, 191–199.

Goldstein M (1988a) Adjusting belief structures. *Journal of the Royal Statistical Society, Series B* **50**, 133–154.

Goldstein M (1988b) The data trajectory. In *Bayesian Statistics 3* (ed. Bernardo JM *et al.*), pp. 189–209. Oxford University Press, Oxford.

Goldstein M (1990) Influence and belief adjustment. In *Influence Diagrams, Belief Nets and Decision Analysis* (ed. Smith J and Oliver R). John Wiley & Sons, Ltd, Chichester.

Goldstein M (1991) Belief transforms and the comparison of hypotheses. *Annals of Statistics* **19**, 2067–2089.

Goldstein M (1994a) Belief revision: subjectivist principles and practice. In *Logic and Philosophy of Science in Uppsala* (ed. Prawitz D and Westerstahl D), pp. 117–130. Kluwer Academic, Dordrecht.

Goldstein M (1994b) Revising exchangeable beliefs: subjectivist foundations for the inductive argument. In *Aspects of Uncertainty: A Tribute to D.V. Lindley* (ed. Freeman P and Smith AFM), pp. 201–222. John Wiley & Sons, Ltd, Chichester.

Goldstein M (1997) Prior inferences for posterior judgements. In *Structures and Norms in Science. Volume Two of the Tenth International Congress of Logic, Methodology and Philosophy of Science, Florence, August 1995* (ed. Chiara MLD, Doets K, Mundici D and van Benthem J), pp. 55–71. Kluwer, Dordrecht.

Goldstein M (1999) Bayes linear analysis. In *Encyclopaedia of Statistical Sciences, update volume 3* (ed. Kotz S *et al.*), pp. 29–34. John Wiley & Sons, Inc., New York.

Goldstein M (2000) Bayes linear methods I – Adjusting beliefs: concepts and properties. *Journal of Stat istical Software*. http://www.stat.ucla.edu/journals/jss/v05/i02.

Goldstein M (2001) Avoiding foregone conclusions: geometric and foundational analysis of paradoxes of finite additivity. *Journal of Statistical Planning and Inference* **94**(1), 73–87.

Goldstein M (2006) Subjective Bayesian analysis: principles and practice (with discussion). *Bayesian Analysis* **1**(3), 403–420.

Goldstein M and O'Hagan A (1996) Bayes linear sufficiency and systems of expert posterior assessments. *Journal of the Royal Statistical Society, Series B* **58**, 301–316.

Goldstein M and Rougier JC (2005) Probabilistic formulations for transferring inferences from mathematical models to physical systems. *SIAM Journal on Scientific Computing* **26**, 467–487.

Goldstein M and Shaw SC (2004) Bayes linear kinematics and Bayes linear Bayes graphical models. *Biometrika* **91**, 425–446.

Goldstein M and Wilkinson DJ (2000) Bayes linear analysis for graphical models: the geometric approach to local computation and interpretative graphics. *Statistics and Computing* **10**(4), 311–324.

Goldstein M and Wilkinson DJ (2001) Restricted prior inference for complex uncertainty structures. *Annals of Mathematics and Artificial Intelligence* **32**, 315–334.

Goldstein M and Wooff DA (1994) Robustness measures for Bayes linear analyses. *Journal of Statistical Planning and Inference* **40**(2–3), 261–277.

Goldstein M and Wooff DA (1995) Bayes linear computation: concepts, implementation and programming environment. *Statistics and Computing* **5**, 327–341.

Goldstein M and Wooff DA (1997) Choosing sample sizes in balanced experimental designs: a Bayes linear approach. *The Statistician* **46**, 167–183.

Goldstein M and Wooff DA (1998) Adjusting exchangeable beliefs. *Biometrika* **85**(1), 39–54.

Goldstein M, Farrow M and Spiropoulos T (1993) Prediction under the influence: Bayes linear influence diagrams for prediction in a large brewery. *The Statistician* **42**, 445–459.

Guttman I (1982) *Linear Models: An Introduction*. John Wiley & Sons, Inc., New York.

Hamilton JD (1994) *Time Series Analysis*. Princeton University Press, Princeton, NJ.

Hartigan JA (1969) Linear Bayes methods. *Journal of the Royal Statistical Society, Series B* **31**, 446–454.

Harville DA (1997) *Matrix Algebra from a Statistician's Perspective*. Springer-Verlag, New York.

Jensen FV (2001) *Bayesian Networks and Decision Graphs*. Springer-Verlag, New York.

Kadane JB and Wolfson LJ (1998) Experiences in elicitation. *The Statistician* **47**, 3–19.

Krzanowski WJ and Marriott FHC (1994) *Kendall's Library of Statistics. Multivariate Analysis Part 1: Di stributions, Ordination and Inference*. Edward Arnold, London.

Kuo L (1988) Linear Bayes estimators of the potency curve in bioassay. *Biometrika* **75**, 91–96.

Kyburg HE and Smokler HE (1964) *Studies in Subjective Probability*. John Wiley & Sons, Inc., New York.

Lad F (1996) *Operational Subjective Statistical Methods: A Mathematical, Philosophical and Historical Introduction*. John Wiley & Sons, Inc., New York.

Lad F, Dickey JM and Rahman MA (1992) Numerical application of the fundamental theorem of previson. *Journal of Stat. Comput. Sim.* **40**, 135–152.

Lauritzen SL (1996) *Graphical Models*. Clarendon Press, Oxford.

Lindley DV (1965) *Introduction to Probability and Statistics from a Bayesian Viewpoint*. Cambridge University Press, Cambridge.

Little JD, Goldstein M and Jonathan P (2004) Efficient Bayesian sampling inspection for industrial processes based on transformed spatio-temporal data. *Statistical Modelling* **4**, 299–313.

Marsaglia G and Styan GPH (1974) Rank conditions for generalized inverses of partitioned matrices. *Sankhya, Series A* **36**, 437–442.

Mitchell P, Arthur W and Farrow M (1992) An investigation of population limitation using factorial experiments. *Journal of Animal Ecology* **16**, 591–598.

Moler CB and Stewart GW (1973) An algorithm for generalized matrix eigenproblems. *SIAM Journal on Numerical Analysis* **10**, 241–256.

Mouchart M and Simar L (1980) Least squares approximation in Bayesian analysis (with discussion). In *Bayesian Statistics* (ed. Bernardo JM *et al.*), pp. 207–222. Valencia University Press, Valencia.

Mukhopadhyay S and Vidakovic B (1995) Efficiency of linear Bayes rules for a normal-mean skewed priors class. *The Statistician* **44**, 389–397.

O'Hagan A (1987) Bayes linear estimators for randomized response models. *Journal of the American Statistical Association* **82**, 580–585.

O'Hagan A (1998) Eliciting expert beliefs in substantial applications. *The Statistician* **47**, 21–35.

O'Hagan A and Forster J (2004) *Kendall's Advanced Theory of Statistics. Volume 2b: Bayesian Inference*. Edward Arnold, London.

O'Hagan A, Glennie EB and Beardsall RE (1992) Subjective modelling and Bayes linear estimation in the UK water industry. *Applied Statistics* **41**, 563–577.

Pearl J (1988) *Probabilistic Reasoning in Intelligent Systems*. Morgan Kaufmann, San Mateo, CA.

Pukelsheim F (1994) The three sigma rule. *American Statistician* **48**, 88–91.

R Development Core Team (2006) *R: A Language and Environment for Statistical Computing*. R Foundation for Statistical Computing, Vienna. ISBN 3-900051-07-0.

Rao CR and Mitra SK (1971) *Generalized Inverse of Matrices and Its Applications*. John Wiley & Sons, Inc., New York.

Rayner AA and Livingstone D (1965) On the distribution of quadratic forms in singular normal variates. *South African Journal of Agric. Science* **8**, 357–369.

Robert CP (2001) *The Bayesian Choice*, 2nd edn. Springer-Verlag, New York.

Savage LJ (1971) *The Foundation of Statistics*, 2nd edn. Dover, New York.

Searle SR (1982) *Matrix Algebra Useful For Statistics*. John Wiley & Sons, Inc., New York.

Shaw SC and Goldstein M (1999) Simplifying complex designs: Bayes linear experimental design for grouped multivariate exchangeable systems. In *Bayesian Statistics 6* (ed. Bernardo JM *et al.*), pp. 839–848. Oxford University Press, Oxford.

Smith JQ (1990) Statistical principles on graphs. In *Influence Diagrams, Belief Nets and Decision Analysis* (ed. Smith JQ and Oliver RM). John Wiley & Sons, Ltd, Chichester.

Spiropoulos T (1995) Decision support for management using Bayes linear influence diagrams. PhD thesis, University of Sunderland.

Stone M (1963) Robustness of non-ideal decision procedures. *Journal of the American Statistical Association* **58**, 480–486.

Stuart A and Ord K (1994) *Kendall's Advanced Theory of Statistics. Volume I: Distribution Theory*. Edward Arnold, London.

Walley P (1991) *Statistical Reasoning with Imprecise Probabilities*. Chapman & Hall, London.

Ward RC (1975) The combination shift QZ algorithm. *SIAM Journal on Numerical Analysis* **12**, 835–853.

West M and Harrison PJ (1997) *Bayesian Forecasting and Dynamic Models*, 2nd edn. Springer-Verlag, New York.

Wickramasinghe LSP, Chazan BI, Farrow M, Bansal SK and Basu SK (1992) C-peptide response to oral glucose and its clinical role in elderly people. *Age and Ageing* **21**, 103–108.

Wilkinson DJ (1995) Bayes linear covariance matrix adjustment. PhD thesis, Department of Mathematical Sciences, University of Durham.

Wilkinson DJ (1997) Bayes linear variance adjustment for locally linear DLMs. *Journal of Forecasting* **16**, 329–342.

Wilkinson DJ (1998) An object-oriented approach to local computation in Bayes linear belief networks. In *Proceedings in Computational Statistics* (ed. Green PJ and Payne RW), pp. 491–496. Physica Verlag, Heidelberg.

Wilkinson DJ (2000) *BAYES-LIN*. University of Newcastle, Newcastle, UK.

Wilkinson DJ and Goldstein M (1996) Bayes linear adjustment for variance matrices. In *Bayesian Statistics 5* (ed. Bernardo JM *et al.*), pp. 791–800. Oxford University Press, Oxford.

Wilkinson JH (1965) *The Algebraic Eigenvalue Problem*. Clarendon Press, Oxford.

Wilkinson JH (1979) Kronecker's canonical form and the QZ algorithm. *Linear Algebra and its Applications* **28**(1), 285–303.

Williams DR and Goldstein M (1999) Graphical diagnostics for the Bayes linear analysis of hierarchical linear models with applications to educational data. In *Bayesian Statistics 6* (ed. Bernardo JM *et al.*), pp. 859–867. Oxford University Press, Oxford.

Wooff DA (1992) [B/D] works. In *Bayesian Statistics 4* (ed. Bernardo JM *et al.*), pp. 851–859. Oxford University Press, Oxford.

Wooff DA (2000a) Bayes linear methods II – An example with an introduction to [B/D]. *Journal of Statistical Software*. http://www.stat.ucla.edu/journals/jss/v05/i02.

Wooff DA (2000b) [B/D] Manual. *Journal of Statistical Software*. http://www.stat.ucla.edu/journals/jss/v05/i02.

Wooff DA and Goldstein M (2000a) Bayes linear methods III – Analysing Bayes linear influence diagrams and exchangeability in [B/D]. *Journal of Statistical Software*. http://www.stat.ucla.edu/journals/jss/v05/i02.

Wooff DA and Goldstein M (2000b) The Bayes linear programming language [B/D]. *Journal of Statistical Software*. http://www.stat.ucla.edu/journals/jss/v05/i02.

Wooff DA, Seheult AH, Coolen FPA and Worrall F (1998) Bayesian discrimination with uncertain covariates for pesticide contamination. In *Statistics for the Environment 4: Statistical Aspects of Health and the Environment* (ed. Barnett V, Stein A and Turkman KF), pp. 337–353. John Wiley & Sons, Ltd, Chichester.

Index

Bayes Linear Statistics: Theory and Methods M. Goldstein and D.A. Wooff

WILEY SERIES IN PROBABILITY AND STATISTICS

ESTABLISHED BY WALTER A. SHEWHART AND SAMUEL S. WILKS

The ***Wiley Series in Probability and Statistics*** is well established and authoritative. It covers many topics of current research interest in both pure and applied statistics and probability theory. Written by leading statisticians and institutions, the titles span both state-of-the-art developments in the field and classical methods.

Reflecting the wide range of current research in statistics, the series encompasses applied, methodological and theoretical statistics, ranging from applications and new techniques made possible by advances in computerized practice to rigorous treatment of theoretical approaches.

This series provides essential and invaluable reading for all statisticians, whether in academia, industry, government, or research.

ABRAHAM and LEDOLTER · Statistical Methods for Forecasting
AGRESTI · Analysis of Ordinal Categorical Data
AGRESTI · An Introduction to Categorical Data Analysis
AGRESTI · Categorical Data Analysis, *Second Edition*
ALTMAN, GILL, and McDONALD · Numerical Issues in Statistical Computing for the Social Scientist
AMARATUNGA and CABRERA · Exploration and Analysis of DNA Microarray and Protein Array Data
ANDĚL · Mathematics of Chance
ANDERSON · An Introduction to Multivariate Statistical Analysis, *Third Edition*
*ANDERSON · The Statistical Analysis of Time Series
ANDERSON, AUQUIER, HAUCK, OAKES, VANDAELE, and WEISBERG · Statistical Methods for Comparative Studies
ANDERSON and LOYNES · The Teaching of Practical Statistics
ARMITAGE and DAVID (editors) · Advances in Biometry
ARNOLD, BALAKRISHNAN, and NAGARAJA · Records
*ARTHANARI and DODGE · Mathematical Programming in Statistics
*BAILEY · The Elements of Stochastic Processes with Applications to the Natural Sciences
BALAKRISHNAN and KOUTRAS · Runs and Scans with Applications
BARNETT · Comparative Statistical Inference, *Third Edition*
BARNETT · Environmental Statistics: Methods & Applications
BARNETT and LEWIS · Outliers in Statistical Data, *Third Edition*
BARTOSZYNSKI and NIEWIADOMSKA-BUGAJ · Probability and Statistical Inference
BASILEVSKY · Statistical Factor Analysis and Related Methods: Theory and Applications
BASU and RIGDON · Statistical Methods for the Reliability of Repairable Systems
BATES and WATTS · Nonlinear Regression Analysis and Its Applications
BECHHOFER, SANTNER, and GOLDSMAN · Design and Analysis of Experiments for Statistical Selection, Screening, and Multiple Comparisons
BELSLEY · Conditioning Diagnostics: Collinearity and Weak Data in Regression

*Now available in a lower priced paperback edition in the Wiley Classics Library.

BELSLEY, KUH, and WELSCH · Regression Diagnostics: Identifying Influential Data and Sources of Collinearity
BENDAT and PIERSOL · Random Data: Analysis and Measurement Procedures, *Third Edition*
BERNARDO and SMITH · Bayesian Theory
BERRY, CHALONER, and GEWEKE · Bayesian Analysis in Statistics and Econometrics: Essays in Honor of Arnold Zellner
BHAT and MILLER · Elements of Applied Stochastic Processes, *Third Edition*
BHATTACHARYA and JOHNSON · Statistical Concepts and Methods
BHATTACHARYA and WAYMIRE · Stochastic Processes with Applications
BIEMER, GROVES, LYBERG, MATHIOWETZ, and SUDMAN · Measurement Errors in Surveys
BILLINGSLEY · Convergence of Probability Measures, *Second Edition*
BILLINGSLEY · Probability and Measure, *Third Edition*
BIRKES and DODGE · Alternative Methods of Regression
BLISCHKE and MURTHY (editors) · Case Studies in Reliability and Maintenance
BLISCHKE and MURTHY · Reliability: Modeling, Prediction, and Optimization
BLOOMFIELD · Fourier Analysis of Time Series: An Introduction, *Second Edition*
BOLLEN · Structural Equations with Latent Variables
BOLLEN and CURRAN · Latent Curve Models: A Structural Equation Perspective
BOROVKOV · Ergodicity and Stability of Stochastic Processes
BOULEAU · Numerical Methods for Stochastic Processes
BOX · Bayesian Inference in Statistical Analysis
BOX · R. A. Fisher, the Life of a Scientist
BOX and DRAPER · Empirical Model-Building and Response Surfaces
*BOX and DRAPER · Evolutionary Operation: A Statistical Method for Process Improvement
BOX, HUNTER, and HUNTER · Statistics for Experimenters: An Introduction to Design, Data Analysis, and Model Building
BOX, HUNTER, and HUNTER · Statistics for Experimenters: Design, Innovation and Discovery, *Second Edition*
BOX and LUCEÑO · Statistical Control by Monitoring and Feedback Adjustment
BRANDIMARTE · Numerical Methods in Finance: A MATLAB-Based Introduction
BROWN and HOLLANDER · Statistics: A Biomedical Introduction
BRUNNER, DOMHOF, and LANGER · Nonparametric Analysis of Longitudinal Data in Factorial Experiments
BUCKLEW · Large Deviation Techniques in Decision, Simulation, and Estimation
CAIROLI and DALANG · Sequential Stochastic Optimization
CASTILLO, HADI, BALAKRISHNAN and SARABIA · Extreme Value and Related Models with Applications in Engineering and Science
CHAN · Time Series: Applications to Finance
CHATTERJEE and HADI · Sensitivity Analysis in Linear Regression
CHATTERJEE and PRICE · Regression Analysis by Example, *Third Edition*
CHERNICK · Bootstrap Methods: A Practitioner's Guide
CHERNICK and FRIIS · Introductory Biostatistics for the Health Sciences
CHILÈS and DELFINER · Geostatistics: Modeling Spatial Uncertainty
CHOW and LIU · Design and Analysis of Clinical Trials: Concepts and Methodologies, *Second Edition*
CLARKE and DISNEY · Probability and Random Processes: A First Course with Applications, *Second Edition*
*COCHRAN and COX · Experimental Designs, *Second Edition*
CONGDON · Applied Bayesian Modelling
CONGDON · Bayesian Statistical Modelling
CONGDON · Bayesian Models for Categorical Data
CONOVER · Practical Nonparametric Statistics, *Second Edition*
COOK · Regression Graphics
COOK and WEISBERG · Applied Regression Including Computing and Graphics
COOK and WEISBERG · An Introduction to Regression Graphics
CORNELL · Experiments with Mixtures, Designs, Models, and the Analysis of Mixture Data, *Third Edition*

*Now available in a lower priced paperback edition in the Wiley Classics Library.

COVER and THOMAS · Elements of Information Theory
COX · A Handbook of Introductory Statistical Methods
*COX · Planning of Experiments
CRESSIE · Statistics for Spatial Data, *Revised Edition*
CSÖRGÖ and HORVÁTH · Limit Theorems in Change Point Analysis
DANIEL · Applications of Statistics to Industrial Experimentation
DANIEL · Biostatistics: A Foundation for Analysis in the Health Sciences, *Sixth Edition*
*DANIEL · Fitting Equations to Data: Computer Analysis of Multifactor Data, *Second Edition*
DASU and JOHNSON · Exploratory Data Mining and Data Cleaning
DAVID and NAGARAJA · Order Statistics, *Third Edition*
*DEGROOT, FIENBERG, and KADANE · Statistics and the Law
DEL CASTILLO · Statistical Process Adjustment for Quality Control
DEMARIS · Regression with Social Data: Modeling Continuous and Limited Response Variables
DEMIDENKO · Mixed Models: Theory and Applications
DENISON, HOLMES, MALLICK, and SMITH · Bayesian Methods for Nonlinear Classification and Regression
DETTE and STUDDEN · The Theory of Canonical Moments with Applications in Statistics, Probability, and Analysis
DEY and MUKERJEE · Fractional Factorial Plans
DILLON and GOLDSTEIN · Multivariate Analysis: Methods and Applications
DODGE · Alternative Methods of Regression
*DODGE and ROMIG · Sampling Inspection Tables, *Second Edition*
*DOOB · Stochastic Processes
DOWDY and WEARDEN, and CHILKO · Statistics for Research, *Third Edition*
DRAPER and SMITH · Applied Regression Analysis, *Third Edition*
DRYDEN and MARDIA · Statistical Shape Analysis
DUDEWICZ and MISHRA · Modern Mathematical Statistics
DUNN and CLARK · Applied Statistics: Analysis of Variance and Regression, *Second Edition*
DUNN and CLARK · Basic Statistics: A Primer for the Biomedical Sciences, *Third Edition*
DUPUIS and ELLIS · A Weak Convergence Approach to the Theory of Large Deviations
EDLER and KITSOS (editors) · Recent Advances in Quantitative Methods in Cancer and Human Health Risk Assessment
*ELANDT-JOHNSON and JOHNSON · Survival Models and Data Analysis
ENDERS · Applied Econometric Time Series
ETHIER and KURTZ · Markov Processes: Characterization and Convergence
EVANS, HASTINGS, and PEACOCK · Statistical Distributions, *Third Edition*
FELLER · An Introduction to Probability Theory and Its Applications, Volume I, *Third Edition,* Revised; Volume II, *Second Edition*
FISHER and VAN BELLE · Biostatistics: A Methodology for the Health Sciences
FITZMAURICE, LAIRD, and WARE · Applied Longitudinal Analysis
*FLEISS · The Design and Analysis of Clinical Experiments
FLEISS · Statistical Methods for Rates and Proportions, *Second Edition*
FLEMING and HARRINGTON · Counting Processes and Survival Analysis
FULLER · Introduction to Statistical Time Series, *Second Edition*
FULLER · Measurement Error Models
GALLANT · Nonlinear Statistical Models
GELMAN and MENG (editors): Applied Bayesian Modeling and Casual Inference from Incomplete-data Perspectives
GEWEKE · Contemporary Bayesian Econometrics and Statistics
GHOSH, MUKHOPADHYAY, and SEN · Sequential Estimation
GIESBRECHT and GUMPERTZ · Planning, Construction, and Statistical Analysis of Comparative Experiments
GIFI · Nonlinear Multivariate Analysis
GIVENS and HOETING · Computational Statistics
GLASSERMAN and YAO · Monotone Structure in Discrete-Event Systems
GNANADESIKAN · Methods for Statistical Data Analysis of Multivariate Observations, *Second Edition*

*Now available in a lower priced paperback edition in the Wiley Classics Library.

GOLDSTEIN and WOOFF · Bayes Linear Statistics: Theory and Methods
GOLDSTEIN and LEWIS · Assessment: Problems, Development, and Statistical Issues
GREENWOOD and NIKULIN · A Guide to Chi-Squared Testing
GROSS and HARRIS · Fundamentals of Queueing Theory, *Third Edition*
*HAHN and SHAPIRO · Statistical Models in Engineering
HAHN and MEEKER · Statistical Intervals: A Guide for Practitioners
HALD · A History of Probability and Statistics and their Applications Before 1750
HALD · A History of Mathematical Statistics from 1750 to 1930
HAMPEL · Robust Statistics: The Approach Based on Influence Functions
HANNAN and DEISTLER · The Statistical Theory of Linear Systems
HEIBERGER · Computation for the Analysis of Designed Experiments
HEDAYAT and SINHA · Design and Inference in Finite Population Sampling
HELLER · MACSYMA for Statisticians
HINKELMAN and KEMPTHORNE: · Design and Analysis of Experiments, Volume 1: Introduction to Experimental Design
HINKELMANN and KEMPTHORNE · Design and Analysis of Experiments, Volume 2: Advanced Experimental Design
HOAGLIN, MOSTELLER, and TUKEY · Exploratory Approach to Analysis of Variance
HOAGLIN, MOSTELLER, and TUKEY · Exploring Data Tables, Trends and Shapes
*HOAGLIN, MOSTELLER, and TUKEY · Understanding Robust and Exploratory Data Analysis
HOCHBERG and TAMHANE · Multiple Comparison Procedures
HOCKING · Methods and Applications of Linear Models: Regression and the Analysis of Variance, *Second Edition*
HOEL · Introduction to Mathematical Statistics, *Fifth Edition*
HOGG and KLUGMAN · Loss Distributions
HOLLANDER and WOLFE · Nonparametric Statistical Methods, *Second Edition*
HOSMER and LEMESHOW · Applied Logistic Regression, *Second Edition*
HOSMER and LEMESHOW · Applied Survival Analysis: Regression Modeling of Time to Event Data
HUBER · Robust Statistics
HUBERTY · Applied Discriminant Analysis
HUNT and KENNEDY · Financial Derivatives in Theory and Practice, *Revised Edition*
HUSKOVA, BERAN, and DUPAC · Collected Works of Jaroslav Hajek—with Commentary
HUZURBAZAR · Flowgraph Models for Multistate Time-to-Event Data
IMAN and CONOVER · A Modern Approach to Statistics
JACKSON · A User's Guide to Principle Components
JOHN · Statistical Methods in Engineering and Quality Assurance
JOHNSON · Multivariate Statistical Simulation
JOHNSON and BALAKRISHNAN · Advances in the Theory and Practice of Statistics: A Volume in Honor of Samuel Kotz
JOHNSON and BHATTACHARYYA · Statistics: Principles and Methods, *Fifth Edition*
JUDGE, GRIFFITHS, HILL, LU TKEPOHL, and LEE · The Theory and Practice of Econometrics
JOHNSON and KOTZ · Distributions in Statistics
JOHNSON and KOTZ (editors) · Leading Personalities in Statistical Sciences: From the Seventeenth Century to the Present
JOHNSON, KOTZ, and BALAKRISHNAN · Continuous Univariate Distributions, Volume 1, *Second Edition*
JOHNSON, KOTZ, and BALAKRISHNAN · Continuous Univariate Distributions, Volume 2, *Second Edition*
JOHNSON, KOTZ, and BALAKRISHNAN · Discrete Multivariate Distributions
JOHNSON, KOTZ, and KEMP · Univariate Discrete Distributions, *Second Edition*
JUREČKOVÁ and SEN · Robust Statistical Procedures: Asymptotics and Interrelations
JUREK and MASON · Operator-Limit Distributions in Probability Theory
KADANE · Bayesian Methods and Ethics in a Clinical Trial Design

*Now available in a lower priced paperback edition in the Wiley Classics Library.

KADANE and SCHUM · A Probabilistic Analysis of the Sacco and Vanzetti Evidence
KALBFLEISCH and PRENTICE · The Statistical Analysis of Failure Time Data, *Second Edition*
KARIYA and KURATA · Generalized Least Squares
KASS and VOS · Geometrical Foundations of Asymptotic Inference
KAUFMAN and ROUSSEEUW · Finding Groups in Data: An Introduction to Cluster Analysis
KEDEM and FOKIANOS · Regression Models for Time Series Analysis
KENDALL, BARDEN, CARNE, and LE · Shape and Shape Theory
KHURI · Advanced Calculus with Applications in Statistics, *Second Edition*
KHURI, MATHEW, and SINHA · Statistical Tests for Mixed Linear Models
*KISH · Statistical Design for Research
KLEIBER and KOTZ · Statistical Size Distributions in Economics and Actuarial Sciences
KLUGMAN, PANJER, and WILLMOT · Loss Models: From Data to Decisions
KLUGMAN, PANJER, and WILLMOT · Solutions Manual to Accompany Loss Models: From Data to Decisions
KOTZ, BALAKRISHNAN, and JOHNSON · Continuous Multivariate Distributions, Volume 1, *Second Edition*
KOTZ and JOHNSON (editors) · Encyclopedia of Statistical Sciences: Volumes 1 to 9 with Index
KOTZ and JOHNSON (editors) · Encyclopedia of Statistical Sciences: Supplement Volume
KOTZ, READ, and BANKS (editors) · Encyclopedia of Statistical Sciences: Update Volume 1
KOTZ, READ, and BANKS (editors) · Encyclopedia of Statistical Sciences: Update Volume 2
KOVALENKO, KUZNETZOV, and PEGG · Mathematical Theory of Reliability of Time-Dependent Systems with Practical Applications
KUROWICKA and COOKE · Uncentainty Analysis with High Dimensional Dependence Modelling
LACHIN · Biostatistical Methods: The Assessment of Relative Risks
LAD · Operational Subjective Statistical Methods: A Mathematical, Philosophical, and Historical Introduction
LAMPERTI · Probability: A Survey of the Mathematical Theory, *Second Edition*
LANGE, RYAN, BILLARD, BRILLINGER, CONQUEST, and GREENHOUSE · Case Studies in Biometry
LARSON · Introduction to Probability Theory and Statistical Inference, *Third Edition*
LAWLESS · Statistical Models and Methods for Lifetime Data, *Second Edition*
LAWSON · Statistical Methods in Spatial Epidemiology, *Second Edition*
LE · Applied Categorical Data Analysis
LE · Applied Survival Analysis
LEE and WANG · Statistical Methods for Survival Data Analysis, *Third Edition*
LEPAGE and BILLARD · Exploring the Limits of Bootstrap
LEYLAND and GOLDSTEIN (editors) · Multilevel Modelling of Health Statistics
LIAO · Statistical Group Comparison
LINDVALL · Lectures on the Coupling Method
LINHART and ZUCCHINI · Model Selection
LITTLE and RUBIN · Statistical Analysis with Missing Data, *Second Edition*
LLOYD · The Statistical Analysis of Categorical Data
LOWEN and TEICH · Fractal-Based Point Processes
MAGNUS and NEUDECKER · Matrix Differential Calculus with Applications in Statistics and Econometrics, *Revised Edition*
MALLER and ZHOU · Survival Analysis with Long Term Survivors
MALLOWS · Design, Data, and Analysis by Some Friends of Cuthbert Daniel
MANN, SCHAFER, and SINGPURWALLA · Methods for Statistical Analysis of Reliability and Life Data
MANTON, WOODBURY, and TOLLEY · Statistical Applications Using Fuzzy Sets
MARCHETTE · Random Graphs for Statistical Pattern Recognition
MARDIA and JUPP · Directional Statistics

*Now available in a lower priced paperback edition in the Wiley Classics Library.

MARONNA, MARTIN, and YOHAI · Robust Statistics: Theory and Methods
MASON, GUNST, and HESS · Statistical Design and Analysis of Experiments with Applications to Engineering and Science, *Second Edition*
MCCULLOCH and SEARLE · Generalized, Linear, and Mixed Models
MCFADDEN · Management of Data in Clinical Trials
MCLACHLAN · Discriminant Analysis and Statistical Pattern Recognition
MCLACHLAN, DO, and AMBROISE · Analyzing Microarray Gene Expression Data
MCLACHLAN and KRISHNAN · The EM Algorithm and Extensions
MCLACHLAN and PEEL · Finite Mixture Models
MCNEIL · Epidemiological Research Methods
MEEKER and ESCOBAR · Statistical Methods for Reliability Data
MEERSCHAERT and SCHEFFLER · Limit Distributions for Sums of Independent Random Vectors: Heavy Tails in Theory and Practice
MICKEY, DUNN and CLARK · Applied Statistics: Anaysis of Variance and Regression, *Third Edition*
*MILLER · Survival Analysis, *Second Edition*
MONTGOMERY, PECK, and VINING · Introduction to Linear Regression Analysis, *Third Edition*
MORGENTHALER and TUKEY · Configural Polysampling: A Route to Practical Robustness
MUIRHEAD · Aspects of Multivariate Statistical Theory
MURRAY · X-STAT 2.0 Statistical Experimentation, Design Data Analysis, and Nonlinear Optimization
MURTHY, XIE, and JIANG · Weibull Models
MYERS and MONTGOMERY · Response Surface Methodology: Process and Product Optimization Using Designed Experiments, *Second Edition*
MYERS, MONTGOMERY, and VINING · Generalized Linear Models. With Applications in Engineering and the Sciences
†NELSON · Accelerated Testing, Statistical Models, Test Plans, and Data Analyses
†NELSON · Applied Life Data Analysis
NEWMAN · Biostatistical Methods in Epidemiology
OCHI · Applied Probability and Stochastic Processes in Engineering and Physical Sciences
OKABE, BOOTS, SUGIHARA, and CHIU · Spatial Tesselations: Concepts and Applications of Voronoi Diagrams, *Second Edition*
OLIVER and SMITH · Influence Diagrams, Belief Nets and Decision Analysis
PALTA · Quantitative Methods in Population Health: Extensions of Ordinary Regressions
PANKRATZ · Forecasting with Dynamic Regression Models
PANKRATZ · Forecasting with Univariate Box-Jenkins Models: Concepts and Cases
*PARZEN · Modern Probability Theory and It's Applications
PEÑA, TIAO, and TSAY · A Course in Time Series Analysis
PIANTADOSI · Clinical Trials: A Methodologic Perspective
PORT · Theoretical Probability for Applications
POURAHMADI · Foundations of Time Series Analysis and Prediction Theory
PRESS · Bayesian Statistics: Principles, Models, and Applications
PRESS · Subjective and Objective Bayesian Statistics, *Second Edition*
PRESS and TANUR · The Subjectivity of Scientists and the Bayesian Approach
PUKELSHEIM · Optimal Experimental Design
PURI, VILAPLANA, and WERTZ · New Perspectives in Theoretical and Applied Statistics
PUTERMAN · Markov Decision Processes: Discrete Stochastic Dynamic Programming
QIU · Image Processing and Jump Regression Analysis
*RAO · Linear Statistical Inference and Its Applications, *Second Edition*
RAUSAND and HØYLAND · System Reliability Theory: Models, Statistical Methods and Applications, *Second Edition*
RENCHER · Linear Models in Statistics
RENCHER · Methods of Multivariate Analysis, *Second Edition*
RENCHER · Multivariate Statistical Inference with Applications

*Now available in a lower priced paperback edition in the Wiley Classics Library.

†Now available in a lower priced paperback edition in the Wiley-Interscience Paperback Series.

RIPLEY · Spatial Statistics
RIPLEY · Stochastic Simulation
ROBINSON · Practical Strategies for Experimenting
ROHATGI and SALEH · An Introduction to Probability and Statistics, *Second Edition*
ROLSKI, SCHMIDLI, SCHMIDT, and TEUGELS · Stochastic Processes for Insurance and Finance
ROSENBERGER and LACHIN · Randomization in Clinical Trials: Theory and Practice
ROSS · Introduction to Probability and Statistics for Engineers and Scientists
ROSSI, ALLENBY and MCCULLOCH · Bayesian Statistics and Marketing
ROUSSEEUW and LEROY · Robust Regression and Outlier Detection
RUBIN · Multiple Imputation for Nonresponse in Surveys
RUBINSTEIN · Simulation and the Monte Carlo Method
RUBINSTEIN and MELAMED · Modern Simulation and Modeling
RYAN · Modern Regression Methods
RYAN · Statistical Methods for Quality Improvement, *Second Edition*
SALTELLI, CHAN, and SCOTT (editors) · Sensitivity Analysis
*SCHEFFE · The Analysis of Variance
SCHIMEK · Smoothing and Regression: Approaches, Computation, and Application
SCHOTT · Matrix Analysis for Statistics
SCHOUTENS · Levy Processes in Finance: Pricing Financial Derivatives
SCHUSS · Theory and Applications of Stochastic Differential Equations
SCOTT · Multivariate Density Estimation: Theory, Practice, and Visualization
*SEARLE · Linear Models
SEARLE · Linear Models for Unbalanced Data
SEARLE · Matrix Algebra Useful for Statistics
SEARLE, CASELLA, and MCCULLOCH · Variance Components
SEARLE and WILLETT · Matrix Algebra for Applied Economics
SEBER · Multivariate Observations
SEBER and LEE · Linear Regression Analysis, *Second Edition*
SEBER and WILD · Nonlinear Regression
SENNOTT · Stochastic Dynamic Programming and the Control of Queueing Systems
*SERFLING · Approximation Theorems of Mathematical Statistics
SHAFER and VOVK · Probability and Finance: Its Only a Game!
SILVAPULLE and SEN · Constrained Statistical Inference: Inequality, Order, and Shape Restrictions
SINGPURWALLA. Reliability and Risk: A Bayesian Perspective
SMALL and MCLEISH · Hilbert Space Methods in Probability and Statistical Inference
SRIVASTAVA · Methods of Multivariate Statistics
STAPLETON · Linear Statistical Models
STAUDTE and SHEATHER · Robust Estimation and Testing
STOYAN, KENDALL, and MECKE · Stochastic Geometry and Its Applications, *Second Edition*
STOYAN and STOYAN · Fractals, Random Shapes and Point Fields: Methods of Geometrical Statistics
STYAN · The Collected Papers of T. W. Anderson: 1943–1985
SUTTON, ABRAMS, JONES, SHELDON, and SONG · Methods for Meta-Analysis in Medical Research
TANAKA · Time Series Analysis: Nonstationary and Noninvertible Distribution Theory
THOMPSON · Empirical Model Building
THOMPSON · Sampling, *Second Edition*
THOMPSON · Simulation: A Modeler's Approach
THOMPSON and SEBER · Adaptive Sampling
THOMPSON, WILLIAMS, and FINDLAY · Models for Investors in Real World Markets
TIAO, BISGAARD, HILL, PEÑA, and STIGLER (editors) · Box on Quality and Discovery: with Design, Control, and Robustness

*Now available in a lower priced paperback edition in the Wiley Classics Library.

TIERNEY · LISP-STAT: An Object-Oriented Environment for Statistical Computing and Dynamic Graphics
TSAY · Analysis of Financial Time Series
UPTON and FINGLETON · Spatial Data Analysis by Example, Volume II: Categorical and Directional Data
VAN BELLE · Statistical Rules of Thumb
VAN BELLE, FISHER, HEAGERTY, and LUMLEY · Biostatistics: A Methodology for the Health Sciences, *Second Edition*
VESTRUP · The Theory of Measures and Integration
VIDAKOVIC · Statistical Modeling by Wavelets
VINOD and REAGLE · Preparing for the Worst: Incorporating Downside Risk in Stock Market Investments
WALLER and GOTWAY · Applied Spatial Statistics for Public Health Data
WEERAHANDI · Generalized Inference in Repeated Measures: Exact Methods in MANOVA and Mixed Models
WEISBERG · Applied Linear Regression, *Second Edition*
WELSH · Aspects of Statistical Inference
WESTFALL and YOUNG · Resampling-Based Multiple Testing: Examples and Methods for *p*-Value Adjustment
WHITTAKER · Graphical Models in Applied Multivariate Statistics
WINKER · Optimization Heuristics in Economics: Applications of Threshold Accepting
WONNACOTT and WONNACOTT · Econometrics, *Second Edition*
WOODING · Planning Pharmaceutical Clinical Trials: Basic Statistical Principles
WOOLSON and CLARKE · Statistical Methods for the Analysis of Biomedical Data, *Second Edition*
WU and HAMADA · Experiments: Planning, Analysis, and Parameter Design Optimization
YANG · The Construction Theory of Denumerable Markov Processes
*ZELLNER · An Introduction to Bayesian Inference in Econometrics
ZELTERMAN · Discrete Distributions: Applications in the Health Sciences
ZHOU, OBUCHOWSKI, and McCLISH · Statistical Methods in Diagnostic Medicine

*Now available in a lower priced paperback edition in the Wiley Classics Library.